concrete

Civil Engineering and Engineering Mechanics Series

N. M. Newmark and W. J. Hall, editors

concrete

Sidney Mindess
University of British Columbia

J. Francis Young
University of Illinois at Urbana-Champaign

PRENTICE-HALL, INC. Englewood Cliffs, New Jersey 07632

Library of Congress Cataloging in Publication Data

MINDESS, SIDNEY.
  Concrete.

  Bibliography: p.
  Includes index.
  1. Concrete. I. Young, J. Francis, joint author.
II. Title.
TA439.M49   620.1′834   80-17389
ISBN 0-13-167106-5

*Editorial production supervision
and interior design: Karen Winkler.
Cover design: Creative Impressions.
Manufacturing buyer: Anthony Caruso.*

Printed in the United States of America
10 9 8 7 6 5 4 3 2 1

Prentice-Hall International, Inc., *London*
Prentice-Hall of Australia Pty. Limited, *Sydney*
Prentice-Hall of Canada, Ltd., *Toronto*
Prentice-Hall of India Private Limited, *New Delhi*
Prentice-Hall of Japan, Inc., *Tokyo*
Prentice-Hall of Southeast Asia Pte. Ltd., *Singapore*
Whitehall Books Limited, *Wellington, New Zealand*

contents

preface

Portland cement concrete is foremost among the construction materials used in civil engineering projects around the world. The reasons for concrete's preeminence are varied, but among the more important are the economic and widespread availability of its constituents; its versatility and adaptability, as evidenced by the many types of construction in which it is used; and the minimal maintenance requirements during service. As is the case with any material, its successful use depends upon an intelligent application of its properties in design, and the supply of a uniform, high quality product. Concrete is unique among major construction materials in that it is generally designed specifically for a particular project using locally available materials. Therefore, the project engineer has full control and responsibility over the final material used in construction. If concrete is not properly designed for the service conditions and is not properly handled and cured, it will result in sub-standard performance. For example, when concrete bridge decks and pavements commonly require extensive maintenance five to ten years after placement, it is a clear indication that the material is not being used to its full potential. It is thus essential that engineers acquire a thorough understanding of the material properties of concretes and the procedures that are essential to providing a material of the required quality and durability.

In the past, concrete technology has been taught largely as an empirical science. However, there is a body of chemical and physical principles underlying the behavior of concrete which are now relatively well-understood. It is the aim of this text to present a unified view of concrete behavior in light of these principles, rather than as a series of more or less unrelated facts. For instance, the material on the workability of concrete is prefaced by a general discussion of the principles of rheology; mechanical properties are discussed from the point of view of concrete as a composite (or multiphase) material; and the underlying chemistry of hydration and microstructure of the hardened cement paste are emphasized.

This book is designed primarily for use at the undergraduate level, but it should also serve as a guide for the professional engineer who did not take a formal course on concrete in college. The text is based on the authors' considerable experience in teaching the principles of concrete at the undergraduate level. It includes the most up-to-date information available on new concrete materials, and considerable attention is given to the role of specifications for concrete and concrete materials and the use of test methods for determining concrete properties.

In order to provide a comprehensive treatment it has been necessary to include more material than can be covered in detail in one semester. Therefore, the instructor will have to choose to omit certain topics. Chapter 4 contains more chemistry than might be considered desirable for an undergraduate course and could be treated in less detail by concentrating only on the reactions of the principal cement compounds. Also, those sections in Chapter 18 dealing with mechanisms of creep and shrinkage and those in Chapter 20 covering deterioration processes could be omitted if desired, although they contribute to a more basic under-standing of the material. Chapters 12 and 22 could be passed over without loss of continuity, while parts of Chapters 17, 20 and 21 could also be omitted, depending on the instructor's interests and the emphasis of the curriculum.

The book is divided into three main parts:

1. The properties of the constituent materials: cements, hardened cement paste, aggregates, water and admixtures.

2. Proportioning of concrete mixes and construction practices: mixing, transporting, placing and consolidating, and curing.

3. The properties of hardened concrete: strength and fracture, fatigue, creep and drying shrinkage, and durability.

In view of the fact that SI units are already in use in Canada and the United Kingdom, and are being adopted in the United States, we have chosen to use SI units as the primary system. Conversion to the English system is given throughout the text, as appropriate. Numerical problems have been given using both sets of units.

Throughout the text, reference is made to three sets of national standards: the American Society for Testing and Materials (ASTM); the Canadian Standards Association (CSA); and the British Standards Insti-tution (BSI). Also, considerable use is made of the recommendations and reports of the American Concrete Institute (ACI). ASTM and ACI documents are the principal standards in the United States, but are also widely used elsewhere. In preparing the text, reference was made to the most up-to-date editions of the standards and recommendations available.

These are subject to frequent revision and the reader should therefore refer to the most recent editions, which may differ in some details from those referred to in the text.

A selected bibliography is provided with each chapter as a guide to further reading, and to provide a point of entry to the original scientific literature. Some exercises are given at the end of each chapter. Selection of these has been difficult since the subject does not lend itself to numerical problems, but they have been chosen to emphasize the important aspects of each chapter.

Sidney Mindess

J. Francis Young

acknowledgments

The authors owe a great deal to Professor C. E. Kesler, University of Illinois at Urbana-Champaign, who gave freely of his advice at all stages of the work, offered many helpful suggestions, comments, and criticisms, and thoroughly read the complete manuscript. Our thanks are due also to Professor S. P. Shah, University of Illinois at Chicago Circle; to Mr. P. J. Hawkins, California Portland Cement Company; and to Mr. R. Philleo, U.S. Army Corps of Engineers, who all reviewed portions of the manuscript. We are indebted to Mr. Bryant Mather, U.S. Army Corps of Engineers, who provided many photographs from his files, and to the Portland Cement Association, who kindly provided a number of excellent photographs (see especially Chapter 12), and have allowed us to reproduce material from their publications. We would also like to thank the following individuals who helped at various stages in the preparation of the manuscript: Susan McLintock, Lynn Browning, Lorraine Dowdall, Cathy Martinsen, Ruth Worner, Ginny Ragle, and Cathy Cassels.

Material taken from ASTM standard specifications and tests, as well as from other ASTM publications, occurs frequently throughout the text. These tables and figures are reprinted, with permission, from the American Society for Testing and Materials, 1916 Race Street, Philadelphia, PA 19103.

Tables 9.5, 9.6, 9.7 and 9.8 are reproduced from CSA Standard CAN3-A23-M77, *Concrete Materials and Methods of Concrete Construction,* which is copyrighted by the Canadian Standards Association, and copies of which may be purchased from the Association, 178 Rexdale Boulevard, Rexdale, Ontario M9W 1R3.

Tables 9.12 and 9.13, and Figures 9.2, 9.3 and 9.4 are taken from D. C. Teychenné, R. E. Franklin, and H. C. Erntroy, "Design of Normal Concrete Mixes," Building Research Establishment, Transport and Road Research Laboratory, 1975; they are reproduced with the permission of the Controller of Her Britannic Majesty's Stationery Office.

Special mention should be made of the co-operation of the American Concrete Institute, who have allowed us to draw freely on the material in their publications, especially their Recommended Practices and the reports of their Technical Committees.

The Cement and Concrete Association (Wexham Springs, Slough, U.K.) have also been very helpful in providing information and in permitting us to reproduce material from their publications.

Acknowledgment is made to the McGraw-Hill Book Company Australia Pty. Limited for the use of Figure 2.9 from "Concrete Technology and Practice," 3rd edition, by W. H. Taylor.

Many other organizations and individuals have given us permission to adapt material from their publications for this book, and we thank them for this courtesy. They are: Academic Press, Inc. (London Ltd.); American Ceramic Society; American Society of Civil Engineers; Applied Science Publishers; Bauverlag GmbH; Brookhaven National Laboratory; Building Research Establishment (U.K.); British Standards Institution; Cement Association of Japan; Cambridge University Press; Canadian Government Publishing Center (Ministry of Supply and Services, Canada); Concrete Construction Publications, Inc. (329 Interstate Road, Addison, IL 60101); The Cement-Gun Co. Ltd. (U.K.); Professor S. Diamond; Elsevier Scientific Publishers (The Netherlands); Federal Highway Administration (U.S. Department of Transportation); John Grist Ltd. (U.K.); The Institution of Civil Engineers (U.K.); Institute Technique du Batiment et des Travaux Publiques (France); IPC Science and Technology Press, Ltd.; Dr. R. A. Helmuth; Dr. D. L. Kantro; McGraw-Hill Book Company (U.K.) Ltd.; Mr. H. A. Newlon, Jr.; National Gallery of Art (Washington, D.C.); National Bureau of Standards; Professor A. M. Neville; National Ready-Mixed Concrete Association; Pergamon Press, Inc.; Dr. S. Popovics; Proceq, S.A. (Switzerland); The Society of Materials Science (Japan); RILEM (France); Transportation Research Board; University of Illinois; U.S. Bureau of Reclamation; John Wiley & Sons, Ltd.; Mr. H. Wenander.

1

concrete as a material

Concrete is a material that literally forms the basis of our modern society. Scarcely any aspect of our daily lives does not depend directly or indirectly on concrete. We may live, work, study, or play in concrete structures to which we drive over concrete roads and bridges. Our goods may be transported by trucks traveling on concrete superhighways, by trains that run on rails supported on concrete crossties, by ships that moor at concrete piers in harbors protected by concrete breakwaters, or by airplanes landing and taking off on concrete runways. Water for drinking and for raising crops is stored behind massive concrete dams (Figure 1.1) and is distributed by systems of concrete waterways, conduits, and pipes. The water thus stored may also be used to generate electric power. Alternatively, electricity can be generated by burning coal in power stations built from concrete, or by harnessing the power of the atom within massive reinforced concrete pressure vessels.

We take concrete for granted in our everyday activities and tend to be impressed by the more dramatic impacts of technology. However, it can be truly said that many of the achievements of our modern civilization have depended on concrete, just as many of the enduring achievements of the earlier civilization of Rome were made possible by the use of the forerunner of modern concrete. The word *concrete* comes from the Latin term "concretus," which means to grow together.

1.1 THE NATURE OF CONCRETE

Concrete is a composite material (see Table 1.1) composed of coarse granular material (the aggregate or filler) embedded in a hard matrix of material (the cement or binder) that fills the space between the aggregate

1

Figure 1.1 The Grand Coulee Dam, located in the State of Washington, provides both electric power and water for irrigation.

particles and glues them together. Aggregates can be obtained from many different kinds of materials, although we mostly make use of the materials of nature—common rocks. They are essentially inert, filler materials which, for convenience, are separated into fine and coarse fractions. Similarly, the cement can be formulated from many diverse chemicals. "Cement" is a generic term that can apply to all binders. Therefore, descriptors must be used to qualify this term when referring to specific materials. A civil engineer may have cause to use portland cement concrete, calcium aluminate cement concrete, or epoxy concrete, where the binders are portland cement, calcium aluminate cement, or epoxy resin, respectively. In concrete construction the engineer will use portland

Table 1.1

Definitions for Concrete

Concrete	=	*Filler*	+	*Binder*
Portland cement concrete	=	Aggregate (fine and coarse)	+	Portland cement paste
Mortar	=	Fine aggregate	+	Paste
Paste	=	Cement	+	Water

cement concrete 95% of the time. (The name refers to a family of calcium silicate cements whose origins are described in Chapter 2.) Thus, for convenience, we will often drop the name *portland* throughout the text and use a qualifying descriptor only when dealing with other kinds of cement and concrete.

1.2 ADVANTAGES OF CONCRETE

As a construction material, concrete is exceeded in use only by timber (see Table 1.2). It should be remembered, however, that a large amount of timber is used for construction of formwork and falsework during the fabrication of concrete structures. Concrete competes directly with all major construction materials—timber, steel, asphalt, rock, plastics, and so on—because of its versatility in applications. To meet the demand for concrete and concrete products, nearly 100 million tons[1] of cement were produced in the United States in 1976; Canadian production was 11 million tons. These figures point clearly to inherent advantages attending the use of concrete. The major advantages and disadvantages of concrete are summarized in Table 1.3. Typical properties of concrete are given in Table 1.4 and compared to other construction materials in Table 1.5. It should be remembered, however, that concrete properties can vary significantly from the figures given, depending on the choice of materials and proportions for a particular application.

Table 1.2

Volume of Materials Used in U.S. Construction, 1974

	Volume		Weight	
Material	*$10^6 m^3$*	*$10^6 ft^3$*	*10^6 metric tons*	*10^6 tons*
Timber	170	6000	—	—
Concrete	200	7000	450	500
Cement	24	850	75	85
Steel	6	200	35	40
Brick and clay products	—	—	40	45
Building stone	0.3	12	1	1
Nonferrous metals	—	—	2	2
Asphalt	—	—	28	31

[1]This is equivalent to about 500 million barrels, which was the measure traditionally used by the cement industry until quite recently. A barrel weighs 396 lb and is equivalent to four sacks or bags. A U.S. sack weighs 94 lb and is equivalent to 1 ft^3 of cement on a dry, loose basis. In Canada cement is sold in 40-kg sacks.

Table 1.3

Advantages and Disadvantages of Concrete as
a Construction Material

Advantages	*Disadvantages*
Ability to be cast	Low tensile strength
Economical	Low ductility
Durable	Volume instability
Fire resistant	Low strength-to-weight
Energy efficient	ratio
On-site fabrication	
Aesthetic properties	

The ability of concrete to be cast to any desired shape and configuration is an important characteristic that can offset other short-comings. Concrete can be cast into soaring arches and columns, complex hyperbolic shells, or into massive, monolithic sections used in dams, piers, and abutments. On-site construction means that local materials can be used to a large extent, thereby keeping costs down. Cement costs only about 5 to 7 cents/kg (2 to 3 cents/lb) (1978) and aggregates only a few dollars per ton. Furthermore, by fabricating concrete on site, its properties may be tailored for the specific application. Concrete is made even more economical by the fact that unskilled or semiskilled workers can largely be employed and that relatively unsophisticated equipment is needed. On the other hand, on-site production is a mixed blessing because the quality of concrete must be carefully controlled. Environ-

Table 1.4

Typical Engineering Properties of Structural Concrete

Compressive strength	$= 35$ MPa (5000 lb/in.²)
Flexural strength	$= 6$ MPa (800 lb/in.²)
Tensile strength	$= 3$ MPa (400 lb/in.²)
Modulus of elasticity	$= 28$ GPa (4×10^6 lb/in.²)
Poisson's ratio	$= 0.18$
Tensile strain at failure	$= 0.001$
Coefficient of thermal expansion	$= 10 \times 10^{-6}$/°C (5.6×10^{-6}/°F)
Ultimate shrinkage strain	$= 0.05$–0.1%
Density	
Normal weight	$= 2300$ kg/m³ (145 lb/ft³)
Lightweight	$= 1800$ kg/m³ (110 lb/ft³)

Table 1.5

Typical Properties of Construction Materials[a]

Material	Density (kg/m³)	Tensile Strength (MPa)	Elastic Modulus (GPa)	Coefficient of Thermal Expansion (10⁻⁶/°C)	Thermal Conductivity (W/m·K)	Energy Requirement (GJ/m³)
Aluminum						
Pure	2800	100	70	23	220	360
Alloy	2800	300	70	23	125	360
Steel						
Mild	7800	300	210	12	50	300
High strength	7800	1000	210	11	45	—
Glass	2500	60	65	6	3	50
Wood						
Soft	350	50	5.5	—	0.2–0.6	—
Hard	700	100	10	—	0.2–0.6	—
Plastic (polystyrene)	1000	~50	~3	72	0.1	—
Rock (granite)	2600	~20(~25ᵇ)	~50	7–9	3	—
Concrete	2300	3(35ᵇ)	~25	10	3	3.4

[a]Conversion factors: $kg/m^3 \times 0.062 = lb/ft^3$; $MPa \times 145 = lb/in.^2$; $GPa \times 0.145 = 10^6\ lb/in.^2$; $10^{-6}/°C \times 0.556 = 10^{-6}/°F$; $W/m·K \times 0.578 = Btu/ft·h·°F$; $GJ/m^3 \times 26.9 \times 10^3 = Btu/ft^3$.

[b]In compression.

mental conditions fluctuate, so that it is difficult to assure uniform processing of concrete throughout a job. Constituent materials are less carefully characterized than they might be and can have undesirably high variations in properties. The use of an unskilled or semiskilled work force means that in the absence of proper supervision on the job site, undesirable practices may be adopted and tolerated.

Casting of concrete can also be adapted to factory-controlled production. Precast building elements for standardized low-cost building systems are more common in European countries but have also been developed in the United States. Precast concrete block has become a very popular building element, and precast concrete pipe is widely used in drainage, sewage, and water supply projects. Precast, prestressed concrete beams, girders, and panels in various configurations are used increasingly in many structures. Precast concrete can be produced more uniformly with closer tolerances compared to concrete cast on site, but requires a more skilled work force and generally more sophisticated equipment.

Good-quality concrete is a very durable material and should remain maintenance-free for many years when it has been properly designed for the service conditions and properly placed. Unlike structural steel, it does not require protective coatings except in very corrosive environments. It is also an excellent material for fire resistance. Although it can be severely damaged by exposure to high temperatures, it can maintain its structural integrity for a considerable period—long after steel buildings would have suffered irreparable damage.

Because of rapidly rising energy costs, more consideration is being given to the energy costs of materials. In this regard concrete comes out ahead of most other construction materials (see Table 1.5). In the first place, concrete requires less energy to produce than does steel. This is because steel is made by high-temperature processes (300 GJ/m^3), whereas only a small component of the concrete, the cement, requires pyro-processing (22 GJ/m^3). The major energy costs of concrete are in cement and reinforcing steel, but the energy consumption of an equivalent steel structural element can be considerably greater. Second, concrete buildings can be more energy-efficient to operate because of the thermal properties of concrete. Concrete conducts heat slowly and is able to store considerable quantities of heat from the environment that can be released during cool periods. It is possible to design buildings to take advantage of this thermal inertia. Last, but not least, concrete has considerable aesthetic possibilities that can be expressed through the use of color, texture, shape, and so on. All of these advantages combine to make concrete very versatile and adaptable.

1.3 LIMITATIONS OF CONCRETE

However, concrete does have weaknesses which may limit its use in certain applications and which must be allowed for when designing structures. Concrete is a brittle material with very low tensile strength. Thus, concrete should generally not be loaded in tension (except for low bending stresses which may be permitted in unreinforced slabs on grade) and reinforcing steel must be used to carry tensile loads; inadvertent tensile loading causes cracking. The low ductility of concrete also means that concrete lacks impact strength and toughness compared to metals. Even in compression concrete has a relatively low strength-to-weight ratio, and a high load capacity requires comparatively large masses of concrete, although, since concrete is low in cost, this is economically possible.

The volume instability of concrete must also be allowed for in design and construction. It shows volume stability that is more characteristic of timber and quite unlike that of steel, which is a volume-stable material under normal conditions of service. Concrete undergoes considerable irreversible shrinkage due to moisture loss at ambient temperatures, and also creeps significantly under an applied load even under conditions of normal service.

Awareness of these problems with concrete enables them to be compensated for by suitable designs and to be controlled in part by a suitable choice of materials and construction practice. A great deal of research effort has been devoted to ameliorating these problems and has led to the development of new types of concrete, such as fiber-reinforced concrete, expansive cement concrete, and latex-modified concretes. Further developments can be anticipated in the future.

Bibliography

BEIJER, O., "Energy Consumption Related to Concrete Structures," *Journal of the American Concrete Institute,* Vol. 72, No. 11, pp. 598–600 (1975).

Concrete Construction (July 1978). Special issue on energy conservation with concrete.

POMEROY, C.D., "Concrete, an Alternative Material," *Proceedings, Institution of Mechanical Engineers,* Vol. 192, pp. 135–144 (1978).

2

historical development of cement and concrete

From very early times, builders have tried to find materials that could be used to cement stones or bricks together. It was clear to them that this mode of construction would provide much more flexibility in construction than the older method of carefully setting stone blocks one above the other without the use of any cementing material between them. Perhaps the oldest cementing material was simply mud, sometimes mixed with straw, to bind dried bricks together, as used in ancient Egypt. Such construction will only work in very dry climates, as the unburnt bricks and clay have no resistance to water. The Babylonians and Assyrians sometimes used naturally occurring bitumens to bind stones or bricks together. In our own time, a variety of organic cements (epoxies) are similarly used to bind structural elements together. However, in this chapter we only follow the development of cements based on compounds of lime (i.e., *calcareous* cements). In particular, we focus on those cements that are capable of hardening under water (i.e., *hydraulic* cements).

2.1 NONHYDRAULIC CEMENTS

The first calcareous materials to be used as cements in mortars were gypsum and lime. The Egyptians used gypsum mortars in the construction of the Pyramid of Cheops (~3000 B.C.); these were prepared by calcining impure gypsum. When mixed with a small quantity of water, the setting of this material is due to the recombination of the calcined gypsum with the water of crystallization that had been driven off during the burning process:

$$2CaSO_4 \cdot 2H_2O \xrightarrow[\sim 130°C]{heat} 2CaSO_4 \cdot \frac{1}{2}H_2O + 3H_2O \qquad (2.1)$$

This material was used instead of lime because it required only a relatively low burning temperature ($\sim 130°C$). Gypsum-based mortars are nonhydraulic; that is, hardening will not take place under water, because gypsum is quite soluble.

Lime mortars were used in Egypt only in the Roman period, but they were used much earlier in Crete, Cyprus, and Greece. These materials were prepared by calcining limestone:

$$CaCO_3 \xrightarrow[\sim 1000°C]{heat} CaO + CO_2 \qquad (2.2)$$

Lime mortar is sometimes known as *air mortar,* since it hardens in air. The hardening is due essentially to the evaporation of the mixing water. A secondary reaction is then the subsequent carbonation:

$$CaO + H_2O \rightarrow Ca(OH)_2 \xrightarrow{CO_2} CaCO_3 + H_2O \qquad (2.3)$$

This helps to solidify the mass, but only on the surface, and if the lime mortar is well compacted, CO_2 is prevented from penetrating beyond a thin surface layer. It should be noted that although the hardened mortar is quite impervious to water, the hardening will not take place *under* water, and thus ordinary limes are not hydraulic. Nevertheless, some excellent mortars were produced by this method—so hard that at one time it was assumed that some secret, now lost, was involved in the production of the mortar. Later study has shown, however, that the excellent performance of these mortars was not due to a secret in the making of the lime, but to thorough mixing and ramming (a lesson that modern engineers would do well to remember). The Colosseum in Rome and the Pont du Gard at Nîmes (Figure 2.1) are tributes to the quality of at least some Roman mortars.

2.2 HYDRAULIC LIMES

Both the Greeks and the Romans also produced hydraulic limes, by calcining limestones containing argillaceous (clayey) impurities. Moreover, they both knew that certain volcanic deposits, when finely ground

Figure 2.1 The Pont du Gard at Nimes.

and mixed with lime and sand, yielded mortars that were not only stronger than ordinary lime mortars but were also resistant to water. The Greeks used a volcanic tuff from the island of Santorin (Santorin earth). The Romans used a volcanic tuff found around the Bay of Naples, which was generally called *pozzolana* because the best variety was found near the village of Pozzuoli, near Mt. Vesuvius. This material is mentioned by Vitruvius,[1] writing in the second half of the first century B.C.:

> There is also a kind of powder which from natural causes produces astonishing results. It is found in the neighborhood of Baiea and in the country belonging to the towns round about Mt. Vesuvius. This substance, when mixed with lime and rubble, not only lends strength to buildings of other kinds, but even when piers of it are constructed in the sea, they set hard under water.

In addition, Vitruvius knew that if natural pozzolanas were not available, a similar effect could be obtained by using finely crushed burnt brick.

[1]Vitruvius, *The Ten Books of Architecture,* Bk. II, Ch. VI (New York: Dover, 1960), pp. 46–47.

Figure 2.2 The interior of the Pantheon. (Giovanni Paolo Panini; National Gallery of Art, Washington, D.C., Samuel H. Kress Collection.)

The Romans also used these hydraulic mortars to make a form of concrete. Perhaps the best preserved building of the ancient world, the Pantheon (Figure 2.2), dating from the second century A.D., was built largely of concrete. The dome, 141 ft 6 in. in diameter, was constructed by pouring concrete into ribbed sections and letting it harden.

The quality of cementing materials generally declined through the Middle Ages. The art of burning lime was almost lost, and ground tiles (or pozzolans) were not added. High-quality mortars did not appear again until after the fourteenth century A.D., at which time the use of pozzolans was reintroduced. It was not until the eighteenth century that work first began on trying to understand the nature of these cementing materials. In 1756, John Smeaton (who was, incidentally, the first person to style himself a "civil engineer") was commissioned to rebuild the Eddystone Lighthouse off the coast of Cornwall, England. Recognizing that the normal lime mortars then in general use would not withstand the action of the water, Smeaton carried out an extensive series of experiments with different limes and pozzolans. He found that the best limestones for use in mortars were those containing a high proportion of clayey material.

Figure 2.3 Smeaton's Eddystone Lighthouse. (Photograph courtesy of H. A. Newlon, Jr.)

This was the first recognition of the factors that control the formation of hydraulic lime. Eventually, Smeaton used a mortar prepared from a hydraulic lime mixed with pozzolan imported from Italy. The Eddystone Lighthouse, thus rebuilt (Figure 2.3), stood for 126 years before it was replaced with a more modern structure.

2.3 DEVELOPMENT OF PORTLAND CEMENT

After this pioneering work, a number of other discoveries followed quite rapidly. James Parker in England took out a patent in 1796 on a natural hydraulic cement (misleadingly called Roman cement), produced by calcining nodules of impure limestone containing clay. A similar process began in France 6 years later. In 1813, also in France, Vicat (who developed the needles we still use to determine the time of setting of cement) prepared artificial hydraulic lime by calcining synthetic mixtures of limestone and clay. James Frost introduced the same approach in England in 1822.

Finally, in 1824, Joseph Aspdin, a Leeds builder, took out a patent on ''Portland'' cement. This cement was prepared by calcining some finely ground limestone (taken, when available, ''from the roads after it is reduced to a puddle or powder''), then mixing this with finely divided clay, and calcining the mixture in a kiln until the CO_2 was driven off. This mixture was then finely ground and used as cement. It seems unlikely that true portland cement could be made according to this patent; the temperatures were not high enough to cause real clinkering. Isaac Johnson claimed to have first burned the raw materials to the clinkering temperature in 1845. The name ''Portland'' cement was coined by Aspdin because of the real or fancied similarity of the hardened cement to a popular, naturally occurring building stone quarried at Portland, England. It might be noted that Aspdin also tried to make his product as mysterious as possible, by pretending to sprinkle some secret salts into the kilns before burning. Whatever the exact nature of Aspdin's cement, however, he made it a commercial success.

The use of hydraulic cements spread rapidly through Europe and North America. Natural cements are made by burning an argillaceous limestone that has the right composition of lime and silica. However, the burning temperature is only high enough to cause partial chemical combination. Thus, natural cements are intermediate between hydraulic limes and portland cement. Although natural cement was produced in the United States as early as 1818, the first American patent on portland cement (by David Saylor) was not taken out until 1871. In Canada, limes and hydraulic cements were probably first produced in 1830. In the late

Table 2.1

Cement Production in the United States and Canada (Short Tons)

	United States			Canada		
		Portland Cement			Portland Cement	
Year	Natural Cement	Imported	Domestic	Natural Cement	Imported[a]	Domestic
1880	500,000	20,000	5,000	—	1,800	—
1890	1,400,000	400,000	60,000	16,000	22,000	—
1900	2,000,000	300,000	1,700,000	22,000	66,000	51,000
1915	100,000	~1,000	15,500,000	—	5,000	1,000,000
1930	N.A.[b]	N.A.[b]	33,600,000	—	25,000	1,900,000
1950	600,000	45,000	46,400,000	—	250,000	2,900,000
1976	—	3,000,000	74,500,000	—	N.A.[b]	10,900,000

[a]Both natural and portland cement.
[b]N.A., not available.

nineteenth century considerable quantities of portland cement were imported by the United States from England and Europe. After domestic production was started, by about 1890, local portland cement quickly displaced the imports and captured the major construction markets from natural cement (Table 2.1).

Great advances were also made in the equipment necessary to produce portland cement. Originally, it had been made in vertical kilns. The disadvantage was obvious; the kiln had to be completely discharged before a new batch of cement could be made. The first rotary kiln was introduced in 1886, and soon Ransome in England was producing satisfactory kilns of this type. In the United States, Thomas Edison was issued a series of patents for rotary kilns in 1909. It was in the United States also that gypsum was first interground with the clinker to increase the time of setting.

With the production of portland cement, work began on testing the material and characterizing its properties. Originally, tensile tests were carried out by building a cantilever beam of bricks held together by mortar and noting the length of beam that would support itself; this was taken to be proportional to the tensile strength of the mortar. The first systematic tests of tensile and compressive strength were carried out in Germany in 1836. After a great deal of experimentation in many countries, the basic cement tests were largely standardized by about 1900, although they have of course been modified since and new tests added as necessary.

2.4 CONCRETE ADMIXTURES

Historically, admixtures are almost as old as concrete itself. It is known that the Romans used animal fat, milk, and blood to improve their concrete. Although these were probably added to improve the workability, blood (due to hemoglobin) is a very effective air-entraining agent. The Romans were certainly unaware of this, but the blood might well have improved the durability of their concrete. In more recent times, calcium chloride was often used as an accelerator. But the systematic study of admixtures began with the introduction of air-entraining agents in the 1930s. This was also accidental—one of the grinding aids used in cement production was beef tallow, and it was found that where tallow was used, the resulting concrete was much more resistant to freezing and thawing. The leakage of oil into the cement in older grinding units had a similar effect. From these beginnings, it is unusual now to find concrete to which some admixture has not been added.

Summary

This has, necessarily, been a very brief and selective outline of the history of cement and concrete. Although a great deal has been learned about cement and concrete since Roman times, a great deal still remains to be discovered. Much of our current research was foreshadowed in the last century by the pioneering studies of Vicat, Michaelis, Le Châtelier, and others, who first began a systematic study of the basic properties of cement, with a view to determining the underlying physical and chemical causes for cement behavior. It is hoped that this brief account of cement and concrete has shown that any real advances in their use were based on fundamental studies, whether by Vitruvius or by Smeaton. Only by continuing this scientific approach will we be able to improve the properties of these materials to meet the challenge of the future.

Bibliography

COHEN, E., AND R. C. HEUN, "100 Years of Concrete Building Construction in the United States," *Concrete International: Design and Construction,* Vol. 1, No. 3, pp. 38–46 (1979).

DRAFFIN, J. O., "A Brief History of Lime, Cement, Concrete, and Reinforced Concrete," *Journal of the Western Society of Engineers,* Vol. 48, pp. 14–47 (March 1943).

FRANCIS, A. J., *The Cement Industry 1796–1914: A History,* David & Charles, Inc., N. Pomfret, Vt., 1977.

GONNERMAN, H. F., *Development of Cement Performance Tests and Requirements,* Bulletin of Research Department No. 93. Portland Cement Association, Skokie, Ill., 1958.

KESLER, C. E., "Reinforced Concrete Materials—A Remarkable Heritage," *Proceedings ASCE, Journal of the Structural Division,* Vol. 103, No. ST4, pp. 747–757 (1977).

LESLEY, R. W., *History of the Portland Cement Industry in the United States.* International Trade Press, Inc., Chicago, 1924.

MALINOWSKI, R. "Concretes and Mortars in Ancient Aqueducts," *Concrete International: Design and Construction,* Vol. 1, No. 1, pp. 66–76 (1979).

MEHTA, P. K., ed., *Cement Standards: Evolution and Trends,* ASTM STP 663, American Society for Testing and Materials, Philadelphia, Pa., 1978.

STONEHOUSE, D. H., *Cement in Canada,* Mineral Bulletin MR 133. Mineral Resources Branch, Department of Energy, Mines and Resources, Ottawa (1973).

3

cements

There is a wide variety of cements that are used to some extent in the construction and building industries, or to solve special engineering problems. The chemical compositions of these cements can be quite diverse, but by far the greatest amount of concrete used today is made with portland cements. This chapter will therefore discuss the composition of portland cements in considerable detail and describe the more specialized cements briefly. The name *portland* originated as a trade name and thus it gives no indication of composition or properties. The name now applies to a family of closely related cements that have an overall similarity of properties. Different portland cements can be made for particular applications by modifying certain properties and later we examine how these modifications are achieved by suitable manipulation of the basic composition of portland cement.

3.1 MANUFACTURE OF PORTLAND CEMENT

In principle, the manufacture of portland cement is very simple and relies on the use of abundant raw materials. An intimate mixture, usually of limestone and clay, is heated in a kiln to 1400 to 1600°C (2550 to 2900°F), which is the temperature range in which the two materials interact chemically to form the calcium silicates. In practice, because of the large amounts of materials being processed and the high temperatures required, considerable attention must be paid to the various stages of processing (see Figure 3.1) if adequate quality control is to be maintained.

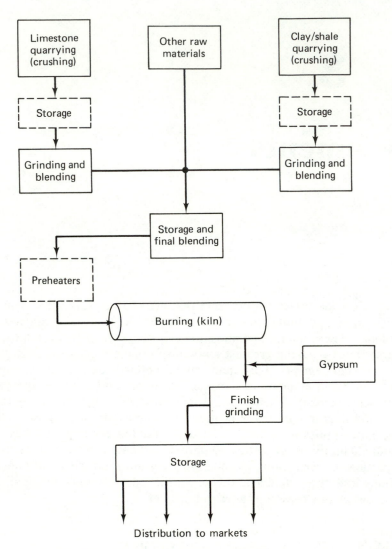

Figure 3.1 Schematic outline of cement production.

Raw Materials

High-quality cements require raw materials of adequate purity and uniform composition. Limestone (calcium carbonate) is the most common source of calcium oxide, although other forms of calcium carbonate, such as chalk, shell deposits, and calcareous muds, are used. The location of cement plants is most often determined by the occurrence of suitable calcareous deposits, since a satisfactory source of silica can usually be obtained within close proximity. Iron-bearing aluminosilicates are invariably used as the primary source of silica. Clays or silts are preferred since they are already in a finely divided state; but shales, schists, and other argillaceous rocks are also used.

The compounds that are derived from the iron and alumina content of siliceous raw materials contribute little to portland cement as regards strength and can lead to problems of durability and abnormal setting behavior. It might be thought, therefore, that the use of pure silica would lead to a better cement. There are, however, good reasons for using aluminosilicates other than the scarcity of economic sources of pure silica. Quartz, the major form of pure silica in nature, is a relatively unreactive material and, moreover, pure lime–silica mixes have very high fusion temperatures (above 2000°C or 3600°F), so that reactions between the two components can only occur by a slow process of sintering. Aluminum and iron oxides act as *fluxing agents*, lowering the fusion temperature of a portion of the raw mix to a practical firing temperature. Quartz and iron oxide are quite commonly added (especially in the western United States) in small amounts to make up for deficiencies of SiO_2 and Fe_2O_3 in the main mixture.

Preparation of Materials

The object of processing the raw materials is to ensure that the raw feed entering the kilns is of constant composition and in a thoroughly comminuted and blended state. Failure to do this would result in a cement with variable composition and unpredictable properties. For example, if the composition of the raw feed varied widely, the percentages of the various chemical compounds in the finished cement would vary widely, making performance difficult to predict with certainty. The optimum burning temperature of the mix will also vary with composition. In addition, if the particle size were too large, complete chemical combination might not occur during the time the material is in the kiln and would result in a cement of inferior performance.

The exact sequence of operations at this stage may vary considerably from plant to plant, depending on the raw materials, equipment, and plant design. In the *wet process* each raw material is quarried, crushed, and stockpiled separately to provide feedstocks of known composition. Blending of the materials may take place during fine grinding or after each material has been ground separately. For example, clay can be broken down to a finely divided slurry merely by good mechanical agitation. Wet grinding of limestone is carried out separately in other mills to form a second slurry. When slurries are used, blending and final proportioning can be handled very conveniently. However, evaporation of large amounts of water consumes a large quantity of heat in the kiln and leads to substantially higher costs. Thus in the final stages of the wet process, the water content of the slurry may be considerably reduced.

If the source of silica is an argillaceous rock, mechanical grinding is required and the advantages of the wet process are lost. Improvements in grinding mills have made the wet process uneconomic and obsolete, and modern cement plants use dry grinding processes. Additional energy that may be required for dry grinding is more than compensated for by savings in energy in the kiln. But even in the *dry process,* water may sometimes be added to the raw feed to granulate it for easy handling and good contact in the kiln. The nature of the raw materials will determine whether the calcareous and argillaceous rocks are ground together or separately. In the dry process care must be taken to ensure adequate mixing and blending and to avoid excessive loss of fines. Closed-circuit grinding with classifying equipment is commonly used to avoid overgrinding and control of dust. Ground materials are stored in silos and final mixing may be accomplished during transfer to a final storage silo before reaching the kiln.

In some plants the *semi-dry* process is used. In this case the raw mix is prepared as in the dry process, but then 12 to 14% of water is added to form nodules, which are fed into the kiln by means of a moving grate.

The Burning Process

Once the raw feed has been satisfactorily ground and blended, it is ready to enter the kiln. This heat treatment is termed *clinkering,* to distinguish it from *sintering* (where no melting occurs) and *fusion* (where complete melting occurs.) In the cement kiln, partial melting takes place; only about one-fourth of the charge is in the liquid state at any time, but it is in this fraction that the necessary chemical reactions proceed. The kiln, in the form of a long steel cylinder lined with refractory brick inclined a few degrees from the horizontal, is rotated at about 60 to 200 rev/h about its axis. Modern kilns may reach 6 m (20 ft) in diameter and over 180 m (600 ft) in length with a production capacity exceeding 5000 tons/day using the dry process. The raw feed enters at the high end and the combination of rotation and inclination slowly moves the material the length of the kiln. In the wet process the material may stay in the kiln for 2 to 2½ h; this is reduced to 1 to 1½ h for dry kilns and to as little as 20 min for some kilns with modern heat exchangers.

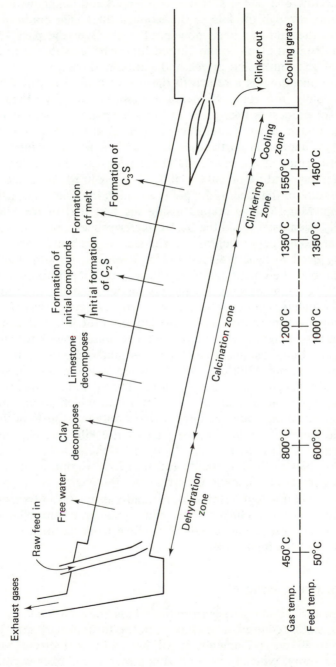

Figure 3.2 Schematic outline of conditions and reactions in a typical cement rotary kiln (dry process).

21

At the low end of the kiln fuel is injected and burnt, with the hot gases passing through the kiln to the exhaust flues. The traditional fuel used to fire the kiln has been powdered coal. Over the past 15 to 20 years many cement plants in the United States have switched to fuel oil and natural gas, but rising prices and periodic shortages in these fuel supplies are encouraging a return to the use of coal. The raw feed thus moves gradually into zones of increasing temperature (see Figure 3.2). Four distinct processes take place in the kiln: evaporation, calcination, clinkering, and cooling. Careful operation of the kiln is the key to successful production of portland cement, and each stage is important to the overall burning of the raw feed. The exit gases at the top end of the kiln are at a temperature of 240 to 450°C (480 to 840°F) and quickly raise the temperature of the feed to the point at which the free water is lost by evaporation. Metal chains draped inside the first part of the kiln have traditionally been used to aid the heat-transfer process, but more sophisticated heat exchangers (such as the suspension preheater) are increasingly being used to heat the feed before it enters the kiln. Once free moisture is lost, the charge is quickly heated to temperatures in excess of 600°C (1100°F), at which point the raw materials begin to decompose (calcine) through loss of bound water and carbon dioxide. The *calcination zone* extends over more than half of the length of the kiln and transforms the charge into a reactive mixture that can enter into new chemical combinations. Initial chemical combinations take place in the latter part of this zone around 1200°C (2200°F), when calcium aluminates and ferrites form through solid-state reactions. These compounds act as the fluxes, which melt at around 1350°C (2450°F), at which point clinkering begins. The *clinkering zone* occupies perhaps only one-fourth of the total kiln length, but it is at the hottest part of the kiln and is where the calcium silicates form in the liquid phase of the charge. The charge temperature rises to 1400 to 1600°C (2550 to 2900°F) during the 15 to 45 min the raw material spends in this zone. As the charge moves past the flame in the final few feet of the kiln, it rapidly drops off in temperature. The rate at which the clinker cools can significantly alter the rate at which the new cement compounds crystallize from the melt and hence the reactivity of the final cement.

Final Processing

The material that emerges from the kiln is known as *clinker*; further processing is still required. The clinker, in the form of dark gray porous nodules (6 to 50 mm in diameter) is still hot and is further cooled by an air or water spray, typically on a moving grate. The clinker is conveyed to ball mills, where it is ground to the fine powder that is so familiar to us. A small amount of gypsum is interground with the clinker to control the early reactions of tricalcium aluminate (C_3A); without the addition of gypsum, C_3A can cause *flash setting* of the clinker. Portland cement is clinker interground with gypsum; without the gypsum it is only ground clinker. The ground cement is subsequently stored in large silos until ready for distribution. At this stage a final blending of the cement will improve the uniformity of performance of the product by averaging out small differences in chemical composition and burning conditions that inevitably occur during manufacture.

Quality Control

The manufacture of portland cement can be seen to involve complex chemical reactions, and all stages of production should be closely monitored and controlled. Plant chemists analyze the raw materials at all stages of production, as well as the finished products, and on-line automated analytical controls have been developed. The burning process is complex and not yet fully understood; thus, although automated control of kilns is progressing, the optimum operating conditions are still, by and large, based on trial-and-error experience and require manual supervision.

Economic Considerations

The manufacture of portland cement is an energy-intensive process, due primarily to the high temperatures that must be maintained in the rotary kilns (see Table 3.1). The wet process is energy-wasteful because

Table 3.1

Energy Consumption in the Cement Industry
(National U.S. Average, 1973 Figures)[a]

	Energy Requirements, MJ/kg (10^6 Btu/ton)		
Operation	Wet Process	Dry Process	Suspension Preheater
Raw materials			
Raw grinding	0.34 (0.29)	0.37 (0.32)	—
Total processing	0.84 (0.72)	0.87 (0.75)	0.87 (0.75)
Kiln operation			
Fuel	7.55 (6.49)	6.34 (5.45)	5.30 (4.56)
Total operation	7.88 (6.78)	6.68 (5.74)	5.64 (4.85)
Finish grinding	0.60 (0.52)	0.60 (0.52)	0.58 (0.50)
Total operations	9.33 (8.02)	8.15 (7.01)	7.19 (6.18)

[a]Data taken from "Energy Conservation in the Cement Industry," Conservation Paper 29, Federal Energy Administration, 1976.

of the extra heat required to vaporize the slurry water. The use of the suspension preheater, which is a very efficient heat exchanger that not only dries the raw feed before it enters the rotary kiln but may also partly calcine it, is more energy-efficient; it is being installed in modern plants. The theoretical fuel requirement is about 1.7 MJ/kg (1.5×10^6 Btu/ton). Coupled with other technological improvements, the average fuel consumption is 4.5 MJ/kg in Japan and 4.2 MJ/kg in West Germany, compared to 7.0 MJ/kg in the United States. The use of suspension preheaters can also increase the production capacity of a given kiln by 50% or more. To avoid rising fuel prices, cement companies are increasing their use of coal, but high levels of impurities in coal such as sulfur can cause numerous problems in production that the cement industry is ill-equipped to deal with.

3.2 COMPOSITION OF PORTLAND CEMENT

Compound Composition

The chemical composition of a general-purpose (ordinary) portland cement such as you would buy at the local building supply store is given in Table 3.2. It will be noted that the quantities do not add up to 100%, the missing percentages being accounted for by impurities. Since the calcium silicates account for three-fourths of a portland cement and are responsible for its cementing qualities, it might be more logically called calcium silicate cement.

The chemical formulas of these compounds are traditionally written in an oxide notation frequently used in ceramic chemistry. This gives rise to a unique, but useful, shorthand notation which has universal use

Table 3.2

Typical Composition of Ordinary Portland Cement[a]

Chemical Name	Chemical Formula	Shorthand Notation	Weight Percent
Tricalcium silicate	$3CaO \cdot SiO_2$	C_3S	50
Dicalcium silicate	$2CaO \cdot SiO_2$	C_2S	25
Tricalcium aluminate	$3CaO \cdot Al_2O_3$	C_3A	12
Tetracalcium aluminoferrite	$4CaO \cdot Al_2O_3 \cdot Fe_2O_3$	C_4AF	8
Calcium sulfate dihydrate (gypsum)	$CaSO_4 \cdot 2H_2O$	$C\bar{S}H_2$	3.5

[a]See p. 32 however, concerning cements produced in the western states.

Table 3.3

Typical Oxide Composition of a General-Purpose
Portland Cement

Oxide	Shorthand Notation	Common Name	Weight Percent
CaO	C	lime	63
SiO_2	S	silica	22
Al_2O_3	A	alumina	6
Fe_2O_3	F	ferric oxide	2.5
MgO	M	magnesia	2.6
K_2O	K ⎫	alkalis	0.6
Na_2O	N ⎭		0.3
SO_3	$\overline{S}$	sulfur trioxide	2.0
CO_2	$\overline{C}$	carbon dioxide	—
H_2O	H	water	—

among cement scientists. The common notation is summarized in Table 3.3. Each oxide formula is abbreviated to a single capital letter, with the number of oxide formulas in the compound designated by a subscript placed after the letter. In order to write carbonates and sulfates (which are important in cement chemistry) in the shorthand notation, it is necessary to make use of CO_2 and SO_3; the symbols $\overline{C}$ and $\overline{S}$ have been adopted. Thus, calcium carbonate ($CaCO_3$) is written as $CaO \cdot CO_2$ or as $C\overline{C}$; gypsum ($CaSO_4 \cdot 2H_2O$) is written as $CaO \cdot SO_3 \cdot 2H_2O$ and becomes $C\overline{S}H_2$ in shorthand notation. Although it is possible to determine the *compound composition*, as given in Table 3.2, by direct analysis, the methods employed are complex and require special skills and expensive equipment. Chemical analysis of portland cement is routinely done using standard methods, and each element present is reported as its oxide. Such analyses are available from the cement plant on request; a typical oxide analysis is given in Table 3.3. It can be seen that an exhaustive analysis reveals a number of impurity oxides that are present in the original raw materials. The significance of these impurities will be discussed later. Using these oxide analyses and the compound stoichiometries, it is a simple matter to calculate a compound composition. This is known as the *Bogue calculation* (after R. H. Bogue, who first emphasized the advantages of knowing the compound composition of a portland cement). A simple Bogue calculation, suitable for most purposes, is given in ASTM C150, although more sophisticated procedures have been developed. The necessary equations are given below; the quantities C_2S, A, F, C_3S, and so on, represent the percentages by weight of the various components.

Case A: A/F $\geqslant$ 0.64
$C_3S =$ $4.071C - 7.600S - 6.718A - 1.430F - 2.852\bar{S}$
$C_2S =$ $2.867S - 0.7544C_3S$
$C_3A =$ $2.650A - 1.692F$
$C_4AF = 3.043F$

Case B:*   A/F $<$ 0.64
$C_3S =$ $4.071C - 7.600S - 4.479A - 2.859F - 2.852\bar{S}$
$C_2S =$ $2.867S - 0.7544C_3S$
$C_3A =$ 0
$C_4AF^* = 2.100A + 1.702F$

ASTM Classifications

A knowledge of the compound composition of portland cement makes it possible to predict the properties of the cement. More important, manipulation of compound composition can be used to modify certain properties of the cement in order that the cement will perform more satisfactorily in particular applications. In order to understand and appreciate this, it is necessary to anticipate some material that will be discussed in more detail in Chapter 4.

Hydration of Cement

When portland cement is mixed with water, its constituent compounds undergo a series of chemical reactions which are responsible for the eventual hardening of concrete. Reactions with water are designated *hydration*, and the new solids formed on hydration are collectively referred to as *hydration products*. All chemical reactions, including hydration, can be described by reaction stoichiometries, rates of reaction, and heats of reaction. Also, in the case of cement chemistry, it is of interest to know whether the hydration products contribute to the strength of the hydrated cement. The hydration characteristics of the cement compounds are summarized in Table 3.4. The stoichiometries are described by chemical equations and need not be discussed at this time.

The rates of hydration for the pure cement compounds are given in Figure 3.3a. It can be seen that C_3A and C_3S are the most reactive compounds, whereas C_2S reacts much more slowly. The presence of gypsum slows the early rate of hydration of C_3A. Quantitative data are not available for C_4AF, but the reaction of C_4AF–gypsum–water is believed to be somewhat slower than C_3S, whereas the hydration of

---

*Note: In case B, C_4AF has a different composition from case A. C_2F may also be present and is included in the quantity calculated.

Table 3.4

Characteristics of Hydration of the Cement Compounds

		Amount	*Contribution to Cement*	
	Reaction	*of Heat*		*Heat*
Compounds	*Rate*	*Liberated*	*Strength*	*Liberation*
C_3S	Moderate	Moderate	High	High
C_2S	Slow	Low	Low initially, high later	Low
$C_3A + C\bar{S}H_2$	Fast	Very high	Low	Very high
$C_4AF + C\bar{S}H_2$	Moderate	Moderate	Low	Moderate

C_4AF without gypsum is faster. The rate of hydration of the compounds in portland cement is plotted in Figure 3.3b, where it can be seen that C_3S (alite) and C_2S (belite) react more rapidly than they do in their pure pastes, and C_4AF hydrates more slowly than C_3S. The actual rate of hydration depends on the particular cement. It is to be noted that both C_3A and C_4AF react with gypsum as well as water during cement hydration. Figure 3.3 shows C_3A and C_3S are the most reactive compounds, whereas C_2S reacts much more slowly. Rates of reactions bear no relation, however, to strength development, as can be seen in Figure 3.4. Clearly, the calcium silicates provide most of the strength developed by portland cement; C_3S provides most of the early strength (in the first 3 to 4 weeks); and both C_3S and C_2S contribute equally to ultimate strength.

The hydration reactions of portland cement are all *exothermic*; that is, they liberate heat. Thus, during the hardening process the concrete is being continually warmed by the internal heat generated. The extent of temperature rise in a concrete section will depend on how quickly the heat is liberated and how quickly it is lost from the concrete to the surroundings. Thus, the *rate* of heat evolution is the important quantity. The contribution of each compound to the overall rate of heat evolution is a function of the heat of hydration, the rate of hydration, and the fraction of the compound in the cement. Multiple regression analyses give the following relationships for the heat of hydration in J/g of a typical cement:

$$H_{3 \text{ days}} = 240(C_3S) + 50(C_2S) + 880(C_3A) + 290(C_4AF) \qquad (3.1a)$$

$$H_{1 \text{ year}} = 490(C_3S) + 225(C_2S) + 1160(C_3A) + 375(C_4AF) \qquad (3.1b)$$

The quantities C_3S, C_2S, and so on, are expressed as weight fractions of the cement. It is to be noted that the coefficients in Eq. (3.1b) are similar

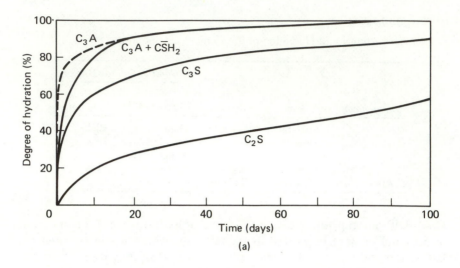

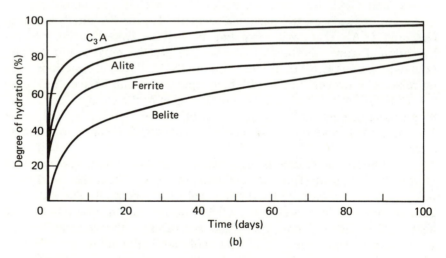

Figure 3.3 Rate of hydration of the cement compounds: (a) in pastes of the pure compounds; (b) in a Type I cement paste.

to the heats of hydration for complete hydration of the pure compounds (see Table 4.3).

ASTM Types

Once the general role of the cement compounds is known, it should be possible, by suitable adjustment of compound composition, to modify the properties of portland cements. This is now done commercially in

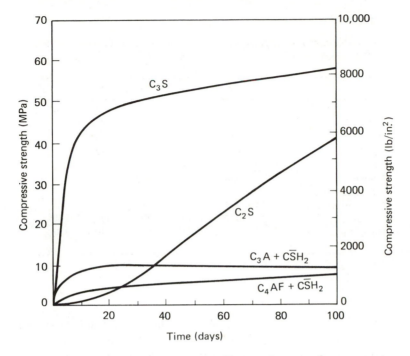

Figure 3.4 Compressive strength development in pastes of pure cement compounds.

most countries. In the United States, five distinct portland cements are recognized by ASTM, differing only in the relative amounts of the cement compounds. These cements are generally known by their ASTM designation or the equivalent description; typical compositions for these cements are given in Table 3.5.

ASTM Type I is the cement most commonly used in general construction where no special properties are needed or specified. If, however, a more rapid rate of hardening is desirable, as, for example, in precast work where forms are reused or when concreting at low temperatures a Type III cement could be specified. Increasing the C_3S content will achieve this aim, but grinding the cement more finely is even more effective. The resultant increased surface area that will be in contact with water means faster hydration and more rapid development of strength. The amount of strength gain of a Type III cement over the first 24 h is about double that of Type I.

The high C_3S content and rapid rate of hydration result in a high rate of heat evolution. If this heat is not allowed to escape (i.e., if *adiabatic conditions* exist), the temperature rise can be quite large as shown in Figure 3.5. In mass concrete, true adiabatic conditions can be

Table 3.5

Typical Chemical Composition and Properties of Portland Cements,
ASTM Types I to V

	I^a	II	III	IV	V
C_3S	50	45	60	25	40
C_2S	25	30	15	50	40
C_3A	12	7	10	5	4
C_4AF	8	12	8	12	10
$C\bar{S}H_2$	5	5	5	4	4
Fineness	350	350	450	300	350
(Blaine, m²/kg)					
Compressive strength^b	7	6	14	3	6
[1 day, MPa (lF in.²)]	(1000)	(900)	(2000)	(450)	(900)
Heat of hydration	330	250	500	210	250
(7 days, J/g)					

^{a}CSA designations are 10, 20, 30, 40, and 50, respectively.
^{b}ASTM C109 2 in. mortar cubes.

approached when concrete volumes are large in comparison to the surface from which the heat can be dissipated to the surroundings. The generation of high internal temperatures occurs primarily over the first few days when the concrete is still relatively plastic, and will not cause immediate damage. However, trouble may occur during subsequent cooling when the concrete has gained rigidity, since the resulting thermal stresses may well lead to tensile cracking. For this reason Type III cement should not be used when placing thick (>0.5 m) concrete sections such as piers and abutments. Even the more modest rate of heat evolution for Type I may cause problems when the heat is not rapidly dissipated or when cement contents are high. Temperature rises approaching 30°C have been measured within massive placements.

Thermal cracking was a frequent problem during the construction of some of the earlier concrete dams, even though low cement contents were generally used. This problem was largely solved by the development of ASTM Type IV, low heat of hydration cement. Since C_3S and C_3A are responsible for most of the early liberation of heat, reduction in the amounts of these compounds substantially reduces the amount of heat produced (see Figure 3.5). With the use of Type IV cements, thermal cracking can be eliminated provided that reasonable care is taken in controlling the initial temperature of the concrete and ensuring adequate opportunity for heat dissipation. In mass concrete applications, the slow rate of hardening that results from low C_3S contents is of little consequence.

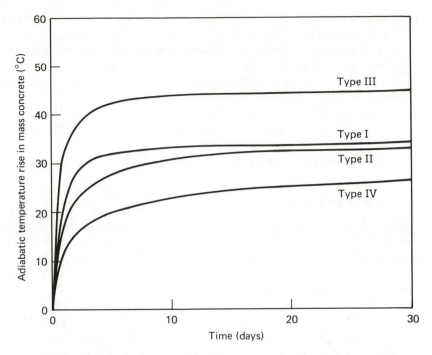

Figure 3.5 Rate of heat evolution measured as a rise in temperature of mass concrete stored under adiabatic conditions. (Based on data in *Concrete Manual*, 8th ed., U.S. Bureau of Reclamation, Denver, Colo., 1975.)

Another problem that was troublesome during the earlier days of concrete construction was the deterioration of portland cement concrete when exposed to waters or soils containing sulfates. An obvious example of such an aggressive medium is seawater, particularly when a structure is in the intertidal zone and is also exposed to wetting and drying. But many groundwaters contain high concentrations of sulfate and are more aggressive than seawater under conditions of total immersion. Sulfate attack involves interactions between the hydration products formed from C_3A, as will be discussed more fully in Chapter 4. Lowering the C_3A content by converting it to C_4AF, which is not so susceptible, has been found to be an effective means of combating sulfate attack. Experience has shown that keeping the C_3A content below 5% by weight, as is done in a Type V cement, will provide satisfactory performance in the most aggressive media (see Table 9.7). Somewhat higher amounts of C_3A, up to 8% by weight, can be tolerated for lesser amounts of sulfate. Type IV cement is a moderately sulfate resistant cement, but because its low rate of strength development is a disadvantage, Type II cement was developed. This cement has a higher C_3S content than Type IV, and therefore better strength development, but it has a higher heat of hydration. It is similar in performance to Type V except for its lower sulfate resistance. Type II is also often used when a lower heat of hydration is required, but with a more normal rate of strength gain.

In the United Kingdom, four main types of portland cement are commonly available: ordinary (BS 12), rapid-hardening (BS 12), low heat (BS 1370), and sulfate-resisting (BS 4027). They are similar to the corresponding ASTM Types I, III, IV, and V, respectively.

Specifications and Properties

It should be noted that the compositions for the various ASTM types given in Table 3.5 are only representative values and that actual compositions can vary widely within each type. For example, a cement company may produce a Type I cement with a C_3S content close to 60% by weight and market the same cement as a Type III merely by grinding it more finely. Although ASTM specifications (see Section 3.5) list chemical requirements for all types, it will be seen (Table 3.11) that there are few limitations on compound composition. ASTM specifications are in essence performance specifications which ensure that certain physical requirements (see Table 3.12) are met regardless of actual chemical composition. Other national standards have adopted the same philosophy and allow considerable latitude in chemical composition. It should also be remembered that the existence of an ASTM specification does not guarantee the commercial availability of a particular cement type. The manufacture of special types of cement requires a change in raw feed composition and burning conditions and the provision of extra storage facilities. Cement companies are reluctant to invest the time and money required for the production of a different cement unless an adequate market exists to allow them to recoup costs. Thus, only Types I and III are available throughout the United States, and Types IV and V are only sold in localities where special markets exist. Type II is in common use in some parts of the country (e.g., the western states), where it is the common cement type rather than Type I. This is dictated by the composition of the most readily available raw materials. Where the properties of a special cement are needed but unobtainable, the performance of Type I can often be modified by the use of admixtures, as will be discussed in Chapter 7.

The ASTM types were developed to control particular aspects of cement hydration, but since all contain the same compounds, although in different proportions, it might be expected that the general properties

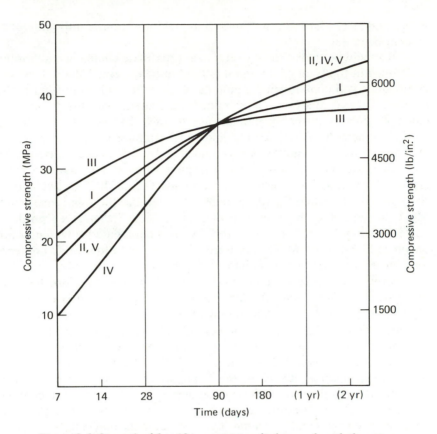

Figure 3.6 Strength of 6 × 12 in. concrete cylinders made with the same aggregates, but different cements. (Adapted from *Concrete Manual*, 8th ed., U.S. Bureau of Reclamation, Denver, Colo., 1975.)

of the hardened concrete would be very similar; and this is indeed the case. Furthermore, in all five types the sum total of the C_3S and C_2S contents is about the same: 75% by weight. Thus, the ultimate strengths of concretes and mortars should be similar even though the initial rates of strength gain are quite different. This is seen to be the case in Figure 3.6, although those cements that gain strength more slowly generally have slightly higher ultimate strengths.

Impurity Oxides

In Table 3.3 a number of impurity oxides were listed. Some, such as MgO, Na_2O, and K_2O, can be of considerable significance with regard to cement performance. Smaller amounts of other oxides, such as Mn_2O_3,

P_2O_5, and TiO_2, are often found, but seldom significantly influence cement behavior.

It should be said at the outset that approximate chemical equilibrium is attained in the rotary kiln and it is thus to be expected that all cement compounds will contain small amounts of the other oxides present in the clinker. Special analyses have indeed confirmed that this is so. The calcium silicates probably contain about 3.0% by weight of impurity oxides, principally Al_2O_3, Fe_2O_3, and MgO. Impure C_3S, as it exists in portland cement, is known as *alite* and impure C_2S as *belite*. Both alite and belite are more reactive than the pure silicates and hydrate more rapidly. C_3A contains considerable amounts (about 10% by weight) of SiO_2 and Fe_2O_3, while C_4AF contains considerable SiO_2 and also much MgO. In fact, it is the presence of MgO in C_4AF that gives portland cement its familiar gray color; pure C_4AF is a chocolate brown color but becomes black when MgO is present. Cements with low MgO contents may have a distinct brownish hue.

Even when free of impurities, however, C_4AF does not have an exact stoichiometry as do the other cement compounds. It is found that C_4AF is the compositional midpoint of a solid solution series between the end members C_2F and (hypothetical) C_2A. In most portland cements the composition lies in the range C_6A_2F to C_6AF_2 and seldom will occur exactly as C_4AF. For this reason it is often called the *ferrite phase* or *ferrite solid solution*, to avoid the implication of a fixed composition. The reactivity of the ferrite phase decreases with increasing iron content. Iron-rich ferrite phases are to be found in cements with a high ferrite content, such as Type V; some Type V cements may contain C_2F in addition to the ferrite phase.

It has been found from experience that limitations on the MgO content of the cement are required; otherwise, the mortar or concrete will start to expand and eventually crack some time after hardening. This is known as *unsoundness*, and has been found to be caused by the fact that much of the MgO is concentrated as *periclase* (crystalline MgO). Ordinarily, MgO hydrates quite rapidly to $Mg(OH)_2$, but because it has been subjected to high temperatures in the kiln, it is much less reactive (dead-burned), and hydration takes place after the cement paste has hardened. The reaction

$$M + H \rightarrow MH \tag{3.2}$$

involves an expansion in solid volume, and hence the periclase particles act as expansive centers that cause stresses within the matrix and eventual cracking.

Even if the MgO content is within specifications (less than 6%), unsoundness may develop if the crystals of periclase in the clinker are large. Thus, the cement must also pass the autoclave expansion test, which accelerates the potentially deleterious expansion by curing at a high temperature. This test also assesses the effect of *free lime* (uncombined lime), which can cause expansion in an analogous manner. Although not specifically limited in the ASTM specifications, free lime contents are generally less than 0.5% in most U.S. cements. Values greater than 1.5% indicate poor control of the burning process and the likelihood of overall inferior performance, apart from unsoundness.

Another form of unsoundness is that due to bulk expansions resulting from high gypsum levels, which can reduce the strength of the paste. This is due to the formation of ettringite (discussed in Chapter 4). Thus, the total SO_3 content is limited in ASTM specifications. There is an optimum addition of gypsum for each cement which will ensure maximum strength and minimum drying shrinkage. Most optimum gypsum levels are less than the equivalent amount of SO_3 permitted by ASTM. However, in some clinkers with high sulfate contents, observance of the ASTM limit may preclude the addition of enough gypsum to reach the optimum level, since the ASTM specification does not distinguish between different origins of SO_3.

The presence of *alkalis* (Na_2O and K_2O) generally does not cause any special problems except when certain aggregates are used that can participate in the alkali-aggregate reaction (see Chapter 6). Under such circumstances, a low-alkali cement should be used (see Table 3.11). It is increasingly difficult, however, to produce low-alkali cements economically. Not only are sources of low-alkali raw materials becoming more scarce, but also modern processes tend to concentrate the alkalis in the clinker. Much of the alkalis in the raw feed volatilizes, and in older plants it is vented up the stack. (The alkalis recondense as the stack gases cool and must be removed to comply with Environmental Protection Agency regulations.) In kilns with preheaters, the alkalis condense as the gases are circulated through the incoming feed, thereby increasing its alkali content or clogging up the preheater ducts or vessels. The use of high-sulfur coals also increases the amount of alkalis in the cement. Most of the additional alkalis are in the form of alkali sulfates (*free alkalis*) rather than as impurities in the cement compounds.

Limits are also placed on the insoluble residue that remains when portland cement is dissolved in concentrated hydrochloric acid. This is to avoid excessive dilution with inert materials. Insoluble residues are derived mostly from the added gypsum, although a part may be unreacted silica. While the limit in ASTM C150 is 0.75%, in BS 12 it is set at 1.5%; thus, the limit appears to be an arbitrary figure.

3.3 MODIFIED PORTLAND CEMENTS

The five ASTM types discussed above all contain the four major clinker compounds —C_3S, C_2S, C_3A, and C_4AF—in varying proportions. There are a number of other portland cements which are formed either by adding other materials to the clinker or by forming other compounds during burning. These cements are collectively referred to as modified portland cements, since they are basically calcium silicate cements.

Portland –Pozzolan Cements

Portland cements can be blended with pozzolans as specified in ASTM C595. A pozzolan is a finely divided form of reactive silica which reacts with calcium hydroxide in cement paste to form additional C-S-H. The pozzolanic reaction is discussed in more detail in Chapter 7. The most important blends are designated Type IP (15 to 40% pozzolan) and Type I-PM (0 to 15% pozzolan). The addition of a pozzolan generally reduces both the heat of hydration and the early strength, but can also confer significant improvements in resistance to sulfate attack and alkali–aggregate reaction, as well as increases in ultimate strength. A portland–pozzolan cement can be designated as moderate or low heat of hydration, or moderate sulfate resistant. A Type IP is similar in properties to a Type IV cement, and a Type I-PM to a Type II. However, low additions of pozzolan will not significantly change the properties of the original cement.

Slag Cements

Blast-furnace slag, a by-product of the iron and steel industry, has potential as a cementitious material. Slag is composed of lime, silica, and alumina, with minor amounts of magnesia, alkali oxides, and iron oxides. The composition depends on the raw materials and industrial processes used but should always have a high lime content (~40%) when used as a cement. Furthermore, the physical structure of the slag depends on the rate of cooling; for use as a cement rapid cooling is necessary to quench the material to form a reactive glass and to prevent the crystallization of unreactive chemical compounds. Slag is generally quenched with water, a process known as *granulation*. Even when prepared in glassy form, slag will not hydrate unless it is activated by adding other compounds: calcium hydroxide or calcium sulfate are generally used. There are various kinds of slag cements, depending on the kind and amount of activator (Table 3.6).

Table 3.6

Types of Slag Cements

Description[a]	Activator	Percent Slag	Comments
Lime–slag cement	CH	70	Obsolete
Portland blast-furnace (slag) cement	CH	20–85	CH from portland cement hydration
Supersulfated cement	$C\bar{S}$	80–85	Small amount of portland cement added

[a]Terminology is not universally established.

Portland Blast-Furnace Cements

Lime for activation is most conveniently supplied by the hydration of portland cement. Table 3.6 shows that wide variations in slag content are allowable in portland blast-furnace cements. These cements are widely used in European countries but are not as popular in the United States. The amounts of slag and portland cement allowable vary from country to country; in the United States ASTM C595 for Type IS cements allows 25 to 65% slag to be blended with portland cement. BS 146 allows any slag content up to 65%, while BS 4246 (Low Heat Portland Blast-Furnace Cement) specifies a slag content of 50 to 90%. This wide variation in slag content means that properties of the cement may vary widely. Slags react slowly to form C–S–H, which is the same hydration product formed by the hydration of C_3S and C_2S. Thus, early-strength development is slower than that of portland cement (see Figure 3.7), depending on the amount of slag present, and heats of hydration are lower also. Slag cements with high slag contents generally have better sulfate resistance than a Type I portland cement, due, it is thought, to a lower calcium hydroxide content. Thus, Type IS cements have comparable properties to a Type IV portland cement.

Supersulfated Cements

When slag is activated with calcium sulfate in the form of anhydrite ($CaSO_4$), together with a small amount of lime or portland cement, the product is known as supersulfated cement. This is not strictly a modified portland cement, but it forms similar hydration products. This cement is not available in the United States but is produced in Europe (BS 4248, for example), although it is less widely used than slag cements. Supersulfated cements have lower heats of hydration than most portland blast

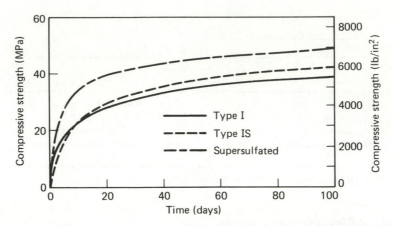

Figure 3.7 Strength development of concretes made with different slag cements.

furnace slag cements and generally show better sulfate resistance. This is caused both by a low CH content and also by the fact that most of the alumina stays combined as ettringite. It is believed that ettringite is responsible for much of the early strength of a supersulfated cement (although it is not a rapid hardening cement).

Expansive Cements

One of the major disadvantages of portland cement concrete is its high drying shrinkage and its susceptibility to tensile cracking when volume contraction is wholly or partially restrained (Figure 3.8a). Shrinkage cracking is unsightly and destroys the integrity of the concrete; thus, special allowance must be made in design and construction. On the other hand, measurable volume expansions during moist curing in ordinary portland cements are very small. If there were greater expansion during hardening, this could offset the contraction that occurs on drying (Figure 3.8b). The development of expansive cements dates back about 50 years. Commercial production began in the United States in the late 1960s, although total production remains quite small, about 500,000 tons annually. A tentative standard specification, ASTM C845-76T, covers expansive cements.

Composition

All three variants of present-day expansive cements are based on the formation of ettringite in considerable quantities during the first few days of hydration. The materials from which ettringite is formed differ sub-

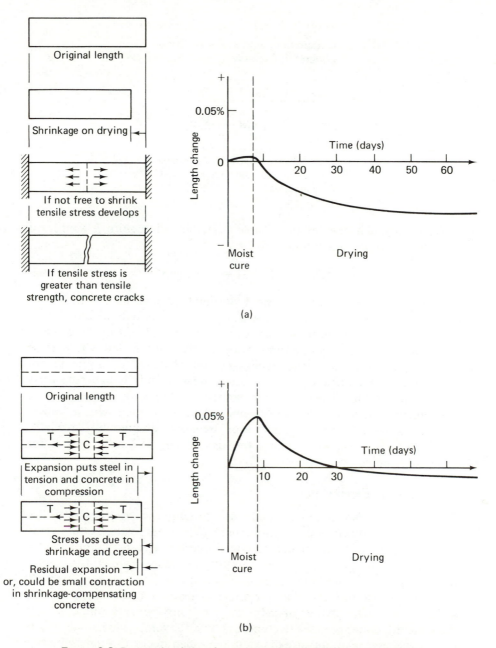

Figure 3.8 Drying shrinkage of concretes made with: (a) Type 1 portland cement; (b) expansive cement. (Adapted from M. Polivka and C. Willson, in *Klein Symposium on Expansive Cement Concretes,* SP-38, American Concrete Institute, Detroit, Mich., 1973, p. 235.)

Table 3.7

Composition of Expansive Components in Expansive Cements

Type K	Type M	Type S
$C_4A_3\overline{S}$	CA	C_3A
$C\overline{S}$	$(C_{12}A_7)^a$	$C\overline{S}H_2$
$(C)^a$	$C\overline{S}H_2$	

[a]Generally contained in the expansive component, but not necessary for expansion.

stantially in each cement (see Table 3.7), but all require a source of calcium aluminate and sulfate ions.

$$(C, A) + \overline{S} + H \rightarrow \text{ettringite} \tag{3.3}$$

Type S is basically a Type I portland cement with a high C_3A content. Type K contains the anhydrous calcium sulfoaluminate $C_4A_3\overline{S}$, which can be formed in the rotary kiln as an integral part of the cement. Alternatively, $C_4A_3\overline{S}$ can be formed separately and blended with a Type I cement together with the necessary anhydrite ($C\overline{S}$). Lime is not necessary for expansion, but it is often present and changes the early rate of expansion. The aluminate source of Type M is the primary constituent of calcium aluminate cement and is blended with a Type I cement. Several expansive admixtures are now commercially available which can confer expansive characteristics on a Type I cement when the concrete is being proportioned.

Mechanism of Expansion

The potential expansion associated with ettringite formation is controlled by the use of ordinary steel reinforcement. The steel restrains the overall bulk expansion of the cement, thereby converting expansion into a slight prestress within the concrete. By resisting expansive forces, the steel is placed into tension and the concrete into compression. About 170 to 700 kPa (25 to 100 lb/in.^{2}) compressive stress is generated, which is generally sufficient to ensure that drying shrinkage will not cause tensile cracking: the concrete either remains in compression or develops only small tensile stresses. Some restraint of expansion may also occur through subgrade friction in slabs or from formwork. The exact prestress will depend on the amount of expansive components and the amount of reinforcement (Figure 3.9). Lack of sufficient restraint will provide little protection

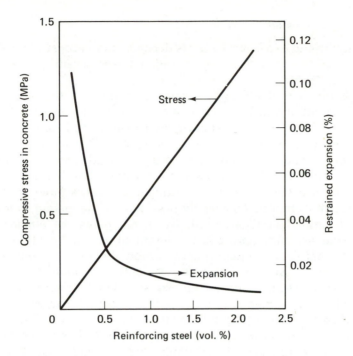

Figure 3.9 Effect of reinforcing steel on expansion of concrete and prestress developed. (Adapted from *New Materials in Concrete Construction*, University of Illinois at Chicago Circle, Chicago, 1972, p. 13–II.)

against shrinkage cracking, and may, in extreme cases, cause the concrete to literally self-destruct; but beyond a certain point additional restraint becomes wasteful. Prestress is developed only in the direction and vicinity of the reinforcing steel. Thus, correct positioning of the steel is important to provide correct restraint, and misplaced reinforcement could lead to lack of adequate prestress or complications such as warping due to differential expansions.

The amount of expansion before drying depends on the amount of expansive component, which should be chosen to ensure that the concrete will remain under a slight compression even after extended drying. Some expansion should be allowed to occur in order to develop maximum residual prestress. It should be emphasized that the development of adequate prestress is the key to shrinkage compensation; as can be seen in Figure 3.8b, concretes made with expansive cements shrink as much, if not more, than ordinary portland cement concretes, but the net tensile stress generated is much less because of the initial compressive prestress.

Control of Expansion

Successful use of expansive cements depends upon proper control of the expansion of the cement during hydration and is sensitive to a number of variables (Table 3.8).

The amount of ettringite formed is controlled by adjusting the cement content or varying the amount of expansive admixture (if one is being used). This assumes that adequate moist curing is provided during the time when ettringite is forming—during the first 7 days. Moist curing is even more critical than it is for ordinary portland cement concrete and requires the supply of additional water for hydration during the curing period, because the formation of ettringite combines large amounts of water, which cannot be completely provided by the mixing water. This is well illustrated in Figure 3.10, which compares the efficiencies of various standard curing procedures (described in Chapter 11).

The other factors influencing expansion affect the amount of *useful* ettringite that is formed. Ettringite can only contribute to an overall volume expansion, and hence to prestress, when it is formed after the concrete has gained rigidity through hydration of C_3S. Any ettringite that forms while the concrete is still plastic is of little value. The effect of temperature on expansion illustrated in Figure 3.11 is an example. At 38°C hydration proceeds faster than at 21°C, and thus the rate of expansion is faster, but ultimate expansion is lower because more ettringite is formed when the concrete is still plastic. The reduction in expansion that occurs with an increase in cement fineness is also a result of the faster initial rate of hydration. An increase in the water/cement (w/c) ratio decreases the amount of expansion. Admixtures may also interfere with the initial formation of ettringite and hence expansion. The

Table 3.8

Variables Affecting Expansion in Expansive Cements

General Category	Specific Variable
Reinforcement	Amount of steel; positioning of rebars
Mix design	Cement content; water/cement ratio; fineness of cement; admixtures
Handling and curing	Time of mixing, handling, and placing; time of moist curing; temperature of curing

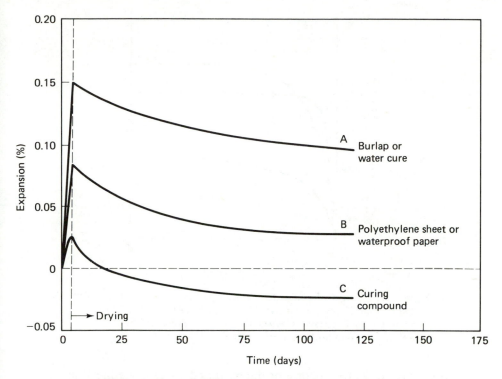

Figure 3.10 Effect of different curing procedures on the expansion of Type M cement concrete. (Adapted from C. E. Kesler, Proceedings ASCE, *Journal of the Construction Division,* Vol. 102, No. CO1, 1976, pp. 41–49.)

effect of a particular admixture cannot be predicted; thus, tests should be run when admixtures are to be used.

Properties

The physical and engineering properties of expansive cement concretes are comparable to Type I cement concretes. Thus, in the absence of any specific data the usual empirical relationships regarding strength, creep, shrinkage, and so on, may be used. Some differences have been observed in (1) compressive strength, (2) rate of heat of hydration, (3) slump loss, and (4) resistance to sulfate attack. Concretes made with Type K cement may have compressive strengths 3.5 to 7.0 MPa (500 to 1000 lb/in.^{2}) higher than comparable Type I cement concrete, particularly at lower w/c ratios and higher cement contents. This has been attributed to the formation of the large quantities of ettringite, but it could also be due to

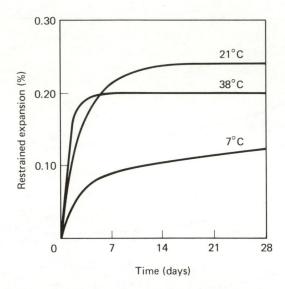

Figure 3.11 Effect of curing temperature on expansion during moist curing of Type K cement concrete. (Adapted from M. Polivka, in *Klein Symposium on Expansive Cement Concretes*, SP-38, ed. S. P. Shah, American Concrete Institute, Detroit, Mich., 1973, p. 250.)

a more uniform microstructure of the hydrated paste since bleeding (see Chapter 8) does not occur. Reduction in bleeding is due to the early crystallization of ettringite, and this can also result in greater slump loss and higher water requirements. Expansive cements are generally not sulfate-resistant because of their high aluminate contents. However, certain Type K cements with high $C_4A_3\bar{S}$ and low C_3A contents may be sulfate-resistant since ettringite remains the stable phase in the mature paste.

Applications

Expansive cements have been used in a wide variety of concrete structures. One of the more frequent uses in the United States has been in parking structures, to prevent water leaks that can cause damage to cars. The largest such structure is the parking building at O'Hare International Airport in Chicago, which used 90,000 m³ (120,000 yd³) of shrinkage-compensating concrete. Shrinkage control joints can be combined with thermal expansion control joints and other construction joints, and the elimination of shrinkage control joints is an attractive advantage in the laying of pavements. The largest paving job to date using this material is probably at the Love Field Airport at Dallas–Fort Worth, where more than 115,000 m³ (150,000 yd³) of shrinkage-compensating cement concrete was used in taxiways. Shrinkage-compensating cement can be used in concrete structures where watertightness is an important requirement, such as water storage tanks, swimming pools, and ice rinks. The use of shrinkage-compensating cement has also been found to be useful in tilt-up construction, where residual prestress can help the building elements to withstand stress imposed during lifting, as well as minimizing separations between elements.

Self-stressing Cements

We have seen that the mechanism of shrinkage compensation is the development of a *chemical prestress* sufficient to overcome tensile stresses induced by differential shrinkage strains. If greater amounts of prestress were developed, the residual compressive stress, after shrinkage is allowed for, could be used in design just as is done in mechanical prestressing. The origin of the prestress is of no concern provided that the required magnitude can be guaranteed. This is more difficult when the prestress is generated internally by a self-stressing cement than when it is applied externally by mechanical means. A self-stressing cement has high potential expansion and the factors affecting expansion become even more critical. Close attention must be paid to the amount and positioning of restraining steel, mix design, and curing conditions for successful use of self-stressing cements. Further, the level of prestress that can be realistically attained will be relatively low, less than 3.5 MPa (500 lb/in.^{2}). For these reasons self-stressing has not advanced much beyond laboratory studies, although it has potential for use in concrete pipe, precast building elements, and ferrocement construction. Precast-concrete elements were once manufactured in Indiana using self-stressing Type M cements. Engineers in Japan and the USSR are actively evaluating the use of self-stressing cements.

Rapid Setting and Hardening Cements

The formation of large quantities of ettringite is also responsible for the properties of two special cements that develop considerable strength within 1 to 2 hours after casting. These are *regulated-set* cement (or *jet cement*) and *VHE* (very high early strength) *cement*. Regulated-set cement, developed in the 1960s, is not presently being produced in North America but is being manufactured in Japan and West Germany. VHE cement is commercially available in North America. In contrast to expansive cements, the ettringite formation gives rise to rapid strength development, and expansions are not as great. In these cements ettringite formation is much more rapid, occurring largely before the paste has gained strength and well before the hydration of the calcium silicates.

Regulated-Set Cement

Regulated-set cement is a modified portland cement (see Table 3.9) in which C_3A is replaced by a new compound, calcium fluoroaluminate, $C_{11}A_7 \cdot CaF_2$[1]. The cement can be produced directly in a rotary kiln or the fluoroaluminate can be blended with a Type I clinker. $C_{11}A_7 \cdot CaF_2$ is even more reactive with water than is C_3A and, in fact, hydrates so rapidly that flash setting will always occur unless sulfate ions are present. The reaction with sulfate ions to form ettringite is also very vigorous, so much so that gypsum will not successfully retard setting because it cannot dissolve fast enough to maintain an adequate supply of sulfate ions. The use of a more soluble salt, such as calcium sulfate hemihydrate (plaster) or sodium sulfate, is needed to avoid flash setting. The time of setting (handling time) can be controlled from about 2 to 40 min using a soluble sulfate (see Figure 3.12) and further extended with an organic retarder such as citric acid. The subsequent supply of sulfate ions required to form additional ettringite for strength development is maintained by slightly soluble anhydrite ($C\overline{S}$).

Strength development occurs very rapidly after setting occurs, and strengths of 7 MPa (1000 lb/in.²) can be attained within 1 h (Figure 3.13). The initial strength rise is due to ettringite forming from $C_{11}A_7 \cdot CaF_2$. Once this reaction slows down, strength development also slows down, until C_3S begins to hydrate in the usual way.

Properties. The level of the early-strength plateau depends on the amount of fluoroaluminate used, which can be controlled either by the cement content of the concrete or by the fluoroaluminate content of the cement.

[1] This is one of the few cement compounds that has a hybrid formula: $C_{11}A_7$ is cement notation, but CaF_2 is in conventional chemical symbols. It is not possible to express calcium fluoride by an oxide formulation.

Table 3.9

Typical Compositions of Rapid-Setting Cements (percent)

Cement Compounds	Regulated-Set Cement	VHE Cement	Type I Portland (for comparison)
C_3S	60	—	50
C_2S	<5	50	25
C_4AF	5	20	8
C_3A	—	—	12
$C_4A_3\overline{S}$	—	20	—
$C_{11}A_7 \cdot CaF_2$	20	—	—
Total SO_3	10	10	3

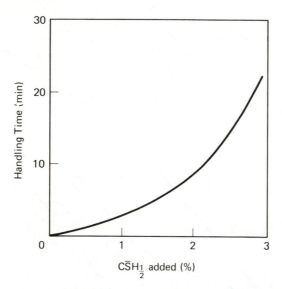

Figure 3.12 Effect of calcium sulfate hemihydrate on the handling time of regulated-set cement. (Adapted from P. Klieger, in *New Materials in Concrete Construction*, ed. S. P. Shah, University of Illinois at Chicago Circle, Chicago, 1972, p. 10 –VI.)

Strengths in excess of 20 MPa (3000 lb/in.^{2}) after 1 h have been achieved in the laboratory using cements with 50% by weight of $C_{11}A_7 \cdot CaF_2$. High early strength is generally accompanied by a relatively short handling time. If the handling time is extended by, for example, the use of citric acid, it will take longer to reach a given strength. A balance between an adequate handling time and a minimum rate of strength development may impose definite limitations on the cement composition.

The reaction of calcium fluoroaluminate is accompanied by the evolution of a large amount of heat, which exceeds that of a typical Type III cement. Even a small pat of regulated-set cement paste quickly becomes warm to the touch. This high rate of heat evolution can help maintain an adequate concrete temperature in winter; if the concrete temperature falls below 4°C, rapid strength development does not occur. Most other concrete properties are much the same as those for ordinary portland cement concrete; but the high aluminate content means that a low sulfate resistance is to be expected, and this has been confirmed.

Applications. One of the first applications visualized for regulated-set cement was in lightweight insulation for roof decks using expanded vermiculate aggregate. This type of concrete would match the rapid set of gypsum plaster, which is widely used, thereby permitting rapid

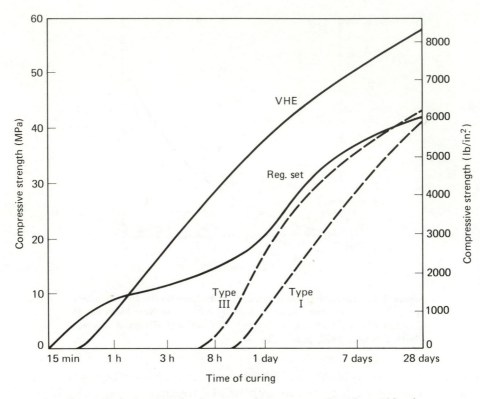

Figure 3.13 Strength developments of concretes made with rapid-hardening cements. (Adapted from W. Perenchio, in *New Materials in Concrete Construction,* ed. S. P. Shah, University of Illinois at Chicago Circle, Chicago, 1972, p. 12–VI.)

placement and have the advantage of better water resistance. But the very rapid strength gain of the cement suggests many other applications in which the properties of a portland cement are desired: pavement and bridge-deck repair, precasting operations, shotcreteing, and slip forming. It is unfortunate that regulated-set cement is not currently available in the U.S., but the interesting properties of the cement will no doubt ensure its reappearance.

VHE Cement

In the production of VHE cement, calcium sulfate is added to the raw mix so that $C_4A_3\overline{S}$ is formed in the rotary kiln. This is the same compound that is present in Type K expansive cements, but the quantities are greater in VHE cement. Calcium sulfate ($C\overline{S}$, insoluble anhydrite) is also formed or is added during grinding. A typical composition is given in Table 3.9, from which it can be seen that C_2S rather than C_3S is the calcium silicate compound.

Properties and composition. Like regulated-set cement, the setting time of VHE can be controlled in the range of 2 to 45 min. Strength development is initially similar to regulated-set cement (see Figure 3.13), and strengths in excess of 7 MPa (1000 lb/in.^{2}) can be attained within 1 h of set. Later strengths are higher than regulated-set cement or Types III and I. As with regulated-set cement, the formation of ettringite will be accompanied by a high rate of heat evolution, but if this early heat can be dissipated, the subsequent hydration of C_2S will not cause additional temperature rise. The presence of C_2S will also improve the durability of the cement paste because of the lower content of calcium hydroxide in the mix, but the long-term sulfate resistance of the cement may not be high. If the concrete is air-entrained, frost resistance should be satisfactory. Creep and drying shrinkage are stated to be lower than for Type III cement concrete.

Miscellaneous Cements

Other Rapid-Hardening Cements

This category covers cements that develop strength faster than a typical Type III cement; regulated-set cements belong in this category. In Britain an *extra-rapid-hardening* cement is sold which is a Type III cement interground with calcium chloride to accelerate hydration. An *ultra-rapid-hardening* cement is also available, which is a very finely ground portland cement: 700 to 900 m^2/kg. In Japan a *super-high-early-strength cement* has been developed with C_3S contents exceeding 70% and traces of chromium, manganese, and fluoride, which are said to increase the rate of hardening.

White Cement

This is a popular cement with architects because it expands the opportunities for creating aesthetic effects. It is especially suitable for exposed aggregate finishes and for making colored cements with pigment additions. The white color is achieved by eliminating iron from the cement; it is thus a Type I or Type III cement with a high C_3A content and no C_4AF. Iron-free clay (kaolinite or china clay) must be used, and bauxite (aluminum oxide) is often needed to achieve the required alumina content. Special ball mills must be used to prevent iron contamination

during grinding. The higher cost of raw materials and changes in manufacturing procedures make white cement expensive.

Masonry Cement

Mortars used for laying brick and block have special requirements for
workability, plasticity, and water retention. To ensure uniform characteristics masonry cements conforming to ASTM C91 or CSA A8 are
generally specified. They are essentially a Type I portland cement mixed
with a finely divided material such as lime or ground limestone together
with an air-entraining agent. These additives provide the "fattiness" and
cohesiveness of a good mortar, improve its adherence to the brick or
block, prevent bleeding, and minimize water loss due to absorption. The
air-entraining agent also provides frost resistance.

Oil-Well Cements

Cements used for sealing oil wells must be slow setting even at high
temperatures and pressures, and stable under highly corrosive conditions.
The American Petroleum Institute has written specifications for six
different classes of cements applicable at certain well depths. Generally,
low-C_3A portland cements are used together with admixtures that retard
setting.

    Cements with no C_3A at all are used as depths increase, and the
improved sulfate resistance of these cements is an advantage since
groundwaters frequently contain sulfate. In the deepest wells, where
temperatures may exceed 260°C (500°F), cements based largely on C_2S
are used to prolong setting times at these temperatures. Oil-well cements
are pumped as slurries with high water contents. Finely divided materials,
such as clays or bentonite, and organic thickening agents are used to
prevent water loss and segregation while maintaining pumpability. The
important parameter is not setting time as measured by penetration tests,
but *thickening time*, which determines how long a slurry can be pumped.

Natural Cement

Early cement production in the United States used "cement rock,"
which is an argillaceous (clayey) limestone with a composition similar to
the raw mix from which portland cement is made. The cement rock is
burned at lower temperatures than are now used for portland cement;
thus very little C_3S is formed and the cement contains mostly C_2S. The
cement is therefore slow hardening and somewhat variable in perform-

ance, since no compositional blending is done. These natural cements are now little used, but are still available in some areas.

3.4 NON-PORTLAND INORGANIC CEMENTS

There are a number of other cements that are used to some extent in building and construction. Although they will not often be encountered, a brief discussion of their composition and performance is warranted.

High-Alumina Cement

High-alumina cement (HAC), which is also known as calcium aluminate cement, deserves consideration because it is widely available and has been used in structural applications in some countries. HAC was originally developed as a sulfate-resisting cement, but other problems preclude its use for this purpose. Several disastrous roof failures involving precast HAC concrete beams have emphasized the limitation of the material. The central problem is the loss in strength that can occur in these concretes due to adverse chemical reactions.

Composition and Chemistry

Unlike portland cement, the manufacture of HAC usually involves complete fusion of the raw materials in a kiln or furnace, although a sintering process is sometimes used, in which pelletized feed is passed through a specially designed rotary kiln. The major component of HAC, as its alternate name suggests, is monocalcium aluminate. Other compounds which are typically present are listed in Table 3.10, but these may vary considerably from one cement to another. The iron content in

Table 3.10

Typical Composition of High-Alumina Cement (percent)

Major constituent:	CA	60
Intermediate constituents:	C_2S	10
	C_2AS	5–20
	(gehlenite)	
Minor constituents:	$C_{12}A_7$	
	FeO (wüstite)	
	Ferrite phase	10–25
	Pleochrite	

particular may vary widely, from 15% to less than 2%. CA hydrates to form calcium aluminate hydrates. The composition of these hydrates depends on the temperature of hydration.

$$CA + H \begin{cases} \xrightarrow{<10°C} CAH_{10} \\[2mm] \xrightarrow{10-25°C} C_2AH_8 + AH_3 \\[2mm] \xrightarrow{>25°C} C_3AH_6 + 2AH_3 \end{cases} \tag{3.4}$$

Under normal ambient conditions a mixture of CAH_{10} and C_2AH_8 can be expected to form. After an initial period of quiescence, CA hydrates quite rapidly. Thus, although HAC has a setting time comparable to portland cements, its early-strength gain is very rapid. Within 24 h of mixing, strengths in HAC concretes can attain values exceeding the 7-day strengths of portland cement (see Figure 3.14). About three-fourths of the ultimate potential strength is obtained by this time. Furthermore, strength gain at low ambient temperatures is much more satisfactory than it is with portland cement. This may be due in part to the high heat of hydration of HAC.

Figure 3.14 Effect of curing temperature on strength development of high-alumina cement concretes.

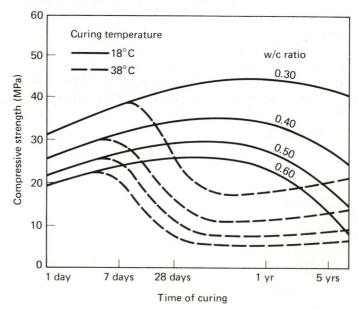

Time of curing

Conversion of HAC

Both C_2AH_8 and CAH_{10} are metastable with respect to C_3AH_6, which means that if the temperature rises above 30°C they will be transformed to C_3AH_6. This transformation is known as *conversion*, and it leads to loss in strength because of a concomitant decrease in solid volume. This results in an increase in porosity, which can be measured experimentally, and a disruption of the original paste microstructure. The increase in porosity is the most important factor, as it not only strongly affects strength, but also opens up the structure to possible environmental attack. Although CAH_{10} and C_2AH_8 are not readily attacked by sulfates or chlorides, C_3AH_6 is less resistant to such chemicals and so the cement becomes prone to sulfate attack. Furthermore, corrosion of reinforcing steel is more likely.

Exposure to hot, moist conditions is conducive to conversion, since the conversion reaction is very sensitive to temperature and proceeds rapidly above 30°C, and it was formerly believed that avoidance of such conditions would minimize any potential strength loss. Yet, substantial conversion seems to have occurred in structures where typical service temperatures would not exceed 30°C. Recent studies have found that conversion occurs even at temperatures below 20°C. Neville (see the Bibliography) quotes considerable field and laboratory data from Europe that point to slow conversion at 20°C, although conversion is always rapid above 30°C, regardless of prior temperature history. However, if the temperature of HAC concrete exceeds 25°C *at any time* during its life, considerable conversion is likely to occur with substantial loss of strength. Just a few hours in excess of 25°C while the concrete is still curing is sufficient for this to happen, and the high heat of hydration makes this a distinct possibility if special precautions are not taken. It would appear that nucleation of C_3AH_6 is the important step in conversion and occurs at about 25°C. Once nucleation has occurred, conversion will take place slowly at lower temperatures.

Strength loss is severe at high *w/c* ratios but can be reduced if the *w/c* ratio is less than 0.40. At lower *w/c* ratios, strength appears to pass through a minimum. This is believed to result from the fact that after initial conversion, residual unhydrated cement can continue to hydrate, thereby removing the additional porosity that is created during conversion. If this explanation is correct, strength loss can be prevented only as long as unhydrated cement is present.

C_3AH_6 is capable of developing strength if it is formed directly, without the intermediate formation of CAH_{10} and C_2AH_8. This can occur during steam curing of HAC when satisfactory and stable strengths can be attained. When C_3AH_6 is formed in the conversion reaction, the disruption of the paste microstructure as well as the increase in porosity prevents the development of intercrystalline bonds between C_3AH_6 crystals. The situation is analogous to strength loss in steam curing of portland cement, which is discussed in Chapter 11.

Structural Use

The early promise of the cement has steadily been eroded by a succession of failures in various countries where it has been used in structural concrete. Not only is strength lost, but the desired sulfate resistance is also diminished, and strength loss due to conversion has been aggravated by subsequent sulfate attack. Structural failures in France, Germany, the United Kingdom, and elsewhere have lead to its ban for structural use in most countries. It is fair to argue that HAC can be successfully used if design strengths are based on residual strengths after conversion that can be *accurately predicted*. We have seen that accurate prediction depends strongly on *w/c* and early temperature history. It might be anticipated that these factors would be adequately controlled during precast operations, but apparently this cannot be guaranteed.

Applications

HAC is now mostly used for refractories, since it is an excellent binder in the production of refractory concretes and castables (such as fire brick) for use at high temperatures. If suitable aggregate is used, the decomposition products that form during the loss of *hydraulic bonding* in the cement paste react with the aggregate to form a new *ceramic bond* which may be stronger than the original bond. The process is analogous to the formation of bonds in fired ceramic ware; but the high green strengths, the ability to successfully fire massive and complex configurations, and the wide choice of aggregate makes it a very versatile industrial refractory. Products may be fired before use, or refractory concretes may be designed to be exposed to high temperatures and ambient temperature simultaneously. The strong ceramic bond at the hot face and the strong hydraulic bond at the cold face ensure good performance, even though at some point there will be a minimum strength during bond transition. There is also considerable scope for the use of HAC in situations of a nonstructural nature. The rapid-strength-gain characteristics can be used to advantage in patching and repair, particularly at low temperatures. (Note that it is inadvisable to mix HAC with portland cement because the mixture has poor strength and durability.)

Gypsum Plaster

Plaster products have a widespread use in the building industry as a surface finish on interior walls and in the production of drywall products for interior lining and partitioning where structural requirements are low. Plasters are calcium sulfate cements which rely on the formation of calcium sulfate dihydrate (gypsum) for strength development. Plaster is β-calcium sulfate hemihydrate ($CaSO_4\cdot\frac{1}{2}H_2O$) formed by controlled heating of natural gypsum. On mixing with water, gypsum is reconstituted by rehydration. If calcination temperatures are too high, anhydrite is formed, which is much less reactive. Dehydration occurs slowly at temperatures below 150°C,

$$\underset{\text{gypsum}}{C\bar{S}H_2} \underset{H_2O}{\overset{150°C}{\rightleftharpoons}} \underset{\substack{\beta\text{-hemihydrate} \\ \text{or plaster}}}{C\bar{S}H_{1/2}} \overset{300°C}{\longrightarrow} \underset{\text{anhydrite}}{C\bar{S}} \tag{3.5}$$

but this temperature is chosen to ensure rapid and complete decomposition.

Gypsum plasters have the advantage of being quick-setting materials that develop strength rapidly. Set time can be controlled through the manufacturing process or by use of special admixtures. The major disadvantage of gypsum plaster as a structural material is its lack of water resistance. Gypsum is quite soluble in water, which makes its use impractical where contact with water is to be expected. It should be noted that leachates from plaster are rich in sulfate ions and the proximity of plaster and portland cement concretes when leaching occurs can lead to sulfate attack.

Other Cements

There are a number of other non-portland, inorganic cements, such as magnesium oxychloride cement (Sorel cement) and phosphate-bonded cements. A discussion of these cements, however, is beyond the scope of this book.

3.5 SPECIFICATIONS AND TESTS OF PORTLAND CEMENT

The production of portland cement requires strict quality control which has made necessary the establishment of specifications for both the chemical and physical requirements of cement. This, in turn, has involved the development of a number of "standard" tests which can be carried out relatively easily and rapidly to ensure that the cement is indeed of the desired quality. It should be noted that different countries have established different tests and specifications, and that even within a country these are subject to frequent revision. Unfortunately, it is usually not possible to compare directly the results obtained by different test methods, though correlations can often be established between them, because the tests do not measure any *fundamental* properties of the cement. Rather, they simply provide information as to whether the cement meets certain (often arbitrary) standards.

Since different test methods for the same property often give quite different results, the test methods used must also be indicated when describing the properties of cement. In North America the most common tests are those developed by ASTM and it is primarily these tests that will be outlined below, though other test methods will be described if they differ substantially from the ASTM procedures.

Chemical Requirements

The chemical requirements for portland cement given in ASTM C150 are shown in Table 3.11. These are similar to BS 12 and BS 1370. These specifications are not very strict, because it has been found that cements with quite different chemical compositions may have suitable physical behavior.

Physical Requirements

The physical requirements for portland cement given in ASTM C150 and shown in Table 3.12 are more significant. The tests for these (and a few other) requirements will be described below. Again, BS 12 and BS 1370 have similar provisions.

Fineness

As has been pointed out, the last step in the production of portland cement is the grinding of clinker and gypsum. The fineness to which the cement is ground can have a considerable effect on the behavior of the cement during hydration. Although it is true that if a cement does meet normal specifications, changing the cement fineness alone will not solve concrete problems that arise in practice, the fineness is nonetheless an important parameter.

Table 3.11

Chemical Requirements of ASTM C150 Standard Specification for Portland Cement[f]

Cement Type	I and IA	II and IIA	III and IIIA	IV	V	Remarks
Standard requirements						
Silicon dioxide (SiO_2), min., %	—	21.0	—	—	—	
Aluminum oxide (Al_2O_3), max., %	—	6.0	—	—	—	
Ferric oxide (Fe_2O_3), max., %	—	6.0	—	6.5	—	
Magnesium oxide (MgO), max., %	6.0	6.0	6.0	6.0	6.0	
Sulfur trioxide (SO_3), max., %						
When ($3CaO \cdot Al_2O_3$) is 8% or less	3.0	3.0	3.5	2.3	2.3	
When ($3CaO \cdot Al_2O_3$) more than 8%	3.5	N.A.[a]	4.5	N.A.	N.A.	
Loss on ignition, max., %	3.0	3.0	3.0	2.5	3.0	
Insoluble residue, max., %	0.75	0.75	0.75	0.75	0.75	
Tricalcium silicate (C_3S), max., %[b]	—	—	—	35	—	
Dicalcium silicate (C_2S), min., %[b]	—	—	—	40	—	
Tricalcium aluminate (C_3A), max., %[b]	—	8	15	7	5	
Tetracalcium aluminoferrite + 2(tricalcium aluminate) (C_4AF + $2C_3A$) or solid solution (C_4AF + C_2F) as applicable, max., %	—	—	—	—	20	
Optional requirements[c]						
Tricalcium aluminate (C_3A), max., %	—	—	8	—	—	For moderate sulfate resistance
Tricalcium aluminate (C_3A), max., %	—	—	5	—	—	For high sulfate resistance
Sum of tricalcium silicate and tricalcium aluminate, max., %	—	58[d]	—	—	—	For moderate heat of hydration
Alkalies (Na_2O + $0.685K_2O$), max., %	0.60[e]	0.60[e]	0.60[e]	0.60[e]	0.60[e]	Low-alkali cement

[a]N.A., not applicable.

[b]The expressing of chemical limitations by means of calculated compounds does not necessarily mean that the oxides are entirely present as such compounds, or that the compounds have the exact stoichiometry implied.

[c]These apply only when specifically requested.

[d]This limit applies when tests for heat of hydration are not requested.

[e]This limit may be specified when the cement is to be used in concrete with aggregates that may be deleteriously reactive.

[f]Reprinted, with permission, from the American Society for Testing and Materials, 1916 Race Street, Philadelphia, PA 19103. Copyright.

Table 3.12

Physical Requirements of ASTM C150 Standard Specification for Portland Cement[f]

Cement Type	I	IA	II	IIA	III	IIIA	IV	V
Standard requirements								
Air content of mortar,[a] vol. %								
Max.	12	22	12	22	12	22	12	12
Min.	—	16	—	16	—	16	—	—
Fineness, specific surface, m²/kg (alternative methods)								
Turbidimeter test, min.	160	160	160	160	—	—	160	160
Air permeability test, min.	280	280	280	280	—	—	280	280
Autoclave expansion, max., %	0.80	0.80	0.80	0.80	0.80	0.80	0.80	0.80
Strength, not less than the values shown for the ages indicated below								
Compressive strength, lb/in.² (MPa)								
1 day	—	—	—	—	1800 (12.4)	1450 (10.0)	—	—
3 days	1800 (12.4)	1450 (10.0)	1500 (10.3) 1000[b] (6.9)	1200 (8.3) 800[b] (5.5)	3500 (24.1)	2800 (19.3)	—	1200 (8.3)
7 days	2800 (19.3)	2250 (15.5)	2500 (17.2) 1700[b] (11.7)	2000 (13.8) 1350[b] (9.3)	—	—	1000 (6.9)	2200 (15.2)
28 days	—	—	—	—	—	—	2500 (17.2)	3000 (20.7)

	I	IA	II	IIA	III	IIIA	IV	V
Time of setting (alternative methods)								
Gillmore test								
Initial set, min., not less than	60	60	60	60	60	60	60	60
Final set, h, not more than	10	10	10	10	10	10	10	10
Vicat test								
Initial set, min., not less than	45	45	45	45	45	45	45	45
Final set, h, not more than	8	8	8	8	8	8	8	8
Optional requirements[c]								
False set, final penetration, min., %	50	50	50	50	50	50	50	50
Heat of hydration[d]								
7 days, max., cal/g (J/g)	—	—	70 (290)	70 (290)	—	—	60 (250)	—
28 days, max., cal/g (J/g)	—	—	80 (330)	80 (330)	—	—	70 (290)	—
Strength, not less than values shown								
Compressive strength, $lb/in.^2$ (MPa)								
28 days	4000 (27.6)	3200 (22.1)	4000 (27.6) 3200[b] (22.1)	3200[b] (22.1) 2560[b] (17.7)	—	—	—	—
Sulfate expansion[e], 14 days, max., %	—	—	—	—	—	—	—	0.045

[a] Compliance with the requirements of this specification does not necessarily ensure that the desired air content will be obtained in concrete.

[b] When the optional heat of hydration or the chemical limit on ($C_3S + C_3A$) is specified.

[c] These apply only when specifically requested.

[d] When the heat of hydration is specified, the ($C_3S + C_3A$) is not limited.

[e] When sulfate expansion is specified, the C_3A and ($C_4AF + 2C_3A$) are not limited.

[f] Reprinted, with permission, from the American Society for Testing and Materials, 1916 Race Street, Philadelphia, PA 19103. Copyright.

1. Of greatest importance, the rate of hydration increases with increasing fineness. This leads to both a higher rate of strength gain (see Chapter 15) and a higher rate of evolution of heat.

2. Since hydration takes place at the surface of the cement particles, and further hydration is hindered by the formation of the reaction products, finer particles will be more completely hydrated than coarser particles. Larger cement particles probably never hydrate completely.

3. Increasing fineness tends to decrease the amount of bleeding, but at high fineness the amount of water required for workability for non-air-entrained concrete is increased, which results in increased drying shrinkage.

4. A high cement fineness reduces the durability of concrete to freeze–thaw cycles.

5. An increased fineness requires a greater amount of gypsum for proper set control, owing to the increased availability of C_3A for reaction.

There are really two things that are of interest: the specific surface of the particles (i.e., the summation of the surface areas of all of the particles in 1 g of cement) and the particle-size distribution. Originally, the fineness of cement was measured by sieve analysis. This is a very awkward technique, because most cement particles are very small, and attempts to use very fine mesh sieves result in clogging. Some specifications still define fineness in terms of the percent passing the No. 200 sieve (75-μm openings), usually requiring at least 80% of the particles to pass through the sieve. (In Canada, the fineness requirement is 82% passing a 75-μm sieve.) However, since this gives no information about the distribution of the fine grains, this method is no longer recommended except for rapid quality control in cement mills.

It is general practice now to describe fineness by a single parameter, the specific surface area. Although cements of quite different particle-size *distributions* might have the same specific surface area, this is still considered to be the most useful measure of cement fineness. And although it is possible to measure particle-size distributions, there is still no agreement on what would constitute a "best" grading curve for cement. It must be noted that for the irregularly shaped cement particles, no two methods of measuring specific surface will give the same results. This is partly because the different methods tend to measure somewhat different properties, and partly because the theories behind the methods are not well-enough developed. Therefore, wherever specific surface is given, the method used to measure it must also be given. The real use of any given method is to allow a comparison between different cements, and thus provide a tool for both quality control and research. Two methods of determining the fineness of cement are recognized by ASTM: the turbidimeter test and the air permeability test.

Wagner Turbidimeter

The Wagner turbidimeter (ASTM C115) method of measuring specific surface involves preparing a suspension of cement in kerosene in a tall glass container. Parallel rays of light are then passed through the container onto a photoelectric cell, and the cross-sectional area of the particles intersecting the beam can be determined by measuring the light intensity. This test is based on *Stokes' law*, which states that when a small sphere falls through a viscous medium under the action of gravity, it ultimately acquires a constant velocity:

$$V = \frac{2ga^2(d_1 - d_2)}{9\eta}$$

(3.7)

where g is the acceleration due to gravity, a the radius of the sphere, d_1 the density of the sphere, d_2 the density of the viscous medium, and η the viscosity. Readings are taken at given times and heights below the surface of the kerosene. From these data, a specific surface area and a particle-size distribution can be obtained. In these calculations the assumptions are made that (1) the particles are spherical, (2) the flow is truly "viscous," and (3) particles in the range of 0 to 7.5 μm have a uniform size of 3.75 μm.

Although the turbidimeter values do not correlate well with the strength of cement, this method does provide a fairly rapid and reproducible measure of the relative fineness of cement.

Blaine Air Permeability Apparatus

The air permeability method of determining the specific surface is based on the relationship between the surface area of the particles in a porous bed and the rate of fluid flow through the bed. This test is described in BS 4550: Part 3: Section 3.1. The basic equation developed by Carman is

$$S = \frac{14}{D(1 - \epsilon)} \frac{\sqrt{\epsilon^3 A i}}{vQ}$$

(3.8)

where S is the specific surface (cm²/g), D the powder density, ϵ the porosity of the bed, A the cross-sectional area, i the hydraulic gradient, ν the kinematic viscosity, and Q the rate of flow. In the Blaine method (ASTM C204), rather than passing air through the bed at a constant rate, a given volume of air is passed through a bed of standard porosity at a steady diminishing rate, and the time (t) required is measured. The specific surface (S) is then calculated from the relationship

$$S = K \sqrt{t} \qquad\qquad (3.9)$$

where K is a constant. In practice, K is not determined directly; rather the sample is compared to a standard sample of known surface area issued by the U.S. Bureau of Standards.

In summary, both the Wagner and Blaine methods provide an acceptable way of measuring surface areas. The Blaine value is generally about 1.8 times the Wagner value, probably because of the different theories involved. The Blaine method is more commonly used, but in cases of dispute the Wagner method is deemed to govern. Various modifications of the turbidimeter and air permeability methods have been developed.

Tests on Cement Paste

Normal Consistency

Two of the common physical requirements for cement paste, time of setting and soundness, depend on the water content of the neat cement paste. Therefore, it is necessary to define a water content at which to do these tests. This is defined in terms of the *normal consistency*, measured according to ASTM C187. This test is based on the depth of penetration of a 10-mm-diameter needle under a load of 300 g into fresh cement paste, using the Vicat apparatus shown in Figure 3.15. A paste is said to have normal consistency when the plunger penetrates 10 ± 1 mm below the original surface in 30 s. The amount of water required is expressed as a percentage by weight of the dry cement, the usual range being about 24 to 33%. (In BS 4550: Part 3: Section 4.5, normal consistency is defined in terms of a 35 ± 1 mm plunger penetration.) This test is a measure of the plasticity of the cement paste, although it does not correlate particularly with the quality of the cement. It is very sensitive to the conditions under which it is carried out, particularly temperature and the way the cement is compacted into the mold.

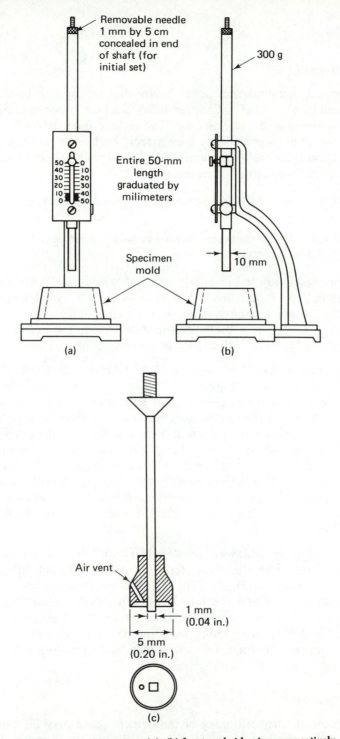

Figure 3.15 Vicat apparatus: (a), (b) front and side view, respectively, of apparatus set up for consistency tests; (c) enlarged view of needle used to determine final set. (Adapted from ASTM C187 and C191. Reprinted, with permission, from the American Society for Testing and Materials, 1916 Race Street, Philadelphia, PA., 19103. Copyright.)

Time of Setting

The chemical reactions that occur during the setting and hardening of cement will be discussed in Chapter 4. We are here concerned only with the measurement of the setting time. The setting times defined below are of no fundamental significance; they merely define two arbitrary points in the general relationship between the time of the addition of water and strength gain. Two setting times are defined:

1. *Initial set,* which indicates that the paste is beginning to stiffen considerably.

2. *Final set,* which indicates that the cement has hardened to the point at which it can sustain some load.

These tests are used primarily for quality control. Generally, initial set occurs in 2 to 4 h, and final set in 5 to 8 h. Concretes, which generally have higher water contents, tend to set more slowly. There are two commonly accepted test methods for determining the setting time of cement paste. They are both carried out on pastes of normal consistency.

Time of setting by the Vicat needle (ASTM C191). In this test, the Vicat apparatus illustrated in Figure 3.15 is used, except that the 1-mm-diameter needle is used for penetration. The initial setting time is defined as the time at which the needle penetrates 25 mm into the cement paste (35 mm in BS 4550: Part 3: Section 3.6). The final set occurs when the needle does not sink visibly into the paste. A tentative revision to this standard is the use of the needle shown in Figure 3.15c for use in determining the final setting time. With this needle, final set is defined as the time when only the needle penetrates the surface, while the attachment fails to do so. This latter method is specified in BS 4550: Part 3: Section 3.6.

Time of setting by Gillmore needle (ASTM C266). This test is less commonly used than the Vicat test and gives different results. The apparatus is shown in Figure 3.16. The cement is considered to have acquired initial set when the initial Gillmore needle, weighing 113.4 g and with a diameter of 2.12 mm, fails to penetrate. Final set is reached when the final Gillmore needle, weighing 453.6 g and with a 1.06 mm diameter, fails to penetrate. Gillmore times tend to be longer than Vicat times.

Early Stiffening

The problem of early stiffening of the cement paste may be manifested by two different phenomena (which are discussed in more detail in Chapter 8).

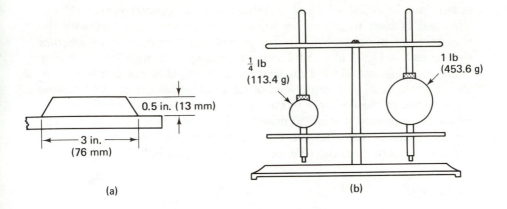

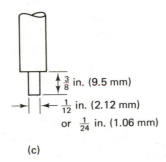

Figure 3.16 Gillmore apparatus: (a) test pat; (b) weighted needles; (c) detail of needle tips. (Adapted from ASTM C266. Reprinted, with permission, from the American Society for Testing and Materials, 1916 Race Street, Philadelphia, PA., 19103. Copyright.)

1. *False set.* This refers to the rapid development of rigidity in cement paste, without the evolution of much heat. The rigidity can be overcome and plasticity regained by further mixing, but with no addition of water.

2. *Flash set (or quick set).* This refers to the rapid development of rigidity in cement paste, *with* the evolution of considerable heat. The rigidity cannot be overcome, and the plasticity cannot be regained, without both further mixing and the addition of water.

The test for early stiffening is described in ASTM C451, using the Vicat apparatus. A paste is prepared with enough water so that the initial penetration 20 s after the completion of mixing is 32 ± 4 mm. The final penetration is determined 5 min after the completion of mixing. The results are reported as (final penetration/initial penetration) $\times$ 100%. This test is used to determine whether the cement complies with the requirements of Table 3.12. It also provides information as to whether the cement is likely to exhibit early stiffening. Although false set may make handling and placing the cement difficult, it is not deleterious to the concrete quality. The problem of false set will not occur with transit mixing, where the concrete is mixed for a long time before placing, or where the concrete is remixed prior to placement, as with pumping. The consequences of flash set are more severe and will usually cause the cement to fail to meet the time of set requirements in Table 3.12.

Unsoundness

Unsoundness in cement paste results from excessive volume change after setting. If there is any appreciable expansion, however slow, cracking and failure of the concrete will result. Unsoundness in cement is caused by the slow hydration of MgO or free lime, and by the reactions of gypsum with C_3A (see Sec. 3.1). These expansive reactions take place very slowly, and so unsoundness will only appear after many months, or even years. Therefore, it is necessary to use some form of accelerated test, so that tendencies toward unsoundness can be detected as a quality control measure. Although a number of soundness tests have been developed over the years, only two are in common use.

Le Châtelier test. This test will detect only unsoundness due to excessive free lime. It is commonly used in Great Britain (BS 4550: Part 3: Section 3.7), where little magnesia is usually present in the raw materials. The test apparatus is shown in Figure 3.17. A cement paste of normal consistency is prepared and is used to fill the split brass cylinder on a piece of glass. The cylinder is then covered with another glass plate, and the whole assembly is immersed in water at 20 $\pm$ 1°C for 24 h. The distance between the indicator points is measured, and then the assembly is returned to the water, brought to the boil in 25 to 30 min, and boiled for 1 h. The assembly is then cooled and the distance between the indicator points measured again. The difference in readings is proportional to the amount of expansion, and must not exceed 10 mm. If this is exceeded, the cement may be aerated for 7 days and the test repeated; in this case, expansions may not exceed 5 mm.

Autoclave expansion. The autoclave expansion test (ASTM C151) is more severe than the Le Châtelier test. It will detect unsoundness due to both excess CaO and excess MgO, and is the test designated in ASTM C150. A cement of normal consistency is molded and cured normally for 24 h. The specimens are then removed from the molds, measured, and

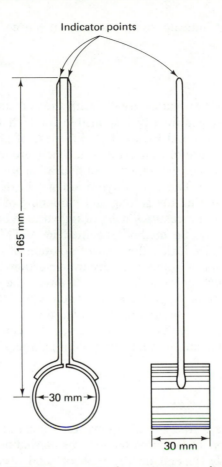

Figure 3.17 Le Châtelier's apparatus for measurement of unsoundness in cement pastes.

placed in an autoclave. The temperature is raised so that the steam pressure in the autoclave rises to 2 MPa (295 lb/in.^{2}) in 45 to 75 min; this pressure is maintained for 3 h, and the autoclave cooled so that the pressure is relieved in 1½ h. The specimen is then cooled in water, to 23°C in 15 min. Its length is again measured after a further 15 min; the expansion must be less than 0.80% if the cement is to be acceptable.

These accelerated tests clearly do not simulate service conditions, and the amount of expansion may be affected by a number of factors, such as the fineness of the cement and the C_3A content. Thus, these tests can only serve as a guide and cannot provide an absolute indication of a tendency toward unsoundness for a given cement. However, they provide a valuable quality control measure, since changes in the expansion tests would indicate some anomaly in the cement manufacturing process.

Heat of Hydration

The heat of hydration varies greatly with the cement composition, with C_3A and C_3S being primarily responsible for high heat evolution [see Eqs. (3.1a and 3.1b) and Figure 3.5]. Fineness of grinding is important with regard to the *rate* of heat evolution. Finely ground cements increase the hydration rate, but the total heat of hydration at very long ages is not particularly affected. The heat of hydration may be defined as the amount of heat evolved during the setting and hardening of portland cement at a given temperature measured in J/g of unhydrated cement. This is most commonly done by the method specified in ASTM C186, the heat of solution method. Basically, the heat of solution of dry cement is compared to the heats of solution of separate portions of the cement that have been partially hydrated for 7 and 28 days. The heat of hydration is then the difference between the heats of solution of the dry and partially hydrated cements for the appropriate hydration period. As may be seen from Table 3.12, only Type II and Type IV cements have heat-of-hydration requirements. A similar procedure is specified in BS 4550: Part 3: Section 3.8.

Tests on Mortar

While some of the specifications for cement require tests done on neat pastes, others require that the tests be carried out on mortars. Since such tests clearly depend on the type of sand used, the mortar tests defined by ASTM all use a graded standard sand. This is a natural silica sand from Ottawa, Illinois, with a specific grading between No. 100 (150-μm) and No. 16 (1.18-mm) sieves as defined in ASTM C109. Sands from different sources may give quite different results even though they meet this grading requirement.

Mortar Flow

As with cement pastes, many of the properties of mortars also depend on the consistency, or w/c ratio. While some specifications for mortar tests are written in terms of a fixed w/c ratio, others are written in terms of consistency. The consistency of mortars is expressed as mortar flow, determined according to the procedures of ASTM C109. Mortar is prepared with a ratio of 2.75 parts Ottawa sand to 1 part cement (by weight). It is compacted in a mold in the form of a truncated cone 2 in.

(50.8 mm) deep, with a bottom diameter of 4 in. (102 mm) and a top diameter of 2¾ in. (69.8 mm). The cone is placed on a flow table, that is, a table whose top can be raised and dropped through a height of 0.5 in. (12.7 mm) by means of a rotating cam. The mold is removed from the mortar, and the table is dropped 25 times in 15 s. The *flow* is then the resulting increase in the average base diameter of the mortar mass, measured as a percentage of the original diameter.

Strength Tests

Since cement is used primarily as a structural material, its strength properties are of prime importance. Therefore, a number of strength tests have been developed to try to answer two questions:

1. What will be the strength of concrete made with a particular cement?

2. How do different cements compare with one another?

Unfortunately, strength is not a very easy property to define, as will be discussed in more detail in Chapter 15. The factors that can influence the measurement of strength include the *w/c* ratio, cement/sand ratio, type and grading of sand, manner of mixing and molding specimens, curing conditions, size of specimen, shape of specimen, moisture content at time of test, loading conditions, and age. In view of these many variables, it should be clear that any strength test must follow the specified testing procedure very closely. Before we examine the test methods themselves, two of these variables must be discussed further: sand and age.

Ever since the testing of cement began, the question has been raised as to whether strength tests should be carried out on neat pastes or on mortars. Although it would appear to be most logical to carry out the test on a neat paste, since the use of sand introduces a number of extra variables, such tests are rarely used today. The problem with neat cement paste is that it is difficult to handle and test, and thus more variability is introduced into the results. It has also been found that cements which appear to be the same when tested neat may behave quite differently when used in mortar, and so mortar tests provide a more reliable indication of cement quality. Tests on neat cement pastes are now used only for research purposes. However, having decided to use mortar, the sand must be very carefully specified, as is done in ASTM standards.

Since cement strength increases with time, it is also necessary to specify the age at which tests should be carried out. Normally, minimum strengths are specified for 3, 7, and 28 days. In addition, high-early-strength cement (Type III) has a 1-day requirement, and the low heat cements (Type IV) which hydrate slowly may sometimes have a 90-day requirement as well. The requirements of ASTM C150 are given in Table 3.12. The point to remember is that it is the shape of the strength–time relationship that is most important, not simply the strength at the single given time. Strength can be measured in compression, tension, or flexure.

Compressive strength. This is by far the most common measure of strength required by cement specifications. The test prescribed in ASTM C109 uses as the test specimen a 2-in. mortar cube. The sand/cement ratio is 2.75:1, using the standard Ottawa sand. The w/c ratio is 0.485 for all portland cements, and 0.460 for air-entraining portland cements. (However, some other specifications express the water requirement in terms of mortar flow.) The mortar is mixed according to a certain schedule, compacted into the molds, and placed in a moist storage room for 24 h. The specimens are then removed from the molds and stored in saturated lime water at 23°C until tested. They are tested wet, using a testing machine with a spherically seated block. The loading rate is such that the specimen will fail in 20 to 80 s.

As an alternative to this test, compressive strength determinations may also be made on portions of test prisms broken in flexure (see below), as described in ASTM C349 (modified cube method). The two broken portions of the flexural specimen (originally 40 × 40 × 160 mm) are both tested. The test is carried out by loading each portion through bearing plates in the form of 40.32 × 50.8 mm rectangles. The strength measured on these modified cubes is not directly comparable to the standard cube strength. The compressive strength S_c of the modified cubes is calculated as

$$S_c = 0.40P \qquad (\text{lb/in.}^2)$$
$$S_c = 0.62P \qquad (\text{kPa})$$

$$(3.10)$$

where P is the maximum load in pounds or newtons.

On the other hand, BS 4550: Part 3: Section 3.4 permits a choice between two different compressive strength tests. One of these is carried out on 70.7 mm mortar cubes, made with w/c ratio of 0.4. The other is performed on 100 mm concrete cubes, with a w/c ratio of 0.6, and made with specified coarse and fine aggregates. These two tests will yield quite different strength values.

Tensile strength. The direct tension test used to be the most common strength test, primarily because in the early years of cement manufacture, testing machines with the capacity for breaking mortar in compression

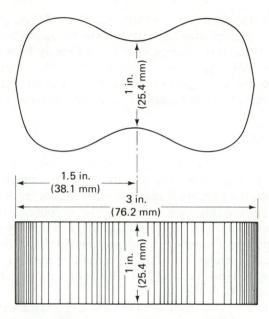

Figure 3.18 Briquet specimen used for tensile strength measurement of mortars. (Adapted from ASTM C190. Reprinted, with permission, from the American Society for Testing and Materials, 1916 Race Street, Philadelphia, PA., 19103. Copyright.)

were not readily available. However, although it has long been known that the tensile test does not provide a useful criterion of the concrete-making properties of cements, the test still persists in some places and is still described by an ASTM Standard, C190. Briefly, the test is carried out on mortar specimens of the shape shown in Figure 3.18, with the load applied through specifically designed grips. Since it is very difficult to apply a truly axial load to these specimens, and since the tensile properties of cement are of little interest, the results are not of much value.

Flexural strength. The flexural test (which is really a measure of the tensile strength in bending) has gained considerable popularity lately, particularly in Europe. This is partly because the same specimen may then be used for compressive strength determinations using the modified cube method described above. In ASTM C340, which is somewhat different from the European method, specimens are cast in the form of $40 \times 40 \times 160$ mm prisms. The proportioning and mixing of the mortar are the same as for the compression test. The specimens are then tested in flexure in center-point loading, and the flexural strength calculated.

It must be remembered, however, that the strengths of mortar as determined by these tests cannot be related directly to the strengths of concrete made with the same cements. Thus, the strength tests on mortars serve primarily as quality control tests. The strength of concrete can only be determined from tests done on the concrete itself.

Air Content of Mortar

The purpose of the test for the air content of mortar (ASTM C185) is to determine the air-entraining potential of a given cement sample. For ordinary cement, the purpose is to ensure that the cement will not entrain undesired air; for air-entraining cements, the purpose is to ensure that the additions are present in the correct quantity. Since the air content of concrete depends on a number of factors, the results of tests on mortar cannot be closely correlated to the air content of concrete made with a particular cement. However, ASTM C150 does specify maximum and minimum air contents of the mortar.

In the test, mortar is made with cement and a standard sand, graded so that all the particles lie between the No. 16 and No. 30 sieves (ASTM C778). The water content is chosen to give a flow of 87.5 ± 7%. The mortar is then compacted lightly into a 400-ml cup and the weight of mortar determined. The volumetric air content is calculated from the measured density and that of the air-free mixture, determined from the mixture proportions and the separate densities of the constituents. However, difficulty may be encountered in carrying out this test, in the form of apparent high air contents that occur occasionally with certain lots of sand.

Sulfate Expansion

There are no really suitable tests available for determining the resistance of a cement to sulfate attack, but ASTM C452 can provide useful information and is required for Type V cements. This test measures the expansion of mortar bars made from a mixture of cement and gypsum such that the total SO_3 content is 7.0% by weight. Mortar specimens are prepared in the form of $1 \times 1 \times 11\frac{1}{4}$ in. ($25 \times 25 \times 285$ mm) bars, using graded standard sand, a sand/(cement + gypsum) ratio of 2.75 and a water/(cement + gypsum) ratio of 0.485. After casting, the specimens are stored in water at 23°C, and the lengths determined at different times. The expansion is then a measure of the sulfate resistance of the cement and should not exceed 0.045% after 14 days.

Bibliography

Portland Cements

DUDA, W. H. *Cement Data Book,* 2nd ed. Bauverlag GmbH, Wiesbaden, Germany, 1976.

LEA, F. M., *The Chemistry of Cement and Concrete,* 3rd ed. Chemical Publishing Co., New York, 1971.

PERAY, K., AND J. J. WADDELL, *The Rotary Cement Kiln.* Chemical Publishing Co., New York, 1972.

POPOVICS, S., *Concrete-Making Materials.* McGraw-Hill Book Company, New York, 1979.

SOROKA, I., *Portland Cement Paste and Concrete.* The Macmillan Press Ltd., London, 1979.

Special Cements

Expansive Cements

A.C.I. COMMITTEE 223, "Expansive Cement Concretes—Present State of Knowledge," *Journal of the American Concrete Institute,* Vol. 67, No. 8, pp. 583–610 (1970).

A.C.I. COMMITTEE 223, "Recommended Practice for the Use of Shrinkage-compensating Cement," *Journal of the American Concrete Institute,* Vol. 73, No. 6, pp. 319–339 (1976).

KESLER, C. E., "Control of Expansive Concretes during Construction," *Proceedings ASCE, Journal of the Construction Division,* Vol. 102, No. CO1, pp. 41–49 (1976).

Klein Symposium on Expansive Cement Concretes, SP-38. American Concrete Institute, Detroit, Mich., 1973.

Rapid-Hardening Cements

KLIEGER, P., "Regulated-Set Cements—Material Properties," and W. PERENCHIO, "Regulated-Set Cements—Applications and Field Problems," in *New Materials in Concrete Construction* (proceedings of a conference, December 1971), ed. S. P. Shah. University of Illinois at Chicago Circle, Chicago, 1972.

UCHIKAWA, H., AND K. TSUKIYAMA, "The Hydration of Jet Cement at 20°C," *Cement and Concrete Research,* Vol. 3, No. 3, pp. 263–277 (1973).

High-Alumina Cement

NEVILLE, A. M., *High Alumina Cement Concrete,* John Wiley & Sons, Inc., New York, 1975.

Slag Cements

NURSE, R. W., "Slag Cements," in *The Chemistry of Cements,* ed. H. F. W. Taylor, Vol. 2, pp. 37–68. Academic Press, Inc., New York, 1964.

SCHRODER, F., "Blastfurnace Slags and Slag Cements," *Proceedings, Fifth International Symposium on the Chemistry of Cement, Tokyo, 1968,* Vol. 4, pp. 149–199. Cement Association of Japan, Tokyo, 1969.

Oil-Well Cements

SMITH, D. K., *Cementing, Monograph Volume 4,* Society of Petroleum Engineers of AIME, American Institute of Mining, Metallurgical and Petroleum Engineers Inc., New York, 1976.

Test Methods

Significance of Tests and Properties of Concrete and Concrete-making Materials, ASTM STP 169B. American Society for Testing and Materials, Philadelphia, Pa., 1978.

Problems

3.1 Calculate the compound composition for a portland cement having the following oxide analysis:

(a) C = 64.15%, S = 21.87%, A = 5.35%, F = 3.62%, $\bar{S}$ = 2.53%

(b) C = 64.15%, S = 21.37%, A = 5.35%, F = 3.62%, $\bar{S}$ = 2.53%

(c) C = 64.15%, S = 21.87%, A = 6.02%, F = 2.63%, $\bar{S}$ = 2.84%

(d) C = 63.54%, S = 23.09%, A = 3.61%, F = 6.38%, $\bar{S}$ = 2.29%

3.2 Calculate the heat of hydration after 3 days for a portland cement with the following compound composition:

(a) C_3S = 55%, C_2S = 24%, C_3A = 10%, C_4AF = 9%

(b) C_3S = 27%, C_2S = 51%, C_3A = 7%, C_4AF = 13%

3.3 Assign each of the following cements to one of the ASTM Types I to V:

(a) C_3S = 55%, C_2S = 21%, C_3A = 11%, C_4AF = 8%, fineness 480 m²/kg

(b) C_3S = 55%, C_2S = 22%, C_3A = 10%, C_4AF = 8%, fineness 380 m²/kg

(c) C_3S = 44%, C_2S = 34%, C_3A = 4%, C_4AF = 13%, fineness 370 m²/kg

(d) C_3S = 30%, C_2S = 45%, C_3A = 6%, C_4AF = 12%, fineness 320 m²/kg

3.4 Why is gypsum interground with portland cement clinker?

3.5 Why is controlled blending an important step in the manufacture of portland cement? (See Problem 3.1.)

3.6 Why must particular attention be paid to curing expansive cements?

3.7 Why must expansive cement concretes be restrained during hardening?

3.8 What cements would be appropriate for pavement repair under winter conditions?

3.9 Why is high-alumina cement not allowed to be used structurally?

3.10 How would you ensure the serviceability of an existing HAC structure?

3.11 Why is it important to follow standard tests exactly when testing portland cement?

3.12 Can the behavior of concrete be accurately predicted from the results of the standard tests used to evaluate the performance of portland cement?

3.13 Is C_3A a useful component of portland cement?

3.14 What role does ettringite play in cement technology?

4

hydration of portland cement

4.1 CHEMISTRY OF HYDRATION

The setting and hardening of concrete are the result of chemical and physical processes that take place between cement and water. An adequate understanding of the chemistry of hydration is necessary for a full appreciation of the properties of cements and concretes and is discussed in detail in this chapter. The underlying chemistry has already been applied in Chapter 3, and will be referred to again in later chapters, particularly those dealing with admixtures and durability.

Hydration of Pure Cement Compounds

The chemical reactions describing the hydration of the cement compounds have been worked out through the study of the hydration of the pure cement compounds. It is assumed that the hydration of each compound takes place independently of the others that are present in portland cement. This assumption is not completely valid, since interactions between hydrating compounds can have important consequences, but in most cases it is reasonable.

Calcium Silicates

The hydration reactions of the two calcium silicates [Eqs. (4.1) and (4.2)] are stoichiometrically very similar, differing only in the amount of calcium hydroxide formed.

$$2C_3S \; + \; 6H \longrightarrow C_3S_2H_3 \; + \; 3CH \qquad (4.1)$$

tricalcium   water   C–S–H   calcium
silicate   hydroxide

$$2C_2S \; + \; 4H \longrightarrow C_3S_2H_3 \; + \; CH \qquad (4.2)$$

dicalcium   water   C–S–H   calcium
silicate   hydroxide

The principal hydration product is a calcium silicate hydrate. The formula $C_3S_2H_3$ is only approximate because the composition of this hydrate is actually variable over quite a wide range. It is a poorly crystalline material which forms extremely small particles in the size range of colloidal matter (less than 1 μm) in any dimension. Its name, C–S–H, reflects these properties (an earlier name, tobermorite gel, is not now used). The properties of this hydration product will be discussed in more detail later. In contrast, calcium hydroxide is a crystalline material with a fixed composition.

The reaction sequence is most conveniently described by reference to a calorimetric curve (see Figure 4.1) which measures the rate of heat evolution with time. The heat flow is proportional to the rate of reaction and is easily measured in the case of C_3S, which is the compound of major interest. When first mixed with water, a *period of rapid evolution of heat* (stage 1) occurs, which ceases within about 15 min. There follows

Figure 4.1 Rate of heat evolution during the hydration of tricalcium silicate.

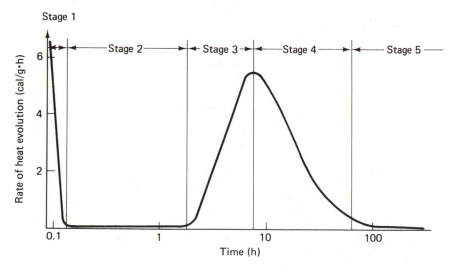

a period of relative inactivity, the *dormant period* (stage 2), which is the reason why portland cement concrete remains in the plastic state for several hours. Initial set occurs in 2 to 4 h, about the time C_3S has begun to react again with renewed vigor at the end of the dormant period. The silicate continues to hydrate rapidly, reaching a maximum rate at the end of the *acceleration period* (stage 3), which corresponds with the maximum rate of heat evolution. By this time (4 to 8 h) final set has been passed and early hardening has begun. Thereafter the rate of reaction again slows down (stage 4) until it reaches a *steady state* (stage 5) within 12 to 24 h.

The following processes occur at each stage (Table 4.1). On first contact with water, calcium ions and hydroxide ions are rapidly released from the surface of each C_3S grain; the pH rises to over 12 within a few minutes, which indicates a very alkaline solution. This hydrolysis slows down quickly but continues throughout the dormant period. When the calcium and hydroxide concentrations reach a critical value, the hydration products, CH and C–S–H, start to crystallize from solution and the reaction of C_3S again proceeds rapidly. The dormant period is apparently caused by the need to achieve a certain concentration of ions in solution before crystal nuclei form from which the hydration products grow. This is a requirement for many chemical reactions and is known as nucleation

Table 4.1

Sequence of Hydration of the Calcium Silicates

Reaction Stage	*Kinetics of Reaction*	*Chemical Processes*	*Relevance to Concrete Properties*
1 Initial hydrolysis	Chemical control; rapid	Initial hydrolysis; dissolution of ions	—
2 Dormant period	Nucleation control; slow	Continued dissolution of ions	Determines initial set
3 Acceleration	Chemical control; rapid	Initial formation of hydration products	Determines final set and rate of initial hardening
4 Deceleration	Chemical and diffusion control; slow	Continued formation of hydration products	Determines rate of early strength gain
5 Steady state	Diffusion control; slow	Slow formation of hydration products	Determines rate of later strength gain

control. CH crystallizes from solution, while C–S–H apparently develops at the surface of the C_3S and forms a coating covering the grain. As hydration continues, the thickness of the hydrate layer increases and forms a barrier through which water must flow to reach the unhydrated C_3S and through which ions must diffuse to reach the growing crystals. Eventually, movement through the C–S–H layer determines the rate of reaction and hydration becomes diffusion-controlled. Reactions that are diffusion-controlled are quite slow and become slower as the thickness of the diffusion barrier increases. Thus, hydration tends to approach 100% completion asymptotically.

C_2S hydrates in a similar manner, but is much slower because it is a less reactive compound than C_3S. The amount of heat liberated by the hydration of C_2S is also lower than it is with C_3S, and thus the calorimetric curve in Figure 4.1 is not easy to measure experimentally.

Chemical reactions are sensitive to temperature, the rate of reaction increasing with temperature. This is also true for hydration, but the temperature dependence is related to the extent of reaction. Hydration is most sensitive to temperature through stage 3 when the reaction is chemically controlled. Once hydration is completely diffusion-controlled

Figure 4.2 Effect of temperature on the hydration of tricalcium silicate. (Adapted from L. E. Copeland and D. L. Kantro, in *Proceedings, Fifth International Symposium on the Chemistry of Cement, Tokyo, 1968,* Vol. 2, pp. 387–419.)

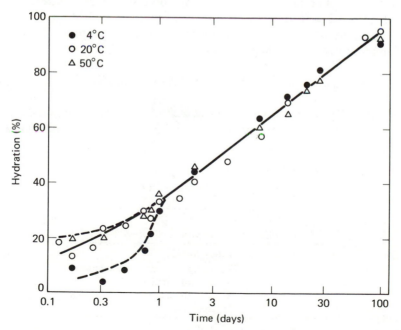

in stage 5, it is much less temperature-sensitive, although the diffusion coefficient of the hydrate barrier will vary with temperature. The overall effect of temperature is illustrated in Figure 4.2, which compares the amount of hydration of C_3S at different temperatures. The stoichiometry of hydration remains essentially the same up to about 100°C, although the composition of the C–S–H changes somewhat. At higher temperatures, which are used in autoclaving, the chemistry begins to change and this will be discussed in Chapter 11.

Tricalcium Aluminate

In portland cement the hydration of C_3A involves reactions with sulfate ions which are supplied by the dissolution of gypsum. The reactions are summarized in Table 4.2. The primary initial reaction of C_3A is:

$$C_3A \quad + \quad 3C\bar{S}H_2 + 26H \ \rightarrow C_6A\bar{S}_3H_{32} \qquad (4.3)$$

tricalcium   gypsum   water   ettringite
aluminate

This calcium sulfoaluminate hydrate, whose correct name is 6-calcium aluminate trisulfate-32-hydrate, is commonly called "ettringite," which is the name given to a naturally occurring mineral of the same composition. The formula is often written $C_3A \cdot 3C\bar{S} \cdot H_{32}$. Ettringite is a stable hydration product only while there is an ample supply of sulfate available (see Table 4.2). If the sulfate is all consumed before the C_3A has completely hydrated, then ettringite transforms to another calcium sulfoaluminate hydrate containing less sulfate.

$$2C_3A + C_6A\bar{S}_3H_{32} + 4H \rightarrow 3C_4A\bar{S}H_{12} \qquad (4.4)$$

This second product is called tetracalcium aluminate monosulfate-12-hydrate, or simply monosulfoaluminate (there is no known mineral of

Table 4.2

Formation of Hydration Products from C_3A

$C\bar{S}H_2/C_3A$ Molar Ratio	*Hydration Products Formed*
3.0	Ettringite
3.0–1.0	Ettringite + monosulfoaluminate
1.0	Monosulfoaluminate
<1.0	Monosulfoaluminate solid solution
0	Hydrogarnet

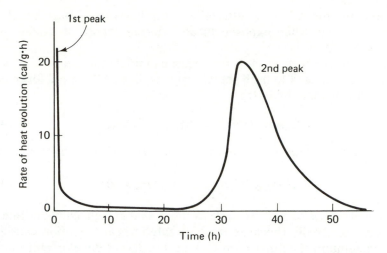

Figure 4.3 Rate of heat evolution during the hydration of tricalcium aluminate with gypsum.

this composition). It is often written $C_3A \cdot C\overline{S} \cdot H_{12}$. Monosulfoaluminate may sometimes form before ettringite if C_3A reacts more rapidly with the sulfate ions than they can be supplied by the gypsum to the mix water. A certain concentration of sulfate ions is required for the formation of ettringite.

Both steps in the hydration of C_3A [Eqs. (4.3) and (4.4)] are exothermic. The formation of ettringite slows down the hydration of C_3A by creating a diffusion barrier around C_3A analogous to the behavior of C–S–H during the hydration of the calcium silicates. This barrier is broken down during the conversion to monosulfoaluminate and allows C_3A to react rapidly again. Thus, the calorimeter curve for hydrating C_3A (Figure 4.3) looks qualitatively much like the curve for C_3S, although the underlying reactions are quite different and the amount of heat evolved is much greater. The first heat peak is completed in 10 to 15 min, but the time at which the second occurs depends on the amount of sulfate available. The more gypsum there is in the system, the longer the ettringite will remain stable. Conversion to monosulfoaluminate will occur in most cements within 12 to 36 h, after all the gypsum has been used to form ettringite.

The formation of monosulfoaluminate occurs because of a deficiency of sulfate ions necessary to form ettringite from all the available aluminate ions. When monosulfoaluminate is brought into contact with a new source of sulfate ions, then ettringite can be formed once again.

$$C_4A\overline{S}H_{12} + 2C\overline{S}H_2 + 16H \rightarrow C_6A\overline{S}_3H_{32} \qquad (4.5)$$

This potential for reforming ettringite is the basis for sulfate attack of portland cements when exposed to an external supply of sulfate ions (see Chapter 20).

Gypsum is added to curb the vigorous initial reaction of C_3A with water, which can lead to flash set (Chapter 8), due to the rapid formation of calcium aluminate hydrates.

$$C_3A + 21H \rightarrow C_4AH_{13} + C_2AH_8 \tag{4.6}$$

These hydrates are not stable and later convert to C_3AH_6 (hydrogarnet).

$$C_4AH_{13} + C_2AH_8 \rightarrow 2C_3AH_6 + 9H \tag{4.7}$$

This conversion is exactly the same as that found in high alumina cement, but occurs so rapidly (because the high rate of heat liberation causes a large temperature rise) that pure C_3A pastes do not develop substantial strength. Even with gypsum present, the formation of sulfate-free hydrates may not be entirely suppressed if the C_3A is very reactive (although not enough will form to cause flash set), and small amounts of hydrogarnet may be found in a hydrated cement.

When quite small amounts of gypsum are present, there may still be unreacted C_3A present when all of the ettringite has been converted to monosulfoaluminate. In such cases a solid solution between $C_4A\bar{S}H_{12}$ and C_4AH_{13} is formed, the two hydrates having the same crystal structure. This solid solution is written $C_3A(C\bar{S},CH)H_{12}$.

$$C_4A\bar{S}H_{12} + C_3A + CH + 12H \rightarrow 2C_3A(C\bar{S},CH)H_{12} \tag{4.8}$$

Ferrite Phase

C_4AF forms the same sequence of hydration products as does C_3A, with or without gypsum. The reactions are slower and involve less heat; C_4AF never hydrates rapidly enough to cause flash set, and gypsum retards C_4AF hydration even more drastically than it does C_3A. Changes in the composition of the ferrite phase affect only the rate of hydration; as the iron content is raised, hydration becomes slower. Iron oxide apparently plays the same role as alumina during hydration (i.e., F can substitute for A in the hydration products). As can be seen in Eqs. (4.9a) and (4.9b), there is insufficient lime to form the calcium sulfoaluminates unless amorphous hydrous oxides of iron or aluminum form also.

$$C_4AF + 3C\bar{S}H_2 + 21H \rightarrow C_6(A, F)\bar{S}_3H_{32} + (A, F)H_3 \tag{4.9a}$$

$$C_4AF + C_6(A, F)\bar{S}_3H_{32} + 7H \rightarrow 3C_4(A, F)\bar{S}H_{12} + (A, F)H_3 \tag{4.9b}$$

In these equations the use of a formula such as $C_6(A, F)\overline{S}_3H_{32}$ indicates that iron oxide and alumina occur interchangeably in the compound, but the A/F ratio need not be the same as that of the parent compound.

Practical experience has shown that cements low in C_3A but high in C_4AF are resistant to sulfate attack. This means that the formation of ettringite from monosulfoaluminate, as in Eq. (4.5), does not occur. It has not been established why this is so; it may be that an iron-substituted monosulfoaluminate cannot react to form ettringite. Alternatively, the presence of the amorphous $(A, F)H_3$ may in some way prevent the reaction described in Eq. (4.5) from occurring.

Special Aluminates

The formation of ettringite in regulated-set cements and expansive cements also proceeds according to specific chemical reactions. As in the hydration of C_3A, a deficiency of sulfate ions ensures the eventual formation of monosulfoaluminate in most cases. With regulated-set cement, this occurs after several hours when the hemihydrate has largely reacted and sulfate is supplied solely by less-soluble anhydrite. In the case of expansive cements, conversion does not take place until after expansion is completed, which may be several weeks, although in some Type K cements ettringite remains the stable phase. Whenever substantial amounts of monosulfoaluminate are present, the possibility exists for a reversion to ettringite according to Eq. (4.5). Such cements are therefore prone to sulfate attack and should not be exposed to sulfate-bearing environments.

Hydration of Portland Cement

Kinetics

The rate of hydration during the first few days is in the approximate order $C_3A > C_3S > C_4AF > C_2S$. It must be remembered, however, that no two preparations of any of these compounds will hydrate at exactly the same rate because their reactivities will be affected by differences in fineness and the rate of cooling of the clinker. There will be additional factors, such as the presence of impurities and the presence of the other cement compounds. For example, alite and belite hydrate faster than do pure C_3S and C_2S because of the impurity atoms contained in the structure. The hydration of C_3A and the ferrite phase will also be affected by impurity oxides.

Compound Interactions

The assumption made earlier that the cement compounds hydrate independently is a reasonable one for most purposes but is not entirely true. For example, C_3A and C_4AF both compete for sulfate ions, but the more reactive C_3A will consume more sulfate than does C_4AF. The effect is to increase the reactivity of C_4AF, since it forms less ettringite than would be expected. Gypsum increases the rate of hydration of the calcium silicates, which also compete for sulfate during hydration. Apparently C–S–H incorporates significant amounts of sulfate, and also alumina and iron, into its structure. It has been estimated that the quantity of sulfoaluminates that form in a paste may be less than half the theoretical quantity calculated from the compound composition of the cement.

Another example is the existence of an optimum gypsum content for the development of maximum strength in a cement paste. One explanation is that too high a gypsum content results in the formation of excessive amounts of ettringite after the paste has hardened, causing unrestrained expansion and disruption of the paste microstructure, whereas too low a gypsum content allows the monosulfoaluminate–solid solution [Eq. (4.8)] to form before the end of stage 2 of C_3S hydration, so that the resulting consumption of lime prevents the nucleation of the hydration products of C_3S, delaying stage 3. Another viewpoint, based on more recent studies, explains the optimum gypsum content by the fact that gypsum accelerates C_3S hydration, but at the same time lowers the intrinsic strength of C–S–H due to the presence of sulfate ions in its structure. Since there is a different optimum gypsum content for maximum strengths at different ages, both explanations may contribute to the phenomenon. The interaction between C–S–H and sulfate ions may help explain the observation that there is also an optimum gypsum content for minimum drying shrinkage, which is not the same as for strength.

Heat of Hydration

The quantity ΔH, which is shown in Table 4.3, is the heat of hydration (or enthalpy) and is a measure of the amount of heat evolved for each unit mass of anhydrous compound that has reacted. It represents the residual energy in the system after the redistribution of energy has occurred with the breaking and making of chemical bonds during hydration. ΔH is a thermodynamic quantity that can either be determined by calculation or measured experimentally (as described in Chapter 3).

We have already discussed the practical importance of the evolution of heat that accompanies hydration. It is possible in theory to calculate the heat of hydration of portland cement at any given time if the compound composition, the amount of hydration that has occurred for

Table 4.3

Heats of Hydration of the Cement Compounds

		ΔH (J/g) for Complete Hydration[a]			
		Pure Compounds		*Clinker[b]*	*Cement[c]*
	Reaction	*Calculated*	*Measured*	*Measured*	*Measured*
C_3S	$\rightarrow$ C–S–H + CH	~380	500	570	490
C_2S	$\rightarrow$ C–S–H + CH	~170	250	260	225
C_3A	$\rightarrow C_4AH_{13} + C_2AH_8$	~1260	—	—	—
	$\rightarrow C_3AH_6$	900	880	840	—
	$\rightarrow$ monosulfoaluminate	—	—	—	~1340
C_4AF	$\rightarrow C_3(A,F)H_6$	520	420	335	—
	$\rightarrow$ monosulfoaluminate	—	—	—	460

[a]These values should be negative since they refer to exothermic reactions, but they are customarily written without the negative sign. To convert from J/g to cal/g, divide by 4.19.

[b]One-year-old pastes of ground clinker (no added gypsum).

[c]One-year-old pastes assumed to be completely hydrated.

each compound, and ΔH for reaction are known. A typical calorimetric curve is shown in Figure 4.4; generally, the contributions of both C_3S and C_3A can be distinguished. In cases where sufficient data have been available, excellent agreement between measured and calculated heats of hydration has been found. However, it is not always easy to determine the amount of hydration and ΔH is not known for every reaction, as can

Figure 4.4 Rate of heat evolution during the hydration of portland cement.

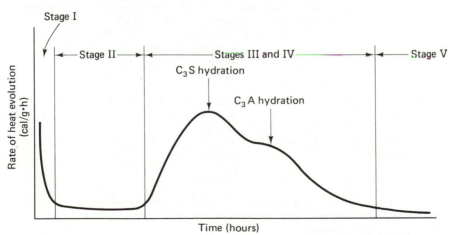

be seen in Table 4.3. Effective values of ΔH for each compound as it hydrates in cement have been determined by multiple regression analyses of heat of hydration data for portland cements of known compound composition. There is good agreement between the two values of ΔH where comparisons can be made.

4.2 HYDRATION PRODUCTS AND MICROSTRUCTURE

Properties of the Hydration Products

The various hydration products that are formed in hydrated cement pastes have quite diverse properties, and the behavior of each compound will contribute to the overall behavior of the paste. In this section we will briefly review the properties of the individual hydration products; a review of their physical interrelationships can be found in the section on microstructure. The important properties are summarized in Table 4.4.

C–S–H: Compositional variations. This calcium silicate hydrate is not a well-defined compound. The formula $C_3S_2H_3$ is only an approximate description, as the stoichiometry is quite variable. The C/S ratio is not

Table 4.4

Summary of Properties of the Hydration Products of Portland Cement Compounds

Compound	Specific Gravity	Crystallinity	Morphology in Pastes	Typical Crystal Dimensions in Pastes	Resolved by:[a]
C–S–H	2.3–2.6[b]	Very poor	Spines; Unresolved morphology	1×0.1 μm (Less than 0.01 μm thick)	SEM
CH	2.24	Very good	Nonporous striated material	0.01–0.1 mm	OM, SEM
Ettringite	~1.75	Good	Long slender prismatic needles	10×0.5 μm	OM, SEM
Monosulfo-aluminate	1.95	Fair–good	Thin hexagonal plates; irregular "rosettes"	$1 \times 1 \times 0.1$ μm	SEM

[a]OM, optical microscopy; SEM, scanning electron microscopy.
[b]Depends on water content.

Table 4.5

Standard Drying Conditions Used to Determine Water Contents of
C-S-H Gel and Hydrated Cement Paste

Type of Drying	Temperature	Water Vapor Pressure (Pa)[a]	Other Conditions
P-drying	Ambient	1.1	Vacuum ($<0.001 Pa$) over $Mg(ClO_4)_2 \cdot 2H_2O - Mg(ClO_4)_2 \cdot 4H_2O$
D-drying	Ambient	0.07	Vacuum ($<0.001 Pa$) over dry ice at $-78°C^b$
Oven drying	105°C	—	Atmospheric pressure; heating in oven

[a]$Pa \times 0.0075 = 1$ torr (mm Hg).
[b]Sublimation temperature of solid CO_2.

exactly 1.5 but varies between 1.5 and 2.0, and may be even higher. It
depends on many factors: the age of the paste, the temperature of
hydration, the *w/c* ratio, and the amount and kind of impurity oxides that
can be incorporated into the product.

The water content varies even more drastically and depends on the
extent of drying as well as the factors mentioned above. Unlike most
hydrated compounds, there are no definite hydration states and when
C–S–H is dried, water is not lost at discrete partial pressures of water
vapor (relative humidities), but there is a continuous loss of water as the
relative humidity is lowered from 100% RH to strong drying in vacuum
or on heating. This is because the water associated with C–S–H exists
in several different states. Because of the continuous loss of moisture
with decreasing water vapor pressures, it is necessary to define standard
drying conditions when determining water contents. The one most widely
adopted is *D-drying* (see Table 4.5), which is the most rigorous vacuum
drying condition. Under these conditions loss of residual moisture is
very slow, and several days drying is required, even when the paste is
ground to a fine powder. Engineers find oven drying at 105°C to be more
convenient and to be satisfactory for most purposes. It gives about the
same water content and is much quicker, water loss being complete after
24 h of heating. The stoichiometry of C–S–H was determined for D-
dried material.

Physical behavior. Because of its variable composition, C–S–H is not a
well-crystallized material; in fact, it is very nearly amorphous. As a
result, it develops as a mass of extremely small irregular particles of
indefinite morphology. The particles are so small that they can be studied

only by electron-optical techniques, and even then cannot be completely resolved. As a consequence of this very finely divided state of C–S–H, hydrated cement pastes have very high surface areas. Measurements using physical adsorption of water vapor on D-dried calcium silicate pastes indicate that most C–S–H preparations formed at ambient temperatures have surface areas (S_{H_2O}) in the range 250 to 450 m^2/g. This is an extraordinarily high surface area (three orders of magnitude higher than unhydrated cement) and indicates that surface effects must play a prominent role in determining the properties of C–S–H. There is controversy over the determination of surface areas, however, since when nitrogen adsorption is used to measure surface areas (S_{N_2}) the values are found to be much lower: 10 to 100 m^2/g. This is still a very high surface area compared to the unhydrated paste. The difference between H_2O and N_2 adsorption data becomes important when using adsorption data to calculate pore-size distributions and when developing structural models of the material.

Structural models of C–S–H. Because of its poor degree of crystallinity, compositional variability and finely divided state, C–S–H is a difficult material to study. Various structural models have been proposed to explain the properties and behavior of C–S–H, and these proposals remain the subject of active debate. The following description is adopted for the purposes of this text, although it is not fully accepted by all research workers. The features discussed below have not yet been resolved by modern microscopes.

C–S–H can be considered to have a degenerate clay structure, by which it is meant that it is based on a layer structure (see Figure 4.5). A well-defined clay mineral has the structure shown in Figure 4.5a. It can be thought of as being composed of layers of bread and filling to make a generous club sandwich. The "bread" is composed of silicoaluminate sheets that are stacked in a specific orientation; the "filling" is made up of metal ions that hold the sheets together with comparatively weak electrostatic attractions between positive charges on the metal ions and residual negative charges on the sheets. Water is also present between the layers. In some clays the layers can be expanded to accommodate additional water, thereby expanding the crystal. Loss of *interlayer water* on drying allows the layers to collapse again and the crystal to contract. This results in the large observed volume changes of clays on wetting and drying.

In C–S–H the "bread" is composed of calcium silicate sheets and the "filling" of calcium ions and water. Unlike a well-crystallized clay mineral, however, the sheets are crumpled and randomly arranged (owing to random variations in their composition) so that they do not fit together neatly (see Figure 4.5b). As a result, the spaces between the sheets are

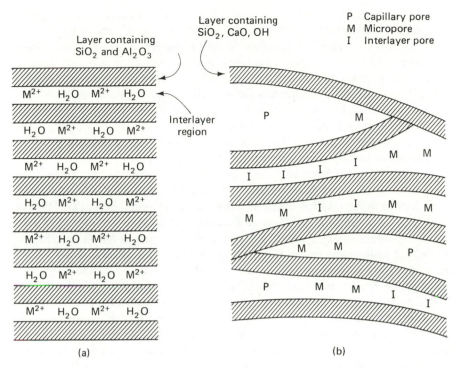

Figure 4.5 Schematic model of C−S−H in cement paste: (a) well-crystallized clay mineral; (b) poorly crystallized C−S−H.

irregular and vary considerably in size. We can distinguish between two kinds of spaces: micropores *(M)* and capillary pores *(P)*. Capillary pores are spaces in which water can behave as bulk water and menisci are created as the pores are filled or emptied. In micropores the adjoining surfaces are so close together that water cannot form menisci and consequently has a different behavior from bulk water. Water in micropores acts to keep the layers apart by exerting a *disjoining pressure*. The disjoining pressure depends on the relative humidity and disappears below 50% RH. When the sheets forming the micropores approach closely in a specific orientation, they may form claylike interlayer spaces *(I)* which bond the sheets together at this point. Interlayer bonding can be regarded as a special case of van der Waals' bonding. In addition, sheets will from time to time be bonded directly by strong ionic–covalent bonds which do not involve the weaker interlayer bonding.

The foregoing description draws on features of three of the major models of C–S–H that have been proposed. These are the *Powers–Brunauer* model, the *Feldman–Sereda* model, and the *Munich* model. A brief outline of the main ideas of each view will be presented.

It should be remembered that a model is merely a theoretical description proposed to explain observed experimental facts and cannot be considered to be a "correct" description in any absolute sense. A good model provides additional insights into behavior of a material and will suggest additional properties not hitherto recognized. More than one model may adequately interpret the known facts until additional data are obtained. In science, theories and models may undergo many changes and modifications as the subject is more thoroughly understood and newer, more complex theories eventually replace the earlier, more simple models. The process is a dynamic, evolutionary one.

The Powers–Brunauer and Feldman–Sereda models stress the chemical structure of C–S–H and both are based on the idea of a layer structure, as discussed above. Here the similarity ends, however. Powers and Brunauer consider C–S–H to be colloidal particles made up of two or three layers bonded together as in a clay (Figure 4.6a). There is not sufficient long-range order to consider the material crystalline. C–S–H is made up of a random arrangement of these particles bonded together by surface forces with occasional strong ionic–covalent bonds linking adjacent particles. Water vapor can penetrate all the spaces between the particles to provide a measure of their surface area, whereas nitrogen can only penetrate larger spaces and does not measure the whole surface area. The smaller pores that the water molecules penetrate are called *micropores* and are characterized by "ink-bottle" shapes with narrow entrances that exclude nitrogen. The water between the layers is held until strong drying occurs, when it is lost irreversibly.

Feldman and Sereda visualize the structure of C–S–H as developing a completely irregular array of single layers (Figure 4.6b) which may come together randomly to create interlayer space, as is observed in clay minerals, but in no regular, ordered way. In contrast to Powers and Brunauer, they consider that water can move reversibly in and out of the interlayer space. Thus, nitrogen adsorption most closely measures the surface area of pores within the material, since adsorption of water molecules within the interlayer region distorts the measurements with water vapor. Bonding between layers is considered to be through solid–solid contacts which are visualized as bonds intermediate in character between weak van der Waals' and strong ionic–covalent bonds. The solid–solid contacts form on drying but are disrupted on wetting. Interlayer bonding is a special kind of chemical bonding and cannot be considered as interactions between free surfaces.

The Munich model was developed by Wittmann and coworkers and is primarily a physical model that considers C–S–H as a xerogel (which is a three-dimensional network of colloidal size particles as shown in Figure 4.6c), the actual chemical structure being of secondary interest.

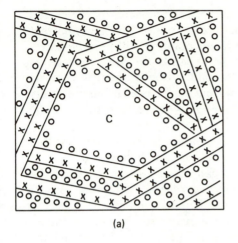

(a)

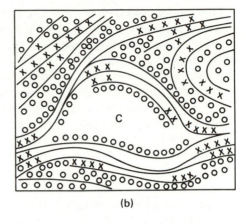

(b)

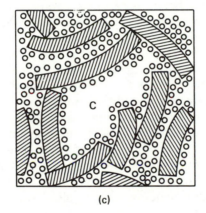

(c)

x Water in interlayer regions	——— C-S-H sheets
o Water adsorbed on surfaces	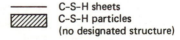 C-S-H particles
C Capillary pore	(no designated structure)

Figure 4.6 Schematic representation of various models of C−S−H. (a) Powers-Brunauer, adapted from T. C. Powers, *The Physical Structure and Engineering Properties of Concrete*, Bulletin No. 90, Portland Cement Association, Skokie, Ill., 1958, and S. Brunauer, *American Scientist*, Vol. 50, No. 1, 1962, pp. 210−229; (b) Feldman-Sereda, from R. F. Feldman and P. J. Sereda, *Engineering Journal (Canada)*, Vol. 53, No. 8/9, 1970, pp. 53−59; (c) Munich, adapted from F. H. Wittman, *Cement Production and Use*, Publication No. 79-08, Engineering Foundation, New York, 1979, pp. 143−161.

This model considers the binding energy between adjacent particles and concludes that van der Waals' forces between surfaces have an important contribution, but that strong ionic–covalent bonding still predominates. The extent of van der Waals' bonding depends on the presence of moisture. Water is strongly attracted to solid surfaces, thereby creating disjoining pressures which can force adjacent surfaces apart and severely reduce their interactions. The influence of water adsorption on the surface free energy is also important.

What each of these theories has in common is the recognition that water plays a very important role in the structure and behavior of C–S–H and can interact to a greater extent than can other adsorbates. Also, the material is in a high-energy state because of the additional energy associated with the extensive free surface. The material can be expected to respond to physical changes in such a way that it will annihilate surface area, thereby lowering its potential energy and increasing the ionic–covalent bonding between particles. Water is held within C–S–H in a variety of ways, ranging from bulk water in capillary pores, through water physically adsorbed on surfaces or between surfaces (micropores) and water structurally associated with the solid (interlayer water), to hydroxyl water in the solid lattice. There appears to be no sharp distinction between the various forms but a gradual transition between them. Thus, there are no universally accepted experimental methods for distinguishing between the different kinds of water.

As water is removed from C–S–H, rearrangement of the particles or layers is possible. For example, loss of water from capillary pores is believed to induce compressive stress on the system due to surface tension effects, while loss of water between the layers or from surfaces changes the potential for bonding: either ionic–covalent, van der Waals', or interlayer. There is evidence to suggest that when it is first formed fully saturated C–S–H may have a surface area exceeding 750 m²/g, but that this is greatly reduced by the extent and method of drying. Changes in bonding may cause permanent changes in the structure; exposure to higher temperatures may give similar changes. However, the description of C–S–H is still in a state of flux.

Calcium Hydroxide

In contrast to C–S–H, calcium hydroxide is a well-crystallized material with a definite stoichiometry. Crystals large enough to be seen with the naked eye can sometimes grow inside voids formed in concrete. These crystals have a distinctive hexagonal prism morphology. Within the body of the paste, the crystals do not grow so large, but can still be seen under a light microscope. For reasons discussed in the section on microstructure that follows, calcium hydroxide does not form homogeneous crystals in a cement paste.

Calcium Sulfoaluminates

Ettringite also crystallizes as hexagonal prisms but with a much greater aspect ratio (ratio of length to diameter) than CH crystals. The exact morphology depends on the available space and supply of ions for crystal growth. In the common portland cements they are typically seen as long slender needles (Figure 4.7a), typically 10×0.5 μm. In expansive cements, where large quantities form rapidly, it forms as stubby crystals, well intergrown (Figure 4.7b). Generally, the crystals are not large

(a)

(b)

Figure 4.7 SEM micrographs of calcium sulfoaluminate hydrates in hydrated cement pastes: (a) ettringite as formed from C_3A in portland cement paste; (b) ettringite as formed in expansive cement paste; (c) monosulfoaluminate formed from C_3A.

(c)

enough to be seen under the optical microscope, but in some concretes that have deteriorated from sulfate attack, large clusters of ettringite needles radiating from a center can often be seen on petrographic examination. Ettringite loses considerable amounts of its crystal water on drying.

When first formed, monosulfoaluminate or its solid solution tends to form clusters or "rosettes" of irregular plates. Later, these tend to grow into well-developed, but very thin, hexagonal plates (Figure 4.7c) which are too small to be seen by the optical microscope. In pure systems these plates are well crystallized, but in portland cement pastes the incorporation of impurities decreases the degree of crystallinity to some extent. $C_4A\overline{S}H_{12}$, C_4AH_{13}, and C_2AH_8 are all structurally related and form the same kinds of crystals, and all lose water on drying. Their crystal structure is also based on a layer arrangement.

Microstructure of Hydrated Cement Pastes

Although the properties of the hydration products will clearly influence the properties of the hydrated cement paste, the behavior of hardened cement paste cannot be properly understood without an appreciation of how the hydration products fit together to form the cementing matrix. The following description of microstructural development is based on SEM observations. Figure 4.8 shows schematically the sequence of structure formation as hydration proceeds. This involves the replacement of water that separates individual cement grains in the fluid paste (Figure 4.8a) with solid hydration products that form a continuous matrix and bind the residual cement grains together over a period of time, as illustrated in Figure 4.8b–d. This happens because the hydration products occupy a greater volume than the original cement compounds, owing to their lower specific gravity ($\sim$2.3 compared to 3.2).

C–S–H

C–S–H makes up about one-half to two-thirds of the volume of the hydrated paste and must therefore dominate its behavior. Little evidence of hydration of the calcium silicates can be seen in SEM micrographs until the paste has set and C_3S has entered stage 3 of hydration. In Figure 4.7a the calcium silicate grains are seen to be covered with a coating of C–S–H, which gives them a spiny appearance like a burr. On close examination (Figure 4.9a) the spines are seen to radiate outward into the surrounding pore space. (The pores would be filled with water in the saturated state, but the paste must be dried in a high vacuum for

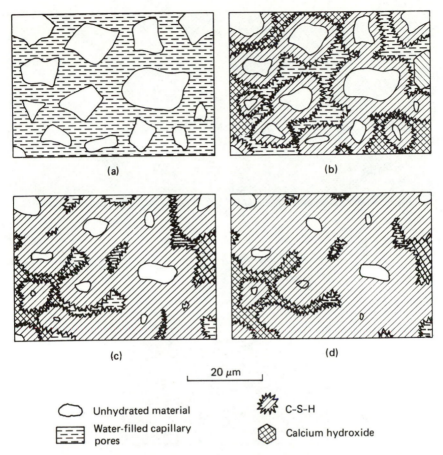

Figure 4.8 Schematic outline of microstructural development in portland cement pastes. (Calcium sulfoaluminates are included as part of C–S–H for convenience, although they will crystallize as separate phases.) (a) Initial mix; (b) 7 days; (c) 28 days; (d) 90 days.

observation.) The spines do not have a unique morphology, being variously pointed, blunt, flat, long and thin, or branched. They are particularly well grown in pastes of the pure calcium silicates. At other times, the spines are scarcely developed, forming more of a "honeycomb morphology" or reticular network (Figure 4.9b). This type of C–S–H is particularly prevalent when certain admixtures have been added to the paste, and it may form when the impurities reach a certain level. It is thought that the varying morphologies reflect the varying compositions of C–S–H and the conditions of formation within the paste.

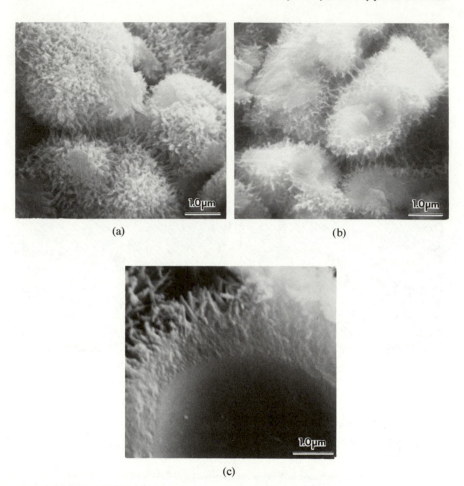

(a)   (b)

(c)

Figure 4.9 Morphologies of C−S−H in hydrated cement pastes (SEM micrographs): (a) spines in 3-day-old paste; (b) honeycomb morphology in paste containing calcium chloride; (c) coating of C−S−H around an unhydrated core.

However, the indications are that this outermost C–S–H may not be directly involved in providing strength. The spines do not grow appreciably after the first day or two; the bulk of additional C–S–H forms below the spines. Figure 4.9c shows how the underlying C–S–H forms a coating around the grain, creating the diffusion barrier described earlier. This material cannot be completely resolved by the SEM; its morphology is not known for certain and may well be variable, as is that of the spines. It is likely that its exact form is not important. As hydration proceeds, the underlying C–S–H grows in thickness and becomes the dominant form of C–S–H in mature pastes. As this hydrate layer grows, its larger specific volume causes each cement grain to increase effectively in size and the spines of C–S–H begin to intermesh, as can be seen at the center of Figure 4.9a. This is the beginning of the formation of a solid bond between two cement grains. As this bond develops with continued hydration, the spines appear to transform to the underlying C–S–H. Exactly how this happens and what the true relationships are between the kinds of C–S–H gel are not known. Thus, it seems a reasonable interpretation of present-day observations to say that cement grains are cemented together by underlying C–S–H (as is shown schematically in Figure 4.8b) and that these "points of contact" grow in area and number as hydration proceeds.

Calcium Hydroxide

Calcium hydroxide crystals occupy about 20 to 25% of the paste volume. During stage 3 hydration of C_3S, many calcium hydroxide crystals nucleate and grow within the capillary pore space. Their size is such that they are readily studied by optical microscopy, which shows more vividly than SEM how the isolated crystals gradually spread throughout the paste (Figure 4.10a and b). Calcium hydroxide will only grow where free space is available. If it is impeded by another calcium hydroxide crystal, it may stop growing or grow in another direction; if it encounters a hydrating cement grain it may well grow right around it, as can be seen in Figure 4.10c. The characteristic striated appearance of the crystal is a consequence of the way the crystal fractures within the paste. Often a calcium hydroxide crystal may completely engulf (occlude) a cement grain. Occlusion of hydrating cement grains increases the effective volume the crystal occupies in the paste and, what is more important, means that calcium hydroxide grown *in situ* may behave quite differently from pure crystals. Calcium hydroxide may vary in morphology, being found as small equidimensional crystals; large flat, platy crystals; large, thin, elongated crystals; and all variations in between. Morphology is particularly affected by admixtures and by the temperature of hydration.

Calcium Sulfoaluminates

The calcium sulfoaluminates are a relatively minor constituent of a mature paste, making up perhaps 10 to 15% by volume. Thus, they play a correspondingly minor role in the microstructure of the hydrated cement paste (although not necessarily in its properties) and are therefore omitted in Figure 4.8. Ettringite is seen in SEM micrographs of very

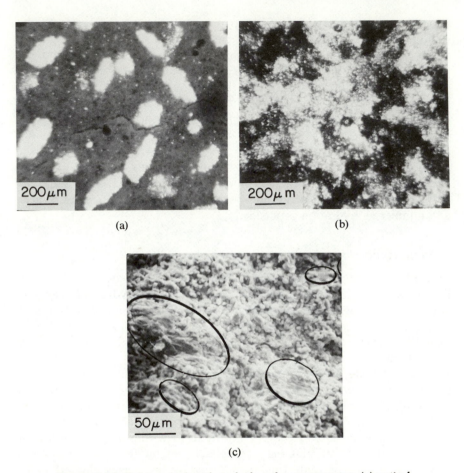

(a)   (b)

(c)

Figure 4.10 Calcium hydroxide in hydrated cement pastes: (a) optical micrograph, 1-day hydration; (b) optical micrograph, 21 days hydration; (c) SEM micrograph, 3 days hydration; circles enclose CH crystals .

young pastes as needles growing into the capillary pores between cement grains (Figure 4.7a). These later convert to the platy morphology of the monosulfoaluminate (Figure 4.7b). Pastes of regulated-set or expansive cements have their early microstructure dominated by ettringite crystals closely packed together (Figure 4.7c).

Minor Components

Unhydrated residues of the cement grains may persist even in well-hydrated cements. Occlusion by calcium hydroxide may preclude complete hydration, or there may be insufficient free space within the paste (Section 4.3). There will also be small amounts of magnesium hydroxide.

These components are not likely to amount to more than 5% by volume in mature pastes, although unhydrated residues will be more prevalent in young pastes.

Porosity

Porosity is another major component of the microstructure that deserves separate discussion. We have been emphasizing the decrease in capillary porosity as seen by SEM observations, but pore-size distribution is also of importance. However, measurement and classification of pore sizes is surrounded by experimental difficulties and by controversy. There are two main problems. First, the paste must be dried in order to make measurements, and drying almost certainly changes the pore structure. Second, a definite geometrical shape of pore must be assumed while SEM micrographs show that the pores form a network of very irregularly shaped pores. Nevertheless, some useful assessment of pore structure has been obtained.

Classification. A classification of porosity in cement paste is given in Table 4.6; it can be seen that there is an enormous range of pore sizes, from 10 μm to less than 0.0005 μm (0.5 nm) in diameter. Thus, the water

Table 4.6

Classification of Pore Sizes in Hydrated Cement Pastes

Designation	Diameter	Description	Role of Water	Paste Properties Affected
Capillary pores	10–0.05 μm (50 nm)	Large capillaries	Behaves as bulk water	Strength; permeability
	50 ~10 nm	Medium capillaries	Moderate surface tension forces generated	Strength; permeability; shrinkage at high humidities
Gel pores	10–2.5 nm	Small (gel) capillaries	Strong surface tension forces generated	Shrinkage to 50% RH
	2.5~0.5 nm	Micropores	Strongly adsorbed water; no menisci form	Shrinkage; creep
	<~0.5 nm	Micropores "interlayer"	Structural water involved in bonding	Shrinkage; creep

that occupies the pores plays many different roles. It is useful to make a distinction between *capillary pores* and *gel pores* in a hydrated paste. The capillary pores are the remnants of water-filled space that exists between the partially hydrated cement grains; the gel porosity can be regarded as part of the C–S–H. The porosity seen in SEM micrographs is capillary porosity; the gel porosity cannot be resolved by the SEM and would be included in the volume occupied by C–S–H. In mature pastes the bulk of the porosity resides within the C–S–H. The size division between capillary and gel porosity is to a large extent arbitrary, as the spectrum of pore sizes is a continuous one. It should also be noted that gel pores include small capillary pores if we use the more accurate definition that a capillary pore is one in which capillarity effects can occur (i.e., a meniscus can form).

As can be seen in Table 4.6, porosity over the whole size range of pores has an influence on paste properties. Yet it is difficult to get an exact assessment of pore-size distributions because no one measurement encompasses the whole size range and because it is difficult to interpret experimental data. Thus, comparisons of porosity should be made with care.

Measurement. There are two main methods that are used to measure the pore-size distribution of hardened cement paste: mercury intrusion porosimetry and physical adsorption of gases. Mercury porosimetry involves forcing mercury into the pore system of the paste by the application of external pressure. Mercury does not wet the paste, and force is needed to overcome surface effects. The pressure required is inversely proportional to the pore radius.

Mercury porosimetry gives a better appreciation of the larger capillary pore system (Figure 4.11), which has an important influence on permeability and on shrinkage at high humidites. Adsorption studies give a more complete measure of the gel porosity: the small capillaries and micropores. In the overlapping range of pore sizes, mercury porosimetry may not always agree very well with data obtained by adsorption experiments. Both methods suffer from the limitation that a pore geometry must be assumed. A look at SEM micrographs (Figure 4.9) very quickly reveals the inadequacy of this assumption. However, there is an even more important problem, which is that the cement paste must be strongly dried before measurements can be made. Since C–S–H undergoes significant changes on drying, it is probable that important changes in the pore-size distributions may also occur on drying. This must always be kept in mind when trying to interpret the behavior of a saturated paste in terms of its porosity distribution. At present, no experimental method is available for determining pore-size distributions directly on saturated pastes.

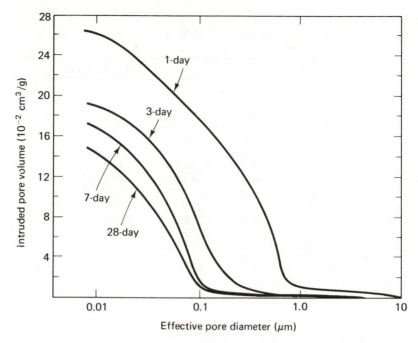

Figure 4.11 Mercury intrusion porosimetry curves for portland cement pastes. (From J. F. Young, *Powder Technology*, Vol. 9, 1974, pp. 173–179.)

4.3 PROPERTIES OF HYDRATED CEMENT PASTES

Volume Changes during Hydration

An important aspect of microstructural development is the decrease in porosity during hydration. All the hydration products of the cement compounds have lower specific gravities and larger specific volumes than the cement compounds themselves (see Table 4.7). Thus, every hydration reaction is accompanied by an increase in solid volume.

Expansive Reactions

It is not only the potential volume change that is of interest, but also the way in which this change is manifested. In our discussion of microstructure, it was noted that CH either grows around solid particles or stops growing in that particular direction when it meets such obstacles. The same is true of C–S–H. Thus, the hydration of the calcium silicates is not accompanied by an increase in the total volume of the paste. The

Table 4.7

Physical Data Determining Volume Changes That Occur during Hydration

Initial Cement Compounds			Hydration Products		
Compound	Specific Gravity	Molar Volume $(GMW/\rho)^b$ $10^{-6}\ m^3$	Compound	Specific Gravity	Molar Volume (GMW/ρ) $10^{-6}\ m^3$
C_3S	~3.5	~72.4	C–S–H	a	a
C_2S	3.28	52.4	CH	2.24	33.2
C_3A	3.03	89.1	$C_6A\bar{S}_3H_{32}$	~1.75	715
C_4AF	~3.73	~128	$C_4A\bar{S}H_{12}$	1.95	313
M	3.58	11.0	MH	2.37	24.2
$C\bar{S}H_2$	2.32	74.2	C_4AH_{13}	~2.02	~260
$C\bar{S}H_{1/2}$	2.74	52.9	C_2AH_8	1.95	165
Portland cement	3.15	—	C_3AH_6	2.52	150

[a]Value depends on the water content of C–S–H, which is related to how much "gel porosity" is included in the structure.

[b]GMW, gram molecular weight; ρ, density in g/m^3.

hydration products will only occupy space that is available to them within the paste, which is the volume originally occupied by the mix water. If this space is filled before complete hydration has occurred, further hydration will virtually cease.

Quite the contrary occurs when ettringite is formed from C_3A or the ferrite phase. Ettringite crystals will make space for themselves when their crystal growth is impeded by solid material. If the impediment is a rigid material, crystal growth pressures will develop at the point where growth has been stopped. It has been estimated that ettringite may develop crystal growth pressures as high as 240 MPa (35,000 lb/in.²). Magnesium hydroxide and calcium hydroxide develop similar crystal growth pressures when formed directly from their respective oxides. This phenomenon is quite common; the expansive growth of ice crystals, for example, can contribute to frost damage.

In a paste that has not yet hardened ettringite has plenty of room in which to form or can make additional room by pushing aside obstructing cement grains. Once the paste has gained rigidity, however, the extra space necessary for continued formation of ettringite is created by expanding the total volume of the paste. We have discussed how this bulk expansion is harnessed to good purpose in expansive cements. If not restrained, however, the internally generated expansions will induce internal stresses that may crack and damage the paste. Ordinarily, almost no bulk expansion occurs during the hydration of portland cement pastes through the crystallization of ettringite, because only small amounts form after the paste has hardened. But if too much gypsum is added, then more ettringite will be formed long after setting which may disrupt the paste sufficiently to lower the compressive strength of the paste or even, in extreme cases, cause cracking. This is another case of unsoundness, exactly analogous to unsoundness caused by MgO and is the reason for the limitation on SO_3 in ASTM C150.

Calculation of Volume Changes

A knowledge of porosity can be very useful since porosity has such a strong influence on paste properties, particularly strength and durability. The total porosity in a paste (but not pore-size distributions) can be calculated quite simply. In theory, it is possible to calculate volume changes, and hence porosities, from the chemical equations describing hydration using the specific gravities of each species. But, apart from the problems of compositional variations, particularly with C–S–H, this would require a knowledge of the compound composition and the rate of hydration of each compound. Clearly, this is not a practical approach, and fortunately a simple set of equations has been worked out by T. C. Powers which can be applied to all portland cement (Types I to V). These equations are empirical, being derived from experimental data. In the form given here, the equations can be easily solved if the amount of hydration is known.

The volume relationships among the constituents are presented schematically in Figure 4.12, ignoring the realities of the microstructure for simplicity. The hydrated cement fraction includes all the hydration products: C–S–H, CH, and sulfoaluminates, without distinguishing them separately. (Powers called this "cement gel" before the individual constituents had been separately identified.)

Two types of water are distinguished: *evaporable water* is lost when a saturated paste is D-dried[1] (or oven-dried) and *nonevaporable water* is lost when a D-dried paste is heated (ignited) to 1000°C. Evaporable water encompasses the water held in both capillary and gel pores (including interlayer pores) as well as some hydrate water from the calcium sulfoaluminates. Nonevaporable water approximately measures the amount of water combined structurally in the hydration products. The nonevaporable water (w_n) is proportional to the amount of hydration that has occurred.

$$w_n = 0.24\alpha \text{ g/g of original cement} \tag{4.10}$$

where α is the degree of hydration (i.e., the fraction of cement that has hydrated). When the cement is fully hydrated ($\alpha = 1.0$), 0.24 g of

---

[1]Equations (4.10) to (4.18) are based on data for D-dried pastes, whereas Powers originally determined values for P-dried pastes.

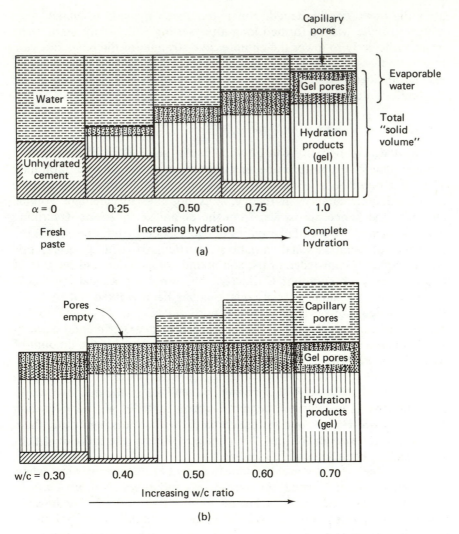

Figure 4.12 Volume relationships among constituents of hydrated pastes: (a) constant *w/c* ratio = 0.50; (b) changing *w/c* ratio (α = 1.0).

nonevaporable water are combined with each gram of cement. Equation (4.10) is used to determine α experimentally. The amount of water held in the "gel pores" (which includes some hydrate water from the calcium sulfoaluminates) is given by the relationship

$$w_g = 0.18\alpha \text{ g/g of original cement} \tag{4.11}$$

where w_g is the *gel water* and is primarily associated with the C–S–H.

The total volume of hydration products (cement gel) is given by

$$V_g = 0.68\alpha \text{ cm}^3/\text{g of original cement}^2 \tag{4.12}$$

The volume of gel porosity (P_g) is also determined by Eq. (4.11) as $\text{cm}^3/$ g of original cement, since the specific gravity of water in the gel pores is effectively 1.0. Gel porosity is included in V_g and P_g can also be expressed as a fraction of V_g:

$$\text{gel porosity} = \frac{W_g}{V_g} = 0.26 \tag{4.13}$$

This fraction is a constant value for all normally hydrated cements. Thus, about one-fourth of the volume of C–S–H is pore volume.

The capillary porosity (P_c) is given by Eq. (4.14):

$$P_c = \frac{w}{c} - 0.36\alpha \qquad \text{cm}^3/\text{g of original cement} \tag{4.14}$$

where w is the original weight of water used to form the paste. w/c is therefore the familiar water/cement ratio (by weight) and capillary porosity is strongly dependent on w/c. As the w/c ratio of a paste is increased, its capillary porosity increases. Powers also used another quantity as a measure of capillary porosity. This is the gel/space ratio (X), which is defined as

$$X = \frac{\text{volume of gel (including gel pores)}}{\text{volume of gel + volume of capillary pores}} = \frac{0.68\alpha}{0.32\alpha + w/c} \tag{4.15}$$

The volume occupied by the unhydrated cement (V_u) is calculated from Eq. (4.16):

$$V_u = (1 - \alpha)v_c \tag{4.16}$$

where v_c is the specific volume of cement and is approximately 0.32 (see Table 4.7). Before hydration has commenced, V_u is simply equal to v_c. Therefore, the original volume of the paste (V_p) is given by

$$V_p = (v_c + w) \text{ cm}^3/\text{g of original cement} \tag{4.17}$$

since the specific gravity of water is 1.0.

---

[2]The units used here are those originally used; the proper SI unit would be m^3/g.

Using Eqs. (4.10) through (4.17) it is possible to calculate the volume of paste constituents for varying values of α and w/c ratios. This was done to construct Figure 4.12. These calculations assume that no overall expansion occurs and that pastes are hydrated at, or near, ambient temperature. It is also assumed that the pastes were continuously moist-cured; if the paste is allowed to dry out, the relationship may change. Figure 4.12 clearly shows the marked effects that w/c and α have on capillary porosity, as indicated by Eq. (4.14).

Minimum w/c ratios. It can be seen that at low w/c ratios there is insufficient space for the hydration products to form so that complete hydration is not possible. Using this criterion, the minimum w/c ratio that can be used and still ensure complete hydration can be determined from Eq. (4.14) by putting $P_c = O$ and $\alpha = 1.0$; the value is 0.36. However, to form the hydration products requires a minimum amount of water:

$$w_{min} = (w_n + w_g) \text{ g/g original cement} \qquad (4.18a)$$

or

$$(w/c)_{min} = 0.42\alpha \qquad (4.18b)$$

Thus, for complete hydration ($\alpha = 1$) the w/c ratio should not fall below 0.42. Complete hydration is not essential to attain a high ultimate strength but it is clear that residual unhydrated cement can be expected to remain indefinitely in pastes made at low w/c ratios. The space requirements for the cement gel are less than the water requirements, so that the available water will be used up while space is still available. This means that below a w/c ratio of 0.42, a paste will *self-desiccate* unless external water is added during the curing period, and the residual capillary pores will be empty of water. Since water is physically lost from the paste by evaporation, absorption by formwork, and so on, during actual concreting, the effective minimum w/c ratio needed to avoid self-desiccation is higher than 0.42. However, self-desiccation is unlikely to prove a serious problem in the field.

Influence of Hydration Products on Paste Properties

It will now be appreciated that hardened cement paste is composed of a complex mixture of diverse compounds. On the microscopic level the paste must definitely be considered as an inhomogeneous, anisotropic, composite matrix. Yet from a macroscopic viewpoint the assumption of a homogeneous, isotropic material is not without validity since the hydration products are closely intergrown and well dispersed throughout the paste. The properties of the paste components must be considered when attempting to understand and predict the properties of the hydrated paste. A general consideration of the effect of the paste composition will be given here, although the subject will be further discussed later. Since C–S–H makes up the bulk of the paste matrix, it is reasonable to assume that it has the dominant effect on paste properties. This is indeed true, the question being to what extent the other components modify the properties of the paste.

Chemical Bonding

This can be illustrated by considering the general mechanical behavior of hydrated cement paste. The initial approach was to ignore all other components and explain the properties in terms of C–S–H. The nature of chemical bonding within a material determines the mechanical response of that material to different states of stress and the relationships between the various parameters describing mechanical response. It was thought that the very low tensile strength of the paste was a result of van der Waals' bonding between the surfaces of the colloidal particles of C–S–H. It is now recognized that hardened cement paste more closely resembles ceramic materials (dominated by ionic–covalent bonding in their mechanical behavior) than it does organic polymers (dominated by van der Waals' bonding). The low tensile strength of concrete is rather the consequence of the presence of flaws on a much larger scale than atomic bonds (see Chapter 14). Nevertheless, there is probably an appreciable van der Waals' component, which may perhaps be one-third to one-fourth of the total bonding energy. The influence of van der Waals' bonding may be reflected in the unique creep and shrinkage behavior of concrete and the influence of moisture on these properties (and only to a lesser extent on mechanical properties).

Porosity

As we have seen, porosity is one of the basic properties of hardened cement paste. In common with other porous materials, the strength of cement paste depends primarily on the porosity; the exact chemical nature of the hydration products is less signficant. As long as there is enough water present (theoretically) to fully hydrate the cement, the *gel porosity* is independent of the *w/c* ratio [Eq. (4.13)]; it depends only on the degree of hydration, or maturity, of the paste. Thus, as may be seen from Figure 4.12, the addition of water in the normal range of *w/c* ratio for concrete (0.40 to 0.55) affects only the *capillary porosity*. Since the gel porosity is an intrinsic part of the paste structure, this means that changes in many concrete properties will be directly related to changes in the capillary porosity. Porosity also dominates the permeability of cement paste. Pastes with high capillary porosities have high permeabilities, as water flows easily through the larger pores. Well-hydrated pastes with low *w/c* ratios have permeabilities that may be three orders of magnitude or more lower than a paste with a high *w/c* ratio. This is because the larger pores become isolated and water movement is controlled by flow through the gel pores, through which it proceeds only with difficulty.

Role of Calcium Hydroxide

The influence of calcium hydroxide on concrete properties is still under debate. Calcium hydroxide must affect strength if for no other reason than that it fills the pores. Any solid material of a comparable stiffness that reduces porosity must contribute to strength, regardless of its nature. The effect of CH on long-term strength is not so clear; it has been suggested that CH may be strength-limiting because of its tendency to cleave under shear. Calcium hydroxide can limit the durability of cement paste because it is more soluble than C–S–H. Leaching of CH can provide a point of entry for aggressive agents. The high alkalinity of CH may also contribute to alkali-aggregate attack, and is obviously important in acid attack. On the other hand, the high alkalinity may protect the reinforcing steel from corrosion.

Role of Calcium Sulfoaluminates

The calcium sulfoaluminate hydrates also affect the properties of the paste. The case of sulfate attack is the most dramatic and we have discussed other consequences of the expansive formation of ettringite. Ettringite and monosulfoaluminate can contribute to early strength. It is known that the C_3A content of a cement can affect the creep and shrinkage of its paste so that calcium sulfoaluminates must have some, as yet undetermined, influence on these properties.

Role of Water

Water is a polar molecule and therefore strong secondary interactions (e.g., hydrogen bonding) with hydroxylated surfaces, such as occur in hydrated paste, are to be expected. Because of its highly polar nature and small molecular size, water appears to play a special role in the structure of C–S–H, as discussed above, and hence has an important influence on concrete properties. Saturated concrete is about 10% weaker in compression than is dry concrete. This can be due to three reasons: (1) as adsorbed water is removed, C–S–H particles can come closer together and produce a stronger system due to an increase in van der Waals' bonding; (2) water may attack Si–O–Si bonds under stress; (3) water may reduce mechanical interlock by acting as a lubricant.

The presence of water in C–S–H is also important in creep and shrinkage, as discussed in Chapter 18. The high degree of hydrogen bonding within water results in high surface tension, which is responsible for the development of large capillary stresses. The strong affinity of water for surfaces must also be important in admixture chemistry, as chemical admixtures must themselves be highly polar molecules.

Bibliography

Hydration, Microstructure, and Paste Properties

COPELAND, L. E., AND D. L. KANTRO, "Hydration of Portland Cement," *Proceedings, Fifth International Symposium on the Chemistry of Cement, Tokyo, 1968,* Vol. 2, pp. 387–419. Cement Association of Japan, Tokyo, 1969.

HANSEN, T. C., F. RADJY, AND E. J. SELLEVOLD, "Cement Paste and Concrete," *Annual Review of Materials Science,* Vol. 3, pp. 233–268 (1973).

Hydraulic Cement Pastes: Their Structure and Properties (Proceedings of a conference held in Sheffield, U.K., 1976). Cement and Concrete Association, Slough, U.K., 1976.

LEA, F. M., *The Chemistry of Cement and Concrete,* 3rd ed. Chemical Publishing Co., New York, 1971.

POWERS, T. C., *The Physical Structure and Engineering Properties of Concrete,* Research and Development Bulletin No. 90. Portland Cement Association, Skokie, Ill., 1958.

SCHRAMLI, W., "An Attempt to Assess the Beneficial and Detrimental Effects of Aluminate in the Cement on Concrete Performance," (2 parts), *World Cement Technology,* Vol. 9, No. 2, pp. 35–36, 39–42; No. 3, pp. 75, 77–78 (1978).

VERBECK, G. J. "Pore Structure," *Significance of Tests and Properties of Concrete and Concrete-Making Materials,* STP 169B, pp. 262–274. American Society for Testing and Materials, Philadelphia, 1978.

VERBECK, G. J. AND R. H. HELMUTH, "Structures and Physical Properties of Cement Paste," *Proceedings, Fifth International Symposium on the Chemistry of Cement, Tokyo, 1968,* Vol. 3, pp. 1–32. Cement Association of Japan, Tokyo, 1969.

Models of C—S—H

BRUNAUER, S. "Tobermorite Gel—The Heart of the Concrete," *American Scientist,* Vol. 50, No. 1, pp. 210–229 (1962).

BRUNAUER, S., I. ODLER, AND M. YUDENFREUND, "New Model of Hardened Portland Cement Paste," *Highway Research Record,* No. 328, pp. 89–101 (1970).

FELDMAN, R. F., AND P. J. SEREDA, "A New Model for Hydrated Portland Cement and Its Practical Implications," *Engineering Journal (Canada),* Vol. 53, No. 8/9, pp. 53–59 (1970).

FELDMAN, R. F., AND P. J. SEREDA, Written Discussion of Paper by Verbeck and Helmuth (see above); *Proceedings, Fifth International Symposium on the Chemistry of Cement, Tokyo, 1968,* Vol. 3, pp. 36–44. Cement Association of Japan, Tokyo, 1969.

WITTMANN, F. H., "Interaction of Hardened Cement Paste and Water," *Journal of the American Ceramic Society,* Vol. 59, No. 8, pp. 409–415 (1973).

WITTMANN, F. H., "The Structure of Hardened Cement Paste—A Basis for Better Understanding of the Material Properties," in *Hydraulic Cement Pastes: Their Structure and Properties.* Cement and Concrete Association, Slough, U.K., 1976.

Problems

4.1   Discuss the effect of temperature on the hydration of portland cement.

4.2   Describe the chemical reactions involved in the sulfate attack of hardened cement paste.

4.3   What are the roles of C_3A and C_4AF in cement hydration?

4.4   Discuss the role of gypsum in the hydration of portland cement.

4.5   Compare the properties of C–S–H and CH, and their roles in determining the properties of hardened cement paste.

4.6   What would you expect the engineering properties of a very high specific surface area (100 to 200 m^2/g) material like C–S–H to be?

4.7   Discuss the problems involved in measuring the porosity of cement paste.

4.8   Describe the classification of porosity in hardened cement paste.

4.9 Calculate (a) the volume of hydration products; (b) the capillary porosity; and (c) the gel/space ratio, given the following experimental data: $\alpha = 0.80$, $w/c = 0.45$.

4.10 Calculate the same quantities as in Problem 4.9 given $w_n = 0.14$ and $w/c = 0.42$.

4.11 What happens in Problem 4.9 if w/c is lowered to 0.27?

4.12 Calculate the same quantities in Problem 4.9 for $\alpha = 1.0$ and $w/c = 0.30, 0.40, 0.50$, and 0.60.

5

water quality

Although water is an important ingredient of concrete, little needs to be written about it. That is because there is much more bad concrete made through using too much good-quality water than there is using the right amount of poor-quality water. The time-honored rule of thumb for water quality is: "If you can drink it, you can make concrete with it;" and a large fraction of concrete is made using municipal water supplies. However, good-quality concrete can be made with water that would not pass normal standards for drinking water. There is no ASTM standard concerning water quality, but BS 3148 addresses this matter.

5.1 IMPURITIES IN WATER

Mixing water can cause problems by introducing impurities that have a detrimental effect on concrete quality. Although satisfactory strength development is of primary concern, impurities contained in the mix water may also affect setting times, drying shrinkage, or durability, or they may cause efflorescence. Water should be avoided if it contains large quantities of suspended solids, excessive amounts of dissolved solids, or appreciable amounts of organic materials. Concentration limits for various impurities are given in Table 5.1.

Suspended Solids

Usually up to about 2000 ppm of suspended clay or silt can be tolerated. Higher amounts may increase water demand, increase drying shrinkage, or cause efflorescence. Muddy water should be allowed to

Table 5.1

Tolerable Levels of Impurities in Mixing Water

Impurity	Maximum Concentration (ppm)	Remarks
Suspended matter (turbidity)	2000	Silt, clay, organic matter
Algae	500–1000	Entrain air
Carbonates	1000	Decrease setting times
Bicarbonates	400–1000	400 ppm for bicarbonates of Ca, Mg
Sodium sulfate	10,000 ⎫	May increase early
Magnesium sulfate	40,000 ⎭	strength, but reduce later strength
Sodium chloride	20,000 ⎫	Decrease setting times,
Calcium chloride	~50,000 ⎬	increase early
Magnesium chloride	40,000 ⎭	strength, and reduce ultimate strength
Iron salts	40,000	
Phosphates, arsenates, borates	500 ⎫	Retard set
Salts of Zn, Cu, Pb, Mn, Sn	500 ⎭	
Inorganic acids	10,000	pH not less than 3.0
Sodium hydroxide	500	
Sodium sulfide	100	Should test concrete
Sugar	500	Affects setting behavior

clear in settling basins before use. The presence of algae or other suspended organic matter may also be a problem, as they will not settle out readily on standing. Organic materials may dissolve during mixing (since the mix water quickly becomes highly alkaline) and subsequently retard setting and strength development by interfering with cement hydration. They may also entrain excessive amounts of air, thereby reducing strength, or conversely, they may interfere with the action of air-entraining agents.

Dissolved Solids

Water containing less than 2000 ppm of dissolved solids can in most instances be used safely, although this depends, of course, on the nature of the dissolved material. As little as 100 ppm of sodium sulfide may cause problems. At the other extreme, seawater [which contains about 35,000 ppm (3.5%) of dissolved salts] can be used to make satisfactory concrete if certain precautions (see below) are taken. Indeed, soluble salts may be added deliberately as admixtures—the most common example being $CaCl_2$, which is used as an accelerating agent. Soluble carbonates and bicarbonates can promote rapid setting; large quantities of carbonates and sulfates may cause a reduction in 28-day strength or long-term strength.

Some soluble inorganic salts may retard the setting and hardening of concrete. Salts of zinc, copper, lead, and, to a lesser extent, manganese and tin fall into this category, as well as phosphates, arsenates, and borates. Up to 500 ppm can generally be tolerated in mixing water. Such compounds are likely to be found at these concentrations only in untreated industrial wastewaters, although waters leached from a mining locale may contain significant quantities of metal salts.

Acidic waters can be used in concrete making; the pH of the water may be as low as 3.0, at which level there are more problems surrounding the handling of the water than will occur in the concrete. Organic acids may affect the setting and hardening of concrete. Alkaline waters, containing sodium or potassium hydroxide, may cause quick setting and low strengths at concentrations above 500 ppm.

Seawater

Seawater is composed mostly of sulfates and chlorides of sodium and magnesium (Table 5.2). Thus, more rapid setting and early strength gain can be expected, owing to the accelerating effect of the chloride ion (see Chapter 7), but the 28-day strength will be lower because of the higher amounts of sulfate, which will prolong the crystallization of ettringite. A strength loss of 10 to 20% will be typical and can be compensated for by using a lower w/c ratio. The presence of chloride ions increases the risk of corrosion of reinforcing steel, and hence

Table 5.2

Typical Composition of Seawater (ppm)

Sodium chloride (NaCl)	27,000
Magnesium chloride ($MgCl_2$)	3,200
Magnesium sulfate ($MgSO_4$)	2,200
Calcium sulfate ($CaSO_4$)	1,100
Calcium chloride ($CaCl_2$)	500
Total dissolved salts	34,000

seawater should *never* be used for making prestressed concrete. The use of seawater is allowed for ordinary reinforced concretes, and if an adequate protective cover of good, dense concrete is provided, problems with corrosion should not occur. However, recent experience has indicated that it may be more difficult to protect reinforcement reliably in concrete made with seawater than has hitherto been assumed. Thus, it is best to avoid using seawater for reinforced and architectural concrete unless it is unavoidable. The use of seawater causes efflorescence and may cause problems with decorative finishes.

Dissolved Organic Material

Colored natural waters generally indicate the presence of dissolved organic material (mostly tannic and humic acids) which may retard the hydration of cements. Many organic compounds that occur in industrial wastes may also severely affect the hydration of cement or entrain excessive amounts of air. Wastes from the pulp and paper industries, the tanning industries, and food-processing industries have been used as a source of chemicals for the formulation of set-retarding or air-entraining admixtures. Thus, untreated industrial wastewaters should be viewed with caution, but if they have passed through a sewage treatment process, organic matter will be reduced to safe levels.

5.2 TESTING OF WATER

There are no ASTM specifications for water quality and no standard tests. However, BS 3148 specifies two methods of assessing the suitability of water for making concrete. These involve comparing both the setting time and the compressive strength of specimens made with the appropriate cement and both the water in question and distilled water. The water is considered to be suitable if it neither changes the setting time by more than 30 min, nor reduces the strength by more than 20% compared to the specimens made with distilled water. Other specifications often used in the United States require that the 7- and 28-day strengths should be at least 90% of those obtained on comparable specimens made with potable water. But other appropriate concrete properties should also be checked to ensure that they are not adversely affected. There should be concern not only for meeting specifications but also for long-term service performance. If there is doubt as to the suitability of a particular water source, an alternative supply should be sought.

Bibliography

Concrete Manual, 8th ed. U.S. Bureau of Reclamation, Denver, Colo., 1975.

McCoy, W. J., "Mixing and Curing Water for Concrete," in *Significance of Tests and Properties of Concrete and Concrete-making Materials,* ASTM STP 169B, pp 765–773. American Society for Testing and Materials, Philadelphia, Pa., 1978.

Problems

5.1 Should seawater be used to make reinforced concrete? Discuss the problems that may arise.

5.2 Is it advisable to use industrial wastewaters for concrete mixing?

5.3 Can wastes containing large amounts of suspended, inorganic materials be used to make concrete?

6

aggregates

Aggregates generally occupy about 70 to 80% of the volume of concrete and can therefore be expected to have an important influence on its properties. They are granular materials, derived for the most part from natural rock, crushed stone, or natural gravels and sands, although synthetic materials such as slags and expanded clay or shale, for example, are used to some extent, mostly in lightweight concretes. In addition to their use as an economical filler, aggregates generally provide concrete with better dimensional stability and wear resistance. The influences that the aggregates can have on the mechanical and physical properties of concrete (see Table 6.1) are discussed in succeeding chapters. Here we will discuss the influence of aggregate properties on mix design and special durability problems associated with their use. Strength of concrete and mix design are essentially independent of the composition of aggregates, but durability may be affected. No particular rock or mineralogical type, in itself, is required for aggregates, but it has been found that certain constituents may cause problems in the field, as will be discussed later. In other instances, a certain kind of rock may be required to attain certain concrete properties (e.g., high density or low coefficient of thermal expansion). But in the absence of special requirements, most kinds of rocks can produce acceptable aggregates that conform to ASTM C33. Thus, aggregates are not generally classified by mineralogy; the simplest and most useful classification is on the basis of specific gravity. (The geology of rocks will not be considered in this text. ASTM C294 describes the principal rock types and constituents of common aggregates.) Aggregates are classed as heavyweight, normal-weight, and light-

Table 6.1

Properties of Concrete Influenced by Aggregate Properties

Concrete Property	*Relevant Aggregate Property*
Durability	
Resistance to freezing and thawing	Soundness, porosity, pore structure, permeability, degree of saturation, tensile strength, texture and structure, clay minerals
Resistance to wetting and drying	Pore structure, modulus of elasticity
Resistance to heating and cooling	Coefficient of thermal expansion
Abrasion resistance	Hardness
Alkali–aggregate reaction	Presence of particular siliceous constituents
Strength	Strength, surface texture, cleanness, particle shape, maximum size
Shrinkage and creep	Modulus of elasticity, particle shape, grading, cleanness, maximum size, clay minerals
Coefficient of thermal expansion	Coefficient of thermal expansion, modulus of elasticity
Thermal conductivity	Thermal conductivity
Specific heat	Specific heat
Unit weight	Specific gravity, particle shape, grading, maximum size
Modulus of elasticity	Modulus of elasticity, Poisson's ratio
Slipperiness	Tendency to polish
Economy	Particle shape, grading, maximum size, amount of processing required, availability

weight. Most of this chapter is devoted to normal-weight aggregates, which make up about 90% of the concrete produced in the United States and Canada, since the same considerations apply generally to the other weight classes.

Aggregates should be hard and strong and free of undesirable impurities. Soft, porous rock can limit strength and wear resistance; it may also break down during mixing and adversely affect workability by increasing the amount of fines. Rocks that tend to fracture easily along specific planes can also limit strength and wear resistance. Therefore, it is best to avoid aggregates that contain a significant proportion of weak or friable materials, or to remove these. Aggregates should also be free of impurities: silt, clay, dirt, or organic matter. If these materials coat the surfaces of the aggregate, they will interfere with the cement–aggregate bond. Silt and clay and other fine materials will also increase the water requirements of the concrete; and organic matter may interfere with cement hydration.

Standard tests used in the United States, Canada, and Britain to evaluate aggregates are listed in the Appendix.

6.1 PROPERTIES REQUIRED FOR MIX DESIGN

In order to be able to proportion suitable concrete mixes, certain properties of the aggregate must be known: (1) shape and texture, (2) size gradation, (3) moisture content, (4) specific gravity, and (5) bulk unit weight. These properties influence the paste requirements for workable fresh concrete.

Shape and Texture

Effect on Workability

Aggregate shape and texture affect the workability of fresh concrete through their influence on cement paste requirements. Sufficient paste is required to coat the aggregates and to provide lubrication to decrease interactions between aggregate particles during mixing. The ideal aggregate particle is one that is close to spherical in shape (well rounded and compact) with a relatively smooth surface (see Figure 6.1). Most natural sands and gravels come close to this ideal. Crushed stone is much more

Figure 6.1 Classification of aggregate shapes.

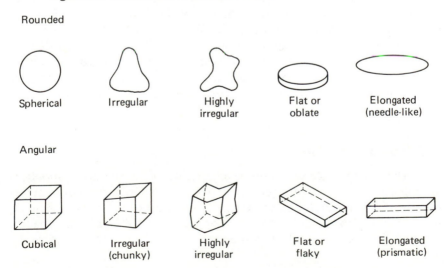

Rounded

| Spherical | Irregular | Highly irregular | Flat or oblate | Elongated (needle-like) |

Angular

| Cubical | Irregular (chunky) | Highly irregular | Flat or flaky | Elongated (prismatic) |

angular (with sharp edges and corners) and may have a rough surface texture. Such particles interfere more severely with the movement of adjacent particles even when nearly uniform (chunky) in shape. They also have a higher surface-to-volume ratio and therefore require more paste to fully coat the surface of each particle. Highly irregular particles with reentrant faces and sharp points will lead to greater interparticle interactions during mixing and handling. Aggregates that are flat or elongated should be avoided because they increase interparticle interaction and the surface-to-volume ratio, thereby increasing the paste requirements. Also, concretes containing aggregates of this shape are more prone to segregation during handling. The presence of flat or elongated particles in crushed rock may be indicative of rock with weak fracture planes. The surface texture of the aggregate is also important for workability, since a rough surface requires more lubrication for movement. Crushed stone has a rougher surface than natural sand and gravel because the surface has not been worn smooth by the effects of water and weather. There are no specified tests or detailed definitions prescribed by ASTM for surface texture and particle shape. However, BS 812: Part 1 does provide classification systems, which, because they do provide some useful reference terms, are given in Table 6.2.

Effect on Mechanical Properties

The shape and texture of the fine aggregate affects only workability, but the characteristics of the coarse aggregate may also affect the mechanical properties of the concrete by affecting the mechanical bond. Shape can favorably influence strength by increasing the amount of surface area available for bonding with the pastes for a given aggregate content. However, extremes in aggregate shape may lead to high internal stress concentrations and hence easier bond failure. Rough, textured surfaces will improve the mechanical component of the bond. It is also thought that the mineralogical character of the aggregate can play a role in determining the strength of the aggregate–paste bond, since a chemical interaction may occur at the interface between certain kinds of rock and cement paste.

Size Gradation

The particle-size distribution or *grading* of an aggregate supply is an important characteristic because it determines the paste requirements for a workable concrete. Since cement is the most expensive component, it is desirable to minimize the cost of concrete by using the smallest amount of paste consistent with the production of a concrete that can be

Table 6.2

Classification of Aggregate Shape and Texture (BS 812: Part 1)

Particle-Shape Classification

Classification	Description	Examples
Rounded	Fully water-worn or completely shaped by attrition	River or seashore gravel; desert, seashore, and windblown sand
Irregular	Naturally irregular, or partly shaped by attrition and having rounded edges	Other gravels; sand or dug flint
Angular	Possessing well-defined edges formed at the intersection of roughly planar faces	Crushed rocks of all types; talus; crushed slag
Flaky	Material of which the thickness is small relative to the other two dimensions	Laminated rock
Elongated	Material, usually angular, in which the length is considerably larger than the other two dimensions	—
Flaky and elongated	Material having the length considerably larger than the width, and the width considerably larger than the thickness	—

Surface Texture of Aggregates

Group	Surface Texture	Characteristics	Examples
1	Glassy	Conchoidal fracture	Black flint, obsidian, vitreous slag
2	Smooth	Water-worn, or smooth due to fracture of laminated or fine-grained rock	Gravels, chert, slate, marble, some rhyolites
3	Granular	Fracture showing more-or-less uniform rounded grains	Sandstone, oolite
4	Rough	Rough fracture of fine- or medium-grained rock containing no easily visible crystalline constituents	Basalt, felsite, porphyry, limestone
5	Crystalline	Containing easily visible crystalline constituents	Granite, gabbro, gneiss
6	Honeycombed	With visible pores and cavities	Brick, pumice, foamed slag, clinker, expanded clay

handled, compacted, and finished and provide the necessary strength and durability. The significance of aggregate gradation is best appreciated by considering concrete as a slightly compacted assembly of aggregate particles bonded together with cement paste, with the voids between particles completely filled with paste. Thus, the amount of paste depends on the amount of void space that must be filled and the total surface area of the aggregate that must be coated with paste. The volume of the voids between roughly spherical aggregate particles is greatest when the particles are of uniform size (Figure 6.2a). When a range of sizes is used, the smaller particles can pack between the larger (Figure 6.2b), thereby decreasing the void space and lowering paste requirements. Using a larger maximum aggregate size (Figure 6.2c) can also reduce the void space. For a given maximum aggregate size, a theoretical grading curve for minimum void space can be worked out using simple particle

Figure 6.2 Schematic representations of aggregate gradations in an assembly of aggregate particles: (a) uniform size; (b) continuous grading; (c) replacement of small sizes by large sizes; (d) gap-graded aggregate; (e) no-fines grading.

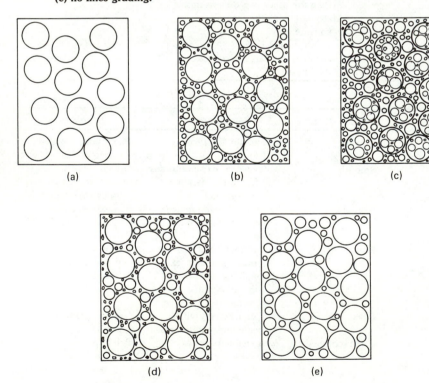

geometries. Such size distributions do not give a workable concrete; a compromise has to be worked out between workability and economy. This is discussed in more detail in Chapter 9.

Size Analysis

The grading of an aggregate supply is determined by a *sieve analysis*. A representative sample of the aggregate is passed through a stack of sieves arranged in order of decreasing size of the openings of the sieve. It is convenient to divide aggregate into coarse and fine fractions. The *coarse aggregate* fraction is that retained on the No. 4 sieve (4.75-mm opening), while the *fine aggregate* fraction is that passing the No. 4 sieve. The common sieve sizes are given in Table 6.3; in the coarse range the sieves are designated by the size of the openings, but in the fine range the sieves are assigned a number that represents the number of openings per inch. It can be seen also that each sieve used to grade the fine aggregate has openings that are half the dimension of the next-larger one. In the coarse sieves the ones that fit in this size sequence are the 9.5 mm, 19 mm, 37.5 mm, and 75 mm (⅜ in., ¾ in., 1½ in., and 3 in.). The sieves in this size sequence are called *standard sieves*; the others are called *half-sizes*.

Table 6.3

ASTM Sieve Sizes Commonly Used for Sieve Analysis of Concrete Aggregates

	ASTM Sieve Designation	*Nominal Size of Sieve Opening*	
		mm	*in.*
Coarse	3 in.	75	3
aggregate	2½ in.[a]	63	2.5
	2 in.[a]	50	2
	1½ in.	37.5	1.5
	1 in.[a]	25	1
	¾ in.	19	0.75
	½ in.[a]	12.5	0.50
	⅜ in.	9.5	0.375
Fine	No. 4 (3/16 in.)	4.75	0.187
aggregate	No. 8	2.36	0.0937
	No. 16	1.18	0.0469
	No. 30	0.60 (600 μm)[b]	0.0234
	No. 50	300 μm	0.0124
	No. 100	150 μm	0.0059

[a]Half-sizes.

[b]1000 μm = 1.0 mm.

Maximum Aggregate Size

The maximum size of the coarse aggregate influences the paste requirements of the concrete, and the optimum grading of the coarse aggregate
depends on the maximum aggregate size. As defined by ASTM C125, the
maximum size of coarse aggregate is the smallest sieve opening through
which the entire sample passes. In practice, it is considered that if only
a small amount of aggregate is retained on a sieve, it will not significantly
affect the properties of the concrete. Thus, it is usual to use a *nominal
maximum size,* which is the smallest sieve opening through which the
entire sample is permitted to pass, but need not do so. A percentage
(usually 5%) of the sample weight may be retained on this sieve. ASTM
grading requirements are based on nominal maximum size.

The choice of nominal maximum aggregate size is determined by
job conditions. If the aggregate size is too large, the concrete at any
given cross-section of a member may not be representative of the entire
material because of the location of an overly large aggregate particle. To
guard against this possibility, the maximum size should not be greater
than one-fifth of the smallest dimension of the member. For slabs on
ground, this figure may be relaxed to one-third of the thickness, because
of the base support. Large aggregate particles may also get obstructed at
narrow openings between reinforcing bars, or between bars and formwork, and cause undesirable segregation during placement. To avoid this
problem, the nominal maximum size should not exceed three-fourths of
the minimum clear spacing between reinforcing bars or between reinforcing bars and forms. These two requirements effectively limit the
maximum size on most jobs to 40 mm (1½ in.) or less. Larger sizes can
be used in unreinforced mass concrete where limitations on size do not
apply. The handling of large volumes of concretes justifies the use of
larger equipment needed to handle large aggregate sizes. However, most
equipment for testing concrete is designed to handle only up to 37.5 mm
(1½ in.) maximum size aggregate, so the larger sizes must be screened
out before testing.

The higher the maximum aggregate size, the lower the paste
requirements for the mix (see earlier discussion). For a given workability
and cement content, the strength of concrete increases with increasing
aggregate size because the *w/c* ratio can be lowered (Figure 6.3).
However, with the larger aggregate sizes, the reduction in bond area and
increased internal stresses tend to lower the strength. This effect is
noticeable only in rich mixes; concretes with low cement contents, which
are typical of mass concrete, show continuing increases in strength.

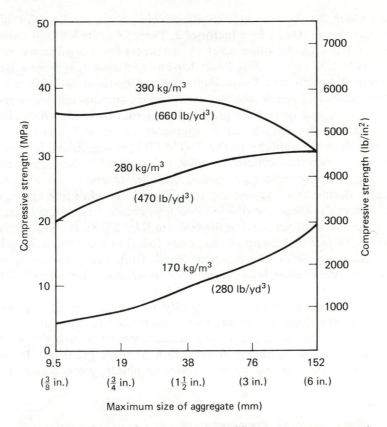

Figure 6.3 Influence of aggregate size on 28-day compressive strength of concretes with different cement contents. (From *Symposium on Mass Concrete*, SP-6, American Concrete Institute, Detroit, Mich., 1963, pp. 219–256.)

Grading Curves

When a sieve analysis has been completed (according to the provision of ASTM C136), the weight of aggregate retained on each sieve is expressed as a percentage of the total weight of the sample. This is then used to calculate the cumulative percentage retained on each successive sieve or the cumulative percentage passing each sieve (Table 6.6). These numbers can then be plotted graphically against sieve size, as in Figure 9.1, to give a grading curve. It is customary to use cumulative percentage passing on the ordinate. The successive standard sieve sizes are plotted linearly along the abcissa, and this effectively gives a semilog plot, since each sieve size is related by a factor of 2. Therefore, the *half-sizes* should be placed between the other sizes as is correct for a logarithmic scale.

ASTM C33 sets grading limits for fine and coarse aggregate based on practical experience. These limits are summarized in Table 6.4 for fine aggregates and coarse aggregates of various nominal maximum sizes. It can be seen that quite wide grading limits are provided, which allows for adjustment for differences in aggregate shape and texture. If an aggregate does not conform to the ASTM C33 grading limits, it does not necessarily mean that concrete cannot be made with the aggregate. It does mean that concrete may require more paste and is more liable to segregate during handling and placing. BS 882 classifies fine aggregates quite differently. These are divided into four zones, based largely on the percentages of material passing the 600-μm (No. 25) sieve. This division is due to the fact that many of the sands found in the United Kingdom can be rationally divided in this way. Sands from any zone can be used in concrete; the grading requirements for these zones are given in Table 6.5.

It is easiest to maintain uniformity of concrete if the aggregate grading remains constant. On large jobs where considerable quantities of concrete are to be handled, it may be advantageous to blend several fractions of coarse aggregate to maintain uniform grading. This is a common procedure in ready-mixed-concrete plants, since concrete can

Table 6.4

ASTM Grading Limits for Concrete Aggregates[c]

Fine Aggregate		Coarse Aggregate				
Sieve Size	% Passing[a]	Sieve Size[b]	\% Passing (Nominal Maximum Size)			
			1½ in.	1 in.	¾ in.	½ in.
⅜ in.	100	1½ in.	95–100	100	—	—
No. 4	95–100	1 in.	—	95–100	100	—
No. 8	80–100	¾ in.	35–70	—	90–100	100
No. 16	50–85	½ in.	—	25–60	—	90–100
No. 30	25–60	⅜ in.	10–30	—	20–55	40–70
No. 50	10–30	No. 4	0–5	0–10	0–10	0–15
No. 100	2–10	No. 8	—	0–5	0–5	0–5

[a]Not more than 45% should be retained between any two consecutive sieves.

[b]1 in. = 25.4 mm.

[c]Data taken from ASTM C33, "Standard Specification for Concrete Aggregates." Reprinted, with permission, from the American Society for Testing and Materials, 1916 Race Street, Philadelphia, PA 19103. Copyright.

Table 6.5

BS 882 Grading Limits for Fine Aggregate

Sieve Size	Percent Passing			
	Grading Zone 1	Grading Zone 2	Grading Zone 3	Grading Zone 4
9.52 mm	100	100	100	100
4.76 mm	90–100	90–100	90–100	95–100
2.40 mm	60–95	75–100	85–100	95–100
1.20 mm	30–70	55–90	75–100	90–100
600 μm	15–34	35–59	60–79	80–100
300 μm	5–20	8–30	12–40	15–50
150 μm	0–10[a]	0–10[a]	0–10[a]	0–15[a]

[a]For crushed-stone sands, the permissible limit is increased to 20%.

be produced more economically when corrections for grading variability are eliminated.

Fineness modulus. The use of a single parameter to describe the grading curve can be useful in checking the uniformity of grading. The *fineness modulus* is such a parameter. It is defined as

$$\text{F.M.} = \frac{\Sigma \text{ (cumulative percent retained on standard sieves)}}{100} \quad (6.1)$$

where the standard sieves used are No. 100, No. 50, No. 30, No. 16, No. 8, and No. 4; and ⅜ in. (9.5 mm), ¾ in. (19.0 mm), 1½ in. (37.5 mm), and larger, increasing in the size ratio 2 to 1. The fineness modulus is usually calculated only for the fine aggregate, but may be calculated for the coarse aggregate as well, based on the assumption that 100% is retained on each of the sieves from No. 8 to No. 100. The fineness modulus for coarse aggregate then becomes

$$\text{F.M.} = \frac{\Sigma \text{ (cumulative percent retained on standard coarse sieves, including No. 4 + 500)}}{100} \quad (6.1a)$$

The fineness modulus for fine aggregate should lie between 2.3 and 3.1 (see the example in Table 6.6). A small number indicates a fine grading; whereas a large number indicates a coarse material. The fineness modulus can be used to check the constancy of grading when relatively small

Table 6.6

Calculation of Fineness Modulus

	Sieve Size	Weight Retained (g)	Amount Retained (wt. %)	Cumulative Amount Retained (%)	Cumulative Amount Passing (%)
Coarse	3 in.	0	0	0	100
fraction	1½ in.	42	4	4	96
	¾ in.	391	39	43	57
	⅜ in.	350	35	78	22
	No. 4	200	20	98	2
	Sample wt.	1000 g		223	
				+ 500	(from fine sieves No. 8 to No. 100)
				Σ = 723	

Nominal maximum size = 1½ in.
Fineness modulus = 723/100 = 7.23

Fine	No. 4	9	2	2	98
fraction	No. 8	46	9	11	89
	No. 16	97	19	30	70
	No. 30	99	20	50	50
	No. 50	120	24	74	26
	No. 100	91	18	92	8
	Sample wt.	500 g		Σ = 259	

Fineness modulus = 259/100 = 2.59

changes are to be expected; but it should not be used to compare the gradings of aggregates from two different sources. The parameter is quite a crude measure of grading; two aggregates with the same fineness modulus can have quite different grading curves.

The fineness modulus of fine aggregate is required for mix proportioning since sand gradation has the largest effect on workability. A fine sand (low fineness modulus) has much higher paste requirements for good workability. ASTM C33 requires that the fineness modulus should not vary by more than 0.2 from the value used for mix design purposes. Larger variations will cause unacceptable changes in workability and will require reproportioning. The fineness modulus of the coarse aggregate is not used for mix design purposes.

Other Grading Requirements

The ASTM specifications permit a wide range of grading, particularly in the case of the fine aggregate. Other specifications may require much narrower limits, depending on whether rounded gravel and sands or angular crushed rock and manufactured sands are available. The amount of fine aggregate passing the No. 50 and No. 100 sieves affects workability, finishability, bleeding, and so on, and adequate fine material is needed for good cohesiveness and plasticity. The 10% passing the No. 50 sieve allowed by ASTM may not provide sufficient fines for hand-finished concrete floors, or when a smooth surface texture is wanted. In such cases it is recommended that a fine aggregate should be used that has at least 15% passing the No. 50 sieve, and 3% or more passing the No. 100. On the other hand, ASTM C33 allows the finer fractions to be reduced or omitted if the concrete contains workability aids: a mineral admixture, an air-entraining agent, or a high cement content. However, ASTM limits the very fine fraction, passing the No. 200 mesh sieve, to not more than 5%. Such material acts as a diluent with a very high water demand. In natural aggregates it is likely to be rich in clayey matter, which may increase the volume instability of the concrete.

Gap Grading

Gap grading is defined as grading in which one or more intermediate-size fractions are omitted (Figure 6.2d and e). Gap grading can be used to produce more economical concrete, particularly when blending of aggregates is required to obtain suitable gradings. Less sand can be used for a given workability, hence allowing the use of less cement or lower *w/c* ratios for a given slump. However, the omission of size fractions can lead to severe segregation problems, particularly in mixes of high workability. Thus, gap grading is recommended primarily for stiff mixes of very low workability that are to be compacted by vibration. Close control of mix proportions must be maintained to avoid segregation. It does not appear that gap-graded concretes are inherently superior to continuously graded concretes. Claims of improved properties can be attributed to decreases in *w/c* ratio at a given cement content and slump. Gap-graded aggregate is often used to give better exposed aggregate finishes.

A special case of gap-grading is *no-fines concrete,* in which the fine aggregate is entirely omitted (Figure 6.2e). As would be expected, this concrete has little cohesiveness in the fresh state and cannot be compacted to a void-free condition. Consequently, no-fines concrete is a low-strength, high-permeability material in which coarse aggregate, lightly compacted by gravity or by light rodding, is cemented together by paste rather than mortar, but without completely filling the voids between the aggregate. The advantages of no-fines concrete are its low density, low drying shrinkage, and high thermal insulation, which can be utilized when structural requirements are not high.

Moisture Contents

Since aggregates contain some porosity,[1] water can be absorbed into the body of the particles. Also, water can be retained on the surface of the particle as a film of moisture. Thus, stockpiled aggregates can have a variable moisture content. It is necessary to have information about the moisture content, since if there is a tendency for the aggregates to absorb water, it will be removed from the paste so that the w/c ratio is effectively lowered and the workability of the concrete decreased. Conversely, if excess water is present at the aggregate surfaces, extra water will be added to the paste and the w/c ratio of the concrete will be higher than desired.

Moisture States

It is convenient to define four moisture states of the aggregate as shown in Figure 6.4. These are:

1. Oven-dry (OD): All moisture removed from the aggregate by heating in an oven at 105°C to constant weight (overnight heating usually suffices). All pores are empty.

2. Air-dry (AD): All moisture removed from surface, but internal pores partially full.

3. Saturated-surface-dry (SSD): All pores filled with water, but no film of water on the surface.

4. Wet: All pores completely filled with water with a film of water on the surface.

Of these four states, only two, the OD and SSD states, correspond to specific moisture contents, and either of these states can be used as reference states for calculating moisture contents. The AD and wet states represent the variable moisture contents that will exist in stockpiled aggregates. The SSD condition is the better choice as a reference state, for the following reasons:

1. It represents the "equilibrium" moisture state of the aggregate in concrete; that is, the aggregate will neither absorb water nor give up water to the paste.

2. The moisture content of aggregates in the field is much closer to the SSD state than the OD state.

[1]Only pores that are connected to the surface are considered in this discussion.

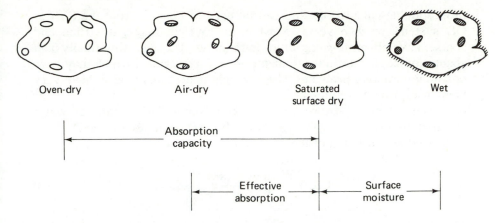

Figure 6.4 Moisture states of aggregates.

3. The bulk specific gravity (BSG) of aggregates is more accurately determined by the displacement method in the SSD condition.

4. The moisture content can be calculated directly from measurements of apparent BSG using the displacement method.

A major disadvantage of using the SSD state is that it is not easy to obtain a true SSD condition and it requires skill and practice to do so. Many people prefer to use the OD state as a reference point because of this. In the rest of this discussion we use the SSD as the reference state, but the equivalent treatment using the OD state is summarized in the appendix to this chapter.

Absorption and Surface Moisture

In order to calculate the amount of water that aggregate will add or subtract to the paste, it is convenient to define three quantities. These are *absorption capacity, effective absorption,* and *surface moisture.*

Absorption capacity (AC or absorption) represents the maximum amount of water the aggregate can absorb. It is calculated from the difference in weight between the SSD and OD states, expressed as a percentage of the OD weight:

$$\text{AC} = \frac{W_{\text{SSD}} - W_{\text{OD}}}{W_{\text{OD}}} \times 100\% \tag{6.2}$$

where W_{SSD} and W_{OD} represent the weight of the aggregate sample in the SSD and OD states, respectively. The absorption capacity is used in mix proportioning calculations and can be used to convert from the SSD to OD system, or vice versa. Most normal-weight aggregates (fine and coarse) have absorption capacities in the range 1 to 2%. Abnormally high absorption capacities indicate high-porosity aggregates, which may have potential durability problems. (Lightweight aggregates tend to have very high absorption capacities.)

The effective absorption (EA) represents the amount of water required to bring an aggregate from the AD state to the SSD state, expressed as a fraction of the SSD weight:

$$EA = \frac{W_{SSD} - W_{AD}}{W_{SSD}} \times 100\% \qquad (6.3)$$

The effective absorption is used to calculate the weight of water absorbed (W_{abs}) by the weight of aggregate (W_{agg}) in the concrete mix:

$$W_{abs} = (EA)W_{agg} \qquad (6.4)$$

If an aggregate is close to the OD condition when batched, it takes the aggregate some time to absorb all the water necessary to reach the SSD condition. In such cases the effective absorption may be taken to indicate the amount of water the aggregate absorbs in 30 min. Further absorption beyond this time will be slow.

The surface moisture (SM) represents water in excess of the SSD state, also expressed as a fraction of the SSD weight:

$$SM = \frac{W_{wet} - W_{SSD}}{W_{SSD}} \times 100\% \qquad (6.5)$$

It is used to calculate the additional water (W_{add}) added to the concrete with the aggregate.

$$W_{add} = (SM)W_{agg} \qquad (6.6)$$

Equations (6.3) through (6.6) can be put in general terms, since both the AD and wet states represent the possible conditions of stockpiled aggregates. The moisture content (MC) of the aggregate is given by

$$MC = \frac{W_{stock} - W_{SSD}}{W_{SSD}} \times 100\% \qquad (6.7)$$

where W_{stock} is the weight of the aggregate in the stockpiled condition. If the moisture content is positive, it is surface moisture; if negative, it is effective absorption. Thus,

$$W_{\mathrm{MC}} = (MC)W_{\mathrm{agg}} \tag{6.8}$$

where W_{MC} is the total moisture associated with the aggregates and is positive (i.e., added) when the moisture content is positive, and negative (i.e., absorbed) when the moisture content is negative.

Bulking of Sand

Stockpiled coarse aggregate is generally in the AD state with an effective absorption of less than 1%. However, fine aggregate is often in the wet state, with a surface moisture typically in the range 0 to 5%. The reason for high surface-moisture values for the fine aggregate is that in addition to thin surface films of moisture, additional water can be held in the interstices between fine particles as the result of formation of menisci (Figure 6.5a). The formation of these menisci creates thicker films of water between the aggregate particles, pushing them apart and increasing the apparent volume of the aggregate (see Figure 6.5b). This phenomenon is known as *bulking* and can cause substantial errors in proportioning by volume. Hence, aggregate is batched by weight and measurements of unit weight are usually made on oven-dry aggregate. When sand is saturated with water, the menisci are destroyed and the volume returns

Figure 6.5 Bulking phenomenon of fine aggregate: (a) dry; (b) partially saturated (menisci formation); (c) fully saturated.

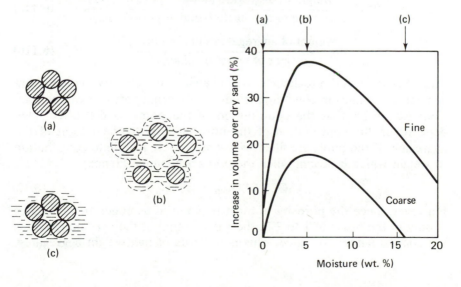

to normal. Coarse aggregates show much less bulking since the particle size is large compared to the thickness of the water film and the effect of meniscus formation is slight.

Specific Gravity

The density of the aggregates is required in mix proportioning to establish weight–volume relationships. The density is expressed as the specific gravity, which is a dimensionless ratio relating the density of the aggregate to that of water:

$$S.G. = \frac{\text{density of solid}}{\text{density of water}} \tag{6.9}$$

Since densities are determined by displacement in water, specific gravities are naturally and easily calculated and can be used with any system of units.

Now, it must be remembered that aggregates contain some porosity and that the specific gravity value depends on whether the pores are included in the measurement. Thus, we can distinguish between absolute specific gravity (ASG), which refers only to the solid material excluding the pores [Eq. (6.10)], and bulk specific gravity (BSG), which includes the volume of pores in the aggregate volume [Eq. (6.11)].

$$\text{ASG} = \frac{\text{weight of aggregate (solid only)}}{\text{volume of aggregate (solid only)}} \cdot \frac{1}{\rho_{\text{water}}} \tag{6.10}$$

$$\text{BSG} = \frac{\text{weight of aggregate (solid + pores)}}{\text{volume of aggregate (solid + pores)}} \cdot \frac{1}{\rho_{\text{water}}} \tag{6.11a}$$

$$= \frac{\text{weight of aggregate (solid + pores)}}{\text{weight of water displaced}} \tag{6.11b}$$

The BSG value is the realistic one to use since the effective volume that aggregate occupies in concrete includes its internal pores. It can be seen from Eq. (6.11) that the contribution of the pores to the BSG value depends on the contribution of the pore medium to the weight of the aggregate. If the pores are filled with water, there is a finite contribution to weight which is absent when the pores are empty. Hence,

$$\text{ASG} > \text{BSG}_{\text{SSD}} > \text{BSG}_{\text{OD}} \tag{6.12}$$

However, since the porosity of most rocks used as concrete aggregates is only of the order of 1 to 2%, the values of all of the specific gravities are approximately the same. This is not true of lightweight aggregates, whose BSGs are strongly dependent on moisture content. The BSG of most rocks is in the range 2.5 to 2.8; a value well below this range is probably indicative of high porosity. However, the BSG of an aggregate cannot be directly related to its performance in concrete, and thus it is not a specified quantity. The only reason for specifying BSG is if a minimum density of concrete is required, since BSG directly relates to concrete density.

Determination of BSG

The BSG_{SSD} of aggregates in the SSD condition is determined by the displacement method given in ASTM C127 and C128. The basic equation used is Eq. (6.11b). In ASTM C127 the coarse aggregate is weighed in air (W_{air}) and in water (W_{water}) and the weight of water displaced (W_{displ}) is given by

$$W_{displ} = W_{air} - W_{water} \qquad (6.13)$$

The weight of water displaced is, of course, numerically equal to the volume of water displaced (in the SI system), which is the same as the volume of aggregate. In ASTM C128 a pycnometer is weighed full of water and then the pycnometer containing the sample of sand is filled with water and again weighed. (Remember that W_{sand} refers to the SSD weight; i.e. W_{SSD}).

$$W_{displ} = (W_{sand} + W_{pyc}) - W*_{pyc} \qquad (6.14a)$$

where W_{pyc} is the weight of the pycnometer containing only water and $W*_{pyc}$ is the weight of the pycnometer containing both sand and water. Equation (6.14a) can be reduced to Eq. (6.14b):

$$W_{displ} = W_{water} - W*_{water} \qquad (6.14b)$$

where W_{water} and $W*_{water}$ are the weights of water in the pycnometer in the absence and the presence, respectively, of the sand.

The displacement method cannot be used successfully on OD aggregate because absorption will take place during the measurement. Since the BSG will be different in the OD case, we write BSG_{SSD} to denote that the quantity pertains to the SSD condition.

Measurement of Moisture Contents

The measurement of the moisture contents of aggregates can conveniently be made using the displacement methods used to determine the BSG of the SSD aggregate. The procedure for determining the surface moisture in fine aggregate is given in ASTM C70, but the principle behind this method can be used to determine effective absorption or surface moisture of both fine and coarse aggregate.

For a given aggregate sample from a stockpile, the following relationship holds true:

$$W_{stock} = W_{SSD} + W_{MC} \qquad (6.15)$$

W_{MC} can be either positive (wet) or negative (AD). The usual measurements are then made as if the BSG of the aggregate was to be determined. (It is assumed that the aggregate reaches the SSD condition during the measurement.) Thus, for the coarse aggregate using Eq. (6.11b), we get

$$BSG_{SSD} = \frac{W_{SSD}}{W_{displ}} \qquad (6.16a)$$

In the experiment measuring BSG_{SSD} the complete weight was W_{SSD}, but in this case this is not true as Eq. (6.15) shows. However, W_{displ} can be expressed in terms of W_{SSD} as follows:

$$BSG_{SSD} = \frac{W_{SSD}}{W_{SSD} - W*_{SSD}} \qquad (6.16b)$$

where $W*_{SSD}$ equals the weight of the *SSD aggregate* in water. This quantity is actually measured experimentally since the aggregate attains the SSD condition before the measurement is complete. (This is always true for wet aggregate, but could be in error if the aggregate is very dry and absorption is slow—not the usual situation.) Therefore, Eq. (6.16b) can be rearranged to obtain W_{SSD} in terms of two known quantities, BSG_{SSD} and $W*_{SSD}$;

$$W_{SSD} = \frac{W*_{SSD} \cdot BSG_{SSD}}{BSG_{SSD} - 1} \qquad (6.17a)$$

Equation (6.17a) can now be used to solve for W_{MC} in Eq. (6.15) and W_{MC} can be expressed as a percentage of W_{SSD} (i.e., as surface moisture and effective absorption).

In the case of fine aggregate, Eq. (6.14a) can be adapted to solve for W_{displ}. Since W_{sand} is now W_{stock}, it can no longer be used in the equation and we must write

$$W_{displ} = (W_{SSD} + W_{pyc}) - W*_{pyc} \qquad (6.17b)$$

$W*_{pyc}$ is measured experimentally and W_{pyc} is known from the experiment used to measure BSG_{SSD}. W_{SSD} is not known but by rearranging Eq. (6.16a) and substituting with Eq. (6.17b) we get

$$W_{SSD} = W_{displ} \cdot BSG_{SSD}$$
$$= [W_{SSD} + (W_{pyc} - W*_{pyc})]BSG_{SSD} \tag{6.18}$$

Unit Weight

Unit weight (UW) can be defined as the weight of a given volume of graded aggregate. It is thus a density measurement and is also known as *bulk density,* but this alternative term is similar to bulk specific gravity, which is quite a different quantity, and perhaps is not a good choice. The unit weight effectively measures the volume that the graded aggregate will occupy in concrete and includes both the solid aggregate particles and the voids between them. The unit weight is simply measured by filling a container of known volume and weighing it (ASTM C29 or BS 812: Part 2). Clearly, however, the degree of compaction will change the amount of void space, and hence the value of the unit weight. The ASTM standard method calls for compaction by rodding. (The unit weight of lightweight aggregate is often determined in a loosely packed condition without any compaction.) Since the weight of the aggregate is dependent on the moisture content of the aggregate, a constant moisture content is required. Oven-dry aggregate is used in ASTM C29. The unit weight of the coarse aggregate is required for the volume method of mix proportioning.

The void space that must be filled with mortar can be calculated using the unit weight, as follows. The unit weight in kg/m³ is equal to the weight of aggregate (W_a) in 1 m³ of volume.

$$\text{total volume} = V_a + V_v = 1 \text{ m}^3 \tag{6.19}$$

Therefore,

$$V_v = (1 - V_a) \text{ m}^3 \tag{6.20}$$

where V_a and V_v are the volume of aggregate and void space, respectively. Also,

$$V_a = \frac{W_a}{BSG \cdot \rho_w} \text{ m}^3 \tag{6.21}$$

where ρ_w is the density of water in kg/m³ and BSG is the bulk specific gravity in the dry state. Therefore,

$$V_v = 1 - \frac{W_a}{\text{BSG} \cdot \rho_w} \text{ m}^3 \qquad (6.22)$$

or

$$V_v = 1 - \frac{\text{UW}}{\text{BSG} \cdot \rho_w} \text{ m}^3 \qquad (6.23)$$

or

$$\text{percent of voids} = \frac{\text{BSG} \cdot \rho_w - \text{UW}}{\text{BSG} \cdot \rho_w} \times 100 \qquad (6.24)$$

The unit weights of both fine and coarse normal-weight aggregates falling within the ASTM grading limits are generally in the range 1450 to 1750 kg/m³ (90 to 110 lb/ft³). However, a plot of unit weights for various aggregate blends of two particular aggregates shows a maximum unit weight when the fine aggregate content is 35 to 40% by weight of total aggregate (Figure 6.6). Using Eq. (6.22) it can be deduced that this maximum corresponds to a minimum percentage of void space (assuming that the BSG of both aggregates is the same, which is a good approxi-

Figure 6.6 Variation of the unit weight of dry-rodded aggregate for blends of fine and coarse aggregate.

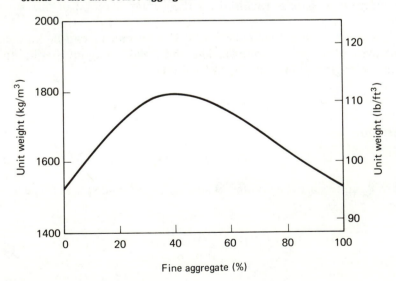

mation) which must be filled with paste. Thus, the most economical concrete is one that has close to 40% of fine aggregate in the total aggregate fraction, since the least amount of cement will be needed.

6.2 DURABILITY OF AGGREGATES

Since aggregates make up the bulk of concrete, any lack of durability of the aggregate will have disastrous consequences for the concrete. Fortunately, in many parts of North America there are few problems concerning the durability of aggregates. In other areas, special screening tests may be required routinely to avoid problem aggregates, or special measures must be taken to counteract the effects of undesirable aggregates. The latter approach will become more important in the future as deposits of high-quality aggregate are worked out and more marginal material is brought into use. The durability of aggregates can be conveniently divided into physical and chemical causes. The former is concerned with susceptibility of aggregates to freezing and thawing or wetting and drying, as well as physical wear. Chemical durability problems are concerned with various forms of cement–aggregate reactions.

Physical Durability

Soundness

Aggregates are said to be unsound if volume changes that accompany environmental changes lead to the deterioration of concrete. Volume changes can arise from alternate freezing and thawing or from repeated wetting and drying. Fortunately, rocks that undergo appreciable volume change on wetting and drying are very rare. Thus, soundness is primarily a question of freeze–thaw resistance and is the basic cause of two fairly widespread forms of concrete deterioration: surface "pop-outs," and D-cracking in pavements. The factors affecting freeze–thaw susceptibilities of aggregates are similar to those affecting hardened cement paste, which are discussed in detail in Chapter 20. Briefly, the resistance of an aggregate depends on whether high internal stresses develop when the water inside the aggregate freezes and causes a volume increase. This stress is a function of the porosity of the aggregate, its permeability, the degree of saturation, and size. A critical aggregate size can be calculated (see Chapter 20) below which freeze–thaw distress will not occur; for most aggregates this is greater than the normal sizes used in practice but for some poorly consolidated sedimentary rocks—cherts, graywackes, sandstones, shales, and limestones—the critical size may be less than the maximum aggregate size (in the range 12 to 25 mm).

Wear Resistance

Wear resistance of concrete is discussed in detail in Chapter 19, but obviously the aggregate must play an important role in determining the resistance of concrete to surface abrasion and wear. A good aggregate will be hard, dense, and strong, and free of soft, porous, or friable particles. The abrasion resistance of aggregates can be tested by the Los Angeles test (ASTM C131), but the resistance of concrete made with the aggregate should also be tested. The Los Angeles abrasion test does not correlate well with concrete wear in the field.

Chemical Resistance

Most chemical durability problems result from a reaction between reactive silica in aggregates and alkalis contained in the cement. The most familiar problem is the *alkali–aggregate reaction,* but other less common reactions have been identified in recent years, and these will also be discussed. Occasionally, other kinds of chemical distress occur. For example, iron pyrites (FeS) may react expansively in the presence of calcium hydroxide to form ferrous sulfate and ultimately ferric hydroxide. This reaction can cause popouts and staining. Natural gypsum will cause sulfate attack if present in significant amounts. Small quantities of zinc or lead are occasionally found in aggregate deposits and may greatly delay setting and early hardening of the concrete.

Alkali –Aggregate Reaction

In the United States there were many reported failures of concrete structures built during the late 1920s to the early 1940s. These failures were the result of overall cracking throughout the structure manifested at the surface as extensive *map cracking* or *pattern cracking* (see Figure 6.7), frequently accompanied by gel exuding from the cracks, or surface popouts and spalling. Such problems were confined mostly to the west and southwest of the country, as well as Kansas, Nebraska, Alabama, and Georgia.

Classic research by Stanton[2] in the early 1940s correctly diagnosed the failures as being due to expansions caused by a chemical reaction

[2]T. E. Stanton, *Proceedings ASCE,* Vol. 66, pp. 1781–1811 (1940).

Figure 6.7 Example of map cracking caused by alkali–aggregate reaction. (Photograph courtesy of Bryant Mather.)

between the alkalis contained in the cement paste and certain reactive forms of silica within the aggregate (Table 6.7). Stanton found opal and chert to be common forms of reactive silica, but the list is considerably broader, and naturally occurring volcanic glasses are a widespread form of reactive silica. It appears that various forms of silica have differing reactivities, depending on the degree of crystallinity, internal porosity, crystallite size, and internal crystal strain. Opal is the most reactive form of natural silica, being both amorphous and porous; container glass (borosilicate) containing sodium or potassium is also a very reactive material. Recent reports indicate that the alkali–aggregate reaction may be more widespread than has hitherto been realized. Most of the structural failures in the 1930s and 1940s occurred within 1 to 10 years, probably involving opaline rocks. In some structures, however, severe deterioration did not occur until after 15 to 20 years, suggesting that less reactive silicas were involved, and there seems no reason why even less-reactive forms of silica may not eventually cause deterioration. In support of this idea is the fact that early signs of deterioration have been found in some structures that are 20 to 25 years old.

Factors affecting expansion. Stanton was able to determine the factors that controlled the alkali–aggregate expansion: (1) nature of reactive silica, (2) amount of reactive silica, (3) particle size of reactive material,

Table 6.7

Forms of Reactive Silica in Rocks That Can Participate in the Alkali-Aggregate Reaction

Reactive Component	Physical Form	Rock Types in Which It Is Found	Occurrence
Opal	Amorphous	Siliceous (opaline) limestones, cherts, shales, flints	Widespread
Silica glass	Amorphous	Volcanic glasses (rhyolite, andesite dacite) and tuffs; synthetic glasses	Regions of volcanic origin; river gravels originating in volcanic areas; container glass
Chalcedony	Poorly crystallized quartz	Siliceous limestones and sandstones, cherts and flints	Widespread
Tridymite, cristobalite	Crystalline	Opaline rocks, fired ceramics	Uncommon
Quartz	Crystalline	Quartzite, sands, sandstones, many igneous and metamorphic rocks (e.g., granites and schists)	Common, but reactive only if microcrystalline or highly strained

(4) amount of available alkali, and (5) amount of available moisture. We have already discussed the nature of the reactive silica, but the amount is also important. Reference to Figure 6.8 shows that maximum expansion occurs at about 5% opal content (the "pessimum percentage"), with decreasing expansion at higher percentages. The particle size of the reactive material is also an important factor (Figure 6.9), and expansion is seen to be greatest with intermediate-sized particles. The alkali content of the cement, and hence of the hydrated paste, is another important factor. Experiments have shown that below 0.6% Na_2O equivalent[3] deleterious expansions usually do not occur (see Figure 6.10). The exact form of the curves in Figures 6.8 to 6.10 will depend on the physical and chemical nature of the reactive component. Finally, there must be sufficient moisture available within the concrete for the alkali–aggregate reaction to proceed.

Mechanism of alkali–aggregate reaction. The experimental data given above have been used as the basis of practical control of the alkali–silica reaction, as will be further discussed shortly, but have not provided a

[3]The Na_2O equivalent is given by $Na_2O + 0.66K_2O$. The factor 0.66 merely accounts for the difference in the molecular weights of Na_2O and K_2O. Thus, $0.66K_2O$ represents the weight of an equal number of moles of Na_2O that is equivalent to the weight of K_2O.

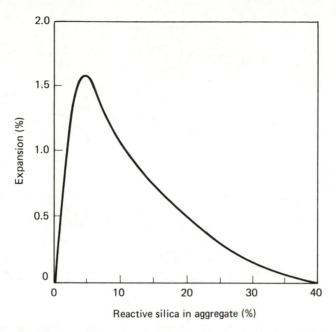

Figure 6.8 Effect of the content of reactive silica in an aggregate on the expansion of concrete due to the alkali–aggregate reaction. (From H. Woods, *Durability of Concrete Construction*, Monograph No. 4, American Concrete Institute, Detroit, Mich., 1968.)

Figure 6.9 Effect of the size of reactive silica constituent of aggregate on alkali–aggregate expansion (particles are retained on the mesh size indicated, but pass through the next larger size). (From H. Woods, *Durability of Concrete Construction*, Monograph No. 4, American Concrete Institute, Detroit, Mich., 1968.)

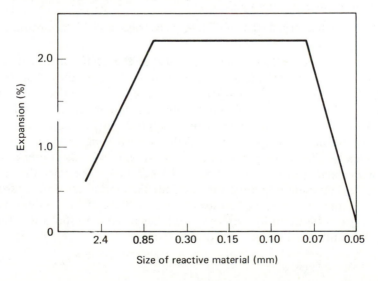

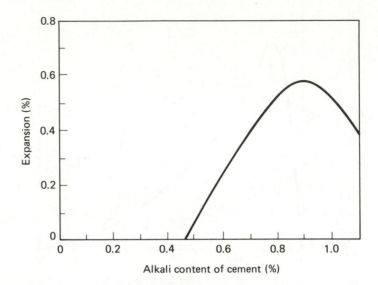

Figure 6.10 Effect of the alkali content of the cement on alkali–aggregate expansion. (From H. Woods, *Durability of Concrete Construction,* Monograph No. 4, American Concrete Institute, Detroit, Mich., 1968.)

detailed explanation for the cause of distress. The following explanation proposed by Diamond visualizes four distinct steps:

1. Initial alkaline depolymerization and dissolution of reactive silica.

2. Formation of a hydrous alkali silicate gel.

3. Attraction of water by the gel.

4. Formation of a fluid sol (a dilute suspension of colloidal particles).

The first step depends on the alkalinity of the solution and does not involve soluble alkali metal ions per se, although they control the alkalinity of the solution contained in the pores of the concrete. As can be seen in Table 6.8, the high-alkali cement can increase the solubility of amorphous silica and also the rate at which it dissolves. Crystalline silica will also be affected by the pH of the surrounding medium, but less dramatically. The initial porosity of the aggregate will also determine the rate and extent of this initial step and whether it can take place throughout the particle or initially only at the surface. This initial alkaline hydrolysis opens up the structure and allows the pore fluid, which is essentially sodium (or potassium) hydroxide, to further hydrolyze the silica fraction to form an alkali–silica gel (or solid gel) *in situ*.

$$S \quad + \quad N(K)H \quad \rightarrow \quad N(K)\!-\!S\!-\!H$$

$$\text{aggregate} \qquad \text{pore fluid} \quad \text{alkali–silica gel}$$

(6.25)

Equation (6.25) does not involve extensive dissolution and is not expansive in itself, but it destroys the integrity of the aggregate particle. The reaction does not necessarily take place from the outside inward, but may proceed throughout the reactive silica particle, depending on its structure. However, the gel has the ability to imbibe considerable amounts of water, which is accompanied by a volume expansion. If this expansion is sufficient, the resulting stress will crack the weakened aggregate and the surrounding cement paste. The final step takes place after the critical expansion has occurred, when further ingestion of water turns the solid gel into a fluid sol which escapes into the surrounding cracks and voids. Secondary reactions with calcium hydroxide in the cement paste may also take place, forming deposits of a calcium–alkali–silica gel at the periphery of the distressed aggregate.

This description of the alkali–aggregate reaction visualizes the localized centers of expansion at the sites of the particles of reactive silica. If the number of reactive particles is relatively small, the soluble alkali metal ions can migrate to these scattered centers to form the alkali silicate gel and cause high, localized expansions that disrupt the matrix. If there are a large number of reactive particles, there are not enough alkali metal ions to cause complete reaction of all particles and therefore expansions are reduced. This explains the occurrence of a "pessimum percentage," as shown in Figure 6.8. Small particle size encourages a rapid reaction without deleterious effects. In these cases the reaction moves rapidly through to step 4 without the large volume changes that would occur during step 3. This mechanism also explains the very high reactivity of opal and container glass. Opal has already proceeded

Table 6.8

Effect of pH on the Hydrolysis of Amorphous Silica in a Cement Paste

Medium	pH	Approximate Solubility of SiO_2 (ppm)
Neutral water	7–8	100–150
Moderately alkaline water	10	<500
Saturated $Ca(OH)_2$	12	90,000
Low-alkali cement paste	~12.5	~500,000
High-alkali cement paste	>13.0	Infinite

through step 1 in nature, while container glass is an alkali silicate that proceeds by alkaline hydrolysis in step 1 directly to step 3.

Control of alkali–aggregate reaction. The knowledge of the factors affecting the alkali–aggregate reaction can be used to control its effects in concrete. The following approaches can be used: (1) control of pH in the pore solution, (2) control of alkali concentrations, (3) control of amount of reactive silica, (4) control of moisture, and (5) alteration of alkali–silica gel.

Pozzolanic admixtures are commonly stated to control the expansions associated with the alkali–aggregate reaction (see, e.g., Figure 6.11a). One reason suggested for the beneficial effects of a pozzolan is that it reacts with the calcium hydroxide in the paste and thereby lowers the pH of the pore solution. However, the highly reactive silica that is characteristic of a finely divided pozzolan may consume the alkali in the cement in a rapid but harmless alkali–pozzolan reaction which proceeds rapidly to step 4 without the deleterious reaction of step 3. This idea has not yet been definitely established.

Alkali concentrations can be most readily controlled by using low-alkali cements with less than 0.6 wt.% Na_2O equivalent (see Figure 6.11b). In the future the alkali contents of cements can be expected to rise due to changes in manufacturing technology and environmental regulations. Thus, the use of low-alkali cements may not remain a feasible means of controlling the alkali–aggregate reaction. It is likely that the use of blended cements such as Types IP and IS may become a reasonable alternative. Higher alkali contents in cement, together with the increasing use of marginal aggregates, are likely to make the alkali–silica problem more widespread than it is at present.

An even better solution is the avoidance of a susceptible aggregate based on petrographic analyses and service records, although this too is not always a practicable or an economical solution. River gravels are often the most dangerous because they may contain relatively small amounts of the reactive rocks. A proper petrographic examination followed by beneficiation of the aggregate can enable the producer to control the amount and particle size of the reactive component to bring expansion within safe levels.

A low *w/c* concrete is very impermeable and may also help to limit the supply of water needed to cause the alkali–silica gel to swell, but will only slow down the reaction. No adverse expansion will occur when external moisture is not available. Lithium and barium salts have been used as additives to control alkali–aggregate expansions, but this is not an efficient or economical solution. The effect of these additions is not exactly clear; it is thought that the metal ions may enter the alkali–silica gel and reduce its ability to imbibe water. Lithium-containing glasses are known not to cause expansion in mortars.

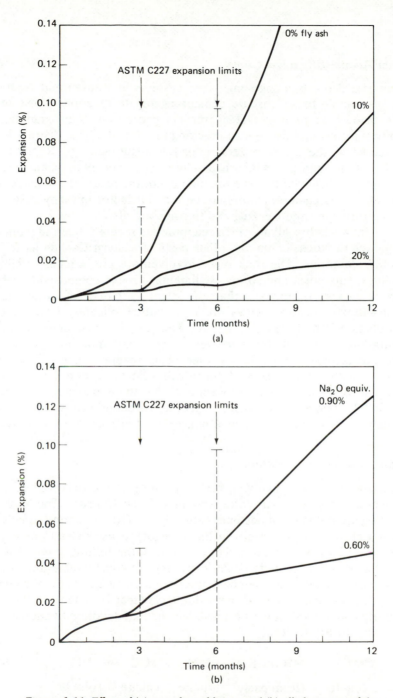

Figure 6.11 Effect of (a) pozzolan additions and (b) alkali content of the cement on the progress of the alkali–aggregate reaction (different aggregate used in each case). (From D. Stark, in *Proceedings, Fourth International Conference on the Effects of Alkalis in Cement and Concrete,* Purdue University, W. Lafayette, Ind., 1978.)

Other Alkali–Silica Reactions

Aggregate deposits along some river systems in Kansas and Nebraska are susceptible to an obscure cement–aggregate reaction. These materials, known as *sand–gravel* aggregates because of their grading, are highly siliceous and cause map cracking in concrete. This distress is not prevented by the use of pozzolans or low-alkali cements. Replacement of 30% of the aggregate with crushed limestone appears to be an effective remedy. It is believed that the reaction is not the result of a single cause. The alkali–aggregate reaction may be a major factor in many cases, but drying shrinkage and freezing and thawing are also involved.

A new kind of alkali–silica reaction has recently been identified in structures in Nova Scotia and other parts of eastern Canada as well as in other countries. The rock types involved are graywackes, argillites, phyllites, siltstones, and so on—sedimentary rocks composed largely of clay minerals. It appears that the primary cause of attack is an *alkali–silicate reaction* (clays are layer-lattice silicates) which causes "exfoliation" of the clay minerals. This can be best described as a separation of flat, platey particles of the clay minerals, which are normally compacted tightly together. This reaction takes place quite slowly, so that the effects of the alkali–silicate reaction may not be apparent for many years, but it appears that the expansions may occur indefinitely. Some normal alkali–aggregate reactions may occur simultaneously because microcrystalline quartz is often present in these rocks.

Alkali–Carbonate Reaction

Expansive reactions involving carbonate rocks have occurred in some midwestern and eastern states and in eastern Canada. The rocks in question are dolomitic limestones ($MgCO_3/CaCO_3$) containing some clay, but not all the reactive rocks in this category cause deleterious expansions. It is known that expansive rocks have the following features: (1) very fine grained dolomite (small crystals), (2) considerable amounts of fine-grained calcite, (3) abundant interstitial clay, and (4) the dolomite and calcite crystals evenly dispersed in a clay matrix. The reaction is not properly understood. It has been established that dedolomitization occurs according to Eq. (6.26).

$$CM\overline{C}_2 \; + \; NH \,(or\, KH) \rightarrow MH \; + \; C\overline{C} \; + \; N\overline{C} \qquad (6.26)$$

dolomite　　(from paste)　brucite　calcite　soluble
　　　　　　　　　　　　　　　　　　　　　　　carbonates

This reaction is expansive and the rate at which it proceeds depends on the crystal size and concentration of the accompanying calcite. Dedolomitization is fastest when there is about 50% calcite in the carbonate fraction, with a particle size of less that 40 μm. The exact role of the clay is unknown; it may be that its role is primarily a question of structural weakening of the carbonate skeleton so that expansion caused by dedolomitization can cause internal cracking. Another suggestion is that dedolomitization exposes clay in a special "active" state which can react with alkali metal ions to form a swelling clay. However, it is also possible that the clay acts as a semi-permeable membrane, thereby setting up an osmotic cell involving the soluble alkali carbonates produced on dedolomitization. The critical amount of clay is dependent on the amount of dolomite in the carbonate fraction.

The alkali–carbonate reaction can be controlled by keeping the alkali content of the cement low or by diluting the reactive aggregate with less-susceptible material. Unlike the alkali–aggregate reaction, there is no pessimum percentage of reactive aggregate, but it appears that the alkali content of the cement should be kept as low as 0.40% for adequate protection. Pozzolans are not effective in controlling the alkali–carbonate reaction.

Deleterious Substances

For satisfactory performance, concrete aggregrates should be free of deleterious substances. ASTM C33 sets down limits for deleterious substances in aggregates (see Table 6.9), which depend on the type of

Table 6.9

Limits for Deleterious Substances in Concrete Aggregates[a]

	Coarse Aggregate		Fine Aggregate	
Material	*Flatwork*[b]	*Structural*	*Flatwork*[b]	*Structural*
Clay lumps and friable particles	3.0	5.0	3.0	3.0
Coal and lignite	0.5	0.5	0.5	1.0
Chert	5.0	5.0	—	—
Material finer than No. 200 sieve (75 μm)	1.0	1.0	3.0	5.0

[a]These figures apply to concrete subjected to severe weathering; limits can be increased in moderate climates or interior concrete. Data taken from ASTM C33, "Standard Specification for Concrete Aggregates." Reprinted, with permission, from the American Society for Testing and Materials, 1916 Race Street, Philadelphia, PA 19103. Copyright.

[b]Concrete subject to abrasion and wear or frequent wetting (e.g., pavements, bridge decks, curbs, etc.).

exposure and the application. We can divide these into two categories: impurities and unsound particles.

Impurities. Impurities can be classified as solid materials or soluble substances. Solid materials are generally present in a very finely divided state, passing the 200-mesh (75-μm) sieve. Such material will appreciably increase the water requirements for workable concrete if present in large amounts. The fine fraction is also likely to adhere to the surfaces of the large aggregate particles and interfere with the cement–aggregate bond. Materials in this class are commonly silt, rock dust, and organic matter. Organic matter may also react chemically with the alkaline cement paste, forming soluble organic compounds that interfere with the setting processes. If sea-dredged aggregates are used, they must be thoroughly washed to avoid serious problems arising from salt contamination. Care should be taken during transportation and storage of aggregates to ensure that inadvertent chemical contamination does not occur.

Unsound particles. A variety of unsound particles can occur in small quantities in aggregates. Soft particles such as clay lumps, wood, and coal will cause pitting and scaling at the surface. Coal may also swell in the presence of moisture or release undesirable organic compounds that interfere with setting and hardening. Weak, friable particles of low density, such as many shales and pumice, should also be avoided if a good wearing surface is needed. Reactive materials such as sulfides, gypsum, and cherts can also lead to problems, as discussed earlier.

6.3 EVALUATION OF AGGREGATES

It is readily apparent from the foregoing discussion that a considerable number of tests may be required in order to evaluate the suitability of an aggregate source for the production of good-quality concrete. The tests required to determine aggregate properties for mix design are quite straightforward and require no further comment. However, many of the tests concerned with aggregate durability and performance have limitations, which will be discussed in some detail. The choice of evaluation tests will depend on the application of the concrete and the service records of local aggregates. Clearly, if the alkali–aggregate reaction is known to occur in the vicinity or in neighboring areas, some tests may be required. A thorough petrographic examination can tell a lot about the potential durability of an aggregate and may eliminate the need for embarking on a lengthy test program. It is important for aggregate evaluation that proper sampling practices be followed as laid down in ASTM C702 and D75. It should also be noted that tests involving aggregates alone are not always satisfactory for predicting aggregate performance. In many cases, comparable tests exist for concrete and these can also be used.

Tests for Physical Durability

These tests can be divided into those measuring abrasion resistance and those concerned with frost resistance of aggregates.

Abrasion Resistance

The Los Angeles test for abrasion resistance (ASTM C131 and ASTM C535) involves ball milling the aggregate with steel balls for a given time and measuring the amount of comminution that takes place. The scratch hardness test (ASTM C235) assumes a relationship between hardness and abrasion resistance. Neither of these methods gives a reliable indication of the abrasion resistance of concrete made with the tested aggregates. It is better to test the abrasion resistance of the concrete itself (see Chapter 19). Then the contribution of the cement–aggregate bond, which depends on the aggregate shape and texture, will also be taken into account. BS 812: Part 3 prescribes more extensive procedures to measure not only the abrasion resistance, but also the impact and crushing resistance of aggregates.

Frost Resistance

The soundness test (ASTM C88) is supposed to simulate the effect of ice crystallization in a saturated aggregate. This is done by soaking the aggregate in a saturated solution of sodium or magnesium sulfate and then drying in an oven. The weight loss occurring after a given number of cycles is determined. The crystallization of the salt in the pores is assumed to simulate the disruption of the aggregate particles by ice. There seems to be no real justification for this assumption and testing an aggregate separately from the cement paste will make the test even less realistic. Correlation between the test and actual service performance is not good. Testing the aggregate in concrete (ASTM C682) is a better approach, but in this case the difficulty of duplicating realistic temperature histories again makes correlations between test data and actual performance less than ideal. This is further discussed in Chapter 20.

Tests for Chemical Durability

These tests are concerned with an evaluation of the alkali–silica or alkali–carbonate reaction. ASTM C289 is a quick chemical test which measured the solubility of silica when the powdered aggregate is treated

with sodium hydroxide. The results are used to assess reactivity. Certain
minerals may interfere with the test, and petrographic examination
should be done to help evaluate the results. Even so, the test is not a
reliable indicator of alkali reactivity. A potentially deleterious aggregate
should be tested for deleterious expansion in the mortar bar test (ASTM
C227). In this test the coarse aggregate is reduced to a fine grading by
crushing and used to cast standard mortar bars (25 × 25 × 285 mm or 1
× 1 × 11¼ in.) using the cement that will be used on the job. The bars
are stored in moist conditions at 38°C to accelerate the alkali–aggregate
reaction. Expansion should not exceed 0.05% after 3 months or 0.1%
after 6 months. This is not always a reliable test since deleterious
expansions can occur after 6 months (see Figure 6.11). Although the
tests can be continued longer if necessary, the time taken to get the
required information is a serious drawback. Grattan-Bellew has found
that a concrete prism test similar to that used in CSA-A23.24 is a better
indicator of potential expansive aggregates. More reliable estimations of
expansivity can be obtained by comparing the initial slopes of the
expansion curves (Figure 6.12), which can be established within 6

Figure 6.12 Expansion of concrete prisms in a test to measure the alkali
reactivity of aggregates (results for two different aggregates with high-
alkali cement). (From P. Grattan-Bellew, in *Proceedings, Fourth International Conference on the Effects of Alkalis in Cement and Concrete*,
Purdue University, W. Lafayette, Ind., 1978).

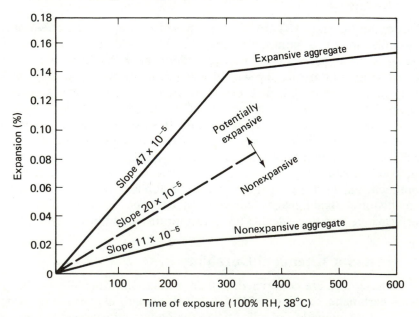

months. However, a rapid, reliable test for alkali–aggregate reactivity has not yet been developed.

The mortar bar test is also used in a modified form for the evaluation of "sand–gravel" reactivity (ASTM C342). Storage is at 55°C (131°F) and a drying period is also included. A mortar bar test, using Pyrex glass as aggregate, is the basis of a test (ASTM C441) to evaluate the effectiveness of a pozzolan in controlling the alkali–aggregate reactions. Expansions should be reduced to 25% of those measured in mortar bars without the pozzolan. Potential reactivity of carbonate rocks is determined by measuring the expansion of rock cylinders stored in a sodium hydroxide solution. Expansions greater than 0.10% will in many cases be evident after 28 days of immersion, and warrant further testing in concrete.

Aggregate Beneficiation

If an aggregate does not meet the specifications required for the job, after the appropriate ASTM or BSI tests have been conducted, the engineer is faced with two courses of action. The engineer can either reject the aggregate or can consider taking measures to bring the aggregate up to specifications. Beneficiation of aggregate may be the only available solution if aggregate supplies are scarce, but it will add to the cost of the aggregate and hence of the concrete. Some possible treatments are summarized in Table 6.10.

Soft, friable particles can be removed by crushing, although this process will also remove some sound material. Soft, lightweight particles can be removed by various forms of separation based on the lower specific gravities of these deleterious substances. Heavy-media separa-

Table 6.10

Beneficiation of Aggregates

Treatment	*Removal*
Crushing	Friable particles
Heavy-media separation	Lightweight particles
Reverse air or water flow	Lightweight particles
Hydraulic jigging	Lightweight particles
Elastic fractionation	Lightweight particles
Washing and scrubbing	Surface coatings, finely divided materials
Selective quarrying, crushing, and blending	Control or removal of deleterious components

tion allows removal of light particles, which float to the top of the liquid. The density cut can be controlled by varying the specific gravity of the liquid. This is a useful method for improving heavyweight aggregates. Reverse water flow or air flow can be used to remove wood and other lightweight material from coarse aggregate. Hydraulic jigging or vertical pulsation of water through a horizontal bed of aggregate stratifies the material in layers of increasing density with the lightest materials at the top. Elastic fractionation is a dry method of density separation. Aggregate is allowed to fall on an inclined steel plate, and the rebound of each particle depends on its elastic modulus. Hard, dense particles with a high modulus will rebound farther than will soft, porous particles with a low modulus. Thus, a suitable location of collection bins can provide a good separation. These methods will remove such material as wood, coal, clay lumps, sandstones, cherts, shale, chalk, and the like. Finely divided material or surface coatings may also be removed, but special washing and scrubbing procedures must be used.

Selective crushing and blending of the aggregate may be required to provide the desired size gradation. If chemical reactivity is suspected, aggregate composition should be carefully monitored. It may be possible to avoid deleterious aggregates by selective quarrying or to obtain safe concentrations or size ranges of deleterious substances by controlled blending and crushing.

Waste Materials as Aggregates

The possibility of using solid wastes as aggregate in concrete has received increasing attention in recent years as one promising solution to the escalating solid-waste problem. The use of solid wastes is not a new concept: industrial wastes are the basis of many concrete admixtures; fly ash has been used as a pozzolanic material in concrete for several decades; blast-furnace slags have been used both as aggregate and as a cementitious material; and cans have been shredded to provide steel fibers for reinforcement. The use of concrete for the disposal of solid wastes has concentrated mostly on service as aggregates, since this provides the only real potential for the utilization of large quantities of waste materials.

A wide variety of materials come under the general heading of solid wastes (see Table 6.11). These range from municipal and household garbage, or building rubble, such as brick and concrete, through unwanted industrial by-products such as slag and fly ash or discarded unused materials such as mine tailings.

When considering a waste material as a concrete aggregate, three major considerations are relevant: (1) economy, (2) compatibility with

Table 6.11

Typical Solid Wastes That Have Been Considered as Aggregate for Concrete

Material	Composition	Industry	Annual Amount (10^6 tons)
Mineral wastes	Natural rocks	Mining and mineral processings	2000
Blast-furnace slags	Silicates or aluminosilicates of calcium and magnesium silicate glasses	Iron and steel	30
Metallurgical slags	Silicates, aluminosilicates, and glasses	Metal refining	30
Fly ash	Silica glasses	Electric power	50
Municipal wastes	Paper, glass, plastics, metals	Commercial and household wastes	200
Rubber	Synthetic rubbers	Scrapped tires	5
Incinerator residues	Container glass and metals and silica glasses	Municipal and industrial	5
Building rubble	Brick, concrete, reinforcing steel	Demolition	25

other materials, and (3) concrete properties. The economical use of a waste material depends on the quantity available, the amount of transportation required, the extent of beneficiation, and the mix design requirements. For instance, many sources of mine wastes are located far from the potential markets for concrete and would entail high transportation costs. Separation of useful materials from undesirable substances (e.g., glass and metals from paper, plastics, and organic wastes in municipal wastes) would be costly unless it is done for other reasons as well. Crushed container glass tends to contain many flat elongated particles, which would lead to a harsh concrete, requiring increased amounts of cement.

Waste materials must not react adversely with other constituents of the mix. Most waste glass will readily take part in the alkali–aggregate reaction and poses a potential durability problem. The use of fly ash (itself a waste material) as a pozzolan can provide the necessary protection. Mine tailings will contain high proportions of lead or zinc if associated with these ores, and these elements will interfere with the setting of cement. Aluminum present in incinerator residues could cause durability problems, particularly in the presence of chlorides.

Finally, the effect of waste materials on concrete properties must be considered. For example, the lower modulus of elasticity of glass compared to that of good-quality rock will lower the elastic modulus of concrete. These effects will be much more pronounced if low-strength, low-modulus materials such as rubber and plastics are used. Scrapped tires have been proposed for use in concretes where high resiliency rather than strength is required. Waste materials can also be used to make special lightweight or heavyweight aggregates, as will be discussed in the next section.

The successful utilization of solid wastes in concrete will depend on anticipating potential problems and the ensuing properties of the concrete, and developing uses that comply with these restraints. For example, the properties of a material may preclude its use in exterior concrete because of durability problems, but it may perform satisfactorily in concrete not exposed to weathering. There are also many applications where structural requirements are low, as in subbases or in interior partition walls. Special properties of aggregates could also be used to advantage for thermal or sound insulation, for example. However, to keep the problem in perspective, it should be remembered that use in concrete is only one possible solution to the disposal of solid wastes and that the solid-waste disposal problem is of too great a magnitude to be solved by any one approach. "Trashcrete" may have its place in the utilization of waste materials, but it will not dominate the field.

6.4 SPECIAL AGGREGATES

Earlier in this chapter it was mentioned that the most common classification of aggregates was on the basis of bulk specific gravity: heavy-weight, normal-weight, and lightweight aggregates. The variability in density can be used to produce concretes of widely different unit weights, as summarized in Table 6.12. The different kinds of lightweight and heavyweight aggregates are described in this section, while the properties of the corresponding concretes are discussed in Chapter 21. A brief discussion of abrasion and skid resistant aggregates is also included here.

Lightweight Aggregates

Lightweight aggregates can be either natural or synthetic materials. The common feature of lightweight aggregates is their high internal porosity, which is the prime reason for their low bulk specific gravity. However, in some cases the solid matrix is also a low-density material. Table 6.13 summarizes some of the different kinds of lightweight aggre-

Table 6.12

Density Classification of Concrete Aggregates

Category	Unit Weight of Dry-Rodded Aggregate (kg/m³)[a]	Unit Weight of Concrete (kg/m³)[a]	Typical Concrete Strengths (MPa)[a]	Typical Applications
Ultra lightweight	<500	300–1100	<7	Nonstructural insulating material
Lightweight	500–800	1100–1600	7–14	Masonry units
Structural lightweight	650–1100	1450–1900	17–35	Structural
Normal weight	1100–1750	2100–2550	20–40	Structural
Heavyweight	>2100	2900–6100	20–40	Radiation shielding

[a]kg/m³ × 0.062 = lb/ft³; MPa × 145 = lb/in.²

gates that have been used in concrete. Most of these are synthetic materials, although many (those designated PN) are natural materials specifically processed to provide concrete aggregates. Expanded clay, shale, or slate is probably the most common type of lightweight aggregate used in structural concrete. The raw material is either crushed to the desired size or ground and pelletized and it is then heated to 1000 to 1200°C (1830 to 2190°F). At these temperatures the material will bloat (i.e., it is puffed up to a greatly expanded size in much the same way as popped popcorn). Bloating occurs because of a rapid generation of gas (from combustion of small quantities of organic matter that often occur naturally, or which may be deliberately added) within the particle, which cannot readily escape. The material is of a composition that causes partial melting at the bloating temperature. The lowered viscosity of the material allows it to expand, and an impervious viscous coating forms on the outside, preventing the gases from escaping too quickly.

Many other synthetic lightweight aggregates are produced by similar pyroprocessing techniques. Perlite (volcanic glass), slags (calcium silicate glasses), or waste glass can be bloated. Water or air can be used as the bloating medium. In the case of perlite, hydrate water in the material is the source of bloating. Pumice, scoria, and tuffs are lavas that have been bloated naturally. Expanded vermiculite is also bloated by loss of hydrate water. Vermiculite is a clay mineral with a platey, layer structure; rapid loss of water between the layers causes the layers to be pushed apart, thereby increasing the volume by as much as 30 times.

Table 6.13

Properties of Concretes Made with Selected Lightweight Aggregates

Aggregate	Aggregate Dry Unit Weight (kg/m³)ᵃ	Originᵇ	Concrete Unit Weight (kg/m³)ᵃ	28-Day Compressive Strength (MPa)ᵃ	Thermal Conductivity K-Factor (W/m·K)ᵃ	Absorption (Wt %)
Expanded shale, clay, slate	550–1050	PN	110–1850	14–42	0.26–0.43	5–15
Foamed slag	650–900	S	110–1850	14–42	0.17–0.34	5–25
Sintered fly ash	600–1000	S	1350–1900	14–42	0.17–0.51	14–24
Exfoliated vermiculite	65–200	PN	400–950	0.67–3	0.07–0.10	20–35
Expanded perlite	65–250	PN	550–800	0.6–3.5	0.07–0.10	10–50
Pumice	—	N	800–1300	4–5	0.15–0.30	—
Expanded glass	250–500	S	1200	9	0.28	5–10
Expanded polystyrene beads	30–150	S	300–900	0.7–12.5	0.07	—
Brick rubble	~750	S	1750–1900	7.7–21	0.40–0.51	19–36
Crushed stoneᶜ	1450–1750	N	2250–2400	21–50	1.0–3.0	0.5–2.0

ᵃkg/m³ = 0.062 = lb/ft³; MPa × 145 = lb/in.²; W/m·K × 0.58 = Btu/ft·h·°F.

ᵇPN, processed natural material; N, natural material; S, synthetic material.

ᶜNatural aggregate listed for comparison.

The shape and structure of a lightweight aggregate can be quite variable and will be a consequence of the processing techniques used in its production. Aggregates may be angular or highly irregular in shape, and this will determine the workability of concrete made with the aggregate. Even more important is the porosity of the aggregate. Lightweight aggregates have high absorption values, because of the large internal interconnected porosity, and this requires a modified approach to concrete proportioning. However, in some cases entry of moisture to the body of the aggregate may be inhibited by a protective outer skin of fused material. Such an aggregate will not have a high effective absorption, although if the aggregate is fractured during mixing or handling, the effective absorption will abruptly increase as the internal pore system is exposed. Slump loss in lightweight concrete due to absorption can be an acute problem, which can be alleviated by wetting the aggregate before batching.

Lightweight aggregates are covered by ASTM C330 (for structural concrete), C331 (for masonry units), and C332 (for insulating concrete). These specifications all contain the requirement that aggregates should be composed "predominately of lightweight-cellular and granular inorganic material," thereby excluding materials such as wood, sawdust, cork, and the like. Organic materials can be used if they are treated to prevent possible retardation of setting and hardening, although they may swell when wet. In the United Kingdom, lightweight aggregates are covered in BS 877, BS 1165, BS 3681, and BS 3797.

Heavyweight Aggregates

Heavyweight aggregates are composed of materials that have a high specific gravity (see Table 6.14). These may be synthetic materials or natural materials, but synthetic aggregates are only used if unit weights greater than 4000 kg/m^3 (250 lb/ft^3) are required. Choices of other materials are determined by availability and cost as well as by physical properties. Heavyweight concrete is mostly used for radiation shielding (see Chapter 19); ASTM C637 (or BS 4619) covers aggregates used for radiation shielding, and BS 4619 also deals with heavy aggregates. It can also be used for counterweights and other applications where a high mass-to-volume ratio is desirable.

Abrasion and Skid-resistant Aggregates

Special aggregates are often used for toppings in heavy-duty industrial floors or other situations where a high resistance to abrasion is required. Hard, dense, strong materials such as calcined bauxite or

Table 6.14

Physical Properties of Some Heavyweight Aggregates[a]

Material	*Chemical Composition*	*Classification*[b]	*Bulk Specific Gravity*	*Unit Weight $(kg/m^3)^c$*	*Unit Weight of Concrete $(kg/m^3)^c$*
Goethite	$Fe_2O_3 \cdot H_2O$	N	3.5–3.7	2100–2250	2900–3200
Limonite	Impure Fe_2O_3	N	3.4–4.0	2100–2400	2900–3350
Barite	$BaSO_4$	N	4.0–4.6	2300–2550	3350–3700
Illmenite	$FeTiO_3$	N	4.3–4.8	2550–2700	3500–3700
Magnetite	Fe_3O_4	N	4.2–5.2	2400–3050	3350–4150
Hematite	Fe_2O_3	N	4.9–5.3	2900–3200	3850–4150
Ferrophosphorus	$Fe_2O_3–P_2O_5$	S	5.8–6.8	3200–4150	4100–5150
Steel	Fe (scrap iron steel punchings)	S	6.2–7.8	3700–4650	4650–6100

[a]See ASTM C638.
[b]N, naturally occurring; S, synthetic products.
[c]$kg/m^3 \times 0.062 = lb/ft^3$.

corundum, emery, metals, flint, or quartz are used for such applications. The strength of the paste and the cement–aggregate bond are more important than aggregate hardness. The aggregate becomes an important factor in the skid resistance of concrete in older pavements, where considerable wear and polishing have occurred. It is desirable to have a mix of hard and soft minerals in the aggregate that wear and polish at different rates. A large proportion of fine aggregates in the range 3 to 10 mm (⅓ to ⅜ in.) improves skid resistance because of the tendency of fine aggregate to contain a higher percentage of harder minerals than does coarse aggregate.

Marginal Aggregates

In the past there has been a natural tendency to avoid aggregates that *might* cause problems in concrete or which would require considerable beneficiation before use. As aggregates become increasingly scarce in some parts of the country, consideration is being given to rock types that have been hitherto rejected (shales, for example). When considering aggregates classed as marginal four major areas of consideration are relevant: (1) the desired concrete properties, (2) the weaknesses of the aggregates, (3) the beneficiation of the aggregate, and (4) the use of protective measures for the concrete.

In the first place it should be remembered that some aggregates may be suitable for some concrete applications but not for others. For example, an aggregate that contains a rather high percentage of porous material may not be suitable for concretes when strengths greater than 35MPa (5000 lb/in.²) are needed, but may be satisfactory for lower-strength concrete when there are no durability problems. Aggregates that have potential durability problems might well be satisfactory for use in concrete that will not be exposed to weathering.

At present, aggregates for concrete are required to meet ASTM C33, or BS 882 and 1201, which means that they must pass various standard tests, including durability assessments. Test results should be considered in the light of the anticipated service conditions and the performance of the aggregate in other regards. We have already discussed the limitations of some of these tests, and the development of improved, more precise tests might be desirable for further evaluation of marginal aggregates that fail one of the tests. For example, it has been suggested that durable shales could be identified by a series of tests involving mineralogical analysis, durability, and mechanical degradation. A thorough petrographic analysis is essential for proper evaluation of marginal aggregates. A skilled and experienced petrographer will be of great help in assessing the suitability of an aggregate for a particular application, and in identifying the potential durability problems.

The discussion of beneficiation applies particularly to marginal aggregates. Density fractionation may be successful at eliminating unsatisfactory fractions, or selective quarrying may be used to advantage. The other remedy is to take protective measures when marginal aggregates must be used. The incorporation of fly ash, for example, will guard against potential alkali–aggregate attack. Crushing aggregate to a smaller maximum size may avoid freeze–thaw problems.

The use of marginal aggregates will require more care, more forethought, more ingenuity, and more cost. There is no doubt that use of marginal materials will require additional testing and closer quality control, but the extra cost involved may be less than the cost of bringing in good-quality aggregate. Necessity is the mother of invention, so we can look forward in the next few decades to the development of better tests, utilizing fundamental aggregate properties, and to the development of new ways of controlling the behavior of components with poor durability.

CHAPTER APPENDIX

Determination of Moisture Contents on an Oven-Dry Basis. Expressing moisture quantities on an OD basis, we have

$$\text{AC} = \frac{W_{\text{SSD}} - W_{\text{OD}}}{W_{\text{OD}}} \times 100\% \qquad \text{same as SSD basis} \qquad (6A\text{-}1)$$

$$EA = \frac{W_{SSD} - W_{AD}}{W_{OD}} \times 100\% \tag{6A-2}$$

$$SM = \frac{W_{wet} - W_{SSD}}{W_{OD}} \times 100\% \tag{6A-3}$$

Equations (6A-2) and (6A-3) are the same as the definitions for the SSD basis, except that the demoninator is now the OD weight. Since the difference between SSD and OD is small (1 to 2%), the actual differences between effective absorption or surface moisture calculated for SSD or OD states can be neglected.

When working on the OD basis, the quantity *total moisture* (TM) is generally used.

$$TM = \frac{W_{stock} - W_{OD}}{W_{OD}} \times 100\% \tag{6A-4}$$

Thus, it can be shown that

$$EA = AC - TM \tag{6A-5}$$

when TM is measured on an air-dry aggregate, and

$$SM = TM - AC \tag{6A-6}$$

when TM is measured on a wet aggregate. Or expressing this in general terms, the moisture content of the stockpiled aggregate is given by

$$MC = TM - AC \tag{6A-7}$$

When the moisture content is positive, the aggregate has surface moisture, and when it is negative, the aggregate is air dry and Eq. (6A-7) now measures effective absorption.

Bibliography

Aggregate Properties

ACI COMMITTEE 221, "Selection and Use of Aggregates for Concrete," *Journal of the American Concrete Institute,* Vol. 58, No. 5, pp. 513–542 (1961).

"Grading of Aggregates," *Highway Research Record,* No. 441, 127 pp. Highway Research Board, Washington, D.C., 1973.

MALISCH, W. R., *Aggregates for Concrete,* Educational Bulletin, E1-78. American Concrete Institute, Detroit, Mich., 1978.

Popovics, S., *Concrete-making Materials*. McGraw-Hill Book Company, New York, 1979.

Significance of Tests and Properties of Concrete and Concrete-making Materials, ASTM STP 169B. American Society for Testing and Materials, Philadelphia, Pa., 1978.

Aggregate Durability

Diamond, S., "A Review of Alkali–Silica Reaction and Expansion Mechanisms. I. Alkalies in Cements and in Concrete Pore Solutions," *Cement and Concrete Research,* Vol. 5, No. 4, pp. 329–426 (1975); "II. Reactive Aggregates," *Cement and Concrete Research,* Vol. 6, No. 4, 549–560 (1976).

"The Effect of Alkalies on the Properties of Concrete," Proceedings of a Symposium held in London, Sept. 1976. Cement and Concrete Association, U.K. 1976.

Grattan-Bellew, P.E., "Study of Expansivity of a Suite of Quartzwackes, Argillites and Quartz Arenites," in Proceedings of the Fourth International Conference on the Effects of Alkalies in Cement and Concrete, pp. 113–140, Publication No. CE-MAT-1-78, School of Civil Engineering, Purdue University, West Lafayette, Indiana, 1978.

Living with Marginal Aggregates, ASTM STP 597. American Society for Testing and Materials, Philadelphia, Pa., 1976.

Proceedings of the Fourth International Conference on the Effects of Alkalies in Cement and Concrete, Purdue University, June 1978. Publication No. CE-MAT-1-78, School of Civil Engineering, Purdue University, West Lafayette, Indiana, 1978.

Symposium on Alkali-Aggregate Reaction: Preventive Measures, Reykjavik, Aug. 1975.

Symposium on Alkali–Carbonate Rock Reactions, *Highway Research Record,* No. 45, Highway Research Board, Washington, D.C. 1963.

Woods, H., *Durability in Concrete Construction,* Monograph No. 4. American Concrete Institute, Detroit, Mich., 1968.

Aggregate Production and Beneficiation

Goldbeck, A. T., "Crushed Stone Production," *Journal of the American Concrete Institute,* Vol. 25, No. 9, pp. 761–772 (1954).

Walker, S., "Production of Sand and Gravel," *Journal of the American Concrete Institute,* Vol. 26, No. 2, pp. 165–178 (1954).

Special Aggregates

Berger, R. L., "Synthetic Aggregates," in *New Materials for Concrete Construction* (proceedings of a conference, December 1971), ed. S. P. Shah. University of Illinois at Chicago Circle, Chicago, 1972.

GUTT, W., "Aggregates from Waste Materials," *Chemistry and Industry,* pp. 439–447 (1972).

HARRISON, W. H., "Synthetic Aggregate Sources and Resources," *Concrete,* Vol. 8, No. 11, pp. 41–44 (1974).

JOHNSTON, C. D., "Waste Glass as Coarse Aggregate for Concrete," *Journal of Testing and Evaluation,* Vol. 2, No. 5, pp. 344–350 (1974).

Problems

6.1 Plot the grading curves for the grading limits of the fine aggregates given in Tables 6.4 and 6.5.

6.2 Plot the grading curves for the grading limits of coarse aggregates given in Table 6.4; maximum aggregate size 1½ in. and 1 in.

6.3 Two sands have the following percentages passing the fine sieves. Calculate the fineness modulus of each sand.
(a) No. 4 = 97%, No. 8 = 95%, No. 16 = 92%, No. 30 = 85%, No. 15 = 30%, No. 100 = 5%
(b) No. 4 = 95%, No. 8 = 75%, No. 16 = 45%, No. 30 = 20%, No. 15 = 10%, No. 100 = 3%

6.4 Two 500-g samples of different sands were found to have the following amounts retained on each sieve. Calculate the fineness modulus of each sand.
(a) No. 4 = 15 g, No. 8 = 60 g, No. 16 = 100 g, No. 30 = 105 g, No. 50 = 130 g, No. 100 = 90 g
(b) No. 4 = 5 g, No. 8 = 55 g, No. 16 = 70 g, No. 30 = 105 g, No. 50 = 200 g, No. 100 = 65 g

6.5 A 1000-g sample of coarse aggregate in the SSD condition weighed 633 g when immersed in water. Calculate the bulk specific gravity of the aggregate.

6.6 A 1000-g sample from the stockpile of the same aggregate as in Problem 6.5 weighed 637 g when immersed in water. Calculate the moisture content of the aggregate in the stockpile.

6.7 A 500-g sample of sand in the SSD condition was placed in a jar, which was then filled with water. The combined weight was 1697 g. The weight of the jar filled with the water only was 1390 g. Calculate the bulk specific gravity of the sand.

6.8 A 500-g sample of the same sand as in Problem 6.7 from the stockpile weighed 1705 g when placed in the jar and topped up with water. Calculate the moisture content of the sand in the stockpile.

6.9 The sample of coarse aggregate in Problem 6.5 weighed only 985 g after drying at 105°C overnight. Calculate the absorption capacity of the aggregate.

6.10 A 1000-g sample of coarse aggregate from the same stockpile as in Problem 6.6 weighed 988 g when dried at 105°C overnight. Calculate the effective absorption on an OD basis. (Compare your answer with the answer to Problem 6.6.)

6.11 The sample of sand in Problem 6.7 weighed 495 g after drying at 105°C overnight. Calculate the absorption capacity of the sand.

6.12 A 1000-g sample of sand from the same stockpile as in Problem 6.8 weighed 949 g when dried at 105°C overnight. Calculate the surface moisture of the sand. (Compare your answer with the answer to Problem 6.8.)

6.13 A mixture of 1800 g of gravel with an effective absorption of 0.90% and 1200 g of sand with a surface moisture of 2.51% was added to a concrete mix. Compute the adjustment that must be made to the added water to maintain a constant *w/c* ratio.

6.14 Can absorption capacity be used as a measure of the total porosity of an aggregate?

6.15 What would be the consequence of using oven-dry aggregate in a concrete mix?

6.16 Why should aggregates be batched by weight rather than by volume?

6.17 Does the fineness modulus completely describe the properties of a fine aggregate?

6.18 Why is it important to control the amount of material passing the No. 200 sieve?

6.19 An aggregate blend of 40% sand and 60% gravel has a unit weight of 1920 kg/m³. If the bulk specific gravity of the sand is 2.60 and of the gravel is 2.70, calculate the volume percent of void space in the blend.

6.20 Suggest a strategy for controlling alkali–aggregate reactions, giving reasons for your choice.

6.21 How would you evaluate a waste material for suitability as a concrete aggregate?

6.22 An Illinois farmer suggests that you use chopped corncobs as an aggregate. Do you think that this would be a suitable material for this purpose, and what potential applications would "corncrete" have?

7

admixtures for concrete

The official definition of an admixture set out in ASTM C125 is "a material other than water, aggregates and hydraulic cement that is used as an ingredient of concrete or mortar and is added to the batch immediately before or during its mixing." This definition encompasses a wide range of materials that are utilized in modern concrete technology. Strictly, it includes materials that act as reinforcing materials—polymer latexes and fibers—which produce concretes with quite different properties, and these are discussed in Chapter 22. The remainder of the large family of admixtures will be discussed here.

7.1 DEFINITIONS AND CLASSIFICATIONS

Readers may encounter two other almost identical terms: addition and additive. The term "additive" belongs to an older terminology, in which it was used synonymously with "admixture." An addition, according to ASTM C219, is "a material that is interground or blended in limited amounts into a hydraulic cement during manufacture either as a 'processing addition' to aid in manufacturing and handling the cement or as a 'functional addition' to modify the properties of the finished product." A *functional addition* would be an admixture added by the manufacturer at the cement plant instead of by the contractor at the job site, or the operator at the ready-mix plant. Thus, a Type IA portland cement has an air-entraining *addition* that should behave identically to a Type I portland cement to which an air-entraining *admixture* is added. Examples of a *processing addition* are a grinding aid or a fluxing agent and these should not, in theory, affect the behavior of the finished cement. There are a bewildering number of products marketed as admixtures for concrete, many of them based on proprietary formulations. This can be confusing, and there is some advantage to simplifying the problem by recognizing some broad categories of admixtures. These groupings will be used in this chapter:

1. *Air-entraining agents* (ASTM C260) are added primarily to improve the frost resistance of concrete.

2. *Chemical admixtures* (ASTM C494 and BS 5075) are water-soluble compounds added primarily to control setting and early hardening of fresh concrete or to reduce its water requirements.

3. *Mineral admixtures* are finely divided solids added to concrete to improve its workability, its durability, or to provide additional cementing properties. Slags and pozzolans are important categories of mineral admixtures.

4. *Miscellaneous admixtures* includes all those materials that do not come under one of the foregoing categories, many of which have been developed for special applications.

7.2 USES OF ADMIXTURES

The enormous variety of admixtures is due in large measure to the fact that almost every property of concrete can be modified to some extent (see Table 7.1). Thus, it is axiomatic that the reasons for using an admixture during the production of concrete will be equally varied. An admixture should be used for good reason: to improve the concrete, whether it be in the handling or consolidation of fresh concrete, in the performance of the hardened concrete, or in the cost of construction. *It should be emphasized that an admixture is not a panacea for poor mix design or sloppy concrete practice, and the potential user should first ensure that the desired result cannot be attained by improvements in these areas.*

Precautions in Use

All engineers who work regularly with concrete have their favorite stories to tell of admixtures preventing concrete from setting or causing some other unfortunate problem. However, the problems that do occur often arise through lack of appreciation of how an admixture interacts

Table 7.1

Beneficial Effects of Admixtures on Concrete Properties

Concrete Property	Admixture Type	Category of Admixture
Workability	Water reducers	Chemical
	Air-entraining agents	Air entraining
	Inert mineral powder	Mineral
	Pozzolans	Mineral
	Polymer latexes	Miscellaneous
Set control	Set accelerators	Chemical
	Set retarders	Chemical
Strength	Water reducers	Chemical
	Pozzolans	Mineral
	Polymer latexes	Miscellaneous
	Set retarders	Chemical
Durability	Air-entraining agents	Air entraining
	Pozzolans	Mineral
	Water reducers	Chemical
	Corrosion inhibitors	Miscellaneous
	Water-repellant admixtures	Miscellaneous
Special concretes	Polymer latexes	Miscellaneous
	Slags	Mineral
	Expansive admixtures	Miscellaneous
	Color pigments	Miscellaneous
	Gas-forming admixtures	Miscellaneous

with concrete. To avoid unnecessary surprises, the concrete engineer would do well to take a few simple precautions when considering the use of admixtures.

1. Require admixtures to conform to relevant ASTM or BSI specifications, where applicable. A reputable manufacturing firm should also be prepared to supply adequate technical data for its product. This should include the following: (a) the main effect of the admixture on concrete performance; (b) any additional influences the admixture may have, whether beneficial or detrimental; (c) physical properties of the material; (d) the concentration of the active ingredient; (e) the presence of any potentially detrimental substances such as chlorides, sulfates, sulfides, phosphates, sugars, nitrates, and ammonia; (f) pH; (g) potential occupational hazards for users; (h) conditions for storage and recommended shelf life; (i) preparation of admixture and procedures for introducing it into the concrete mix; and (j) recommended dosage under identified conditions, maximum permissible dosage, and effects of overdosage.

2. Follow manufacturer's instructions regarding dosage, but run relevant tests to check that the desired effects are being obtained. It is

particularly important that these tests be run using job materials under job-site conditions because the precise effect of an admixture will depend on many factors: cement composition, aggregate characteristics and impurities, mix proportions, presence of other admixtures, type and length of mixing, time of addition and concrete temperature, and curing conditions.

3. Ensure that reliable procedures are established for accurate batching of the admixture. This is particularly important for air-entraining admixtures and chemical admixtures, where typical dosages may be well below 0.1% by weight of cement. In such cases, overdoses can easily occur and the effects may be disastrous.

4. The effects that the admixture may have on other concrete properties should be taken into account, particularly if these effects may be unfavorable. The majority of admixtures do affect several concrete properties (see Table 7.1). Some commercial admixture formulations contain materials that belong to different categories.

Analysis of Admixtures

Many admixtures are derived from waste products of other industries; hence their compositions may be variable and in many cases they may be complex mixtures of compounds that are not well characterized. For these reasons, determining their exact chemical constitution is a difficult, if not impossible task, best left to experts. ASTM specifications for admixtures are therefore primarily performance specifications. An even more difficult task is analyzing hardened concrete to determine if an admixture has been added originally. Such an analysis is fraught with difficulties and is only possible in certain cases. In addition, the complex nature of many admixtures has hindered the study of cement–admixture interactions during hydration. Thus, the way cement composition affects admixture behavior is not fully understood. Two admixtures based on supposedly the same active ingredient may have different effects on concrete properties.

7.3 AIR-ENTRAINING ADMIXTURES

One of the major disadvantages of concrete is its susceptibility to damage during freezing and thawing cycles when it is in a saturated or near-saturated condition. Concrete can be badly damaged after a single winter's exposure if corrective measures are not taken, and this rapid deterioration would effectively preclude the use of concrete in most major applications: pavements, dams, foundations, and so on. All porous

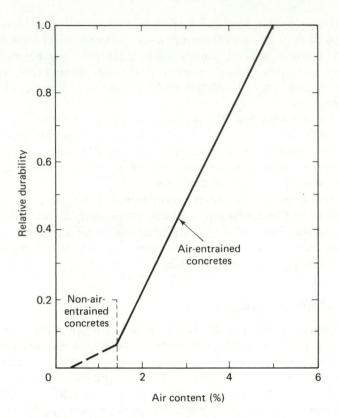

Figure 7.1 Effect of air entrainment on frost resistance of concrete.

materials (brick, ceramic tile, etc.) suffer from the same deficiency unless the material is specially designed to resist freezing and thawing. Fortunately, concrete can be made frost resistant by the addition of air-entraining admixtures (Figure 7.1). All concrete designed for outside exposure should be air-entrained routinely in the northern United States and throughout Canada. The mechanism of frost protection is described in detail in Chapter 20; here, we are only concerned with describing the effect air-entraining agents have on concrete properties.

The Entrained Air-void System

Air-entraining admixtures do what their name suggests: they entrain air in the concrete. The volume of air required to give optimum frost resistance has been found to be about 9% by volume of the mortar

Table 7.2

Characteristics of Air-entrained Concrete at Optimum Frost Resistance (Cement Content 250 lb/yd,[3] or 326 kg/m[3])[a]

| Maximum Aggregate Size, mm (in.) | *Measured Air Contents (% by volume)* | | | | | Bubble Spacing Factor, mm (in.) |
	Non-air-entrained	Recommended Air Content (ACI) ± 1%	Concrete	Mortar	Paste	
63.5 (2½)	0.5	4.0	4.5	9.1	16.7	0.18 (0.007)
38 (1½)	1.0	5.0	4.5	8.5	16.4	0.20 (0.008)
19 (¾)	2.0	6.0	5.0	8.3	16.9	0.23 (0.009)
9.5 (⅜)	3.0	7.5	6.5	8.7	19.7	0.28 (0.011)
Mortar	—	—	—	9.0	23.0	0.30 (0.012)

[a]Based on data from P. Klieger, in *Significance of Tests and Properties of Concrete and Concrete-Making Materials*, ASTM STP 169 B, pp. 787–803, American Society for Testing and Materials, Philadelphia, PA, 1978.

fraction (Table 7.2), although the air is actually entrained within the paste fraction. For routine field tests the easiest quantity to measure is the amount of air expressed as a percentage of the *concrete* volume. The air content must be in the range 4 to 8% by volume of concrete for satisfactory frost protection, the actual amount depending on the maximum size of the coarse aggregate. This is because more paste is required to provide workable concrete with a lower maximum size of coarse aggregate. The air-content measurement includes voids naturally entrapped within the paste during mixing. Intentionally entrained air increases the total volume of air voids by 3 to 4% of the volume of the concrete.

However, total air content is not the only parameter that determines how well concrete will resist freezing and thawing. The nature of the entrained air-void system is also important. Air-entrained concrete cannot be distinguished from non-air-entrained concrete with the naked eye, but when viewed under a microscope (Figure 7.2) it can be seen that literally millions of tiny air bubbles are dispersed uniformly throughout the paste. These bubbles are spherical in shape and are very close together. It looks as if the paste has been mixed with a fine foam, and this is essentially what has happened. The air-entraining admixture causes the mixing water to foam, and this foam is locked into the paste during hardening. The critical parameter of the air-entrained paste is the *spacing factor,* which is defined as the average maximum distance from any point in the paste to the edge of a void. The spacing factor should

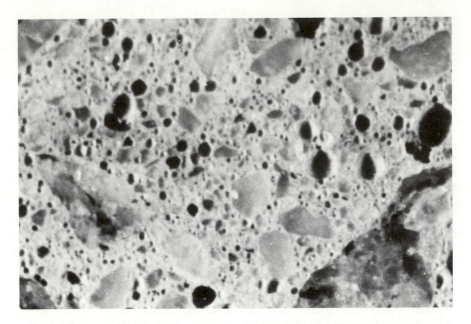

Figure 7.2 Micrograph of air-entrained concrete. (Photograph courtesy of Bryant Mather.)

not exceed 0.2 mm (0.008 in.) to ensure adequate frost protection (Figure 7.3); the smaller the spacing factor, the more durable the concrete. The voids should be small, in the range 0.05 to 1.25 mm (0.002 to 0.05 in.) diameter to ensure that the required spacing factor is obtained at relatively low air contents. The smaller the bubble size, the smaller the spacing factor at a given air content. Furthermore, small voids remain as discrete, isolated bubbles which do not readily fill with water even when the concrete is kept saturated. Since void sizes are not uniform, they are expressed in terms of specific surface (mm^2/mm^3 of concrete); typically, values should be in the range of 16 to 25 mm^{-1} (400 to 625 $in.^{-1}$).

Air-entraining Materials

It follows from the discussion above that air-entraining admixtures must contain compounds that will promote the formation of stable foams. Bubble formation in water is normally a transient phenomenon because the high surface tension of water opposes the creation of the air–water interface that bounds a bubble. Foams can only be formed if the energy barrier represented by surface tension can be overcome in order that

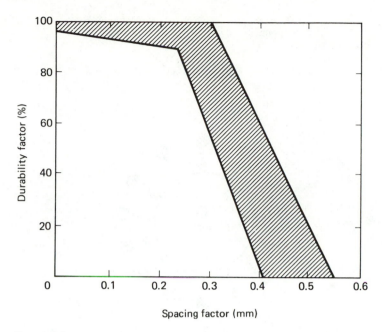

Figure 7.3 Relationship between durability and bubble spacing factor of entrained air.

masses of stable bubbles can be created during agitation. Air-entraining agents contain *surface-active agents* which concentrate at the air–water interface, lower the surface tension so that bubbles can form more readily, and stabilize the bubbles once they are formed. Surface-active agents are molecules which at one end have chemical groups that tend to dissolve in water [*hydrophilic* (water-loving) *groups*], and which at the other end have groups that are repelled by water [*hydrophobic* (water-hating) *groups*]. The molecules tend to align at the interface with their hydrophilic groups in the water and the hydrophobic groups in air (Figure 7.4). Carboxylic acid or sulfonic acid groups are commonly responsible for the hydrophilicity of the molecules, while aliphatic or aromatic hydrocarbons make up the hydrophobic component. Air-entraining admixtures are thus closely related to synthetic detergents, although the foaming capacity of the latter is only a side effect of other "surface-active" properties. Most household detergents cannot be used as air-entraining admixtures, however, because the foams are too coarse (large bubble size) and not sufficiently stable to remain intact while the concrete is hardening.

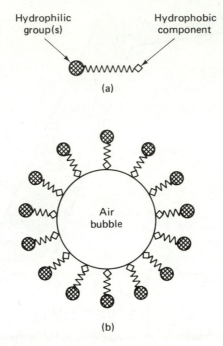

Figure 7.4 Schematic representation of air entrainment by surface active molecules: (a) surface-active molecule; (b) stabilized air bubble.

Because most commercial air-entraining admixtures are prepared from by-products of other industries (pulp and paper, petroleum, and animal processing), the active ingredient may only be a fraction of the total admixture, but since the admixture acts only at surfaces, a little goes a long way. A typical dose is 0.005 to 0.05% of active ingredient by weight of cement, so that admixtures are often prediluted to assist accurate batching. The admixture is added to the concrete with the mix water. Some manufacturers intergrind an air-entraining agent with the cement so that one can obtain air entrainment directly without the need of an admixture, and such cements are designated IA, IIIA, and so on. Air-entraining cements may not always entrain sufficient air and should not be used without being tested under job conditions.

Testing of Air-entraining Agents

The development of a satisfactory entrained air-void system can be determined, according to ASTM C457, by counting bubbles in a polished section of concrete viewed under a microscope and then calculating the spacing factor. This method is tedious, requires trained personnel and expensive equipment, and is clearly not suited for routine analysis. Thus, quality control is based on the measurement of total air content (see Chapter 8), and it is fortunate that commercial admixtures apparently provide satisfactory air-void systems at recommended air contents.

Air-entraining admixtures should conform to ASTM C260, which is essentially a performance specification in which the admixture is tested according to the provisions of ASTM C233. The key test in this method is the performance of the concrete exposed to rapid freezing and thawing (ASTM C666). Admixtures are compared to neutralized vinsol resin, which has been selected as the reference air-entraining admixture. ASTM C226 is the relevant specification for air-entraining additions.

Factors Affecting Air Entrainment

An air-entraining admixture should always be tested under field conditions because the amount of air entrained can be affected by a variety of factors. The recommended admixture dosage can be adjusted to bring the air content to within the recommended limits. Increasing the admixture dosage will invariably increase air content, decrease bubble size, and decrease the spacing factor. The variety of surface-active agents present in commercial admixtures means that job conditions may have different effects on different admixtures.

Materials

The air-entraining potential of an admixture depends on the concrete materials used and their proportions. Finely ground cements entrain less air than do more coarsely ground ones. Similarly, a high proportion of fines in the fine aggregate or the addition of finely divided mineral admixtures (such as fly ash) will reduce air contents. The gradation and particle shape of the aggregates will also affect air. Impurities in the water may have positive or negative effects on air entrainment. The use of other admixtures simultaneously with air-entraining agents should be approached with caution, because they may affect air-entraining abilities. Calcium chloride can be used successfully, but some common chemical admixtures may interact when mixed with air-entraining admixtures and inhibit entrainment. In such cases the second admixture should be added to the concrete after the air has been entrained. Carbon present in fly ashes can suppress air entrainment.

Cement content has a marked effect on the total air entrained. At a given dose, lean concretes entrain more air than do rich mixes (Figure 7.5), and low-*w/c*-ratio concretes entrain less air than do concretes with

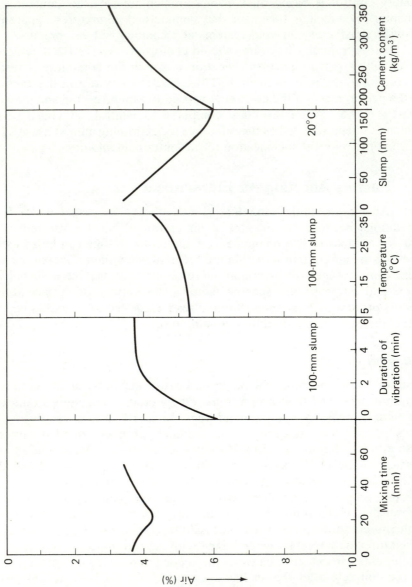

Figure 7.5 Effect of concrete parameters on total volume of entrained air.

high *w/c* ratios. However, such differences in total air content do not necessarily imply that the spacing factor has changed, because the void size may change.

Mixing and Consolidation

Air entrainment occurs during mixing, while the concrete is being agitated. The total volume of entrained air depends on the type of mixer, the rate of mixing, and the amount of concrete being mixed. The time of mixing is also important: the air content will initially increase with time and then gradually decrease during prolonged mixing. Maximum air entrainment is generally achieved during normal mixing times, and if air loss should occur during extended agitation (e.g., during transit of ready-mixed concrete), this can readily be allowed for by adjusting the admixture dose. Increasing the initial slump of the concrete will result in a greater air content over normal slump ranges (25 to 150 mm).

Temperature is also an important parameter that affects air entrainment; air content varies inversely with temperature. Increasing the temperature from 10 to 38°C (50 to 100°F) will approximately halve the air content.

Much concrete is now consolidated using high-frequency internal or external vibration. Vibration significantly reduces the air contained in concrete (Figure 7.5), but it has been shown that if vibration is applied just long enough to effect consolidation (about 15 s) only the larger, accidentally entrapped air voids are eliminated and the entrained air void system is unaffected. Prolonged vibration will, however, lower the amount of entrained air in the concrete.

Effect on Other Concrete Properties

Although air-entraining admixtures are added primarily to improve the durability of concrete, they have other important influences on the behavior of both fresh and hardened concrete.

Fresh Concrete

Air entrainment improves the workability and cohesiveness of fresh concrete, at both high and low slumps. Air-entrained concrete is more workable than a non-air-entrained mix at the same slump and can be placed and compacted more readily. Bleeding and segregation are considerably reduced in wet mixes during handling and transportation. Addition of 5% entrained air will increase the slump of concrete about 15 to 50 mm (½ to 2 in.) at a given paste content, depending on the mix design and materials. This is due partly to the fact that the surface-active ingredient to some extent acts at the solid–water interface in the same way as a water-reducing admixture (see Section 7.4). But the main reason for improved workability is that the tiny bubbles behave as low-friction, elastic, fine aggregate which reduces interactions between conventional solid aggregate particles as the concrete is handled. Thus, air-entrained mixes behave as if the mix were oversanded, and its beneficial effects are most apparent in lean or harsh mixes deficient in fines.

Rather than using air-entraining agents to increase slump, it is more common to reduce the amount of sand and water required to produce a desired slump. Water contents can generally be reduced 20 to 30 kg/m³ (30 to 50 lb/yd³), for a 5% increase in air content. This partially offsets the strength reduction that accompanies air entrainment (see below). The improvements in workability are such that air entrainment may be used even when freeze–thaw protection is not required. This is particularly beneficial where mixes have low cement contents, poor aggregate gradings, or harsh aggregate textures. Air entrainment will help produce a more uniform, well-compacted concrete.

Hardened Concrete

The introduction of additional void space with air entrainment must have a detrimental effect on strength. The increased air voids are partially offset by the decrease in capillary porosity from the lower w/c ratio that can be used. Generally, a strength loss of 10 to 20% can be anticipated for most air-entrained concretes. However, air-entrainment may not affect the strength of lean concrete mixes (cement contents below 300 kg/m³ or 500 lb/yd³), since the lower air contents needed for good frost protection may be entirely offset by decreases in w/c. There is an optimum amount of entrained air for good durability of concrete (Figure 7.6). Excessive amounts of air will drastically lower the strength of concrete and reduce its resistance to stress from freezing. The high air contents generally indicate larger bubble sizes rather than smaller spacing factors. As well as improving freeze–thaw durability, air entrainment helps combat the scaling caused by de-icing salts used on pavements and bridge decks (see Chapter 20). It also improves the sulfate resistance of concrete. The lower w/c ratios that can be used and the better compaction characteristics result in more impermeable concrete and a better overall resistance to aggressive agents.

Recommended Air Contents

The presence of a suitable entrained air-void system acts to protect concrete during freezing by providing a reservoir for water that is expelled from the paste on freezing. This mechanism will be discussed

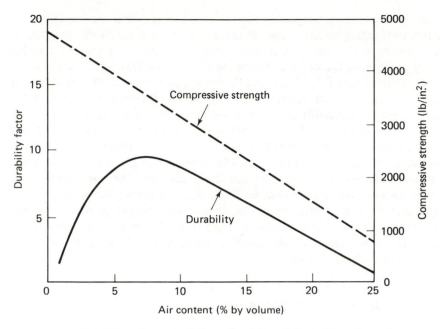

Figure 7.6 Effect of entrained air on durability. (Adapted from *Concrete Manual,* 8th ed., U.S. Bureau of Reclamation, Denver, Colo., 1975.)

in detail in Chapter 20 and is the reason for the importance of the spacing factor. Yet we have to rely on total air contents, such as are used in the ACI recommended practice for mix design (Table 9.2), which depend on many factors and which need not be related directly to the spacing factor. An example of this important point is illustrated in Table 7.3. It can be seen that for a given air content, increasing the *w/c* ratio increases

Table 7.3

Effect of w/c Ratio on Air-void System in Concrete[a]

w/c Ratio	Air Content (%)	Spacing Factor mm (in.)	Linear Expansion per Freeze–Thaw Cycle (%)
0.35	4.8	0.11 (0.0043)	0.00004
0.45	4.7	0.14 (0.0055)	0.00014
0.55	5.2	0.15 (0.0060)	0.00021
0.65	4.9	0.18 (0.0072)	0.00026
0.75	5.3	0.23 (0.0089)	0.00036

[a]Data from R. C. Mielenz, National Ready-Mixed Concrete Association, Publication No. 132 (1970).

bubble size, hence increasing the spacing factor and lowering the resistance to freezing and thawing. It can be seen, as a corollary, that the air content must be increased in concretes with high w/c ratios to maintain a given spacing factor. The recommendations in Table 9.2 should therefore be used sensibly. They should be applied to concretes that will be exposed to severe conditions where freeze–thaw cycles may be frequent, de-icers will commonly be used or sulfate attack is anticipated. For low-slump concretes (less than 25 mm slump) with low w/c ratios and high cement contents, air contents can be safely reduced below the recommended amounts. A suitable air-void system can generally be achieved in such concretes by using the same dose that would be required to attain the recommended air content in Table 9.2 at a higher slump (e.g., 75 to 100 mm or 3 to 4 in.), even though the air content in the low-slump concrete will fall below recommendations. If air entrainment is used only to improve workability and handling, recommended air contents can be reduced by about one-third.

7.4 CHEMICAL ADMIXTURES

This class of admixtures encompasses the total spectrum of soluble chemicals that are added to concrete for the purposes of modifying setting times and reducing the water requirements of concrete mixes. Most of these admixtures come under the provisions of ASTM C494. This document refers to five types of chemical admixtures, each description being self-explanatory:

Type A. Water-Reducing Admixtures

Type B. Retarding Admixtures

Type C. Accelerating Admixtures

Type D. Water-Reducing and -Retarding Admixtures

Type E. Water-Reducing and -Accelerating Admixtures

The three distinct types represented by A, B, and C will each be discussed in detail. (BS 5075: Part 1 also deals with these three types.)

Water-reducing Admixtures

As its name implies, a water-reducing admixture lowers the water required to attain a given slump. This property can be used to advantage in several ways. Achieving the desired slump with less water at a constant cement content means an effective lowering of the w/c ratio with a consequent general improvement in strength, water tightness (impermeability), and durability. Alternatively, the desired slump may be achieved without changing the w/c ratio by lowering the cement content. This may be done for economic reasons (by cutting back on the most expensive ingredient of concrete) or for technical reasons (e.g., to lower the heat of hydration of the concrete). Finally a water-reducing admixture may be used to increase slump without increasing cement and water contents, to facilitate difficult placements. According to ASTM C494, an admixture can be classified as water-reducing if it reduces water requirements by 5%; under this criterion many air-entraining admixtures could also be classed as water-reducing admixtures. Most conventional admixtures achieve water reductions of 5 to 10% at normal dosages, but some newer admixtures, called "superplasticizers," can routinely achieve reductions of 15 to 30% in water requirements. The actual water reduction achieved at any dosage depends on the type of admixture, cement fineness, mix proportions, temperature of the concrete, and time of addition. Even the composition of the cement can be important; it has been found that water reduction is lower for cements high in C_3A or in alkalis. It is believed that during the early hydration of C_3A the admixture can be removed from the cement surface by being absorbed in the hydration products.

Composition and Mechanism of Water Reduction

Water-reducing admixtures can be divided into three categories based on the general composition of their active ingredients. These are (1) salts and derivatives of lignosulfonates, (2) salts and derivatives of hydroxycarboxylic acids, and (3) polymeric materials. Most conventional water-reducing admixtures belong to the first two categories, while superplasticizers belong to (1) or (3). The common feature of all categories is that these compounds are adsorbed primarily at the solid–water interface, and this is believed to be the basis of water reduction. Solid particles carry residual charges on their surfaces, which may be positive, or negative, or both. In cement paste, opposing charges on adjacent particles of cement can exert considerable electrostatic attractions, causing the particles to flocculate (Figure 7.7a). A considerable amount of water is tied up in these agglomerations and adsorbed on the solid surfaces, leaving less water available to reduce the viscosity of the paste and hence of the concrete. Molecules of the water-reducing admixtures interact to neutralize these surface charges or to cause all surfaces to carry uniform charges of like sign. Particles now repel each other rather than attract, and remain fully dispersed in the paste (Figure 7.7b). Thus,

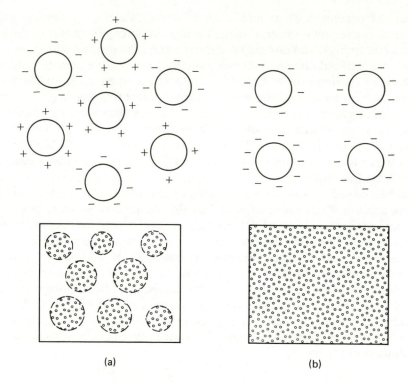

Figure 7.7 Dispersing action of water-reducing admixtures (schematic representation): (a) flocculated paste; (b) dispersed paste.

most of the water is available to reduce the viscosity of the paste and of the concrete. These surface interactions are most noticeable with the fine cement particles but may act to some extent with the finest fractions of the fine aggregate, since the tendency for flocculation is common to all finely divided solids. For example, deflocculating agents or plasticizers are commonly used with clay–water systems to produce free-flowing slurries at low water contents. The use in concrete technology is analogous.

Effects on Other Concrete Properties

Water-reducing admixtures can influence other properties of concrete, in both the plastic and hardened states.

Fresh concrete. Although water-reducing admixtures improve the workability of concrete as measured by slump, they may not necessarily improve cohesiveness. Admixtures based on hydroxycarboxylic acids tend to increase bleeding and should be used with care in high-slump concretes. Lignosulfonate-based admixtures perform better in this respect primarily because they tend to entrain air; indeed, air entrainment can be excessive unless controlled by air-detraining agents. Any water-reducing admixture, regardless of its own air-entraining ability, will reduce the amount of an air-entraining admixture required to attain a given air content. Presumably, this is because the air-entraining admixture is no longer adsorbed on the solid surfaces of the cement and is wholly available to act at the air–water interface.

Most of the conventional water-reducing admixtures will also act as retarding admixtures. If prolonged setting times are not convenient, the admixture can be formulated with an accelerating admixture to counteract the retarding tendencies or even to provide some net accleration of setting. However, nonretarding formulations sometimes show a tendency to cause retardation at low curing temperatures. It has been found that delaying the addition of the admixture until a few minutes after water has been first added will increase potential water reduction, enhance air entrainment, and increase the retardation of set (if this occurs on regular additions).

Although added to increase workability, water-reducing admixtures are often accused of aggravating the problem of loss of workability (slump loss) that occurs in fresh concrete. It is not clear to what extent this accusation is justified, since it seems to be a general rule that the higher the initial slump of concrete, the greater will be the rate of slump loss (see Chapter 8). Certainly, admixtures that also have retarding tendencies can cause workability problems. Retempering concrete with water to restore workability is an undesirable practice because of the resulting increase in the w/c ratio. Using additional water-reducing agents during retempering can reduce or eliminate the need to add further water, provided that excessive set retardation does not occur.

Hardened concrete. When admixtures are used to lower water requirements, increases in compressive strength can be anticipated. These increases can be observed as early as 1 day if excessive retardation does not occur. It is generally agreed that increases in compressive strength are up to 25% greater than would be anticipated from the decrease in w/c ratio. The reason for this is not known, but it may indicate a better uniformity within the paste structure through elimination of early flocculation. Flexural strengths will also be increased, but to a lesser extent. Water-reducing admixtures are often considered to increase drying shrinkage and creep of concrete. It may be that these increases are illusory; if comparisons were made with a properly designed concrete mix that did not contain the admixture, the differences in creep behavior would be less. However, decreases in cement content or w/c ratio that can be obtained by the use of a water-reducing admixture should balance any intrinsic, detrimental influence the admixture may have. When used intelligently, water-reducing admixtures should produce more durable concrete, by lowering w/c ratios and improving the uniformity and density of in-place concrete.

Superplasticizing Admixtures

A new class of water-reducing agents, which can achieve water reductions of 15 to 30%, has been gaining widespread acceptance in recent years. Water reductions of this magnitude have a much greater impact on the properties of concrete than do the reductions obtained with conventional water-reducing admixtures. Superplasticizers are used for two main purposes:

1. To create "flowing" concretes with very high slumps in the range of 175 to 225 mm (7 to 9 in.).

2. To produce high-strength concretes at w/c ratios in the range 0.30 to 0.40.

These materials were initially developed in Japan and West Germany. The name "superplasticizer" is perhaps rather subjective, although graphic, and the ASTM terminology for these materials is "high-range water-reducing admixtures."

Composition. Superplasticizers are linear polymers containing sulfonic acid groups attached to the polymer backbone at regular intervals. Two major polymer types are the basis of most commercial formulations: these are a sulfonated melamine–formaldehyde condensate and a naphthalene sulfonate–formaldehyde condensate. In some countries a third type, based on special lignosulfonate polymers, is marketed. However, the nature of the backbone is relatively unimportant. The sulfonic acid groups are responsible for neutralizing the surface charges on the cement particles and causing dispersion. If added in amounts comparable to normal water-reducing admixtures, water reductions are similar (5 to 10%). The effectiveness of these materials is that the undesirable side effects, air entrainment and set retardation, are absent or at least very much reduced. Thus, they can be used at high rates of addition (see Table 7.4), typically in amounts exceeding 1% of active ingredient by weight of cement, whereas conventional water reducers cannot be used in such large quantities. Superplasticized concretes can be air-entrained for freeze-thaw resistance, although the characteristics of the air void system are different from that of conventional concrete at the same air contents.

Table 7.4

Effect of Dosage on Water Reduction Using a Superplasticizer[a]

Dosage[b]	Cement Content (kg/m³)	Slump (mm)	w/c Ratio	Water Reduction (%)
—	309	50	0.49	—
0.64	309	64	0.43	15
1.73	326	64	0.35	25
2.30	329	108	0.35	24
2.82	332	132	0.35	24

[a]Based on data from W. F. Perenchio, D. A. Whiting, and D. L. Kantro, in *Superplasticizers in Concrete (Proceedings of an International Symposium)*, Vol. 1, pp. 295–324, CANMET, Ottawa, Canada, 1978.

[b]Percent of solids by weight of cement.

Flowing concrete. Concrete with very high slumps can be used to advantage in difficult placements or in placements where adequate consolidation by vibration cannot be readily achieved. Using conventional water-reducing agents, slumps reaching 180 mm (7 in.) can only be achieved with oversanded mixes with very high cement contents. The use of superplasticizers allows even higher slumps to be obtained using more normal mix proportions without the occurrence of excessive segregation or bleeding. Developments in West Germany were directed at providing "flowing" concrete which could be poured like a slurry for self-leveling pouring of slabs. Flowing concrete is particularly well suited to tremie placement and rapid pumping of concrete. Care should be taken in mix design to ensure that segregation will not occur during placement; trial mixes must be checked before use. A reasonable starting point is a conventional mix designed for pumping at a slump of about 75 mm (3 in.), which can be increased to about 200 mm (8 in.) with the addition of a superplasticizer.

One of the problems with flowing concrete is its high rate of slump loss (Figure 7.8). This can be overcome by adding the admixture just prior to placement; alternatively, the concrete can be retempered using very little additional water with the addition of more admixture. Delays in the addition of the admixture do not affect its performance unduly. Slump loss is also increased in concretes of lower slumps (Figure 7.8).

High-strength concretes. When w/c ratios can be lowered below 0.40, very high strengths can be achieved (even though complete hydration may not be possible). After 24 h of normal curing it is possible to obtain compressive strengths that would take 7 days to develop at more usual w/c ratios. Similar improvements can be obtained when superplasticizers

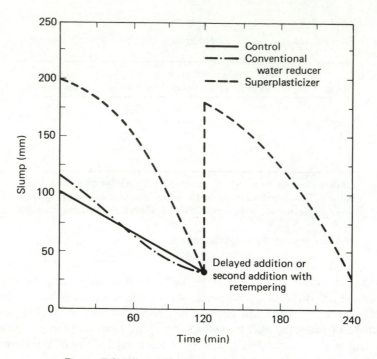

Figure 7.8 Effect of superplasticizers on slump loss.

are used with steam curing. Thus, superplasticizers can modify Type I cement to behave as super-high-early-strength cement exceeding even the strength gain of a Type III. It should be emphasized that this is the result of a greatly reduced w/c ratio; the rate of hydration is not increased. Thus, low cement contents can be used and, unlike Type III cement, the more rapid strength gain will not be accompanied by an increase in the rate of evolution of heat.

Effect on hardened concrete. As yet, there appear to be no detrimental effects on the properties of hardened concrete made with a superplasticizer. A lower w/c ratio would naturally lead to better durability and lower creep and drying shrinkage. Superplasticized concrete should be air entrained for frost resistance where necessary. There is evidence that superplasticizers can modify the air-void system in concretes having the recommended air contents so that the recommended spacing factor is exceeded. Thus there is currently some question about long-term freeze–thaw durability. However, superplasticized concrete has satisfactory field service records, and further testing should resolve this problem.

Set-retarding Admixtures

Use of Retarding Admixtures

Retarders can be used whenever it is desirable to offset the effects of high temperatures which decrease setting times, or to avoid complications when unavoidable delays may occur between mixing and placing. Prolonging the plasticity of fresh concrete can be used to advantage in placing mass concrete. Successive lifts can be blended together by vibration, with the elimination of cold joints that would occur if the first lift were to harden before the next were placed. Retarders can also be used to resist cracking due to form deflection that can occur when horizontal slabs are placed in sections. Concrete that has set but has acquired little strength is liable to microcrack when subsequent pouring alters the amount of form deflection. If the plastic period is prolonged, the concrete can adjust to form deflections without cracking.

Composition

These admixtures can be divided into several categories, based on their chemical composition: (1) lignosulfonic acids and their salts, (2) hydroxycarboxylic acids and their salts, (3) sugars and their derivatives, and (4) inorganic salts. It will be noted that categories 1 and 2 also possess water-reducing properties, and these admixtures can be classified under both groups. Lignosulfonate-based admixtures are prepared from pulp and paper industrial wastes, and studies have indicated that most of the retarding properties of these admixtures may be due to compounds that belong to category 2 or 3. Some inorganic salts (e.g., borates, phosphates, and zinc and lead salts) can act as retarders but are not used commercially.

Mode of Action

Basic research on the effect of retarders has shown that they slow down the rate of early hydration of C_3S by extending the length of the dormant period (stage 2). Thus, the setting time of portland cement, as measured by the penetration tests, is extended. However, subsequent hydration in stages 3 and 4 may be more rapid, so that strength development need not be much slower than an unretarded paste if retardation is not excessive. The extension of the dormant period is proportional to the amount of retarding admixture that is used, and when the dosage exceeds a certain critical point, the C_3S hydration will never proceed beyond stage 2 and the cement paste will never set. Thus, it is important to avoid overdosing a concrete with a retarding admixture.

On the other hand, deliberate over-retarding of concrete has helped many a ready-mix truck driver out of a tight spot. A bag of table sugar added to a batch can take care of those emergencies when the concrete may set up in the truck. Carbonated beverages can also be effective, since these contain considerable amounts of phosphoric or citric acids, both strong retarders, as well as sugar.

Retarders also tend to retard the hydration of C_3A and related aluminate phases. Thus, hydroxycarboxylic acids, such as citric acid or gluconic acid, are used to control the handling time of regulated-set cement. The interaction between C_3A and a retarder is quite complex. There is evidence that (1) the initial reactions between C_3A and water may be accelerated even though the overall hydration is retarded; and (2) the admixture may be incorporated into the hydration products of C_3A during their formation. The C_3A–admixture interactions are responsible for the following observations:

1. The effectiveness of a retarder depends on the C_3A content of the cement. (More retarder is removed from solution during the formation of the hydration products of C_3A, so that less is available to retard C_3S hydration.)

2. The effectiveness of a retarder is increased if its addition to the fresh concrete is delayed for a few minutes. (Less retarder is removed from solution, since some hydration products have already formed before the admixture is added.)

3. A retarder may cause abnormal setting problems with particular cements: both early stiffening and abnormal retardation of setting have been observed. Cements from the same mill have been found to behave differently with a water-reducing admixture, depending on the SO_3 content of the cement. (It appears that the aluminate/sulfate ratio can be thrown out of balance when the retarders change the rate of hydration of C_3A.)

4. An admixture may extend setting times but not the time during which the concrete can be handled and placed. (The early reactions of C_3A affect workability, and accelerating the early hydration may promote slump loss.)

5. A retarding admixture may influence the behavior of an expansive cement. (The rate of formation of ettringite is changed.)

In addition, the content of alkali oxides in a cement may determine the effectiveness of a retarder. It is not known exactly what causes this, but it has been postulated that the alkalis may cause the breakdown of the admixture. Because retarders are so sensitive to cement composition, an admixture should be evaluated with the cement that is to be used on the job.

Effect on Concrete Properties

The influence of an admixture on air entrainment should be considered, particularly if the admixture also has water-reducing properties. Retarders may also increase the rate of loss of workability in fresh concrete (slump loss), even when abnormal setting behavior does not occur. Thus, there may be a decrease in the time during which the concrete can be handled and placed, even though the setting time has been extended.

As mentioned above, the retarding power of an admixture increases when its addition to concrete is delayed a few minutes after the first addition of water. Retardation is increased rapidly up to about 10 min delay (depending on admixture type and dose) and then slowly decreases (Figure 7.9). After about 2 to 4 h delay, the admixture will no longer retard the setting time. Consequently, an additional dose of a water-reducing admixture can be used during retempering of fresh concrete without prolonging setting times unduly. Whenever a retarding admixture is used, some reduction in the 1-day strength of the concrete should be anticipated. Within 7 days, the strength should approach that of an

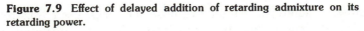

Figure 7.9 Effect of delayed addition of retarding admixture on its retarding power.

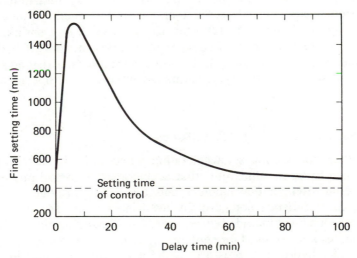

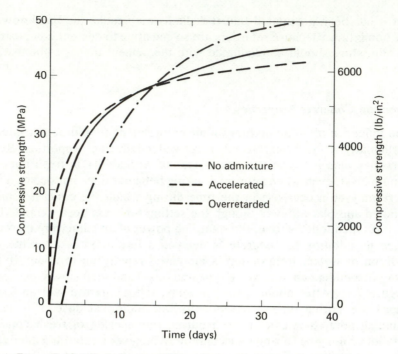

Figure 7.10 Effect of set-modifying admixtures on strength development of concrete.

unretarded concrete unless an overdose has been used (Figure 7.10). Retarding admixtures have been reported to increase ultimate compressive strength and, to a lesser extent, flexural strength. Although set-controlling admixtures are reported to increase drying shrinkage and creep, the effects depend on changes in mix design, time of hydration, and time of drying or loading. Laboratory work indicates that admixtures may increase the rate of drying shrinkage and creep but not the ultimate values.

Set-accelerating Admixtures

When considering set-accelerating admixtures, it is important to distinguish between admixtures that accelerate the normal processes of setting and strength development and those that provide very rapid setting characteristics not normally associated with ordinary portland cements. Quick-setting admixtures may provide setting times of only a few minutes and are used for shotcreting applications, for plugging leaks while under hydrostatic pressure, and for emergency repair in general where very rapid development of rigidity is required. The conventional accelerators are used to speed construction by permitting earlier finishing of surfaces, earlier form removal, and earlier attainment of sufficient strength to carry construction loads. Accelerators are beneficial during winter concreting by partially overcoming the slower rate of hydration caused by low temperatures and shortening the period for which protection against damage by freezing is required.

Composition

We can divide accelerating admixtures into three groups: (1) soluble inorganic salts, (2) soluble organic compounds, and (3) miscellaneous solid materials. Most soluble inorganic salts will accelerate the setting and hardening of concrete to some degree, calcium salts generally being the most effective. Calcium chloride is the most popular choice because it gives more acceleration at a particular rate of addition than other accelerators and is also reasonably inexpensive. Soluble carbonates, aluminates, fluorides, and ferric salts have quick-setting properties. Sodium carbonate and sodium aluminate are the most common ingredients of shotcreting admixture formulations used to promote quick setting. Calcium fluoroaluminate or calcium sulfoaluminate (Chapter 3) can be used as admixtures to obtain rapid-hardening characteristics.

A variety of organic compounds have accelerating properties (although many more act as retarders), but triethanolamine, calcium formate, and calcium acetate account for most commercial uses. They are commonly used in formulations of water-reducing admixtures to offset their retarding action. Although triethanolamine is listed as an accelerator, recent research shows that its reaction with portland cement is rather complex. It can cause retardation or flash setting, depending on the amount used.

Solid materials are not often used for acceleration. Additions of calcium aluminate cements cause portland cements to set rapidly, but strength development is poor. Concrete can be "seeded" by adding fully hydrated cement that has been finely ground during mixing to cause more rapid hydration. Finely divided carbonates (calcium or magnesium), silicate minerals, and silicas are reported to decrease setting times.

Mode of Action

Conventional accelerators have exactly the opposite action that retarders have: they increase the rate of hydration of C_3S. Generally, the dormant period is shortened, but the rate of hydration during stages 3 and 4 may be increased. This is the case with calcium salts. Organic accelerators are believed to act by increasing the rate of hydration of C_3A. Quick-setting admixtures are believed to cause flash setting of C_3A by promoting very rapid hydration on contact with water. Admixtures that affect C_3A hydration can also be expected to affect the expansion characteristics of expansive cements.

Effects on Concrete Properties

Accelerators do not as a rule have adverse effects on air entrainment, although trial batches should be used to check this. It must be remembered that use of an accelerator will cut down the time during which the concrete can be handled and placed, so that reliable scheduling of construction operations is necessary. Accelerating admixtures may require some additional water if early stiffening occurs, but they can be formulated with a water-reducing admixture. Conventional accelerators can be expected to increase 1-day strengths, the increase depending on the type and dosage of the admixture. An optimum dose of calcium chloride (2% by weight of cement[1]) will approximately double the 1-day compressive strength of concrete. These increases diminish with time and later strengths (at 28 days or more) may actually be lower than the strength of concretes without an accelerating admixture (Figure 7.10). This reduction in later strength is more pronounced when the initial accelerating effects are large. The extreme case is that of some quick-setting admixtures, which may have lower strengths at 1-day even though some strength is obtained very rapidly. Flexural strengths are affected to a lesser extent than compressive strengths by accelerating admixtures.

Accelerating admixtures are said to increase drying shrinkage and creep, but, as discussed with retarders, it is the rate of deformation rather than the ultimate value that is affected. Tests have shown that the use of calcium chloride in concrete reduces its resistance to sulfate attack and aggravates the alkali–aggregate reaction if this is a potential problem. Quick-setting admixtures can be expected to reduce the durability of the concrete or mortar if they have an adverse effect on strength.

Corrosion of Reinforcement

A major disadvantage attending the use of calcium chloride is its tendency to increase the rate of corrosion of metals embedded in concrete. Additions of calcium chloride, up to 2% by weight of cement, have generally been considered safe to use in ordinary reinforced concrete provided that an adequate cover of dense concrete is provided. However,

[1]This is 2% of Type I calcium chloride (77% pure); see ASTM D98 specifications for calcium chloride, or BS 3587.

there is not universal agreement about the levels of calcium chloride that should be tolerated, and many people believe that much lower levels should be used. ACI Committee 318 in the Building Code Commentary notes that "chloride ion concentrations [in the mixing water] exceeding 400 to 500 ppm might be considered dangerous." (A dose of 2% gives about 30,000 ppm in the mixing water.) ACI Committee 201 (Durability) recommends that not more than 0.15% chloride ion (~300 ppm) from all sources should be tolerated in reinforced concrete and not more than 0.10% if it is to be exposed to chlorides in service. It should be remembered that if the reinforcement is later exposed by microcracking or scaling, the presence of chloride in the concrete will accelerate subsequent corrosion. There is general agreement that chlorides should never be used in prestressed concrete (or in grouts used in prestressed concrete) or in concrete in which aluminum or galvanized metal is to be embedded. Aluminum, zinc, and steel under high stress are particularly vulnerable to corrosion in the presence of the chloride ion.

Many engineers prefer to avoid the use of calcium chloride altogether, and alternative admixtures based on calcium formate and calcium nitrite are available. Other soluble salts, such as thiosulfates, thiocyanates, and nitrates, will also accelerate cement hydration without causing corrosion. It must be remembered that such accelerators must be added in greater amounts to achieve the same effects as calcium chloride and will be more expensive. An alternative solution is to use Type III cement and higher cement contents in order to get the same early strength gain.

7.5 MINERAL ADMIXTURES

The addition of finely ground solid materials to concrete is an established practice in modern concrete technology. They are used to improve the workability of fresh concrete and the durability of hardened concrete. We may subdivide this class of admixtures into:

1. Materials of low reactivity
2. Cementitious materials
3. Pozzolanic materials

Materials of Low Reactivity

Materials in this category are used solely to improve the workability of fresh concrete. They are used primarily in concretes deficient in fines, such as concretes made with coarsely graded sand, or concretes where low cement contents are desirable. If insufficient fine material is present in the mix, the concrete will be susceptible to excessive segregation and bleeding. Such concretes are not suitable for pumping or placement by tremie and do not finish well. The improvement of cohesiveness and plasticity can be obtained by increasing the cement content, but if this is not desired, because of the additional cost or because of technical reasons (e.g., increased evolution of heat in mass concrete), a mineral admixture can be used. Generally, cementitious or pozzolanic materials are preferred because they eventually contribute to strength and durability of the hardened concrete. Materials such as ground limestone, dolomite, quartz, bentonite, rock dust, and hydrated lime are relatively inert and contribute little to the strength of the concrete. Such additions are used primarily in applications that do not require high strength, such as mortars for brick or block, grouts, oil-well cementing slurries, and the like. Although such finely divided materials are probably never completely inert (hence the use of the term "low reactivity"), any contribution to strength will be masked by the increased water demand that their use requires.

Cementitious Materials

Materials that have hydraulic activity in their own right and can thus contribute to the strength of the concrete include natural cements, hydraulic lime, and blast-furnace slag. The latter is the most common admixture in this category but is usually used as an addition in a blended cement (e.g., Type IS cement). Portland cement–slag blended cements and slag cements were discussed in detail in Chapter 3. Slags may also have pozzolanic properties and so have benefits similar to those of pozzolans.

Pozzolanic Materials

The Pozzolanic Reaction

If a material is able to react with calcium hydroxide, it is said to have pozzolanic activity and is called a *pozzolan* (Chapter 2). The pozzolanic reaction is

$$CH + S + H \rightarrow C\text{-}S\text{-}H \tag{7.1}$$

A pozzolan contains amorphous silica which is sufficiently reactive to combine with calcium hydroxide to form C–S–H. A pozzolan may be mixed directly with calcium hydroxide in lime–fly ash mixtures used in subbases for pavements; but when the pozzolan is mixed with portland cement, it will react with the calcium hydroxide formed during hydration. The effect of the pozzolanic reaction is then to increase the proportion of C–S–H in the hydrated paste at the expense of calcium hydroxide. Pozzolans can be added as an admixture at the job site, but some cement companies market a blended pozzolanic cement, Type IP, which should conform to ASTM C595.

Uses. Beside improving the workability of harsh mixes, pozzolans will lower the total heat of hydration and also the rate of liberation of heat, because the pozzolanic reaction is quite slow. The speed of the pozzolanic reaction is comparable to C_2S hydration, and the result of adding a pozzolan is to effectively raise the C_2S content of the cement. Thus, a Type I cement can be used with a pozzolanic admixture as a substitute for Type IV in mass concrete and, as a result, Type IV cements are now seldom manufactured. The use of a pozzolan will also improve the impermeability and durability of hardened concrete. Expansion resulting from the alkali–aggregate reaction is reduced with a pozzolan, and a Type I cement can be used with a pozzolan that is low in alumina to replace Type V cement in concretes exposed to sulfate attack. It is believed that the reduction of calcium hydroxide in the paste is an important factor in improving chemical durability. Pozzolans will not improve freeze–thaw durability, and air entrainment is still required. Rates of additions depend on the particular application but may be as high as 35% by weight of cement.

Composition. A wide variety of different materials contain silica that has pozzolanic activity. Many of these are naturally occurring materials: volcanic ashes and tuffs, pumicite, diatomaceous earth, opaline cherts, clays, and shales. A natural pozzolan usually requires grinding to cement fineness and may need to be calcined (e.g., clays and shales). These treatments will maximize their pozzolanic activities. Synthetic pozzolans include fly ash, quenched boiler slag, and precipitated silica. Fly ash is the inorganic residue that remains after powdered coal has been burned and is trapped by electrostatic precipitators; most fly ash is obtained from coal-fired power stations. It is a popular pozzolan because it is already very finely divided, and since it is often in the form of tiny spheres, it can improve workability without an undue increase in water demand, unlike most natural pozzolanic materials. Fly ash is widely available, particularly near centers of population where much concrete construction is located, and is thus very cost-competitive. The cost of a pozzolan must be lower than cement if its use as a cement replacement is to be justified.

Pozzolans can have quite variable composition. Frequently, reactive alumina is also present and it can react analogously to silica:

$$CH + A + H \rightarrow C\text{-}A\text{-}H \text{ (calcium aluminate hydrates)} \qquad (7.2)$$

The calcium aluminate hydrates, whose nature depends on the amount and reactivity of alumina, can react expansively with sulfate to form ettringite. Thus, if a pozzolan is to be used to improve sulfate resistance, a material low in alumina should be used. Otherwise, pozzolans should conform to ASTM C618, which puts limits only on the amounts of MgO, SO_3, and Na_2O. Some fly ashes may contain free carbon and lime. Excessive carbon can inhibit air entrainment and its amount is controlled indirectly by limits on loss of ignition. The presence of free lime or MgO can lead to unsoundness, which is measured by the autoclave expansion test. ASTM C618 also specifies a minimum fineness, since fineness determines the rate of the pozzolanic reaction, the improvements in workability, and the control of bleeding.

Other effects on concrete properties. The use of a pozzolan is often accompanied by a low rate of early strength gain such as occurs with Type IV cement. Compressive strengths may actually be improved at later ages (Figure 7.11) if the admixture is nearly 100% pozzolanic (some

Figure 7.11 Effect of a pozzolan on the compressive strength of concrete (30% replacement by weight of cement). (From E. C. Higginson, in *Significance of Tests and Properties of Concrete and Concrete-making Materials,* ASTM STP 169A, 1966, pp. 543–555. Reprinted, with permission, from the American Society for Testing and Materials, 1916 Race Street, Philadelphia, PA 19103. Copyright.)

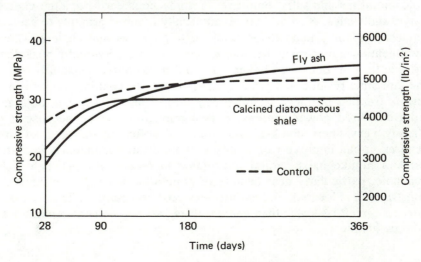

fly ashes) but may be lower if there are considerable inert impurities or the pozzolanic reaction is very slow. A good pozzolan is considered by some to be a necessary addition for high-strength concrete. Pozzolanic activity is estimated by determining the strength developed when the material is reacted with pure calcium hydroxide at 23°C for 1 day and then at 55°C for 6 days (ASTM C595). If a pozzolan increases the water requirements of the concrete, an increase in drying shrinkage and creep can be anticipated. ASTM C618 limits the increase in drying shrinkage to 0.03%.

7.6 MISCELLANEOUS ADMIXTURES

There are many other types of admixtures that are commercially available. The consumption of these various materials added together is less than the amount used of any single type so far discussed. Some brief discussion of the more important kinds is warranted, however.

Bonding Admixtures

Polymer latexes (emulsions) are used to improve the bonding properties of concrete. This can be bonding between old and new concretes in repair work, or bonding between concrete and other materials, such as steel. These admixtures will be discussed in more detail in Chapter 22.

Corrosion Inhibitors

The incorporation of compounds that will prevent or reduce corrosion has been suggested for concretes in which corrosion of reinforcement can be a problem. Generally, these will be salts that contain an oxidizable ion, such as nitrites, thiosulfates, benzoates, stannous salts, and ferrous salts. But it is doubtful if the use of inhibitors is really warranted, since they are not considered to provide protection in the presence of a chloride ion, which is precisely the situation in which protection is desired. However, the foregoing materials are accelerators in their own right and could replace calcium chloride when an accelerating admixture is desired.

Dampproofing Admixtures

The term "dampproofing" implies prevention of water penetration into dry concrete or the transmission of water through concrete. No admixture can actually prevent such movement of water, although it may reduce the rate at which it occurs. Certain formulations based on salts of fatty acids (soaps) or petroleum products (mineral oils and asphalt emulsions) may give the concrete a water-repellent effect. Such materials have been used to prevent the penetration of rain into porous concrete block but are not likely to affect the performance of dense, well-cured concrete.

Dampproofing admixtures are not able to eliminate moisture migration through slabs on ground or foundation walls. Any claims to that effect should be viewed with a healthy scepticism. Moisture migration should be prevented by the construction of an appropriate vapor barrier.

Expansion-producing Admixtures

Admixtures based on calcium sulfoaluminate can be used to convert an ordinary portland cement into an expansive cement. These can be used to offset drying shrinkage in general construction or in specialty applications, such as nonshrink grouts. Gas-forming admixtures can also be used to provide expansion and counteract settlement and bleeding. Aluminum powder is the most widely used gas-producing admixture; it reacts with the alkaline medium of cement paste to produce hydrogen gas. The release of gas before setting causes the mixture to expand by forming gas bubbles dispersed through the mix. These bubbles are much larger than entrained air voids and do not provide freeze–thaw resistance. The rate and amount of gas must be carefully controlled to provide sufficient expansion without seriously weakening the material.

Grouting Admixtures

Cement-based grout used for specialty applications such as cementing oil wells may contain a great variety of different admixtures. Flocculating admixtures, thickeners, and mineral admixtures are used to prevent bleeding and segregation and to increase cohesion and retention of water during pumping. Retarders are commonly used to extend pumping times.

7.7 CONCLUDING REMARKS

It can be seen from this discussion of admixtures that there is virtually no aspect of concrete performance that cannot be modified to some advantage by the use of admixtures. By using admixtures a general purpose Type I cement can be used in place of any of the other four ASTM types (Table 7.5). Much of the present use of admixtures is dictated by economic considerations determined by the relative cost of

Table 7.5

Modification of Type I Cement by Admixtures to Achieve the Properties of Other ASTM Cements

Admixture	Blended Cement	Comparison	Remarks
Pozzolan	IP, I-PM	II	Low additions
Pozzolan	IP	II	C_3A content similar to Type II
Accelerator	—	III	
Water-reducing	—	III	Very low w/c ratios
Pozzolan	IP	IV	Moderate to high additions
Slag	IS	IV, V	Low heat and slow strength development
Pozzolan	IP	V	Low alumina content

materials or the cost of labor. Engineers and contractors may have a reluctance to use admixtures, because of the increased risk of encountering technical problems or a natural distrust of materials not properly understood. A good understanding of how admixtures interact with concrete can do much to encourage the use of admixtures to make better concrete.

Bibliography

General

ACI COMMITTEE 212, "Admixtures for Concrete," *Journal of the American Concrete Institute,* Vol. 60, No. 11, pp. 1481–1524 (1963).

ACI COMMITTEE 212, "Guide for Use of Admixtures in Concrete," *Journal of the American Concrete Institute,* Vol. 68, No. 9, pp. 646–676 (1971).

RILEM COMMITTEE 11A, "Concrete Admixtures," *Materials and Structures (Paris),* Vol. 8, No. 48, pp. 451–472 (1976).

RIXOM, M. R., ed., *Concrete Admixtures: Use and Applications.* The Construction Press, Lancaster, U.K.; Longman, Inc., New York, 1977.

Specific Admixture Categories

MALHOTRA, V. M., E. E. BERRY, AND T. A. WHEAT, "Superplasticizers in Concrete," *Proceedings of International Symposium, Ottawa, 1978* (2 vols.). CANMET, Department of Energy, Mines and Resources, Ottawa, 1978, also published in part as ACI SP-62, American Concrete Institute, Detroit, 1979.

PRICE, W. H., "Pozzolans—A Review," *Journal of the American Concrete Institute,* Vol. 72, No. 5, pp. 225–232 (1975).

RAMACHANDRAN, V. S., *Calcium Chloride in Concrete Science and Technology.* Applied Science Publishers Ltd., London, 1976.

Significance of Tests and Properties of Concrete and Concrete-making Materials, ASTM STP 169B. American Society for Testing and Materials, Philadelphia, Pa., 1978.

SKALNY, J., AND J. N. MAYCOCK, "Mechanisms of Accelerating by Calcium Chloride—A Review," *Journal of Testing and Evaluation,* Vol. 3, No. 4, pp. 303–311 (1975).

Superplasticizing Admixtures in Concrete. Cement and Concrete Association, Slough, U.K., 1976.

YOUNG, J. F., "Reaction Mechanisms of Organic Admixtures with Hydrating Cement Compounds," *Transportation Research Record,* No. 564, pp. 1–9. Transportation Research Board, Washington, D.C., 1976.

Problems

7.1 What tests would you make to determine the suitability of an admixture for concrete?

7.2 What problems may arise when two or more admixtures are used simultaneously?

7.3 Why is it important to have accurate control of the batching of admixtures?

7.4 What are the effects of high ambient temperatures on the behavior of admixtures?

7.5 Not every material that acts as a detergent is suitable as an air-entraining agent for concrete. Why?

7.6 Calculate the gel-space ratio for the following air contents at a w/c ratio of 0.50, a paste content of 20 vol.% of the concrete, and a 75% degree of hydration: (a) 4.0%; (b) 6.0%; and (c) 8.0%.

7.7 When can a water-reducing agent be used to advantage?

7.8 What are the potential disadvantages of superplasticizers?

7.9 Why are superplasticizers more effective than conventional water-reducing agents?

7.10 Are there any circumstances when you might want to use both an accelerator and a retarder?

7.11 Why can cement composition have an influence on the effectiveness of retarding admixtures?

7.12 What are the dangers of using $CaCl_2$ as an accelerating admixture?

7.13 What are the alternatives to the use of $CaCl_2$ as an accelerating admixture?

7.14 Describe the pozzolanic reaction.

7.15 What are the advantages and disadvantages of a pozzolan?

7.16 For what purposes can fly ash be used in concrete?

8

fresh concrete

The properties of fresh concrete are only important inasmuch as they affect the choice of equipment needed for handling and consolidation, and because they may affect the properties of the hardened concrete. There are therefore two sets of criteria that we must consider when making concrete:

1. Long-term requirements of the hardened concrete, such as strength, durability, and volume stability.

2. Short-term requirements, while the concrete is still in the plastic state, which are generally lumped together under the term "workability."

Unfortunately, as will be seen when the problem of proportioning concrete is considered in Chapter 9, these two sets of requirements are not complementary, and a compromise is needed between them. In this chapter only the short-term requirements for concrete are considered.

For hardened concrete to be of an acceptable quality for a given job, the fresh concrete must be capable of satisfying the following requirements:

1. It must be easily mixed and transported.

2. It must be uniform throughout a given batch and between batches.

3. It should have flow properties such that it is capable of filling completely the forms for which it was designed.

4. It must have the ability to be compacted fully without an excessive amount of energy being applied.

5. It must not segregate during placing and consolidation.

6. It must be capable of being finished properly, either against the forms or by means of trowelling or other surface treatment.

8.1 WORKABILITY

Definition of Workability

The aspects of the quality of fresh concrete just mentioned all have different requirements, so a number of terms are in common usage, each emphasizing a different facet of concrete behavior: consistency, flowability, mobility, pumpability, compactibility, finishability, and harshness. These terms are both subjective and qualitative; they mean different things to different people and are therefore not very useful. Here we use the term *workability* to represent all of the properties mentioned above. To be a little more precise, workability is often defined in terms of the amount of mechanical work, or energy, required to produce full compaction of the concrete without segregation. This is a useful definition, since the final strength of the concrete is largely a function of the amount of compaction; a small increase in void content (or decrease in relative density) will lead to a large decrease in strength.

Basic Principles of Rheology

Before discussing the various ways of measuring workability, it is helpful to examine a few basic principles of rheology, which is the science dealing with the deformation and flow of materials under stress. The simplest fluid is one that obeys Newton's law of viscous flow, which is illustrated in Figure 8.1. This law is derived by considering the shearing of adjacent layers of liquid, and it can be written

$$\tau = \eta D \qquad (8.1)$$

where τ is the shear stress, η the coefficient of viscosity, and D the rate of shear or the velocity gradient. From Eq. (8.1) it may be seen that a plot of the rate of shear vs. the shear stress will give a straight line passing through the origin, with a slope of $1/\eta$, as shown in Figure 8.2a. Therefore, if concrete could be approximated as a Newtonian fluid, a single measurement of a corresponding pair of values of τ and D would serve to define the line. In other words, a "single-point" method of determining the workability would be sufficient.

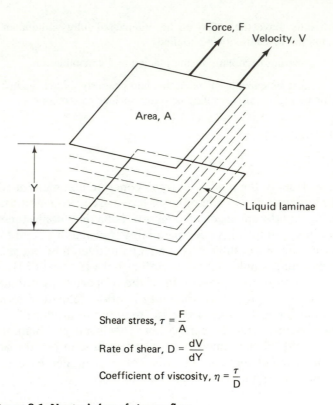

$$\text{Shear stress, } \tau = \frac{F}{A}$$

$$\text{Rate of shear, } D = \frac{dV}{dY}$$

$$\text{Coefficient of viscosity, } \eta = \frac{\tau}{D}$$

Figure 8.1 Newton's law of viscous flow.

For very dilute suspensions of a solid in a liquid, where there are no interparticle forces, Newton's equation holds. The effect of small increases in the amount of suspended solid is merely to increase the coefficient of viscosity (i.e., to decrease the slope of the line in Figure 8.2a). However, the Newtonian model breaks down for fluids in which the volume of suspended solids is large. Fresh concrete can be considered to be a very concentrated suspension. Typically, the ratio of the volume of solids to the volume of water would be about 4.5:1. For such materials, there are forces acting between the particles. This does not merely increase the viscosity (as in a Newtonian liquid), but actually changes the type of flow behavior. In particular, the fresh concrete has a definite *shear strength*, τ_0, which must be exceeded before flow can occur. A common description of materials that exhibit this type of behavior is given by the *Bingham model*,

$$\tau - \tau_0 = \mu D \tag{8.2}$$

where τ_0 is the yield stress and μ is called the plastic viscosity. Such behavior is shown in Figure 8.2b. Clearly, for such a material, which is defined by two constants, τ_0 and μ, two pairs of values of τ and D must be determined in order to define the straight line of Figure 8.2b. A single-point test would not be very useful in trying to describe such a fluid. There is now considerable evidence that the behavior of fresh concrete can be reasonably approximated by the Bingham model. Thus, the single-point tests of workability currently in use do not adequately quantify the behavior of the material.

One other term that must be defined is *thixotropy*. A thixotropic (or shear thinning) material is one that undergoes "a decrease of the apparent viscosity under shear stress (Figure 8.2c), followed by a gradual recovery

Figure 8.2 Rheological models: (a) Newtonian liquid; (b) Bingham model; (c) thixotropy.

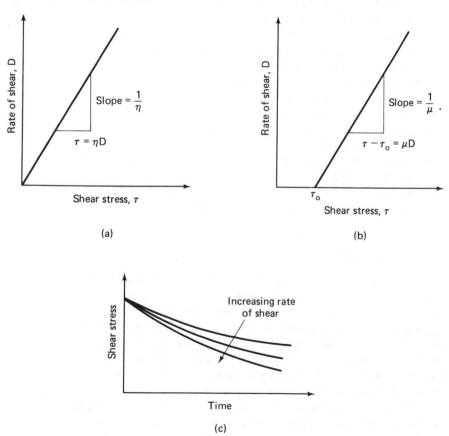

when the stress is removed. The effect is time-dependent." Such behavior is due to interparticle attraction and weak bonding. Concrete will exhibit thixotropic behavior as well; this can be important when considering the placing and vibration of concrete.

Factors Affecting Workability

The workability of concrete is affected by a number of factors: water content of the mix, mix proportions, aggregate properties, time, temperature, characteristics of the cement, and admixtures.

Water Content of the Mix

The single most important factor governing the workability of concrete is the water content. Increasing the amount of water will increase the ease with which concrete flows and can be compacted. However, apart from reducing the strength, increased water may lead to segregation (settling of the coarse aggregate) and to bleeding. In general, any collection of particles requires a certain amount of water to achieve plasticity so that it can be "worked." First, there must be enough water to adsorb on the particle surfaces. Then, water must fill the spaces between particles; additional water "lubricates" the particles by separating them with a water film. From this it follows that finer particles, which have a higher specific surface area, require more water. On the other hand, without some minimum quantity of fine material, the concrete cannot exhibit plasticity. Thus, the water content of the mix cannot be considered in isolation from the aggregate grading. For optimum workability finer aggregate gradings require high water contents.

Influence of Aggregate Mix Proportions

When considering the effect of aggregates on workability, two factors are important: the amount of aggregate and the relative proportions of fine and coarse aggregate. For a constant *w/c* ratio, an increase in the aggregate/cement ratio will decrease the workability; also, more cement is needed when finer aggregate gradings are used. A deficiency in fine aggregate results in a mix that is harsh, prone to segregation, and difficult to finish. (A harsh mix is one that lacks the desired consistency because of a deficiency in mortar or aggregate fines.) On the other hand, an excess of fine aggregate ("oversanded") will lead to a rather more permeable and less economical concrete, although the mixture will be easily workable.

Aggregate Properties

It is not sufficient merely to look at the ratio of coarse/fine aggregate. Different sands will behave differently because of differences in particle-size distribution. This is generally expressed in terms of the fineness modulus, which was described in Chapter 6.

The shape and texture of aggregate particles can also affect the workability. As a general rule, the more nearly spherical the particles, the more workable the resulting concrete will be. This is due partly to the fact the spherical particles will act as "ball bearings" while angular particles will have more mechanical interlock and will therefore need more work to overcome the resulting internal friction. Spherical particles will have a lower surface-to-volume ratio, and less mortar will be needed to coat the particles, leaving more to provide "workability." When flat or elongated particles are contained in the coarse aggregate, the quantities of sand, cement, and water must be increased. Also, smooth particles will tend to be more workable than rough ones, such as are obtained when crushed rock is used as aggregate.

The porosity of the aggregates may also affect workability. If the aggregate can absorb a great deal of water, less will be available to provide workability. Thus, it may be necessary to distinguish between the *total* water content (which may include absorbed water) and the *free* water content which is available to provide workability, and which determines the *w/c* ratio.

Time and Temperature

There is considerable evidence that as the ambient temperature increases, the workability decreases, as shown in Figure 8.3, since higher temperatures will increase both the evaporation rate and the hydration rate. This suggests that more water should be used in very warm weather to maintain the same workability. Also, the quantity of water required to change the slump by some given amount increases as the temperature increases. On the other hand, there is some recent laboratory evidence that, at least for short times, workability is not affected by temperature. Thus, where temperature may be a factor, field tests should be carried out with the specific materials in order to determine the temperature effects for a given job.

Loss of Workability

During the period of fluidity there will be a steady decrease in workability with time, which is caused partially by the hydration of C_3S and C_3A, which continues slowly even during the dormant period, and partially by

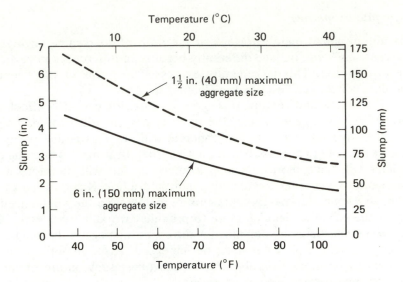

Figure 8.3 Relationship between slump and temperature of concrete made with two maximum sizes of aggregates. Each curve represents the average obtained with twelve different cements. (Adapted from *Concrete Manual*, 8th ed., U.S. Bureau of Reclamation, Denver, Colo., 1975.)

loss of water from the concrete through evaporation or absorption. In addition, particle interactions change because of the presence of hydration products on the surface. This "slump loss" is approximately linear with time, although it is greatest in the first ½ to 1 h after mixing. Since the slump at the time of placement is of greatest importance, the slump loss of concrete must be considered when selecting mix proportions. Slump loss is increased when concrete temperatures are higher and also when accelerating admixtures are used; in both cases, setting times are decreased. Accelerated slump loss can occur when water-reducing and set-retarding admixtures are used, even when the setting times are unchanged or even increased. The rate of loss of workability is less for lean (low cement content) mixes and when high *w/c* ratios are used. The brand of cement is also important (Figure 8.4), since two cements of the same type can have different compound compositions.

If the concrete mix is improperly designed, it may be necessary to remix the concrete with additional water ("retempering") just prior to placing in order to restore workability. Retempering increases the *w/c* ratio and may be highly detrimental to the properties of the concrete unless additional cement is also added. Therefore, retempering is a practice that should be strongly discouraged.

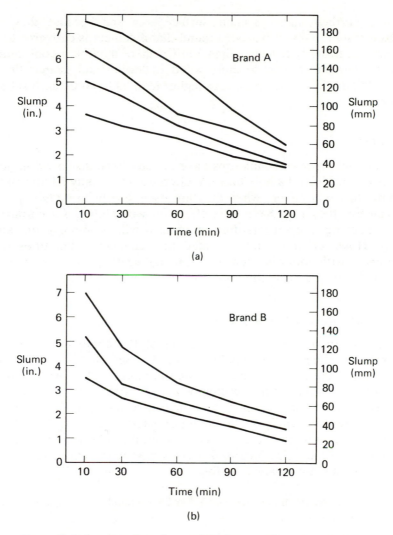

Figure 8.4 Concrete slump loss at 21°C for two different brands of Type I cement. (From R. W. Previte, *Journal of the American Concrete Institute*, Vol. 74, No. 8, 1977, p. 361.)

Cement Characteristics

The cement characteristics are much less important in determining workability than are the aggregate properties. However, the increased fineness of Type III (rapid-hardening) cements will reduce workability at a given *w/c* ratio, as these cements have a higher water requirement because of their higher specific surface area and because they also hydrate more rapidly. It has been found that if cement is delivered to the job hot (particularly if higher than 100°C), there may be a considerable loss of workability, due to more rapid hydration and evaporation of water. In addition, at temperatures higher than 60 to 80°C, flash set may occur.

Admixtures

The different types of admixtures have already been discussed in detail in Chapter 7. Mineral admixtures are often used to supply additional fine material to harsh mixes; when they are used simply to *replace* part of the cement, they may have little effect on workability. Air-entraining, water-reducing, and set-retarding admixtures will all improve the workability. However, it should be noted that chemical admixtures react differently with different cements and aggregates, and can in some circumstances aggravate the loss of workability.

Segregation and Bleeding

Segregation refers to a separation of the components of fresh concrete, resulting in a nonuniform mix. In general, this means some separation of the coarse aggregate from the mortar. This separation can be of two types: either the settling of heavy particles to the bottom of the fresh concrete, or a separation of the coarse aggregate from the body of the concrete, generally due to improper placing or vibration. Although there are no quantitative tests for segregation, it can be seen quite clearly when it does occur. The factors that contribute to increased segregation have been listed by Popovics as follows:

1. Larger maximum particle size (over 25 mm) and proportion of the large particles.

2. A high specific gravity of the coarse aggregate compared to that of the fine aggregate.

3. A decreased amount of fines (sand or cement).

4. Changes in the particle shape away from smooth, well-rounded particles to odd-shaped, rough particles.

5. Mixes that are either too wet or too dry.[1]

---

[1]S. Popovics, in *Proceedings of a RILEM Seminar, March 22–24, 1973, Leeds,* Vol. 3, pp. 6.1-1 to 6.1-37, The University, Leeds, 1973.

The use of finely divided mineral admixtures or air-entraining agents reduces the tendency toward segregation, but careful handling and placing methods are more important. The correct techniques of handling concrete to prevent segregation will be referred to in more detail in Chapter 10.

Bleeding may be defined as the appearance of water on the surface of concrete after it has been consolidated but before it has set. Water, being the lightest component, segregates from the rest of the mix, and thus bleeding is a special form of segregation. This is the most common manifestation of bleeding, although the term may also be used to describe the draining of water out of the fresh mix. Bleeding is generally caused by the fact that as the aggregate particles settle within the mass of fresh concrete, they are unable to hold all the mixing water. Some bleeding is normal for good concrete; it results in a small amount of uniform seepage over the entire surface.

On the other hand, bleeding may occur through distinct, localized channels, and small "craters" may form at the mouth of each channel. The upper layer of the concrete may become rich in cement paste which has a *w/c* ratio that is too high. This leads to weakness, porosity, and a lack of durability. Water pockets may form under large aggregate particles or reinforcing bars, leaving weak zones in the concrete and reducing bond (Figure 8.5). If the bleed water evaporates more quickly than the bleeding rates, which often occurs in hot, dry weather, plastic shrinkage cracks will form (Chapter 18). Sometimes, a scum of fine particles may be carried to the surface, creating a weak and nondurable surface, or salts may crystallize at the surface, leading to the formation of $CaCO_3$.

Figure 8.5 Different types of bleeding in concrete.

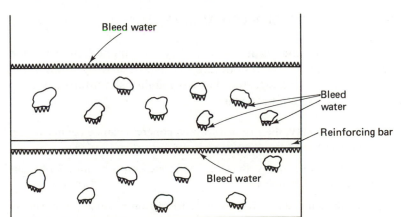

This is referred to as *laitance*. If this occurs at the top of a lift, poor bond to the next lift will be the result. If it occurs at the top of a flat slab, the surface will be prone to *dusting*; that is, a dry, powdery layer will appear on the surface of the hardened concrete, and this will be a permanent feature of the concrete. Deterioration is then a combination of laitance and plastic shrinkage. Thus, laitance should always be removed by brushing and washing the surface.

Bleeding can be reduced in a number of ways:

1. By increasing cement fineness or by using pozzolans or other finely divided mineral admixtures.

2. By increasing the rate of hydration of the cement by using cements with high alkali contents or high C_3A contents (which may, however, have other undesirable effects), or by using $CaCl_2$ as an admixture (which may also be undesirable).

3. Through air entrainment, which is very effective.

4. By reducing the water content, if this can be done while maintaining an acceptable workability.

Both the rate of bleeding and the total bleeding capacity can be measured using the procedures outlined in ASTM C232. This method involves consolidating a sample of concrete in a container with a diameter of 254 mm (10 in.) and a height of 279 mm (11 in.). The container is then covered to prevent evaporation. The bleed water rising to the surface is carefully drawn off every 10 min for the first 40 min, and every 30 min thereafter until the bleeding has stopped. The total bleeding and rate of bleeding may then be determined.

8.2 MEASUREMENT OF WORKABILITY

As we have seen, the term "workability" is applied to a number of different properties of fresh concrete and is affected by a large number of factors. It has been suggested that workability should measure at least three separate concrete properties:

1. Compactibility, the ease with which concrete can be compacted and air voids removed.

2. Mobility, the ease with which concrete can flow into forms, around steel and be remolded.

3. Stability, the ability for concrete to remain a stable, coherent, homogeneous mass during handling and vibration without the constituents segregating.

A large number of workability tests have been proposed over the years, almost all of them completely empirical. Only a few of these tests have been incorporated into standards; the rest have received very limited use. None of the available tests measures workability in terms of fundamental properties of fresh concrete. In addition, they are all single-point tests, and therefore cannot be expected to describe the properties of a material such as concrete which appears to follow a Bingham model (Section 8.1). The available tests cannot even easily be compared to one another, since they tend to measure somewhat different properties of the concrete.

Nevertheless, it *is* important to have some measure of the workability of concrete. Thus, until some better test procedures are developed, we must make do with one or more of the available test methods. Most of the available methods do, at least, provide information as to *variations* in workability for a given mix and are therefore useful as quality control measures. It might also be noted that none of the available test methods can be used over the whole range of very wet to very dry concrete; any individual test method can only be used over a relatively narrow range of workabilities.

The different measures of workability that have been developed over the years will be described under the following categories: (1) subjective assessment, (2) slump test, (3) compaction tests, (4) flow tests, (5) remolding tests, (6) penetration tests, (7) mixer tests, and (8) miscellaneous tests.

Subjective Assessment

The subjective assessment of fresh concrete (by an experienced worker) is, of course, the oldest "measure" of workability and is very widely used, at least as an adjunct to more quantitative tests. Some engineers still prefer to use their own judgment, based on the way the concrete behaves in the mixer and while being placed. Concrete may be described as being of high, medium, or low workability, or as being wet or dry, or plastic, and so on. Unfortunately, these terms mean different things to different people, so they may not be very helpful. Pity the poor engineer who had to work with this description of workability taken from an old textbook: "The mortar was wet enough to quake like liver under moderate ramming."

Slump Test

The slump test is by far the oldest and the most widely used test of workability. It first appeared as an ASTM Standard in 1922, and now is described in ASTM C143. The apparatus for this very simple test consists

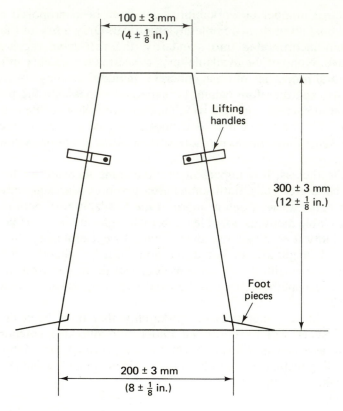

Figure 8.6 Slump cone.

primarily of a hollow mold in the form of a frustrum of a cone, with the dimensions shown in Figure 8.6. The mold is filled with concrete in three layers of equal volume; each layer is rodded 25 times with a 16-mm-diameter steel rod. The mold is then lifted away vertically, and the slump is measured by determining the difference between the height of the mold and the height of the concrete over the original center of the base of the specimen. (It should be noted that there are differences in the procedures for carrying out the slump test in different standards. According to BS 1881: Part 2, the specimens are filled in four lifts, and slump is measured to the *highest* part of the slumped concrete.) If a distinct shearing off of concrete from one side of the cone occurs, the test is disregarded and a new one carried out. If such shearing occurs consistently, this indicates that the test is not suitable for that particular concrete mix.

The slump test may be considered to be a measure of the shear resistance of concrete to flowing under its own weight. Depending on the mix, three distinct types of slump may occur, as shown in Figure 8.7. *"True"* slump consists of a general subsidence of the mass, without any breaking up. *Shear slump* often indicates a lack of cohesion; it tends to occur in harsh mixes, or in mixes prone to segregation. Shear slump may indicate that the concrete is not suitable for placement. On the other hand, normal mixes are sometimes found to display shear slump. *Collapse slump* generally indicates a lean, harsh, or more likely, a very wet mix.

There are a number of difficulties associated with the slump test. The test is completely empirical and is not related to our earlier definition of workability, which involved a measure of the amount of energy required to compact concrete. With different aggregates or mix properties, the same slump can be measured for very different concrete

Figure 8.7 Types of slump of concrete: (a) true slump; (b) shear slump; (c) collapse slump.

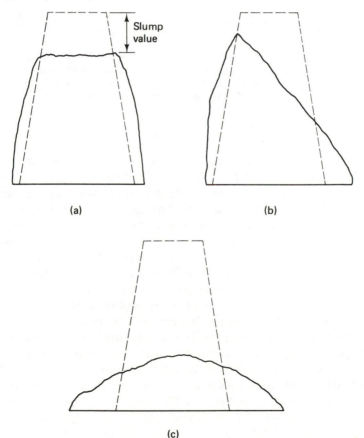

(a)   (b)

(c)

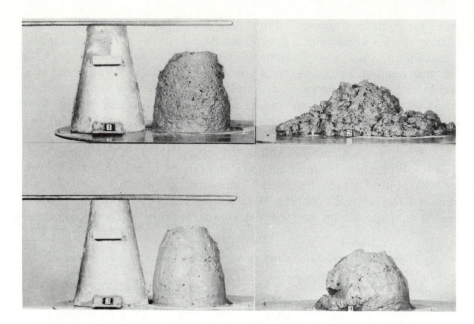

Figure 8.8 Slump test for consistency. By tapping the side of a slump specimen with the tamping rod (see the views at the right), additional information as to the workability of the concrete is obtained. (From *Concrete Manual,* 8th ed., U.S. Bureau of Reclamation, Denver, Colo., 1975.)

consistencies. This is illustrated in Figure 8.8; in the views at the right, the slumped concrete has been gently tapped on the side with the rod (according to the U.S. Bureau of Reclamation procedure) to provide additional information on workability. The upper concrete is a very harsh mix, with a minimum of fines and water. The lower concrete is plastic and cohesive. Clearly, these two concretes could not both be used for the same purposes, even though their slump values are the same. In addition, the slump test cannot differentiate between different low-workability concretes, which may all give "no slump." Concretes with slumps less than 25 mm should be tested by another procedure, preferably one involving vibration, since vibration will be needed to consolidate such stiff concrete. Although the test is reasonably reproducible with a skilled operator, it is quite sensitive to variations in test procedure, and it is not uncommon for different operators to achieve values more than 25 mm apart for the same concrete. As indicated earlier when describing the workability–time relationships, the slump may vary considerably depending on how long after mixing the test is carried out.

In spite of these limitations, slump tests can provide useful information. In general, concretes of similar slump *can* be used the same way. More important, the slump test is a valuable quality control tool. Changes in slump on a given job generally indicate that a change has occurred in the aggregates or in the amount of water or admixture being used. This should provide a warning that something is happening to the mix so that remedial action can be taken if necessary.

Compaction Tests

The strength of concrete is approximately proportional to the relative density. A test to measure the compactibility of concrete for a given amount of work more closely approaches our earlier definition of workability and can provide a very useful measure of concrete properties. A number of different compaction tests have been proposed, the most common being the compacting factor test.

Compacting Factor Test

The compacting factor test was developed in Great Britain in 1947 and is described in BS 1881: Part 2. The apparatus is shown in Figure 8.9. The upper hopper is completely filled with concrete, which is then successively dropped into the lower hopper and then into the cylindrical mold. The excess concrete is struck off, and the compacting factor is defined as the ratio of the concrete in the cylinder to the same concrete fully compacted in the cylinder.

Apart from the fact that the apparatus is not very suitable for field use, it has been found that some mixes stick to the sides of the hoppers (particularly when air-entrained) and must be rodded through. It has also been found that mixes with the same compacting factor do not necessarily require the same amount of work for compaction. Although some authorities feel that the test is suitable for very dry mixes (it is recommended in ACI 211, Recommended Practice for Proportioning No-Slump Concrete), others feel that it should be used only for "standard" mixes. Moreover, as the maximum size of aggregate increases, the size of the apparatus must be increased, making it impractical for aggregates greater than 40 mm (1½ in.) in size.

Flow Tests

Flow tests measure the ability of a concrete to flow under jolting or continuous vibration and provide information as to the tendency for segregation. A number of tests are available, but there are none recog-

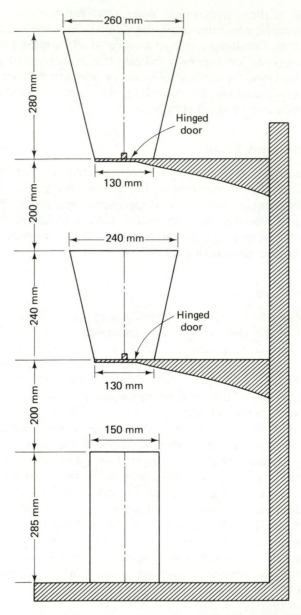

Figure 8.9 Compacting factor apparatus as specified in BS 1881: Part

nized by ASTM. However, until 1973, under ASTM C124, a flow-table test was described which is still sometimes used. In this test, a sample of concrete is cast in the form of a frustrum of a cone, 254 mm (10 in.) in diameter at the base, 171 mm (6.75 in.) in diameter at the top, and 127 mm (5 in.) high. This is done on a drop table, which is then dropped 15 times in 15 s through a height of 12.7 mm (½ in.) by revolving a cam shaft. The flow is defined as the increase in diameter expressed as a percentage of the original diameter. Again, concretes with the same flow may have quite different workabilities in the field. Similar tests are still in use in parts of Europe. For example, a flow test for concrete consistency is described in the German DIN 1048, Part 1, Chapter 3.1.2.

Remolding Tests

Remolding tests were developed to measure the work required to cause concrete not only to flow but also to conform to a new shape. They are intended to try to simulate, in the laboratory, actual field conditions. A number of tests were developed, but only the Vebe test is now used very much.

Vebe Test

The Vebe consistometer (Figure 8.10) was developed in 1940 and is probably the most suitable test for determining differences in consistency of very dry mixes. Although there is no ASTM standard for this test, it is one of the tests recommended in ACI 211. The test is widely used in Europe and is described in detail in BS 1881: Part 2. It is, however, only applicable to concretes with a maximum size of aggregate of less than 40 mm (1½ in.). A standard slump cone is cast, the mold removed, and a transparent disk placed on top of the cone. This is then vibrated at a controlled frequency and amplitude until the lower surface of the transparent disk is completely covered with grout. The time in seconds for this to occur is the Vebe time. The test is probably most suitable for concretes with Vebe times of from 5 to 30 s. The only real difficulty with the test is that the wetting of the disk with mortar is not uniform, and it may be difficult to pick out the end point of the test.

Thaulow Drop Table

This test is also suggested for no-slump concrete in ACI 211. It is similar to the Vebe test in that it involves changing the shape of a sample of concrete from a slump cone to a cylinder. A drop table is used and the number of 10-mm drops to achieve this remolding is counted.

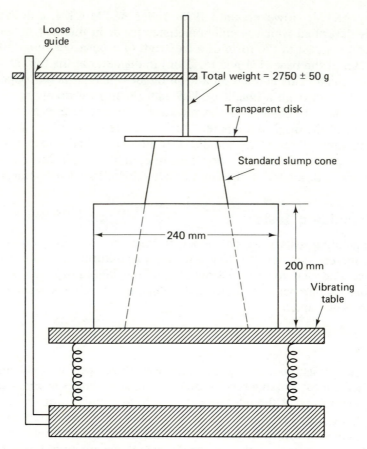

Figure 8.10 Vebe apparatus.

Penetration Tests

Penetration tests measure the depth of penetration of some type of indenter into concrete, under a variety of conditions. A number of tests have been proposed, but only one, the Kelly ball penetration test, is included in ASTM Standards (ASTM C360).

The *Kelly ball penetration test* uses the apparatus shown in Figure 8.11. It measures the depth of penetration into fresh concrete of a 152-mm (6-in.)-diameter hemisphere weighing 13.6 kg (30 lb). This test is very quick and can be performed on concrete in a buggy, open truck, or in the forms if they are not too narrow. It can be compared to the slump test as a convenient measure of the consistency of concrete.

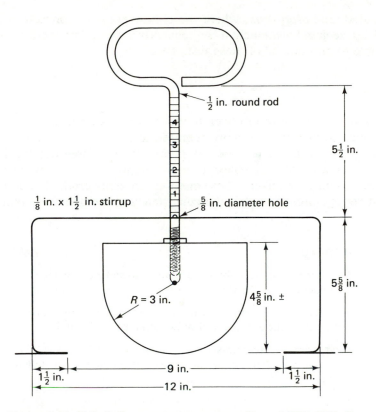

Figure 8.11 Kelly ball penetration apparatus. (From ASTM C360. Reprinted, with permission, from the American Society for Testing and Materials, 1916 Race Street, Philadelphia, PA 19103. Copyright.)

The *K-slump tester* involves pushing a slotted tube into fresh concrete. Slump is determined in terms of the amount of mortar that flows into the tube in 60 s. This test is not very effective in discriminating among stiff mixes.

Miscellaneous Tests

Other Tests

In addition to the tests described above, several dozen other tests have been proposed to measure workability. These include *deformation tests*, where the amount of work required to deform the concrete in a mold is measured, and *drop tests*, where the cohesiveness of concrete is measured by seeing how much it segregates on being dropped. It is not possible to describe all of these tests here.

Mixer Tests

As a practical measure of concrete workability, several tests have been developed to measure the work required to mix concrete in a concrete mixer. This can be done either by measuring the force exerted by the concrete on a vane in the mixer, or more simply by measuring the power required to turn the mixer. These methods are quite crude, but do enable the mixer operator to control the consistency at the point of mixing.

Summary

1. All of the tests described are empirical and do not measure in any fundamental way the rheological properties of concrete.

2. None of the tests will work for *all* concretes, and they may yield the same values for concretes of quite different workabilities.

3. Their primary usefulness is as a quality control measure *for a given concrete mix*; changes in the measurement may then indicate some change in the mix.

8.3 SETTING OF CONCRETE

Setting is defined as the onset of rigidity in fresh concrete. It is distinct from *hardening*, which describes the development of useful and measurable strength. Setting precedes hardening, but it should be emphasized that both are gradual changes which are controlled by the continuing hydration of the cement. We can view setting as a transitional period between states of true fluidity and true rigidity. The penetration tests used to measure the times of setting (which are described in Chapter 3) are purely arbitrary measurements. Figure 8.12 shows that initial set and final set, as measured by ASTM C403, do not correspond exactly to any specific change in concrete properties, although it is useful to consider that initial set represents approximately the time at which fresh concrete can no longer be properly handled and placed, while final set approximates the time at which hardening begins. Fresh concrete will have lost measurable slump prior to initial set, while measurable strength will be achieved sometime after final set.

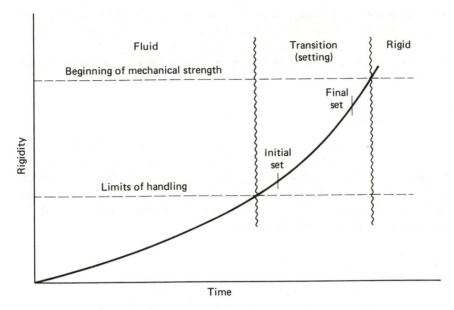

Figure 8.12 Process of setting and hardening.

Effect of Hydration on Setting

Role of C_3S

It should be remembered that the cement hydration begins as soon as water is added at the mixer. Setting is controlled primarily by the hydration of C_3S. The period of fluidity corresponds to the dormant period (stage 2) of C_3S hydration (Chapter 4). Setting occurs when the dormant period is terminated and rapid hydration of C_3S occurs in stage 3. Initial set corresponds approximately to the beginning of stage 3 and final set to its midpoint. Thus, initial set is marked by the beginning of a rapid temperature rise of the concrete, which will reach a maximum rate at final set. Setting is also accompanied by a decrease in electrical conductivity and an increase in the velocity of sound waves propagating through the paste. Measurements of either of these properties could form the basis of an adequate test for setting.

Role of C_3A and Gypsum

When using ordinary portland cements (the five ASTM types), C_3A plays a relatively minor role in determining setting behavior except in the cases of abnormal set discussed below. Gypsum is always interground with modern cements; it reacts with C_3A to help control the setting time. When large amounts of ettringite are formed rapidly, the aluminate phase will decrease initial and final set, as in the case of expansive cement. In the extreme case, regulated-set cements, for example, ettringite formation entirely controls setting.

Abnormal Setting Behavior

Abnormal setting of concrete was a troublesome problem in earlier times but is much rarer now. The different types of abnormal setting behavior are shown in Figure 8.13. They are most likely to be encountered under certain conditions when an admixture is used (usually a set-retarding admixture). Two major types of setting problems may be encountered; *false set* and *flash set*.

False Set

A concrete may stiffen rapidly a short time after mixing is completed (see Figure 8.13). Fluidity is restored by remixing and the concrete will then set normally; thus, false set has more nuisance value than anything else. This phenomenon is sometimes called *plaster set* because it is most often caused by crystallization of gypsum. When gypsum is interground with clinker, the material in the grinding mill can get quite hot because of the high energy input during grinding. The temperature can rise high enough (~120°C or ~250°F) to cause the gypsum to partially dehydrate to calcium sulfate hemihydrate (plaster) according to the first part of Eq. (3.5). When water is added, the hemihydrate will rehydrate back to gypsum and form a rigid crystalline matrix, but because there are only small amounts of plaster in the concrete, very little strength can actually develop and the plaster set can easily be disrupted by further mixing. The formation of the hemihydrate during grinding is minimized by cooling the grinding mills. A small amount of hemihydrate may still be formed but will mostly rehydrate while mixing is continuing and thus is harmless. However, a set-retarding admixture can also delay rehydration of the hemihydrate until after the mixing sequence is completed. In this way, a cement that normally behaves satisfactorily may exhibit false set.

False set may also be caused by the excessive formation of ettringite soon after mixing is completed. Ettringite has a crystal morphology similar to gypsum and thus can cause "plaster set" in an analogous way. Ettringite crystallizes most rapidly during the mixing period, but if its formation is delayed or extended in the presence of an admixture, false set can also occur. There is evidence (Chapter 7) that set-retarding admixtures can accelerate the initial hydration of C_3A, and thus cause

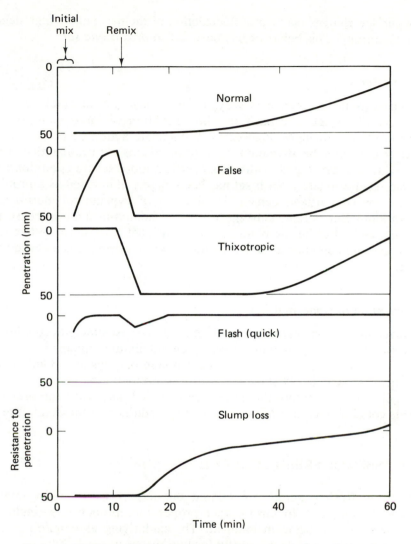

Figure 8.13 Diagrammatic sketches of different types of set of cement mortar, and slump loss of concrete. (Adapted from G. L. Kalousek, *Abnormal Set of Portland Cement. Causes and Correctives,* Rept. REC-OCP 69-2, U.S. Bureau of Reclamation, Denver, Colo., 1969.)

false set by increasing the formation of ettringite. In some high alkali cements the formation of syngenite ($KC\bar{S}_2H$) may cause false set.

It has been found that a concrete may sometimes show abnormal set even when the crystallization of gypsum, ettringite, or syngenite cannot be implicated. It is thought that an abnormally high concentration of surface charges may cause flocculation of the paste and a high degree of thixotropy. This behavior has been called *thixotropic set*.

Flash Set

If the C_3A in the cement is very reactive, flash set (or quick set) may occur. Flash set is caused by the formation of large quantities of monosulfoaluminate or other calcium aluminate hydrates. This is a rapid set that cannot be disrupted by further mixing, indicating that some strength has developed. Thus, flash set is a more severe condition than false set. Fortunately, flash set has been largely eliminated as a problem with normal portland cement by the use of gypsum to control C_3A hydration. But occasionally the use of an admixture may increase the hydration of C_3A to the point at which flash set may occur. When C_3A and gypsum contents are high, the formation of ettringite can cause flash set.

Prevention of Abnormal Set

Correcting a problem of abnormal set may be most simply accomplished by merely changing to another equivalent admixture, eliminating the use of the admixture, or changing the amount of gypsum added to the cement. Research has shown that a little prehydration of cement (i.e., principally C_3A) before the admixture is added may solve the problem. This could be most easily done by delayed addition of the admixture.

8.4 TESTS OF FRESH CONCRETE

Tests on fresh concrete are carried out largely as a quality control measure and to help ensure that proper mix proportions (including admixtures) are being maintained. The underlying assumption is that tests on fresh concrete are useful for two reasons:

1. They permit some estimation of the subsequent behavior of the hardened concrete.

2. *Changes* in the properties of fresh concrete imply that the concrete mix is changing, so that remedial action can be taken if necessary.

By far the most common tests on fresh concrete are slump tests (or some other workability tests), as described. We will here consider the other properties of fresh concrete that may be determined.

Sampling Fresh Concrete

Concrete is a manufactured material, made from cement, aggregate, water, and admixtures. All of these ingredients are subject to variation, both in quantity and quality. Therefore, the first consideration in testing fresh concrete is that the concrete tested be *truly representative* of the batch. This is, of course, very much a matter of judgment and experience; it is difficult to lay down hard and fast rules to define "representative." Also, samples must be taken reasonably often; a common requirement is that at least one sample must be taken for every 115 m³ (150 yd³) of concrete cast.

A typical specification for sampling fresh concrete is the one described in ASTM C172. This covers sampling from stationary and truck mixers, and from equipment used to transport central-mixed concrete. The specification requires that tests be done on *composite* samples, that is, samples taken at several points during the discharge of the concrete and then combined. The main problem in collecting these samples is to ensure that they are not contaminated by the forms or subgrade and that no segregation occurs during sampling. It is also important to carry out the tests as quickly as possible once the sample is obtained (usually within 15 min), as workability, in particular, quickly changes with time. A minimum sample size of 1 ft³ (0.03 m³) is required for strength tests, but smaller samples are permitted for air content and slump tests. Similar requirements are specified in BS 1881: Part 1.

It is important not only to maintain reasonable uniformity between different batches of concrete but also to ensure that the variability of concrete from different parts of the same batch ("within-batch" variation) is low. This becomes significant when using older or worn equipment or where there is inadequate equipment maintenance. There are stringent requirements for the permissible variation of ready-mixed concrete in ASTM C94. Samples are taken from two locations in the batch, between the discharge of about 15 and 85% of the concrete. These samples are tested separately and the uniformity requirements are given in Table 8.1.

It is, of course, normal to expect some variability in concrete tests and quality, and this will be discussed in detail in Chapter 17. However, it may be stated here as a general rule that the less variability there is in concrete properties, the better.

Time of Setting

Although the time of setting of concrete is an arbitrary measurement (Chapter 3), it is an important parameter for a number of reasons. It is used:

Table 8.1

Requirements for Uniformity of Concrete (ASTM C94)[d]

Test	Requirement, Expressed at Maximum Permissible Difference in Results of Tests of Samples Taken from Two Locations in the Concrete Batch
Weight per cubic foot (weight per cubic metre) calculated to an air-free basis, lb/ft³ (kg/m³)	1.0 (16)
Air content, vol. % of concrete	1.0
Slump	
If average slump is 4 in. (102 mm) or less, in. (mm)	1.0 (25)
If average slump is 4 to 6 in. (102 to 152 mm), in. (mm)	1.5 (38)
Coarse aggregate content, portion by weight of each sample retained on No. 4 (4.75-mm) sieve, %	6.0
Unit weight of air-free mortar[a] based on average for all comparative samples tested, %	1.6
Average compressive strength at 7 days for each sample,[b] based on average strength of all comparative test specimens, %	7.5[c]

[a]"Test for Variability of Constituents in Concrete," Designation 26, *Concrete Manual*, 8th Ed, U.S. Bureau of Reclamation, Denver, Colo., 1975.

[b]Not less than three cylinders will be molded and tested from each of the samples.

[c]Tentative approval of the mixer may be granted pending results of the 7-day compressive strength tests.

[d]Reprinted, with permission, from the American Society for Testing and Materials, 1916 Race Street, Philadelphia, PA 19103. Copyright.

1. To help regulate the times of mixing and transit.

2. To gauge the effectiveness of various set-controlling admixtures (either retarding or accelerating).

3. To help plan the scheduling of finishing operations.

As we have seen, there are two tests in use for the setting time of cement, but unfortunately these do not correlate with the setting times of concrete, and this is the reason for the current test procedure outlined in ASTM C403. This penetration test is limited to concretes whose slump is greater than zero, and to cases "when tests of the mortar fraction of the concrete will provide the information required."

Briefly, the test consists of removing the mortar fraction of the concrete by passing it through a No. 4 (4.75-mm) sieve, rodding the mortar into containers, and then measuring the force required to cause a needle to penetrate 25.4 mm (1 in.) into the mortar. Removable needles with bearing areas ranging from 645 to 16 mm^2 (1 to 0.025 in.^{2}) are used. This is done at regular intervals, and a curve of penetration resistance versus time is plotted. Times of initial set and final set are defined as the times at which the penetration resistances are 3.5 MPa (500 lb/in.^{2}) and 27.6 MPa (4000 lb/in.^{2}), respectively. Although these are arbitrarily chosen points, 3.5 MPa corresponds approximately to the point at which the concrete will no longer become plastic under vibration; 27.6 MPa corresponds to a concrete strength of about 700 kPa (100 lb/in.^{2}).

Air Content

Air-entraining agents have already been discussed in Chapter 7, and the mechanisms of air entrainment will be considered in Chapter 20. This section deals only with the techniques available to measure the amount of air in concrete. It must be remembered that in an air-entrained concrete, there will be two types of air, entrained air and entrapped air. The test methods described below are unable to distinguish between entrained air and entrapped air; they simply measure the *total* air content. Also, the air content is expressed in terms of the volume of concrete; clearly, however, air is present only in the cement paste. It should be remembered that we have already described a test for determining the air-entraining potential of cement. Unfortunately, an air-entraining cement meeting those requirements may still not give the required air content in concrete.

There are three principal tests available to measure the total air content of concrete: the *gravimetric, volumetric,* and *pressure* methods.

Gravimetric Method

The gravimetric method of determining the air content (ASTM C138 or BS 1881: Part 2) is the oldest and simplest method. It consists, basically, of comparing the unit weight of concrete containing air with the *calculated* unit weight of air-free concrete, computed from the proportions and specific gravities of the mix components. The air content, A, is then calculated as

$$A = \frac{T - W}{T} \times 100 \qquad (8.3)$$

where T is the theoretical weight of the concrete on an air-free basis and W is the unit weight of concrete. This method can also be used to determine the yield (i.e., the volume of fresh concrete produced from a known quantity of ingredients) and the cement content, as well as the unit weight. This test is not suitable for field use, since it requires an accurate knowledge of the specific gravities of the ingredients. Also, the batch proportions and moisture content of the aggregates must be accurately known. It may not be used with lightweight aggregates, since their specific gravity is very difficult to determine accurately. However, the test is suitable for laboratory work, where the other parameters may be measured accurately.

Volumetric Method

The volumetric method (ASTM C173) is based on comparing the volume of fresh concrete containing air with the volume of the same concrete after the air has been expelled by agitating the concrete under water. An apparatus similar to that shown in Figure 8.14 is used.

A sample of fresh concrete is rodded into the measuring bowl having a volume of at least 0.07 ft³ (0.002 m³). Excess concrete is struck off, the top section clamped in place, and water is added to the zero mark. The unit is then inverted and agitated vigorously until all the air seems to have been removed from the concrete and has risen to the top of the apparatus. One cup of isopropyl alcohol is then added to dispel bubbles, and a direct reading of the air content can be made by reading the amount of liquid remaining in the neck of the apparatus (correcting, of course, for the amount of isopropyl alcohol added). The chief difficulty of this test is the large amount of physical effort required to remove the air from the concrete.

Pressure Method

The pressure method (ASTM C231 or BS 1881: Part 2) is the most common method for measuring the air content of fresh concrete in the field. The basis of the method is the measurement of the change in volume of the concrete when subjected to a given pressure. This change in volume is assumed to be caused entirely by compression of the air, and the principle of Boyle's law can then be used to calculate the air content.

An apparatus similar to that shown in Figure 8.15 is used. The concrete is consolidated into the measuring bowl, struck off level to the rim, and the apparatus assembled. Water is added to the standpipe to the zero mark, and the test pressure applied. (Each apparatus is calibrated for a specific pressure.) The water level is read, the pressure released,

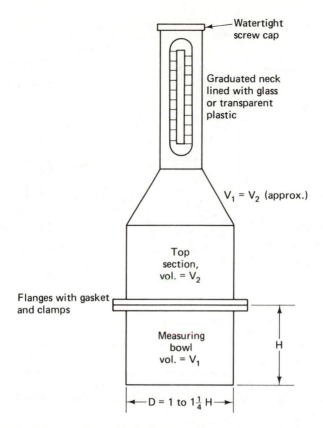

Figure 8.14 Apparatus for measuring air content of fresh concrete by volumetric method. (From ASTM C193. Reprinted, with permission, from the American Society for Testing and Materials, 1916 Race Street, Philadelphia, PA 19103. Copyright.)

and the new water level read. This gives the apparent air content, A_1. (Similarly, the air content of the aggregate itself, A_2, is determined for porous aggregates.) Then the true air content is simply $A_1 - A_2$. The chief advantage of this method is that it is not necessary to determine the specific gravities, moisture contents, or mix proportions of the components. If the air content of the concrete exceeds the range of the apparatus, the determination can be repeated at a lower test pressure.

Other Tests

The previous three methods are the only ones specified by ASTM. However, a number of variations on the volumetric and pressure methods are available. In addition, a pocket-size indicator (Chace Air Meter) is

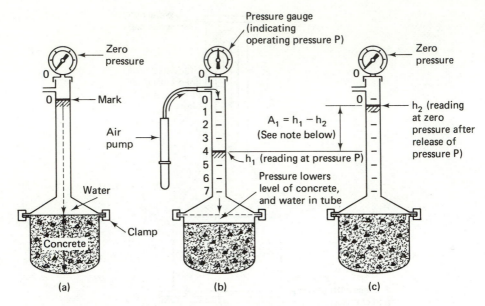

Note: $A_1 = h_1 - h_2$ when bowl contains concrete, as shown in
this figure; when bowl contains only aggregate and water, $h_1 - h_2 = G$
(aggregate correction factor). $A_1 - G = A$ (entrained air content of concrete)

Figure 8.15 Pressure method for air content, Type A meter. (From ASTM C231. Reprinted, with permission, from the American Society for Testing and Materials, 1916 Race Street, Philadelphia, PA 19103. Copyright.)

available which measures the volumetric displacement of air from the mortar by isopropyl alcohol. This method, although quick and easy, is not very accurate or reliable, but may be used as a rapid field check.

It should be noted that the foregoing tests measure only the total volume of air in the concrete. They yield no information about the nature of the air-void system. However, as we have seen in Chapter 7, it is the nature of the air-void system (i.e., the pore size and pore spacing) that determines the durability of the concrete. To determine these, it is necessary to examine the void system of the hardened concrete itself. This may be done using the procedures described in ASTM C457 (see section 7.3).

Unit Weight and Yield

The density (unit weight) of the fresh concrete can be determined by weighing a known volume of concrete. It is usual to weigh the sample of concrete used for the air measurement immediately before the air content is determined, since this is a known volume of concrete. The presence of entrained air affects the unit weight, since air contributes to volume but not to weight. The unit weight can be used as an approximate indication of air content for concretes made with the same materials.

The unit weight can also be used to determine the volume of the concrete batch:

$$V = \frac{w}{\text{unit weight}} \; \text{m}^3 \, (\text{ft}^3) \tag{8.4}$$

where w is the combined weight of the concrete constituents, including water. Thus, the *yield* of the concrete can be determined:

$$Y = \frac{V}{w_{\text{cem}}} \; \text{m}^3/\text{kg} \, (\text{ft}^3/\text{lb}) \tag{8.5}$$

This tells us how much concrete can be made using a given weight of cement.

Rapid Analysis of Fresh Concrete

The tests that we have described so far are really indirect methods of trying to ascertain that the mix proportions in the concrete being tested are really the same as those specified. Clearly, the quality control requirements would be largely met if we could, in fact, determine the proportions of *all* the ingredients in the mix. Then a simple strength test at some age (say, 7 days) would suffice to ensure that some unforeseen accident was not preventing the concrete from hardening properly. In fact, even a test to measure directly the two most important parameters, the cement content and the w/c ratio, would be very useful. However, there is as yet no test method available which provides sufficient accuracy but is still relatively easy to carry out. Several test methods have been proposed. Although none has as yet been adopted as an ASTM Standard, a test method is described in BS 1881: Part 2. In this test the aggregate is separated from the paste by washing the fresh concrete through a set of sieves. If the specific gravity, water absorption, and grading of the fine and coarse aggregates are known, then the water content, cement content, and the other mix proportions can be determined gravimetrically.

Another method under consideration is the Kelly-Vail chemical technique. This is a method of determining the water and cement contents of fresh concrete within 15 min of sampling. A sample of fresh concrete is washed over a nest of sieves until free of cement, using a known volume of wash water. A sample of the cement suspension in water is then obtained and its calcium content determined by flame photometry. The cement content can then be obtained from a previously constructed calibration curve. To obtain the water content, the principle employed is that the water in fresh concrete is all available for intermixing with added aqueous solutions. Two samples of concrete are taken, and a sodium chloride solution is added to one of them. After mixing, the chloride content of the liquid in each sample is determined by titration, and the water content of the concrete can be determined from a comparison of the chloride contents. This test seems to give quite good results, but some types of calcareous aggregate may cause an error, as may variations in the calcium content of the cement used.

A number of similar tests exist, which basically wash the cement out of a concrete sample and determine its quantity. However, these tests are not, as yet, considered to be fully reliable. In addition, a number of other techniques have been proposed which do not require a separation of the constituents of the concrete:

1. *Thermal conductivity tests.* The higher the water content, the slower the temperature rise.

2. *Capacitance tests.* The higher the water content, the greater the dielectric constant.

3. *Electrical resistance tests.* The electrical resistance of fresh concrete is inversely proportional to the water content; this is complicated by the various soluble salts that may be present.

4. *Nuclear methods.* X-ray, gamma ray, and neutron activation analysis can be used to measure the cement or water contents of the fresh concrete.

None of these methods measures the relative quantities of fine and coarse aggregates; these must be determined separately.

In addition, no reliable test is yet available to determine the type and amount of admixtures present (which is complicated further by the fact that the exact chemical composition of many admixtures is a closely guarded secret). Thus, although the rapid analysis techniques can provide some useful information on cement and water contents, they are not yet sufficiently developed to allow the replacement of the indirect tests described earlier.

Bibliography

Concrete Manual, 8th ed. U.S. Bureau of Reclamation, Denver, Colo., 1975.

POWERS, T. C., *The Properties of Fresh Concrete,* Chaps. 4, 10. John Wiley & Sons, Inc., New York, 1968.

Smith, D. T., "Uniformity and Workability," in *Significance of Tests and Properties of Concrete and Concrete-making Materials,* ASTM STP 169B, pp. 74–101. American Society for Testing and Materials, Philadelphia, Pa., 1978.

Tattersall, G. H., *The Workability of Concrete.* Cement and Concrete Association, Wexham Springs, Slough, U.K., 1976.

Problems

8.1 Discuss the implications of the one-point workability tests currently in use.

8.2 Why is the slump test universally used in North America?

8.3 What is slump loss and what factors affect it?

8.4 Without changing the water content, what other strategies could you use to change the workability of concrete?

8.5 What are the practical consequences of bleeding?

8.6 Describe the process of setting.

8.7 Do measurements of initial and final set correspond to fundamental changes in the hydration process?

8.8 Compare false set, flash set, and plaster set.

8.9 What is the importance of the sampling procedures used to obtain fresh concrete for testing?

8.10 Would it be desirable to have an air-content test that distinguishes between entrained air and entrapped air?

9

proportioning concrete mixes

The proportioning of concrete mixtures, more commonly referred to as *mix design,* is a process that consists of two interrelated steps: (1) selection of the suitable ingredients (cement, aggregate, water, and admixtures) of concrete, and (2) determining their relative quantities ("proportioning") to produce, as economically as possible, concrete of the appropriate workability, strength, and durability. These proportions will depend on the particular ingredients used, which will themselves depend on the application. Other criteria, such as designing to minimize shrinkage and creep or for special chemical environments, may also be considered. However, although a considerable amount of work has been done on the theoretical aspects of mix design, it still remains largely an empirical procedure. And, although many concrete properties are important, most design procedures are based primarily on achieving a specified compressive strength at some given workability and age; it is assumed that if this is done, the other properties (except perhaps resistance to freezing and thawing, or other durability problems, such as resistance to chemical attack) will also be satisfactory. But before turning to the mix design methods now in common use, it is worthwhile examining the basic design considerations themselves in more detail.

9.1 BASIC CONSIDERATIONS

Economy

The cost of concrete is made up of the costs of materials, labor, and equipment. However, except for some special concretes, the costs of labor and equipment are largely independent of the type and quality of concrete produced. It is therefore the material costs that are most important in determining the relative costs of different mix designs. Since cement is much more expensive than aggregate, it is clear that minimizing the cement content is the most important single factor in reducing concrete costs. This can, in general, be done by using the lowest slump that will permit adequate placement, by using the largest practical maximum size of aggregate, by using the optimum ratio of coarse to fine aggregates, and, where necessary, by using appropriate admixtures. It should be noted here that in addition to cost, there are other benefits to using a low cement content; shrinkage will in general be reduced and there will be less heat of hydration. However, if the cement contents are too low, they will diminish the early strength of the concrete and will make uniformity of the concrete a more critical consideration.

The economy of a particular mix design should also be related to the degree of quality control that can be expected on a job. As will be discussed in Chapter 17, the mean concrete strength must be higher than the specified minimum compressive strength because of the inherent variability of concrete. At least on small jobs, it may be cheaper to "overdesign" the concrete than to provide the extensive quality control that would be required with a more cost-efficient concrete.

Workability

Clearly, a properly designed mix must be capable of being placed and compacted properly *with the equipment available*. Finishability must be adequate, and segregation and bleeding should be minimized. As a general rule, the concrete should be supplied at the *minimum* workability that will permit adequate placement. The water requirement for workability depends mostly on the characteristics of the aggregate rather than those of the cement. Where necessary, workability should be improved by redesigning the mix to increase the mortar content rather than by simply adding more water or more fine material. Thus, cooperation between the mix designer and the contractor is essential to ensure a good concrete mix. In some cases, a less economical mix may be the best solution. A deaf ear should be turned to the frequent pleas from any job site for "more water."

Strength and Durability

In general, concrete specifications will require a minimum compressive strength. They may also impose limitations on the permissible *w/c* ratios and minimum cement contents. It is important to ensure that these requirements are not mutually incompatible. As we will see in Chapter 17, it is not necessarily the 28-day strength that is most important; strengths at other ages may control the design.

Specifications may also require that the concrete meet certain durability requirements, such as resistance to freezing and thawing, or chemical attack. These considerations may provide further limitations on the *w/c* ratio or cement content and in addition may require the use of admixtures.

The process of mix design, then, involves the satisfactory resolution of all the requirements described above. Since these requirements cannot all be optimized simultaneously, some compromises (as between strength and workability) will be necessary. It must be remembered that even a "perfect" mix will not perform properly unless the proper placing, finishing, and curing procedures are carried out.

9.2 FUNDAMENTALS OF MIX DESIGN

There have been two aspects of mix design in which most of the theoretical work has been carried out: water content and aggregate grading. Most of the modern empirical design methods depend heavily on these two considerations.

Water/Cement Ratio

In 1919, Duff Abrams enunciated his water/cement ratio law for the strength of concrete: "For given materials, the strength depends only on one factor—the ratio of water to cement." [1] This can be expressed by a formula in the form

$$\sigma_c = \frac{A}{B^{1.5(w/c)}} \tag{9.1}$$

where σ_c = compressive strength at some fixed age, A = empirical constant, B = constant that depends mostly on the cement properties, and w/c = water/cement ratio by weight. Usually, A is taken to be 14,000 lb/in.^{2}; B depends on the type of cement, but may be taken to be about

---

[1] Feret in 1896 had formulated a more general "law," expressing concrete strength in terms of the cement/total voids ratio, of which Abrams' law is a special case.

4. This observation, that strength is inversely proportional to the w/c ratio, remains the basis for most mix design procedures. The reason for this was shown by Powers and Brownyard[2] to be the fact that the w/c ratio determines the porosity of the cement paste. This was discussed in terms of the gel/space-ratio concept in Chapter 4. Of course, Eq. (9.1) must be normalized in terms of the maturity (degree of hydration) of the cement. It might be noted as well that since the w/c ratio controls the porosity, it also largely controls the durability of the concrete, as discussed in Chapter 20.

Ideal Aggregate Grading

Theoretical work on the "best" aggregate grading was based on the packing characteristics of granular materials. After a considerable amount of experimental work, Fuller and Thompson in 1907 determined that their "ideal" grading curve (one giving minimum voids) could be represented closely by the expression

$$P_t = \left(\frac{d}{D}\right)^{1/2} \tag{9.2}$$

where P_t = fraction of total solids finer than size d
$\quad\quad D$ = maximum particle size

This was later generalized to the expression

$$P_t = \left(\frac{d}{D}\right)^{q} \tag{9.3}$$

where q may have values between 0 and 1.

It may be recognized that these relationships are *parabolic,* hence the general name "parabolic gradings" for these types of grading curves. It may be shown that when *all* aggregate sizes below D are present, the voids content depends only on q, approaching zero as q approaches zero. Thus, by choosing an arbitrarily low q, an arbitrarily dense grading could be obtained. The assumption was that denser gradings (with the

---

[2]T. C. Powers and T. L. Brownyard, "Studies of the Physical Properties of Hardened Portland Cement Paste," Bulletin 22, Research Laboratories of the Portland Cement Association, 1948.

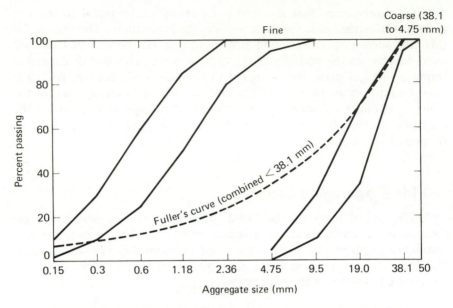

Figure 9.1 Grading curves indicating the specified grading limits (ASTM C33) for fine aggregate and one size of coarse aggregate. The corresponding Fuller–Thompson "ideal" grading curve is also shown.

minimum value of voids requiring the minimum cement content) would produce both more economical and better concrete.

However, in practice, these parabolic gradings simply do not work. From workability considerations, a certain proportion of fine material is required. From 2 to 10% of the fine aggregate must pass the No. 100 sieve, and 10 to 30% must pass the No. 50 sieve. Thus, the recommended grading curves for fine and coarse aggregates (ASTM C33, CSA 23.1) are not truly "parabolic" (Figure 9.1). In addition, Eq. (9.3) does not really work either. The problem is that very fine particles cannot form densely packed aggregates; the packing density decreases as the average particle size decreases. The lowest practical value of q is approximately ½.

Thus, we have now largely given up the idea that there is some "ideal" aggregate grading. The grading limits shown in Figure 9.1 are based on practical experience rather than on theory. In fact, it is possible to make a satisfactory concrete from almost any type of aggregate grading, although gradings outside the limits of Figure 9.1 may be uneconomical and difficult to handle with regard to segregation, finishing, and so on.

9.3 ACI METHOD OF MIX DESIGN

There are a number of different methods of mix design available. Although they are not directly comparable, they do give approximately the same relative proportions of materials, and all are capable of yielding suitable concrete mixes. The most common method used in North America is that established by ACI Recommended Practice 211.1, and this method will be described here in some detail. It must be remembered, however, that any mix design method will provide only a first approximation of proportions. These must be checked by trial batches in the laboratory or in the field and can then be adjusted as necessary to produce the desired concrete characteristics. With any given set of materials, it may be found that considerable deviations from the ACI recommended practice may be necessary. Once sufficient experience with local materials is acquired, the ACI method should be modified to take their properties into account.

As was stated earlier, the job specifications may dictate certain mix requirements, such as minimum cement contents and w/c ratios, slump, air content, maximum aggregate size, strength, the use of admixtures, or other special requirements. But regardless of the specification requirements, the establishment of the batch weights [note that concrete should always be batched by *weight* (Chapter 10) rather than by volume] can best be accomplished by following the sequence of steps laid out below. This will ensure that the characteristics of the available materials are properly considered in combining them into a suitable concrete mixture. In summary, the mix design process consists of (1) determining the job parameters—aggregate properties, maximum aggregate size, slump, w/c ratio, admixtures; (2) calculation of batch weights; and (3) adjustments to the batch weights based on a trial mix made according to these calculations.

Mix Design Procedures

1. Required material information. Before starting the mix design process, the following raw material properties should be determined: sieve analyses of both the fine and coarse aggregates, unit weight of the coarse aggregate, bulk specific gravities, and absorption capacities of the aggregates.

2. Choice of slump. Usually, slump will be specified for a particular job, to take into account the anticipated methods of handling and placing the concrete. However, where the slump has not been specified, appropriate

Table 9.1

Recommended Slumps for Various Types of Construction[a]

	Slump (in.)		Slump (mm)	
Types of Construction	*Max.*[b]	*Min.*	*Max.*[b]	*Min.*
Reinforced foundation walls and footings	3	1	80	20
Plain footings, caissons, and substructure walls	3	1	80	20
Beams and reinforced walls	4	1	100	20
Building columns	4	1	100	20
Pavements and slabs	3	1	80	20
Mass concrete	3	1	80	20

[a]From ACI 211.1. Reproduced with permission.

[b]May be increased 1 in. (25 mm) for methods of consolidation other than vibration.

values can be chosen from Table 9.1, which applies when the concrete is to be consolidated by vibration. As a general rule, the lowest slump that will permit adequate placement should be selected.

3. Maximum aggregate size. Generally, the largest maximum size of aggregate available (and consistent with the limitation listed below) should be used, as this will minimize the required cement content. The limitations on maximum aggregate size (see section 6.1) are:

(a) For reinforced (or prestressed) concrete, the maximum size may not exceed one-fifth of the minimum dimension between forms, or three-fourths of the minimum clear spacing between bars, strands, bundles of bars, or between the steel and the formwork.

(b) For slabs on grade, the maximum size may not exceed one-third the slab depth.

If it is shown by experience that it is possible to place the concrete without honeycombing or voids, these requirements may be relaxed. It has also been found that, at a given w/c ratio, higher strengths can be achieved with smaller maximum sizes of aggregates. There is thus a trend toward the use of reduced maximum aggregate sizes. In many areas, the largest sizes available are 20 or 25 mm (¾ in. or 1 in.). It should also be remembered that, for a given job, it may be unwise to recommend different maximum aggregate sizes for different parts of the structure, as this may lead to confusion and increases the probability of error.

4. Estimation of mixing water and air content. As we have seen in Chapter 8, the workability of concrete is dependent primarily on the paste content of the concrete; the amount of entrained air; and on the maximum size, grading, and particle shape of the aggregate. An estimate of the water requirement to produce different slumps for both air-entrained and non-air-entrained concrete can be obtained from Table 9.2, which is based on experience obtained over many years of practice. It is better to establish these numbers from experience with the actual materials in question rather than resorting to the use of Table 9.2. This table also shows the approximate amount of entrapped air to be expected in non-air-entrained concrete and gives the recommended levels of air entrainment (when required) for different maximum sizes of aggregate and for three different levels of severity of exposure. Since there is frequent pressure from the job site for concrete with "more slump" (which is often obtained by adding more water rather than by adding more paste, leading to a higher *w/c* ratio and lower strength), it is suggested that the trial batches of the mix used to develop the strength relationships be made to represent the most unfavorable combination of air content and water content. That is, both the maximum allowable air content and the maximum allowable slump should be used in these batches rather than using average values. This will help prevent overestimating the strength.

5. Water/cement ratio. The selection of the appropriate *w/c* ratio may be governed not only by strength but also by durability requirements.

(a) *Strength;* In the absence of strength vs. *w/c* ratio data for the specific materials, a conservative estimate can be made for the expected 28-day compressive strength from Table 9.3. Of course, it is possible that the specifications may be based on a required strength at a time other than 28 days (e.g., at the time when forms are to be stripped), or the design may require the use of high-early-strength (Type III) or low-heat cements (Type IV). In these cases, Table 9.3 is not applicable. The designer must develop his or her own data for these cases, or when the design is governed by a flexural strength requirement. It is always more desirable to develop the appropriate strength–time–*w/c* ratio relationships for the materials that are actually to be used on the job. In this way, the effects of admixtures can also be determined.

(b) *Durability:* If there are severe exposure conditions, such as freezing and thawing, or exposure to seawater, or sulfates, the more severe *w/c* ratio requirements of Table 9.4 may govern. It might be noted that other standards may have somewhat different requirements. For instance, CSA Standard CAN3-A23.1-M77 provides a more detailed guide to the selection of maximum permissible *w/c* ratios for severe exposure conditions. These requirements, and the corresponding

Table 9.2

Approximate Mixing Water and Air Content Requirements for Different Slumps and Nominal Maximum Sizes of Aggregates[a,b]

		Water, lb/yd³(kg/m³) of Concrete for Indicated Nominal Maximum Sizes of Aggregate												
Slump		⅜ in.	(10 mm)	½ in.	(12.5 mm)	¾ in.	(20 mm)	1 in.	(25 mm)	1½ in.	(40 mm)	2 in.	(50c mm)	3c in.	(70c mm)
in.	mm														
		Non-air-entrained Concrete													
1–2	30–50	350	(205)	335	(200)	315	(185)	300	(180)	275	(160)	260	(155)	240	(145)
3–4	80–100	385	(225)	365	(215)	340	(200)	325	(195)	300	(175)	285	(170)	265	(160)
6–7	150–180	410	(240)	385	(230)	360	(210)	340	(205)	315	(185)	300	(180)	285	(170)
Approximate amount of air in non-air-entrained concrete, %		3		2.5		2		1.5		1		0.5		0.3	
		Air-entrained Concrete													
1–2	30–50	305	(180)	295	(175)	280	(165)	270	(160)	250	(145)	240	(140)	225	(135)
3–4	80–100	340	(200)	325	(190)	305	(180)	295	(175)	275	(160)	265	(155)	250	(150)
6–7	150–180	365	(215)	345	(205)	325	(190)	310	(185)	290	(170)	280	(165)	270	(160)
Recommended average total air content, %, for level of exposure:															
Mild		4.5		4.0		3.5		3.0		2.5		2.0		1.5d	
Moderate		6.0		5.5		5.0		4.5		4.5		4.0		3.5d	
Extreme[e]		7.5		7.0		6.0		6.0		5.5		5.0		4.5d	

[a]Adapted from ACI 211.1. Reproduced with permission.

[b]These quantities of mixing water are for use in computing cement factors for trial batches. They are maxima for reasonably well-shaped, angular, coarse aggregates graded within limits of accepted specifications.

[c]The slump values for concrete containing aggregate larger than 1½ in. (40 mm) are based on slump tests made after removal of particles larger than 1½ in. (40 mm) by wet screening.

[d]For concrete containing large aggregates which will be wet-screened over the 1½ in. (40 mm) sieve prior to testing for air content, the percentage of air expected in the 1½ in. (40 mm)-minus material should be as tabulated in the 1½ in. column. However, initial proportioning calculations should include the air content as a percent of the whole.

[e]These values are based on the criterion that 9% air is needed in the mortar phase of the concrete. If the mortar volume will be substantially different from that determined in this recommended practice, it may be desirable to calculate the needed air content by taking 9% of the actual mortar volume.

Table 9.3

Relationships between Water/Cement Ratio and Compressive Strength of Concrete[a]

| | Water/Cement Ratio, by Weight | |
Compressive Strength at 28 Days[b]	Non-air-entrained Concrete	Air-entrained Concrete
6000 lb/in.²	0.41	—
5000 lb/in.²	0.48	0.40
4000 lb/in.²	0.57	0.48
3000 lb/in.²	0.68	0.59
2000 lb/in.²	0.82	0.74
45 MPa	0.37	—
40 MPa	0.42	—
35 MPa	0.47	0.39
30 MPa	0.54	0.45
25 MPa	0.61	0.52
20 MPa	0.69	0.60
15 MPa	0.80	0.71

[a]Adapted from ACI 211.1. Reproduced with permission.

[b]Values are estimated average strengths for concrete containing not more than the percentage of air shown in Table 9.2. Strength is based on 6 × 12 in. (150 × 300 mm) cylinders moist-cured in accordance with ASTM C31.

Table 9.4

Maximum Permissible Water/Cement Ratios for Concrete in Severe Exposures[a,b]

Type of Structure	Structure Wet Continuously or Frequently, and Exposed to Freezing and Thawing[c]	Structure Exposed to Seawater or Sulfates
Thin sections (railings, curbs, sills, ledges, ornamental work) and sections with less than 1 in. cover over steel	0.45	0.40[d]
All other sections	0.50	0.45[d]

[a]From ACI 211.1. Reproduced with permission.

[b]Based on report of ACI Committee 201, Durability of Concrete in Service.

[c]Concrete should also be air-entrained.

[d]If sulfate-resisting cement (Type II or Type V of ASTM C150) is used, permissible water/cement ratio may be increased by 0.05.

ranges of air content, are shown in Tables 9.5 and 9.6. Table 9.7 shows the CSA requirements for concretes exposed to different concentrations of sulfates in soils or groundwaters, which include a specification of the type of cement to be used (see Chapter 20).

6. Calculation of cement content. Once the water content and w/c ratio (steps 4 and 5) are determined, the amount of cement per unit volume of concrete is determined simply by dividing the estimated water requirement by the w/c ratio. However, many specifications, in addition, require a *minimum cement content*. Such a requirement may be used to ensure satisfactory finishability, quality of vertical surfaces, or workability; it may also ensure against low strengths due to increased water demands at the job site. In the absence of suitable test data, or when the w/c ratio–strength relationships are not available or are inadequate, the minimum cement contents such as those in Table 9.8 may be used, but only for concretes with a specified compressive strength less than 25 MPa (3600 lb/in.^{2}).

Table 9.5

Maximum Permissible Water/Cement Ratios (Normal-Weight Concrete) and Minimum Specified Compressive Strengths (Structural Lightweight Concrete) for Different Types of Structures and Varying Degrees of Exposure[a]

Class of Exposure	Condition of Exposure	Normal-Weight Concrete Maximum Water/Cement Ratio	Structural Lightweight Concrete Minimum Specified 28-Day Compressive Strength (MPa)	Typical Examples
A	Concrete subjected to frequent cycles of freezing and thawing in a saturated condition in seawater or subjected to applications of de-icing chemicals	0.45	35	Structures exposed at the waterlines of seawater bridge abutments, locks, and docks; surfaces subjected to de-icing chemicals including bridge decks, pavements, sidewalks, curbs and gutters, garage and parking floors

Table 9.5 (*continued*).

Class of Exposure	Condition of Exposure	Normal-Weight Concrete Maximum Water/Cement Ratio	Structural Lightweight Concrete Minimum Specified 28-Day Compressive Strength (MPa)	Typical Examples
B	Concrete subjected to frequent cycles of freezing and thawing in a saturated condition in fresh water or infrequent wetting by seawater; concrete subjected to complete and continuous immersion in seawater	0.50	30	Piles, columns, piers, bridge abutments, locks and docks partially immersed in fresh water and retaining walls or structures located near seawater and exposed to infrequent wetting therefrom
C	Concrete subjected to frequent cycles of freezing and thawing in an unsaturated condition	0.55	25	Exposed columns and beams, retaining walls, bridge piers, and abutments
D	Concrete not exposed to freezing and thawing or the application of de-icing chemicals	Select *w/c* ratio or cement content on the basis of strength, workability, and finishing requirements		Interior concrete slabs on ground, concrete floor, interior walls and columns

[a]This table is reproduced from CSA Standard CAN3-A23.1-M77, *Concrete Materials and Methods of Concrete Construction,* which is copyrighted by the Canadian Standards Association. Copies may be purchased from the Association, 178 Rexdale Boulevard, Rexdale, Ontario M9W 1R3.

Table 9.6

Range of Total Air Content for Various Classes of Exposure As Defined in Table 9.5[a]

	Range in Total Air Content for Concretes with Indicated Nominal Sizes of Coarse Aggregate (%)			
Class of Exposure	*10 mm*	*14 mm*	*20 mm*	*40 mm*
A, Subjected to de-icing chemicals	7–10	6–9	5–8	4–7
A, Not subjected to de-icing chemicals, and B	6–9	5–8	4–7	3–6
C	5–8	4–7	3–6	3–6
D[b]	<5	<4	<3	<3
Approximate amount of entrapped air in non-air-entrained concrete	3	2.5	2	1

[a]Adapted from CSA Standard CAN3-A23.1-M77. See note to Table 9.5.

[b]The use of entrained air is not mandatory for class of exposure D, but under normal conditions it is recommended as an aid to workability and reduced bleeding.

7. Estimation of coarse aggregate content. It has been found empirically that aggregates having the same maximum size and grading will yield workable mixes when used in concrete in the volumes (on a dry-rodded basis[3]) shown in Table 9.9. For the same workability, the volume of coarse aggregate depends only on its maximum size and on the fineness modulus of the fine aggregate. The OD weight of coarse aggregate required per cubic meter of concrete is simply equal to the value from

Table 9.7

Types of Cement and Water/Cement Ratio Requirements of Concrete in Contact with Soils and Groundwaters Containing Various Sulfate Concentrations[a]

Potential Degree of Sulfate Attack	*Total Sulfate (as SO_4) in Soil Sample (%)*	*Water-Soluble Sulfate (as SO_4) in Soil Sample (%)*	*Sulfate (as SO_4) in Groundwater Samples (mg/ℓ)*	*Type of Cement to Be Used[b]*	*Maximum Water/Cement Ratio*
Negligible	Less than 0.10	—	Less than 150	10, 20, 30, 40, or 50	—
Mild	0.10–0.20	—	150–1000	20, 40 or 50	0.50
Considerable	—	0.20–0.50	1000–2000	50	0.50
Severe	—	Over 0.50	Over 2000	50	0.45

[a]From CSA Standard CAN3-A23.1-M77. See note to Table 9.5.

[b]Canadian Types 10, 20, 30, 40, 50 correspond to ASTM Types I, II, III, IV, V, respectively.

[3]Oven-dry rather than air-dry.

Table 9.8

Maximum Permissible Water and Minimum Cement Contents to Be Used When Suitable Test Data Are Not Available[a]

	Minimum Cement Content (kg/m³)[b]					
	Non-air-entrained Concrete			Air-entrained Concrete		
	Nominal Size Aggregate (mm)					
Specified Compressive Strength (MPa)	10	20	40	10	20	40
15	285	250	225	290	255	235
20	325	290	260	335	300	270
25[c]	365	320	290	390	340	315
Maximum water (kg/m³)	200	180	160	170	150	140

[a]Adapted from CSA Standard CAN3-A23.1-M77. See note to Table 9.5.

[b]Minimum cement contents are based on concrete slumps not to exceed 100 mm and air contents according to Table 9.6.

[c]Over 25 not applicable.

Table 9.9

Volume of Coarse Aggregate per Unit of Volume of Concrete[a]

Maximum Size of Aggregate		Volume of Dry-Rodded Coarse Aggregate[b] per Unit Volume of Concrete for Different Fineness Moduli of Sand			
in.	mm	2.40	2.60	2.80	3.00
⅜	10	0.50	0.48	0.46	0.44
½	12.5	0.59	0.57	0.55	0.53
¾	20	0.66	0.64	0.62	0.60
1	25	0.71	0.69	0.67	0.65
1½	40	0.76	0.74	0.72	0.70
2	50	0.78	0.76	0.74	0.72
3	75	0.82	0.80	0.78	0.76
6	150	0.87	0.85	0.83	0.81

[a]Adapted from ACI 211.1. Reproduced with permission.

[b]Volumes are based on aggregates in dry-rodded condition as described in ASTM C29. For less workable concrete, such as required for concrete pavement construction, they may be increased about 10%. For more workable concrete, such as may sometimes be required when placement is to be by pumping, they may be reduced up to 10%.

Table 9.9 multiplied by the dry-rodded unit weight of the aggregate in kg/ m³. (The volume of aggregate in ft³/yd³ of concrete can be obtained by multiplying the appropriate value by 27. This volume can then be converted to an OD weight by multiplying it by the dry-rodded weight per cubic foot of coarse aggregate.) To convert from OD to SSD weights, multiply by $(1 + AC/100)$ as in Eq. (6.2).

8. Estimation of fine aggregate content. We have now established the weights of all of the concrete ingredients except that of the fine aggregate. We can establish fine aggregate content in two ways:

(a) *"Mass" ("Weight") Method:* This requires a knowledge of the weight (per m³ or yd³) of the fresh concrete, which can often be estimated from previous experience with the materials in question; failing this, Table 9.10 may be used as a first estimate.

An exact calculation of the weight of fresh concrete in kg/m³ can be obtained using Eq. (9.4) [or in lb/yd³ using Eq. (9.5)].

$$U_m = 10G_a(100 - A) + C_m\left(1 - \frac{G_a}{G_c}\right) - W_m(G_a - 1) \qquad (9.4)$$

$$U = 16.85G_a(100 - A) + C\left(1 - \frac{G_a}{G_c}\right) - W(G_a - 1) \qquad (9.5)$$

where U_m, (U) = weight of fresh concrete, kg/m³(lb/yd³); G_a = weighted average bulk specific gravity (SSD)[4] of combined fine and coarse aggregate, assuming reasonable weight proportions; G_c = specific gravity of cement (generally 3.15); A = air content, %; W_m, (W) = mixing water requirement, kg/m³ (lb/yd³); C_m, (C) = cement requirement, kg/m³ (lb/yd³).

If the first estimate of the weight of the fresh concrete is not very good, an iterative procedure may be required in order to obtain G_a. The weight of fine aggregate is then the difference between the total weight of the fresh concrete and the weight of the other ingredients.

(b) *"Volume" Method:* This is the preferred method, as it is a somewhat more exact procedure, which requires a knowledge of the volumes displaced by the various ingredients. That is, the volumes of the cement, water, air, and coarse aggregate are subtracted from the total volume; the difference is the volume of fine aggregate. The weight of fine aggregate can then be obtained by multiplying this volume by the density of the fine aggregate.

---

[4]If the aggregate is in a condition other than SSD, the appropriate BSG should be used.

Table 9.10

First Estimate of Weight of Fresh Concrete[a]

		First Estimate of Concrete Weight[b]			
		Non-air-entrained Concrete		Air-entrained Concrete	
Maximum Size of Aggregate					
in.	mm	lb/yd³	kg/m³	lb/yd³	kg/m³
⅜	10	3840	2285	3690	2190
½	12.5	3890	2315	3760	2235
¾	20	3960	2355	3840	2280
1	25	4010	2375	3900	2315
1½	40	4070	2420	3960	2355
2	50	4120	2445	4000	2375
3	70	4160	2465	4040	2400

[a]Adapted from ACI 211-1. Reproduced with permission.

[b]Values calculated by Eqs. (9.4) and (9.5) for concrete containing 550 lb/yd³ (330 kg/m³) of cement, slump of 3 to 4 in. (75 to 100 mm), and aggregate bulk specific gravity of 2.7.

The aggregate calculations given above are best carried out using SSD weights, but they can be done by using OD weights. In this case, the coarse aggregate weight is converted to OD by measuring its absorption capacity and then using the OD bulk specific gravity.

9. Adjustment for moisture in the aggregates. The actual water content of the paste will be affected by the moisture content of the aggregates. If these are air-dry, they will absorb some water, thereby effectively lowering the w/c ratio and reducing the workability. On the other hand, if the aggregates are too wet, they will contribute some of their surface moisture to the paste, increasing both the w/c ratio and the workability, and reducing strength. Therefore, these effects must be estimated and the mix adjusted to take them into account.

10. Trial batch. Having now estimated the proportions of all the ingredients, the next step is to prepare a trial batch using these estimates, using only as much water as is needed to reach the desired slump (but not exceeding the permissible w/c ratio). This batch may be prepared in the laboratory or using a full-sized field batch. The concrete thus produced should be tested for slump, unit weight, yield, and air content, as well as observing segregation tendencies and finishing characteristics. Finally, 28-day (or other age) compressive (or flexural) strength should also be determined. Adjustments can now be made in the batch proportions for those requirements which were not satisfied by the original estimate, as follows:

(a) If the slump is incorrect, a new water content can be estimated from the observation that an increase or decrease of 3.5 kg/m^3 (10 lb/yd^3) of water will increase or decrease the slump by approximately 25 mm (1 in.). If the correct slump is obtained at a lower water content, it is permissible to reduce the cement content to reach the design w/c ratio, consistent with any specified limitations on cement content. However, unless this will achieve a substantial saving in cement (which might indicate that the mix should be entirely redesigned), it is probably advisable not to reduce the cement content. The batch weights should also be recalculated since the concrete volume has now been reduced. If the water content must be increased to obtain the desired slump, then the w/c ratio will also be increased. In this case, additional cement must be added until the design w/c ratio is again achieved (or the entire mix redesigned). New batch weights should also be calculated, since the concrete volume has now been increased.

(b) If the desired amount of air entrainment was not achieved, the amount of air-entraining admixture should be reestimated. The mixing water required should then be increased or decreased by 2 kg/m^3 (5 lb/yd^3) for each decrease or increase of 1% air entrainment, because of the influence of air entrainment on workability.

(c) If the weight method of proportioning is used, and if the estimated weight of fresh concrete is incorrect, this can be reestimated from the unit weight of the trial batch, making allowance for the necessary changes in air content.

(d) Any adjustment will change the yield, and therefore new batch weights must then be calculated, following the foregoing procedure from step 5 on.

Concrete Mixes for Small Jobs

Often, for small jobs, it is not feasible to work through the rather elaborate procedures described above. Also, there may not be time to wait for trial batches to harden so that the strength characteristics can be determined. In these cases, the mixes given in Table 9.11 should provide concrete that is sufficiently strong and durable, as long as only enough water is added to produce a workable consistency. However, such a crude estimation is not a substitute for proper mix design procedures.

Table 9.11

Recommended Proportions for Concrete Mixes for Small Jobs[a]

Procedure: Select the proper maximum size of aggregate. Use mix B, adding just enough water to produce a workable consistency. If the concrete appears to be undersanded, change to Mix A and, if it appears to be oversanded, change to mix C.

			Approximate Weights of Solid Ingredients of Concrete $(lb/ft^3)^b$			
			Sand[c]		*Coarse Aggregate*	
Maximum Size of Aggregate (in.)	*Mix*	*Cement*	*Air-entrained Concrete*	*Concrete without Air*	*Gravel or Crushed Stone*	*Iron Blast-Furnace Slag*
½	A	25	48	51	54	47
	B	25	46	49	56	49
	C	25	44	47	58	51
¾	A	23	45	49	62	54
	B	23	43	47	64	56
	C	23	41	45	66	58
1	A	22	41	45	70	61
	B	22	39	43	72	63
	C	22	37	41	74	65
1½	A	20	41	45	75	65
	B	20	39	43	77	67
	C	20	37	41	79	69
2	A	19	40	45	79	69
	B	19	38	43	81	71
	C	19	36	41	83	72

[a]Adapted from ACI 211.1. Reproduced with permission.

[b]$lb/ft^3 \times 16.0 = kg/m^3$.

[c]Weights are for dry sand. If damp sand is used, increase tabulated weight of sand 2 lb and, if very wet sand is used, 4 lb.

Example Mix Design

To illustrate the mix design procedure described above, consider the following sample problem. Concrete is required for an exterior column to be located above ground level in an area where substantial freezing and thawing may occur. The concrete is required to have an average 28-day compressive strength of 35 MPa (5000 lb/in.²).[5] For the

[5]This value takes into account the difference between the average strength required and the specified design strength, which is based on the variability of the concrete (see Chapter 17).

conditions of placement, the slump should be between 25 and 50 mm (1 and 2 in.), and the maximum aggregate size should not exceed 20 mm (¾ in.). The properties of the materials are as follows:

Cement: Type I, specific gravity = 3.15
Coarse aggregate: Bulk specific gravity (SSD) = 2.70; absorption capacity = 1.0%; total moisture content = 2.5%; dry-rodded unit weight = 1600 kg/m³ (100 lb/ft³)
Fine aggregate: Bulk specific gravity (SSD) = 2.65; absorption capacity = 1.3%; total moisture content = 5.5%; fineness modulus = 2.70

The sieve analyses of both the coarse and fine aggregates fall within the specified limits. With this information, the mix design will now be carried through in detail, using the sequence of steps outlined.

Step 1. *Required material information.* This is already given.

Step 2. *Choice of slump.* The slump is also given, consistent with Table 9.1.

Step 3. *Maximum aggregate size.* The maximum aggregate size, 20 mm (¾ in.), is governed by reinforcing details.

Step 4. *Estimation of mixing water and air content.* Since the concrete will be exposed to freezing and thawing, it must be air-entrained. From Table 9.2, the air content recommended for extreme exposure is 6.0%; the water requirement is 165 kg/m³ (280 lb/yd³).

Step 5. *Water/cement ratio.* From Table 9.3, the (conservative) estimate for the required *w/c* ratio to give a 28-day compressive strength of 35 MPa (5000 lb/in.²) is 0.4. This does not exceed the limits based on durability in Table 9.4 (or Table 9.5).

Step 6. *Calculation of cement content.* The required cement content, based on the results of steps 4 and 5, is 165/0.4 = 413 kg/m³ (280/0.4 = 700 lb/yd³).

Step 7. *Estimation of coarse aggregate content.* Interpolating in Table 9.9 for the fineness modulus of the fine aggregate of 2.70, the volume of dry-rodded coarse aggregate per unit volume of concrete is 0.63. Therefore, the coarse aggregate will occupy 0.63 m³/m³ (or 0.63 × 27 = 17.01 ft³/yd³). The OD weight of the coarse aggregate is 0.63 × 1600 = 1008 kg (17.01 × 100 = 1701 lb). The SSD weight is 1008 × 1.01 = 1018 kg (1718 lb).

Step 8. *Estimation of fine aggregate content.* The fine aggregate content can be established either by the mass (weight) method or by the absolute volume method.

8(a). Mass (weight) method. From Table 9.10, the estimated concrete weight is 2280 kg/m³ (3840 lb/yd³). Although for a first trial it is not generally necessary to use the more exact calculation based on Eq. (9.4), this value will be used here.

$$U_m = (10)\,(2.68)\,(100 - 6) + 413\,(1 - 2.68/3.15) - 165\,(2.68 - 1)$$

$$= 2304 \text{ kg/m}^3$$

Using Eq. (9.5), the equivalent value of 3879 lb/yd³ is obtained. The weights already determined are water = 165 kg (280 lb), cement = 413 kg (700 lb), and coarse aggregate (SSD) = 1018 kg (1718 lb). Therefore, the weight of the fine aggregate (SSD) is 2304 − 165 − 413 − 1018 = 708 kg (3879 − 280 − 700 − 1718 = 1181 lb).

8(b). Volume method. Knowing the weights and specific gravities of the water, cement, and coarse aggregate, and knowing the air volume, we can calculate the volumes per m³ (yd³) occupied by the different ingredients.

Water: 165/1000 (280/62.4)		= 0.165 m³	(4.49 ft³)
Cement: 415/(1000 × 3.15)	[700/(3.15 × 62.4)]	= 0.131 m³	(3.56 ft³)
Coarse aggregate (SSD)[6]:	1018/(1000 × 2.70)	= 0.377 m³	
	[1718/(62.4 × 2.70)] =		(10.20 ft³)
Air: 0.06 (0.06 × 27)		= 0.06 m³	(1.62 ft³)
Total		= 0.733 m³	(19.87 ft³)

Therefore, the fine aggregate must occupy a volume of 1 − 0.733 = 0.267 m³ (27.0 − 19.87 = 7.13 ft³). The required SSD weight of fine aggregate = 0.267 × 2.65 × 1000 = 708 kg (7.13 × 2.65 × 62.4 = 1179 lb). As may be seen, this is essentially the same as the weight calculated according to the weight method.

Step 9. *Adjustment for moisture in the aggregate.* Since the aggregates will be neither SSD or OD in the field, it is necessary to adjust the aggregate weights for the amount of water contained in the aggregate. (Note that very dry aggregates will absorb water from the mix, and this too must be allowed for.) Only surface water need be considered; absorbed water does not become part of the mix water. For the given moisture contents, the adjusted aggregate weights become (see the appendix to Chapter 6)

Coarse aggregate (wet) =	1018(1.025 − 0.01)	= 1033 kg/m³
	[1718(1.025 − 0.01)	= 1744 lb/yd³]

[6]If bulk specific gravities of the aggregates are given on an OD basis, then the OD weight should be used to calculate the solid volume of the coarse aggregate, and the weight of the fine aggregate will be determined on an OD basis also.

$$\text{Fine aggregate (wet)} = 708(1.055 - 0.013) = 738 \text{ kg/m}^3$$
$$[1179(1.055 - 0.013) = 1229 \text{ lb/yd}^3]$$

Surface moisture contributed by the coarse aggregate is $2.5 - 1.0 = 1.5\%$; by the fine aggregate, $5.5 - 1.3 = 4.2\%$. The *additional* mixing water required is then

$$165 - 1018(0.015) - 708(0.042) = 120.0 \text{ kg/m}^3$$
$$[280 - 1718(0.015) - 1179(0.042) = 205 \text{ lb/yd}^3]$$

Thus, the estimated batch weights per m³ (yd³) are:

Water (to be added)	=	120 kg	(205 lb)
Cement	=	413 kg	(700 lb)
Coarse aggregate (wet)	=	1033 kg	(1744 lb)
Fine aggregate (wet)	=	738 kg	(1229 lb)
Total	=	2304 kg/m³	(3878 lb/yd³)

Step 10. *Trial batch.* A trial mix is now made using the proportions calculated. If any of the desired concrete properties are not achieved, these properties must be adjusted as indicated above. If very large adjustments appear to be indicated, it is probably best to redesign the entire mix, perhaps by modifying the materials themselves.

9.4 MIX DESIGN IN THE UNITED KINGDOM

A rather different approach to mix design is the method widely used in the United Kingdom, even though, like the ACI method, it is based on proportioning the ingredients to provide the required workability, compressive strength, and durability. This method depends on the use of tabulated and graphical data, obtained from years of experience and research, to help the designer choose appropriate mix proportions according to the characteristics of the available aggregate. For many years, the method used was that described in *Road Note No. 4*. This was replaced in 1975 by a new procedure, "Design of Normal Concrete Mixes," based, however, on broadly similar principles. These principles may be described as follows:

1. This procedure explicitly recognizes the variability of concrete (see Chapter 17). The target mean strength, f_m, for which the mix is to be designed, depends upon the specified design strength ("characteristic" strength, f_c), the standard deviation of the strength data, and the allowable proportion of test results which may be expected to fall below the level of f_c (usually 5%).

2. Workability may be specified by either slump tests or Vebe tests.

3. It is assumed that water absorbed by the aggregate will not be available for hydration. Thus, *w/c* ratios refer to the *free* water in the mix.

4. The relationship between the *w/c* ratio and the compressive strength of cubes is given in Figure 9.2 (essentially a graphical presentation of Abrams' *w/c* ratio law). These curves do not indicate any particular

Figure 9.2 Relationship between compressive strength and free-water/ cement ratio. (From D. C. Teychenné, R. E. Franklin, and H. C. Erntroy, "Design of Normal Concrete Mixes," Building Research Establishment, Transport and Road Research Laboratory, 1975. Reproduced with permission of the Controller of Her Britannic Majesty's Stationery Office.)

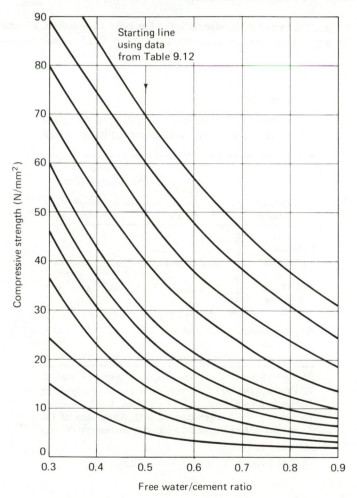

test age; they merely show the type of relationship that exists between the compressive strength and the *w/c* ratio based on tests of many different cements, aggregates, and mixes.

5. In *Road Note No. 4,* three types of aggregate were defined: angular, irregular, and rounded. However, the present procedure considers only two types of aggregate: crushed and uncrushed. Only three different maximum aggregate sizes are considered: 40 mm, 20 mm, and 10 mm.

Mix Design Procedures

Based on these considerations, the mix design process can be briefly described as follows:

1. Once the target mean strength (f_m) has been established, a value for the compressive strength of a mix with *w/c* = 0.5 is obtained from Table

Table 9.12

Approximate Compressive Strengths of Concrete Mixes Made with a Free-Water/Cement Ratio of 0.5[a]

Type of Cement[c]	Type of Coarse Aggregate	Compressive Strengths (MPa)[b] Age (days)			
		3	7	28	91
Ordinary portland	Uncrushed	18	27	40	48
Sulfate-resisting portland	Crushed	23	33	47	55
Rapid-hardening portland	Uncrushed	25	34	46	53
	Crushed	30	40	53	60

[a]Adapted from D. C. Teychenné, R. E. Franklin, and H. C. Erntroy, *Design of Normal Concrete Mixes,* Building Research Establishment, Transport and Road Research Laboratory, 1975. (Reproduced with permission of the Controller of Her Britannic Majesty's Stationery Office.)

[b]MPa × 145 = lb/in.²

[c]ASTM Types I, V, and III, respectively.

Table 9.13

Approximate Free-Water Contents (kg/m³) Required to Give Various
Levels of Workability[a]

Slump (mm): V-B (s):		0–10 >12	10–30 6–12	30–60 3–6	60–180 0–3
Maximum Size of Aggregate (mm)	Type of Aggregate				
10	Uncrushed	150	180	205	225
	Crushed	180	205	230	250
20	Uncrushed	135	160	180	195
	Crushed	170	190	210	225
40	Uncrushed	115	140	160	175
	Crushed	155	175	190	205

[a]Adapted from D. C. Teychenné, R. E. Franklin, and H. C. Erntroy, *Design of Normal Concrete Mixes,* Building Research Establishment, Transport and Road Research Laboratory, 1975. (Reproduced with permission of the Controller of Her Britannic Majesty's Stationery Office.)

9.12, for the age, type of cement, and type of aggregate to be used. This value is plotted on Figure 9.2; a curve is then drawn through this point, parallel to the printed curves, until it intercepts a horizontal line passing through the ordinate representing f_m. The value of the w/c ratio corresponding to f_m can then be obtained.

2. The free water content is then obtained from Table 9.13 according to the type of aggregate and the specified workability. The cement content can then be obtained by dividing the water content by the w/c ratio.

3. The total aggregate content is then obtained by estimating the wet density of the fully compacted mix from Figure 9.3; the total SSD aggregate content is then equal to (wet density − cement content − free water content).

4. The proportion of fine aggregate is then chosen from a series of graphs that plot the percent of fine aggregate versus w/c ratio for different maximum aggregate sizes, workabilities, and the four zones of fine aggregate defined in BS 882 (see Chapter 6). A typical graph is shown in Figure 9.4. The proportion of coarse aggregate is then the total aggregate content minus the amount of fine aggregate.

5. Finally, trial mixes are made, tested, and adjusted if necessary, as in the ACI procedure.

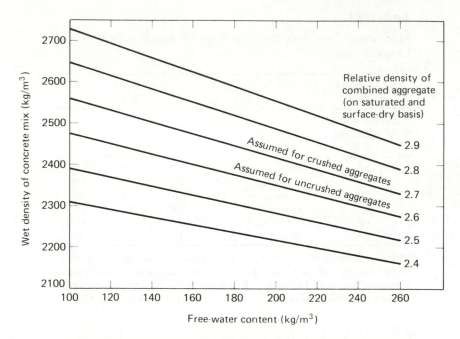

Figure 9.3 Estimated wet density of fully compacted concrete. (From D. C. Teychenné, R. E. Franklin, and H. C. Erntroy, "Design of Normal Concrete Mixes," Building Research Establishment, Transport and Road Research Laboratory, 1975. Reproduced with permission of the Controller of Her Britannic Majesty's Stationery Office.)

In addition, using a family of curves similar to Figure 9.2, the design procedure may also be carried out where the indirect tensile strength is the strength specified (see Chapter 16).

In order to consider the effects of air entrainment (which have so far not been considered), it is assumed that each 1% of entrained air will result in a compressive strength loss of 5.5%. Therefore, for air-entrained concrete, the target mean strength becomes $f_m/(1 - 0.055A)$, where A is the percent of entrained air. The design process described above is also modified slightly to take into account the effects of entrained air on the workability and the density of the concrete.

This method is not directly comparable to the ACI method of mix design, in part because the British Standard for compressive strength is based on cube tests, which are not equivalent to cylinder tests (see Chapter 16). However, when properly used, both methods (as well as several other design methods which are available) will provide concrete of satisfactory quality.

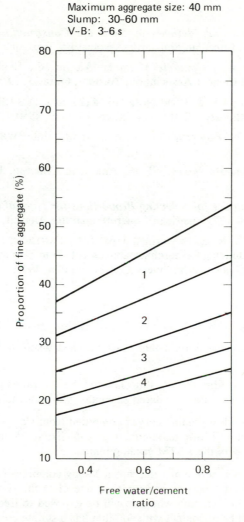

Maximum aggregate size: 40 mm
Slump: 30–60 mm
V–B: 3–6 s

Figure 9.4 Recommended properties of fine aggregate. (From D. C. Teychenné, R. E. Franklin, and H. C. Erntroy, "Design of Normal Concrete Mixes," Building Research Establishment, Transport and Road Research Laboratory, 1975. Reproduced with permission of the Controller of Her Britannic Majesty's Stationery Office.)

Bibliography

Concrete Materials and Methods of Concrete Construction, CSA Standard CAN3-A23.1-M77. Canadian Standards Association, Rexdale, Ontario, 1977.

Design and Control of Concrete Mixtures, Metric ed., Engineering Bulletin. Canadian Portland Cement Association, Toronto, Ontario, 1978.

Design of Concrete Mixes, Road Note No. 4 (2d ed.), D.S.I.R. Road Research Laboratory. Her Majesty's Stationery Office, London, 1950.

POWERS, T. C., *The Properties of Fresh Concrete.* John Wiley & Sons, Inc., New York, 1968.

Proportioning Concrete Mixes, SP-46, American Concrete Institute, Detroit, Mich., 1974.

Recommended Practice for Selecting Proportions for Normal and Heavyweight Concrete, ACI 211.1-77. American Concrete Institute, Detroit, Mich., 1977.

TEYCHENNÉ, D. C., R. E. FRANKLIN, AND H. C. ERNTROY, *Design of Normal Concrete Mixes,* Building Research Establishment, Transport and Road Research Laboratory. Her Majesty's Stationery Office, London, 1975.

Problems

9.1 What are the criteria that must be used when designing a concrete mix?

9.2 Calculate Abrams' w/c ratio law for the w/c ratios given in Table 9.3 and compare the calculated strengths with those tabulated.

9.3 Plot the ideal grading curves computed from Eq. (9.3) with $q = 0.33$, 0.5, and 0.67 for a maximum aggregate size of 25 mm. Compare your results with the ASTM grading limits.

9.4 Design a concrete mix that has a 28-day compressive strength of 4500 lb/in.^{2} and a maximum aggregate size of ¾ in. The concrete will be placed in a column where it will be exposed to freezing and thawing and will be in contact with soil that has a sulfate content of 0.3%. The material properties are as follows:
Cement: all types have a specific gravity of 3.15
Coarse aggregate: bulk specific gravity (SSD) = 2.65, absorption capacity = 1.5%, surface moisture = 1.0%, dry-rodded unit weight = 105 lb/ft$_3$.
Fine aggregate: bulk specific gravity (SSD) = 2.75, absorption capacity = 1.0%, surface moisture = 3.0%, fineness modulus = 2.70.

9.5 Recompute the mix in Problem 9.4 if the bulk specific gravities of the aggregates now represent the OD condition and if the total moisture contents are 1.0% (coarse) and 3.0% (fine).

9.6 Design a concrete mix using a maximum aggregate size of 40 mm to be used for an interior beam having a 28-day compressive strength of 35 MPa. The material properties are as follows:

Cement: all types have a specific gravity of 3.15.

Coarse aggregate: bulk specific gravity (SSD) = 2.90, absorption capacity = 0.5%, effective absorption = 0.3%, dry-rodded unit weight = 1700 kg/m³.

Fine aggregate: bulk specific gravity (SSD) = 2.62, absorption capacity = 1.2%, surface moisture = 2.8%, fineness modulus = 2.80.

9.7 Recompute the mix in Problem 9.6 if the bulk specific gravities of the aggregate now represent the OD condition and if the total moisture contents are 0.8% (coarse) and 2.5% (fine).

9.8 Recompute the mix in Problem 9.6 using the mix design method adopted in the United Kingdom.

10

handling and placing

The production of high-quality concrete does not rest solely on proper proportioning. The concrete placed in a structure must be of uniform quality, free of voids and discontinuities, and adequately cured. Although the design of a concrete mix is critical for strength and durability, lack of sufficient attention to mixing, handling, and placing can result in poor concrete from a well-designed mix. In this chapter we discuss batching and mixing and the various methods of transporting, placing, and consolidating concrete. Curing methods are the subject of the following chapter. The overall objective is to ensure that the concrete within a structure is a uniform blend of the constituent materials in the correct proportions and thus conforms to the specifications.

10.1 BATCHING AND MIXING

Batching

Batching of aggregates and cements should always be by weight; dispensing of solids on a volume basis can lead to gross errors. Only water and liquid admixtures can be measured accurately by volume. Batching by weight also allows rapid and convenient adjustment of batch weights of aggregate and water when changes in aggregate moisture contents occur. Material quantities should be measured with a high degree of accuracy. The tolerances that are frequently used are given in Table 10.1, which is based on ASTM C94, Specifications for Ready-

Table 10.1

Recommended Tolerance for Batching
Concrete Constituents[a]

	Individual[b]	*Cumulative*[b]
Cement	±1%	±1%
Water	±1%	N.R.
Aggregates	±2%	±1%
Admixtures	±3%	N.R.[c]

[a]Batch weights should be greater than 30% of scale capacity.

[b]Individual refers to separate weighing of each constituent. Cumulative refers to cumulative weighing of cement and pozzolan or of fine and coarse aggregate.

[c]Not recommended except for pozzolans.

Mixed Concrete. Weighing equipment should be capable of measuring quantities within these tolerances for the smallest batches that will be used. On small jobs, cement can be dispensed by sack (fractional sacks should be weighed), although the tolerance allowed by ASTM on individual sacks (3%) is greater than that recommended in Table 10.1.

Batching Equipment

Any batching equipment that operates reliably within the specified tolerances can be used provided that it minimizes segregation by allowing free, unobstructed flow of materials. The exact arrangement and type of equipment can vary widely, but for each type of material there should be a storage bin that discharges into a weight batcher and then to the mixer. Typical recommended arrangements are shown in Figure 10.1. Cement is preferably batched separately, but aggregates may be batched separately or cumulatively.

Available materials-handling equipment can be divided into three general categories: manual, semiautomatic, and fully automatic. Manual batching is generally only acceptable for small jobs (up to 400 m³ or 520 yd³) and low output requirements (15 m³/h or 20 yd³/h); otherwise, semiautomatic or automatic batching should be used. In semiautomatic arrangements, the charging and discharging of the batchers are activated manually but are automatically terminated. In a fully automatic system, a single starter switch activates the batching sequence. Both systems require interlocks to maintain tolerances and to prevent the batcher

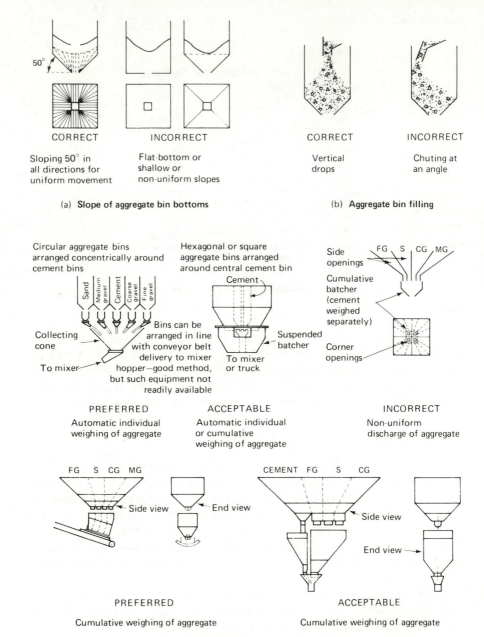

CORRECT

Sloping 50° in all directions for uniform movement

INCORRECT

Flat-bottom or shallow or non-uniform slopes

(a) Slope of aggregate bin bottoms

CORRECT

Vertical drops

INCORRECT

Chuting at an angle

(b) Aggregate bin filling

Circular aggregate bins arranged concentrically around cement bins

Hexagonal or square aggregate bins arranged around central cement bin

Sand, Medium gravel, Cement, Coarse gravel, Fine gravel

Collecting cone

To mixer

Bins can be arranged in line with conveyor belt delivery to mixer hopper—good method, but such equipment not readily available

Cement

Suspended batcher

To mixer or truck

Side openings

FG S CG MG

Cumulative batcher (cement weighed separately)

Corner openings

PREFERRED

Automatic individual weighing of aggregate

ACCEPTABLE

Automatic individual or cumulative weighing of aggregate

INCORRECT

Non-uniform discharge of aggregate

FG S CG MG

Side view

End view

CEMENT FG S CG

Side view

End view

PREFERRED

Cumulative weighing of aggregate

ACCEPTABLE

Cumulative weighing of aggregate

(c) Batching of aggregate and cement
(cement is always batched so that discharge occurs during discharge of all aggregate)

Figure 10.1 Methods of batching: (a) slope of aggregate bin bottoms; (b) aggregate bin filling; (c) batching of aggregate and cement. (Adapted from ACI Committee 304, *Journal of the American Concrete Institute,* Vol. 69, No. 7, 1972, pp. 374–414.)

discharging and charging simultaneously. Automatic systems, when properly designed, can maintain high-speed batching within specified tolerances.

Batching equipment should be kept clean and receive regular maintenance. The accuracy of the weight batchers or other measuring devices should be checked regularly. This is particularly important when batching finely divided solids (cement and pozzolans), since dust from these materials can cause malfunction of the weighing mechanism if it is not properly protected. Avoiding free fall of cement avoids excessive dust and material loss. Proper transfer of such fine material can be a problem, and vibration or aeration are often used to facilitate complete discharge. Dispensing equipment for admixtures requires more frequent maintenance and calibration. Malfunctions can easily lead to severe overdoses, which can cause serious problems with the fresh or hardened concrete. Liquid dispensers should be regularly flushed with water to prevent buildup of gummy deposits or sediments that could clog or stick valves and damage pumps.

Handling of Aggregates

In Chapter 6 the importance of proper sampling was emphasized to obtain aggregate properties from a truly representative sample. However, proper sampling will be of a little value if suitable precautions are not also taken to prevent segregation of aggregate sizes during handling. Figure 10.2 indicates some of the proper ways to handle aggregates to avoid segregation. In some plants the division of coarse aggregate into coarse, medium, and fine fractions can minimize this problem. The proper gradation is then achieved during batching. Protection of aggregates from the weather will eliminate gross variations in aggregate moisture contents. For good batching control, proper allowance for changes in the moisture content should be made. The variations are greatest for the fine aggregate (see Chapter 6), and often equipment for monitoring the moisture content of the sand is incorporated into an automatic batching plant. Such equipment should be regularly checked by standard moisture content tests.

Mixing

Thorough mixing is essential for the complete blending of the materials which are required for the production of homogeneous, uniform concrete. Not only does inadequate mixing result in lower strengths but also in greater batch-to-batch and within-batch variations. However, overly long mixing times do not improve the quality of concrete and

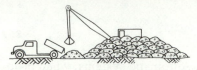

PREFERABLE

Crane or other means of placing material in pile in units not larger than a truck load which remain where placed and do not run down slope.

LIMITED ACCEPTABILITY

Pile built radially in horizontal layers by bulldozer working from materials as dropped from conveyor belt. A rock ladder may be needed in setup.

GENERALLY OBJECTIONABLE

Bulldozer stacking progressive layers on slope not flatter than 3:1 Unless materials strongly resist breakage, these methods are also objectionable.

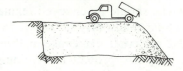

OBJECTIONABLE

Methods which permit the aggregate to roll down the slope as it is added to the pile or permit hauling equipment to operate over the same level repeatedly.

Uniform about center

CORRECT

Chimney surrounding material falling from end of conveyor belt to prevent wind from separating fine and coarse materials, openings provided as required to discharge materials at various elevations on the pile.

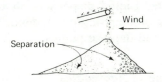

Wind

Separation

INCORRECT

Free fall of material from high end of stacker permitting wind to separate fine from coarse material.

(a) Storage of unfinished aggregate

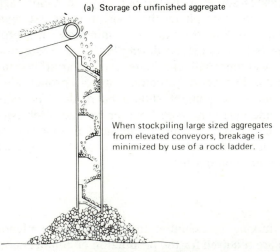

When stockpiling large sized aggregates from elevated conveyors, breakage is minimized by use of a rock ladder.

(b) Finished aggregate storage

Figure 10.2 Handling and storing aggregates: (a) storage of unfinished aggregate; (b) finished aggregate storage. (Adapted from ACI Committee 304, *Journal of the American Concrete Institute*, Vol. 69, No. 7, 1972, pp. 374–414.)

severely limit the output of the batching plant. Unusually long mixing times may cause some breakdown of the aggregate and may decrease the air content.

Mixing Times

The optimum mixing time depends on (1) the type of mixer, (2) the condition of the mixer, (3) the speed of rotation, (4) the size of the charge, and (5) the nature of the constituent materials. Thus, the most efficient mixing time should be assessed in the field by determining batch-to-batch variations using the materials and mixing conditions for the job. Lean, dry, or harsh mixes require longer mixing times; concretes made with angular aggregates need more mixing than those made with rounded gravels. A good rule of thumb is 1 min of mixing time for 1 m³ of concrete plus 1/4 min for each additional 1 m³.

Charging the Mixer

The mixing time refers to the duration of mixing once the mixer is fully charged. The complete mixing cycle includes the charging and discharging process, so that its duration may easily be two to three times that of actual mixing. Charging of the mixer should be an opportunity to preblend the materials. Although the sequence can vary to suit particular applications, it is desirable to add about 10% of the mixing water before the aggregates are added. The water should enter well inside the mixer and be added uniformly during the whole time the solid ingredients are added, leaving about a final 10% to be added at the end. The cement should enter the mix after about 10% of the aggregates have been charged. Mineral admixtures are generally added with the cement, but water-soluble admixtures should be dissolved in the mixing water. If more than one admixture is used, they should be batched separately, not premixed, and added to each batch at the same time in the mixing sequence and in the same order. By allowing one admixture to interact with the solid ingredients before the second is added, adverse interactions between the two admixtures may be avoided.

Types of Mixers

A great variety of concrete mixers are available on the market today. These can generally be divided into three categories: drum mixers, pan mixers, and continuous mixers. The capacity of such mixers can range up to 12 yd^3 (9 m^3). Laboratory mixers are designed to handle as little as $1/10$ ft^3 (0.003 m^3), while planetary mixers are recommended for mixing mortars (ASTM C305.)

A satisfactory type of *drum mixer* has an arrangement of interior fixed blades to ensure end-to-end exchange of material during mixing. In some cases a horizontal shaft with spiral blades may be rotated in the drum. Tilting drums, which are commonly used for small jobs, are now available in large sizes also. They have the advantage of a quick, clean discharge even of dry mixes. Horizontal drums are discharged by inserting a chute to intercept the concrete as it is mixing or by reversing the direction of rotation, which forces the concrete out of the mixer. *Slip-form pavers* generally have drum-type mixers attached to a unit that spreads the concrete by means of a screw and also screeds and vibrates the concrete after it has been placed.

Pan mixers are particularly good for mixing lean and dry mixes. They are thus commonly used in precast concrete plants, where their greater bulk and less convenient discharge are not necessarily disadvantages. The blades are fixed to an assembly so that they agitate the concrete throughout the stationary pan as the vertical drive shaft rotates. In another type an eccentric mounting of blades revolves within a rotating pan. Small-scale pan mixers are popular for trial mixing in the laboratory, as they produce consistent concrete throughout a wide range of mixes.

In a *continuous mixer* the materials are fed into a mixing drum by means of special conveyors. The concrete is mixed as it passes through the drum to the discharge end. The concrete is proportioned by adjusting the relative speeds of the conveyors. Thus, proportioning is actually by volume, so that it is more difficult to produce uniform concrete, unless the aggregate source is uniform and the moisture content of the sand is kept constant to avoid errors introduced by "bulking." The special conveyors should be well maintained to ensure an accurate supply of materials. Continuous mixers are useful where a continuous supply of concrete is desired, as in slipforming, and in placing large quantities of mass concrete.

Ready-Mixed Concrete

Increasing quantities of concrete are being placed using concrete mixed at permanent batching plants, which are to be found in all communities with a sizable population. The advantages of ready-mixed concrete for small jobs are obvious, but it is being used more and more on large jobs also. The advantages are to be found not only in automated equipment and permanent, trained personnel, which results in better quality control, but also in the elimination of materials storage on congested building sites. Since the batch plant must be within reasonable distance of the building site, field batch plants must still be used in more remote locations. There are several ways in which concrete from central batching plants can be handled. These are central-mixed, transit-mixed, shrink-mixed, and truck-mixed.

Central-mixed concrete is completely mixed at the plant and the truck mixer is only used as an agitating conveyance. Slight agitation of the concrete during transportation reduces the amount of slump loss that occurs when the concrete is left standing, and prevents segregation. The truck can also be used to remix the concrete, if necessary, at the job site. *Transit-mixed* concrete is partially or completely mixed during the time the concrete is being transported to the job site. *Shrink-mixed* concrete is partially mixed at the plant to shrink or reduce the overall volume, and mixing is completed in the truck mixer. *Truck-mixed* concrete is completely mixed within the mobile mixer after it has been charged at the central plant. The advantage of truck mixing is that the water can be kept separate from the solid materials and mixed just prior to placement at the construction site. Therefore, problems of delays in transportation or placement are avoided. However, in many cases the concrete is partially mixed in transit and mixing is completed at the job site. ASTM C94 (CSA A23.1) requires the concrete to be placed within 1½ h of mixing or before the drum has passed through 300 revolutions. The drum should provide efficient mixing within 100 revolutions at mixing speed (generally 8 to 12 rev/min). However, it would seem that such requirements are unduly restrictive if the concrete maintains the required slump, air content, and unit weight.

Remixing

When ready-mixed concrete is supplied at the job site, it is often remixed just before placing to ensure that the correct slump is achieved. If this is the case, the earlier period of mixing should be such that at least half the minimum mixing time occurs during the remixing period and that the total number of revolutions does not exceed specifications. Concrete that has been remixed is liable to set more rapidly than that mixed only once. It is common practice to add materials (particularly water) to the concrete after it has been batched, usually at the job site, to ensure that the specified properties are attained before placement. This should be discouraged, since it will only lead to trouble. Accurate batching on the job site is difficult, so that quality control suffers. The concrete mix should be designed to ensure that the correct properties are attained at the time of placement. This means that normal slump loss must be allowed for by designing for a higher slump than that specified for placement.

10.2 TRANSPORTATION

There are many different ways of handling concrete and the choice will depend mostly on the amount of concrete involved, the size and type of construction, the topography of the job site, the location of the batch plant, and relative costs. From the concrete point of view, any method of transportation should protect the concrete from the effects of the weather (heat, cold, or moisture). It should also not cause undue segregation by excessive jarring or shaking. Methods of transportation can be roughly assigned to four general categories (see Table 10.2): wheeled transports, buckets, conveyors and chutes, and pumps.

Transport of concrete from the batching plant to the job is generally by ready-mixed trucks equipped with a revolving drum for agitation. The use of dumpsters should not involve long trip times to avoid excessive slump loss or segregation. Buckets, overhead cableway, or truck can be used to transport concrete from an adjacent batch plant.

Buckets or skips are a versatile means of moving concrete about the job site by the use of a hoist, crane, or overhead cableway. Large quantities [up to 8 yd³ (6 m³) at a time] can be moved horizontally or

Table 10.2

General Methods of Transporting Concrete

Method	Application	Capacity^a	Remarks
Truck (agitated)	Ready-mixed concrete	4–9 m³	Up to 1½ h trip time
Truck or rail cars (nonagitated)	From nearby batch plant to job site		30–45 min trip time; avoid wet mixes or very dry mixes (slump range 12–50 mm or ½–2 in.)
Rail cars (nonagitated)			
Buckets or skips	On site, can be transferred from nearby batch plant	Up to 6 m³	Requires cranes or hoists
Power buggies	On site	Up to 0.75 m³	Requires good terrain; vertical transfer limited
Wheelbarrows	On site	About 0.03 m³	Small jobs
Belt conveyors	On site	Up to 115 m³/h	Avoid wet mixes; use a cohesive mix, 75 mm (3-in.) slump
Pumping	On site	Up to 75 m³/h	Mixes should be designed especially for pumping

^{a}m³ × 1.32 = yd³.

vertically. Buckets should be designed so that discharge can be properly regulated and so that segregation does not occur during discharge. The bucket should slope down to the exit gate and the concrete should be released vertically (Figure 10.3). The bucket should be charged from the mixer in a similar fashion.

Motorized buggies or wheelbarrows can be used to transport small loads around the job site on good terrain. They can only be used to advantage on level ground or slight grades.

Belt conveyors can also be used to transport concrete both horizontally and vertically, but special care should be taken with this form of transportation. Segregation is liable to occur at transfer points between conveyors or during discharge. Wet or lean mixes are liable to segregate particularly if the incline of a conveyor is too great. A large maximum aggregate size should be avoided. Care should also be taken to prevent mortar from adhering to the belt; it should be scraped off at the point of discharge. Since the concrete is spread out relatively thinly, it is susceptible to loss of moisture or increase of temperature during hot weather, which will accelerate slump loss. Thus, belts should be protected from direct sunlight and wind and their use is not recommended in very hot conditions. Chutes can be used when the concrete is being placed at points below the point of delivery. They suffer from the same disadvantages.

Pumping

Today an increasingly large amount of concrete is being moved by pumping. Although developed in the 1930s, this method of moving concrete was still relatively uncommon 15 to 20 years ago; today perhaps over one-fourth of the concrete on building sites is moved through pipelines. This is a very versatile method and is particularly useful where space is limited, such as in tunnels. Concrete can be pumped successfully over quite large distances, typically more than 450 m (1500 ft) horizontally and 150 m (500 ft) vertically. These distances are increasing as pumping equipment and technology improve with experience.

Pumping Equipment

There are three basic types of concrete pumps on the market today (see Figure 10.4): piston pumps, pneumatic pumps, and squeeze pumps. The *piston pump* operates with synchronized inlet and outlet valves. On the backward stroke the inlet valve opens and concrete is drawn into the cylinder from the hopper, while on the forward stroke the concrete is forced into the pipeline through the open outlet valve. Piston pumps can

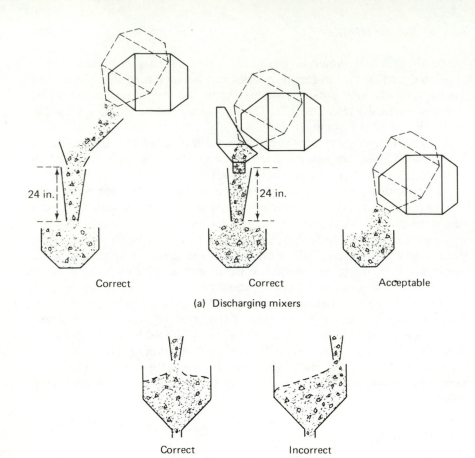

Correct   Correct   Acceptable

(a) Discharging mixers

Correct   Incorrect

(b) Filling hoppers or buckets

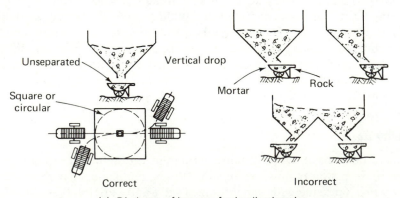

(c) Discharge of hoppers for loading buggies

Figure 10.3 Handling of concrete: (a) discharging mixers; (b) filling hoppers or buckets; (c) discharge of hoppers for loading buggies; (d) control of separation at the end of concrete chutes; (e) control of separation of concrete at the end of a conveyor belt; (f) control at transfer point of two conveyor belts. (Adapted from ACI Committee 304, *Journal of the American Concrete Institute,* Vol. 69, No. 7, 1972, pp. 374–414.)

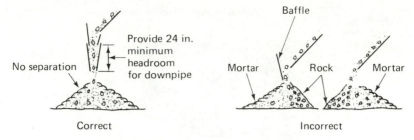

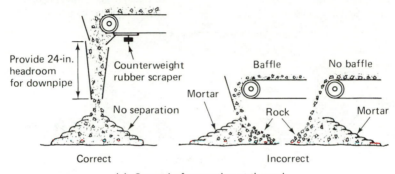

(d) Control of separation at the end
of concrete chutes

(e) Control of separation at the end
of a conveyor belt

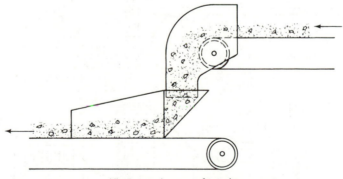

(f) Control at transfer point
of two conveyor belts

Figure 10.3 (*continued*).

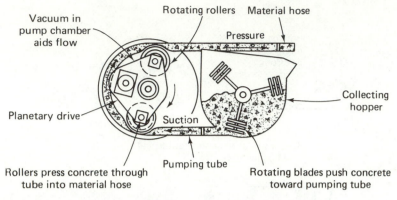

(a) Squeeze-pressure pump

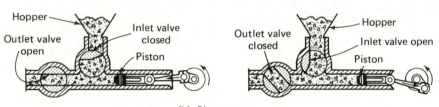

(b) Piston pump

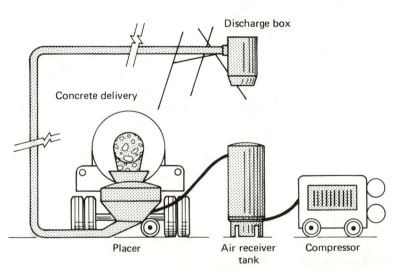

(c) Pneumatic pump

Figure 10.4 Types of concrete pumps: (a) squeeze-pressure pump; (b) piston pump; (c) pneumatic pump. (Adapted from ACI Committee 304, *Journal of the American Concrete Institute,* Vol. 68, No. 5, 1971, pp. 327–344.)

be operated either mechanically or hydraulically, and it is now common for pumps to contain two pistons operating alternately to give a steadier flow of concrete. *Pneumatic pumps* force the concrete through the line by means of compressed air which is supplied to a pressure vessel at the beginning of the line. A discharge box at the end of the line bleeds off the air and controls the delivery of the concrete. The *squeeze pump* forces the concrete through a flexible hose by means of rollers that press against the hose. A vacuum is maintained within the pumping chamber to aid in restoring the shape of the hose and thus ensure a steady supply of concrete. This kind of pump is based on peristaltic pumps used to pump corrosive fluids, since the pumping mechanism never comes in contact with the fluid.

Concrete pipelines can be made either of rigid pipe or heavy-duty flexible hose. The latter does not perform as well as rigid pipe because it presents a greater resistance to the movement of concrete, but it is used in rigid lines for curves and is invaluable for use with moving booms and other situations where flexibility is required. Generally, flexible hose greater than 100 mm (4 in.) i.d. is difficult to handle conveniently, but rigid pipe up to 200 mm (8 in.) i.d. is regularly used. Small-bore pumps are used for mobile pumping units that can be moved rapidly from site to site and will conveniently handle relatively small quantities of concrete. Such units are usually equipped with a hydraulically operated boom for both vertical and horizontal placements. With a 75 to 100 mm (3 to 4 in.) line a good unit can deliver concrete up to 60 m³/h (75 yd³/h), 180 to 250 m (600 to 800 ft) horizontally and 45 to 60 m (150 to 200 ft) vertically.

Pipe or hose materials should be resistant to wear and abrasion, reasonably lightweight, and should not react with concrete. Thus, aluminum, because of its reactivity with alkalis, is not recommended as a pipeline material unless satisfactory performance can be demonstrated. Couplings between pipe sections should be watertight, strong enough to withstand stresses from misalignment or poor pipe support, present no obstruction to concrete flow, and be easily connected or disconnected.

Pumping Distances

The distance concrete can be pumped depends on many job factors: (1) the capacity of the pump, (2) the size of the pipeline, (3) the number of obstructions to uniform flow, (4) the velocity of pumping, and (5) the characteristics of the concrete. The pump must supply enough force to overcome friction between the concrete and the interior surface of the line. A knowledge of actual frictional losses can be used with conventional hydraulic theory in order to calculate pump capacity for a given rate of delivery. Bends in the line and decreases in pipe diameter considerably add to frictional resistance. When concrete is pumped vertically, extra force is needed to overcome gravity, and this requires about 23-kPa/m (1-lb/in.²/ft) lift. Proper maintenance of pipe and couplings is needed to minimize frictional losses.

Concrete moves as a cylindrical plug when it is pumped, separated from the pipe wall by a thin lubricating film of mortar or grout. Thus, the line should be initially primed by starting pumping with a properly designed mortar (or with a batch of the regular concrete with the coarse aggregate removed). For 150 to 200 mm (6 to 8 in.) horizontal lines 0.4 m³ (0.5 yd³) will lubricate about 300 m (1000 ft) of line, and this lubrication will be maintained as long as pumping continues. Once pumping is completed, the pipes are cleaned by blowing compressed air through and then flushing with water until completely clean.

Mix Design

Special attention should be paid to mix design when concrete is to be pumped. Concrete for pumping is not radically different from other concrete, but it should be plastic and cohesive; harsh or dry mixes do not pump well. More emphasis on quality control is needed and materials should be uniform in properties throughout the job. The two major causes of failure to pump concrete successfully are high frictional resistance and a tendency to segregate. As a result, badly proportioned concrete requires a greater pump capacity and is prone to pipe blockage.

The ratio of the maximum size of the coarse aggregate to the smallest inside pipe diameter should not exceed 0.33 for angular aggregates, or 0.40 for well-rounded gravel, in order to minimize the possibility of blockages. Size gradation of both fine and coarse aggregate should be as close as possible to the middle range of the grading limits given in ASTM C33, but uniformity of gradation is even more important. Blending of size fractions may be needed to achieve this. The role of the fine aggregate is particularly important, since it is the fluid mortar that is the pumping medium in which the coarse aggregate is suspended. Special attention should be given to having adequate fines; designation of fineness modulus alone is not enough. For small line systems 15 to 30% should pass the No. 50 (300-μm) sieve and 5 to 10% should pass the No. 100 (150-μm) sieve. Lightweight aggregates, both fine and coarse, should be thoroughly presoaked prior to use to avoid excessive slump loss in the pipeline. Pumping aids are available commercially; these are water-soluble polymers which reduce frictional resistance and bleeding in the pipeline by increasing the viscosity of water. The effectiveness of such admixtures for a particular job should be established by on-site evaluation.

Concrete with slumps in the range 50 to 150 mm can be successfully pumped if properly proportioned. High slumps may cause segregation of coarse aggregate and excessive bleeding, which can lead to loss of lubrication and blocking of the lines. Adding water solely to correct slump may cause more problems by lowering the cohesiveness of the concrete. The desired slump should be achieved by proper proportioning. This will usually require additional cement because of the increased aggregate fines, but mineral admixtures or water-reducing admixtures can be used to reduce cement requirements. Air-entrained concrete can be successfully pumped if air contents are not too high, generally less than 5%, and air entrainment reduces the possibility of segregation. However, the compressibility of the entrained air reduces pumping capacity, particularly over long distances.

10.3 PLACEMENT OF CONCRETE

Proper handling of concrete during placement in forms can prevent problems of segregation of the coarse aggregate, which would result in the formation of rock pockets and honeycombing. Experience has shown that certain methods give satisfactory results, and these are illustrated in Figure 10.5. These principles also apply to the handling of graded aggregates and in the transfer of concrete from the mixer to the placement site.

Concrete should fall vertically but should not be allowed free fall for long distances. Segregation is enhanced as material bounces off from form faces or strikes reinforcing steel. The use of a drop chute or pipe protects concrete during its fall. In deep, narrow forms or in curved forms, placing concrete by pump and hose or by tremie can be used to advantage. Alternatively, concrete can be initially placed in the lower part of the form through openings specially made for this purpose. In such instances the concrete should be allowed to flow slowly into the form; it should not enter the form rapidly at an angle from the vertical. On slopes the concrete should be constrained to fall vertically onto the slope; otherwise, coarse aggregate is likely to roll to the bottom. When placing slabs, both horizontal or sloping, concrete should be dumped into the face of the concrete already placed. Thus, if any segregation has occurred during transportation or occurs during placement, the segregated aggregate can be reworked into the mass. Excessive lateral movement of concrete is not desirable. A strict watch should always be kept for

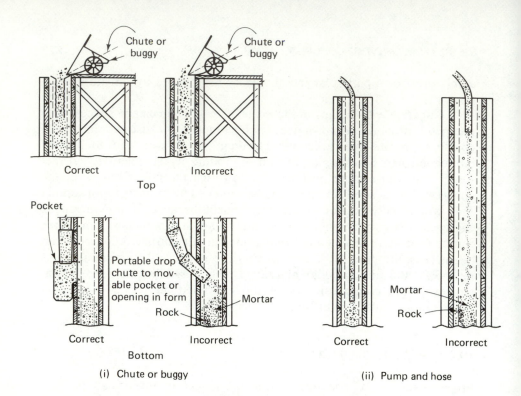

Top

Bottom

(i) Chute or buggy

(ii) Pump and hose

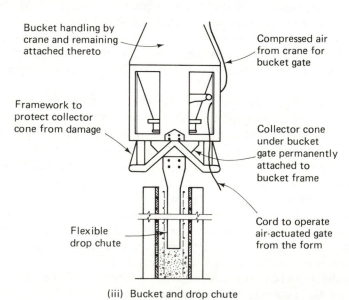

(iii) Bucket and drop chute

(a) Placement in deep and narrow forms

Figure 10.5 Placement of concrete: (a) placement in deep and narrow forms; (b) placement in curved forms; (c) placement on horizontal surfaces; (d) placement on sloping surfaces; (e) consolidation during placement. (Adapted from ACI Committee 304, *Journal of the American Concrete Institute,* Vol. 69, No. 7, 1972, pp. 374–414.)

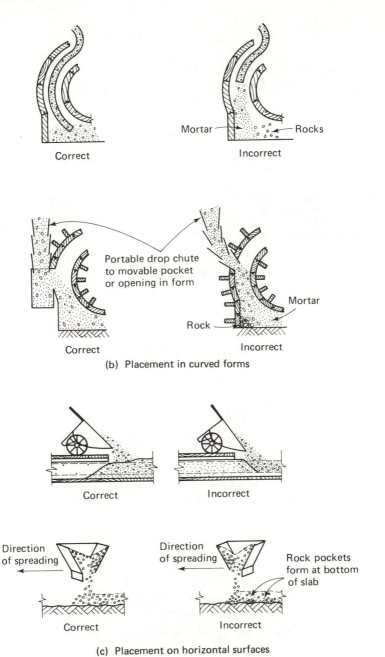

(b) Placement in curved forms

(c) Placement on horizontal surfaces

Figure 10.5 (*continued*).

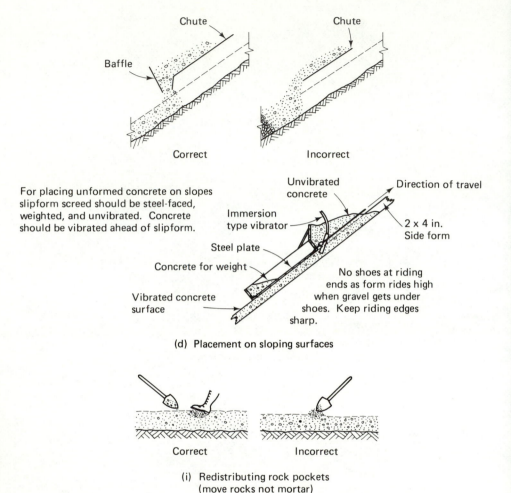

Chute

Baffle

Chute

Correct

Incorrect

For placing unformed concrete on slopes slipform screed should be steel-faced, weighted, and unvibrated. Concrete should be vibrated ahead of slipform.

Unvibrated concrete

Direction of travel

Immersion type vibrator

2 x 4 in. Side form

Steel plate

Concrete for weight

No shoes at riding ends as form rides high when gravel gets under shoes. Keep riding edges sharp.

Vibrated concrete surface

(d) Placement on sloping surfaces

Correct

Incorrect

(i) Redistributing rock pockets (move rocks not mortar)

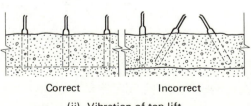

Correct

Incorrect

(ii) Vibration of top lift

(e) Consolidation during placement

Figure 10.5 (*continued*).

282

segregation during placement and action taken to correct problems as soon as they arise.

Special Placements

A number of special placement techniques have been developed for the placement of concrete in certain applications. Those discussed here are slip forming, preplaced aggregate, shotcreteing, and the tremie method.

Slip Forming

Slip forming perhaps should no longer be considered a special technique, since it is being applied to an increasing variety of construction projects. Formerly used primarily in the construction of silos, chimneys, and monolithic tunnel linings, the slip-form method is now used for paving and in some high-rise construction. Slip forming involves continuous placement and consolidation of concrete. Slip-formed paving uses low-slump concrete that retains its shape shortly after it has been placed without the need for formwork support. Concrete can be placed at the rate of several hundred feet per hour. Vertical slip forming requires formwork to confine the concrete until it has gained sufficient strength to support the concrete placed above and perhaps the formwork also. Rates of movement are therefore much slower. A spectacular example of a slip-formed structure is the CN tower in Toronto (Figure 10.6), which was erected to a height of 460 m (1500 ft) in about 100 days during the winter months. Successful slip forming requires careful quality control of materials and production and a properly designed testing program to ensure that strength development is adequate. The increased productivity and higher quality of the finished concrete can more than offset the higher initial capital investment.

A variant of slip forming is the jump-forming technique. Forms are not moved continuously but are repositioned (jumped) for the next lift after each casting. Jump forming thus has similar advantages to slip forming, but should also be accompanied by an adequate quality control and testing program.

Preplaced Aggregate

This method entails packing forms with well-graded coarse aggregate and injecting structural mortar (or grout) into the mass to fill the voids. This method is adaptable to underwater concreting, for repairs of existing structures, or for situations where conventional placements are difficult.

Figure 10.6 View of CN Tower, Toronto.

Because the aggregate can be packed more densely than in ordinary concrete, less cement paste need be used and concrete of low volume change results. However, this method requires skill and experience to ensure complete filling of the void space. The secret of good placement is the use of a high-quality, well-graded, clean aggregate conforming to ASTM C33 and a cohesive, yet fluid, mortar that develops good strength. The sand used in the mortar is much finer than normal. Any type of portland cement may be used, but admixtures are generally required to attain the required grout properties. Concrete made by the preplaced aggregate method will have direct contact between aggregate particles, whereas this does not occur in concrete made by conventional means. This may affect the elastic properties of the concrete and also its mode of fracture.

Shotcreteing

Concrete can be applied pneumatically by spraying it from a nozzle by means of compressed air. This process is known as *shotcreteing* or *gunniting*. The latter term is in more common usage in Europe, but remains the registered trademark of the Allentown Pneumatic Gun

Company, which first developed the process on a commercial basis early in this century. A schematic view of the cement gun is shown in Figure 10.7. It is usual to mix the solid materials in the dry state prior to spraying and to add water just before ejection at the nozzle, but in some cases concrete is sprayed in the wet state after mixing. In the dry-mix process, the amount of water is adjusted by the nozzleman; it requires considerable experience to get a mix that is neither too wet nor too dry. Concrete for shotcreteing is generally made with fairly fine aggregate:

Figure 10.7 Schematic view of shotcreteing equipment: (a) supply chamber; (b) nozzle. (From J. R. Illingworth, *Movement and Distribution of Concrete*, McGraw-Hill Co. (U.K.) Ltd., 1972. Reproduced by permission of The Cement Gun Co. Ltd.)

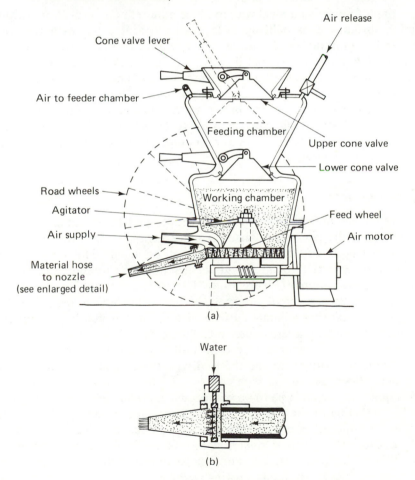

pea gravel (10-mm maximum size) and sand, although aggregates up to 20-mm maximum size can be used. Fiber-reinforced concrete can also be shotcreted.

The concrete is ejected from the nozzle at high velocity so that the concrete is well compacted at the surface. Thus, a dense concrete with a low *w/c* ratio can result with compressive strengths of 40 to 55 MPa (6000 to 8000 lb/in.²) at 28 days, and it forms a good bond to steel and substrate. However, about ¼ to ½ of the material rebounds on impact and is not retained on the surface. Most of this is the coarser material, so the in-place shotcrete is richer than the original mix. Admixtures must be used in shotcreteing, the most common being quick setting admixtures which are claimed to reduce rebound and give very rapid strength development in a matter of hours.

Shotcreteing is an ideal way to place concrete on vertical or steeply sloping surfaces. It is built up in layers by repeated passes across the face and reinforcing steel may be incorporated in the concrete. It has been used to construct tunnel linings, stabilize rock faces, and provide surface supports, without the need for formwork, and can be used in any situation where conventional formwork would be difficult to use. But its most versatile use is perhaps in the repair and restoration of existing structures. Shotcreteing can restore structural integrity to damaged members or provide new, protective coatings over such diverse materials as brick and timber.

Tremie Concrete

The tremie method of placement was developed for pouring concrete under water, but can be used also for placement in deep forms or where normal methods of consolidation cannot be used. The concrete is delivered by gravity through a rigid vertical pipe (the tremie) fed through a funnel-shaped hopper (Figure 10.8). The tremie method is designed to minimize the occurrence of entrapped air and other voids, and it has the advantage of causing minimal surface disturbance, which is particularly important when a concrete–water interface exists. For successful tremie placement a steady, uninterrupted flow of concrete through the tremie pipe should be maintained and the seal with the placed concrete kept at all times. Blockages in the pipe can result from even short delays and will require the tremie pipe to be withdrawn and placement started again. At the beginning of the placement the tremie pipe is sealed with a plug so that the concrete is not disturbed as it displaces the water in the pipe.

Because the concrete must flow into place, it should be designed to have a high slump (150 to 250 mm or 6 to 10 in.) without a tendency to segregate or bleed. To obtain flowing, cohesive concrete it is necessary

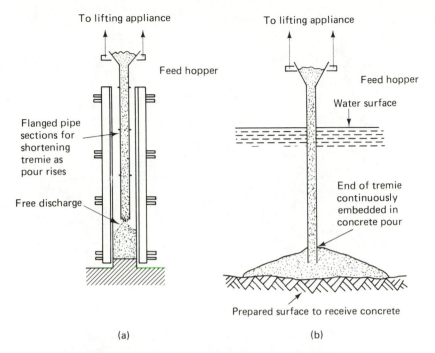

Figure 10.8 Use of the tremie in concrete placement: (a) use in high wall pours; (b) use for underwater placing. (Copyright McGraw-Hill Co. (U.K.) Ltd. From J. R. Illingworth, *Movement and Distribution of Concrete,* 1972. Reproduced by permission.)

to use a high percentage of sand: 40 to 50% by weight of the total aggregate. Consequently, mixes will have high cement contents. Water-reducing admixtures, air-entraining agents, and pozzolans can all be used to advantage in designing tremie concrete. The size of the tremie will depend on the maximum size of the coarse aggregate: a pipe diameter of 200 mm (8 in.) is recommended for 38-mm (1½-in.) aggregate and 150 mm (6 in.) for 19-mm (¾-in.) aggregate.

Underwater Placement

The tremie method is only one method of placing concrete underwater, although the considerations applying to tremie placement also apply to other methods. Pumping methods can be adapted to underwater placement or the preplaced aggregate method used. Special bottom-dump buckets have been designed for underwater placement (Figure 10.9). Such a bucket has "skirts" which are lowered while the concrete is discharged to protect it from the surrounding water.

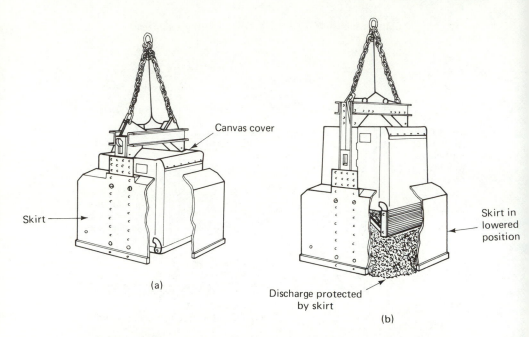

Figure 10.9 Placement with the bottom dump bucket: (a) filled; (b) discharging. (From J. R. Illingworth, *Movement and Distribution of Concrete*, McGraw-Hill Co. (U.K.) Ltd., 1972. Reproduced by permission of John Grist Ltd.)

Consolidation

After placement, the concrete should be worked to eliminate voids and entrapped air and to consolidate the concrete into the corners of the forms and around the reinforcing steel. Most concrete now placed is consolidated by vibration. Proper vibration allows stiffer mixes to be used and generally leads to better consolidation and a superior finish. Concrete slumps may be as little as one-third of those consolidated by hand, and indeed concretes that can be easily worked by hand should not be vibrated, as they are prone to segregation. Vibration is needed for proper compaction of concrete with less than a 50-mm slump. Overvibration brings excess paste to the surface, enhances bleeding, and causes loss of entrained air (Figure 10.10).

Concrete Vibrators

Vibrators apply periodic forces to the concrete with an eccentric rotating mass. The concrete flows (or "liquefies") under the shear forces accompanying the vibration, and the concrete is compacted away from the

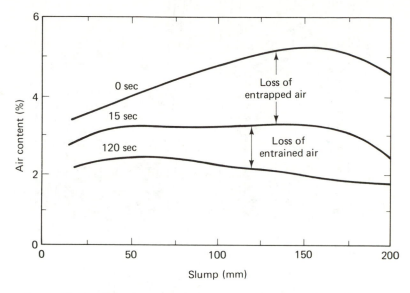

Figure 10.10 Effect of vibration on air content of concrete. (From *Design of Concrete Mixes,* 11th ed., Portland Cement Association, Skokie, Ill., 1968.)

vibrator. Internal or immersion vibrators (often called spud or poker vibrators) operate at frequencies in the range 4000 to 12,000 rev/min, while external vibrators are generally run at 2000 to 6000 rev/min. The response of concrete depends not only on its workability but also on both the frequency and amplitude of vibration. Higher frequencies are generally attained only at the expense of amplitude (dependent on the size of the vibration mass), which may offset any potential improvement in efficiencies. High-frequency vibrators tend to require more maintenance.

External vibrators can be clamped to formwork, when the proper use of internal vibrators is not possible (congested reinforcement, narrow spaces, curved sections, or in slip forming). Vibrating screeds or pan-type vibrators are commonly used on flatwork, especially slip-formed paving. Consolidation on vibrating tables is used in precast work. External vibrators require more power as they impart energy to the formwork also and are generally less effective than internal vibrators. Forms must be strong and rigid and remain watertight, and for this reason metal forms are best used with the external vibrators.

Internal vibrators are most suitable in general construction. The energy imparted by the head of the vibrator excites the solid particles in the concrete mix, causing it to flow. However, the concrete does not move uniformly, as can be seen in Figure 10.11. The coarse aggregate

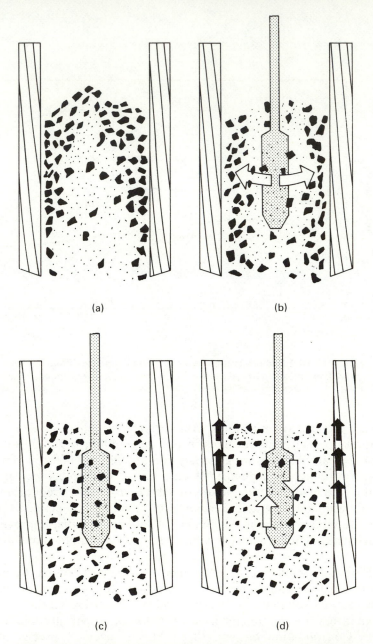

Figure 10.11 Idealized representation of the influence of a high-frequency vibrator on concrete consolidation: (a) The mix is introduced into the form. (b) The vibrator moves aggregate closer together at the form face and cement−sand mortar begins to move outward. Air pockets collect on the faces of forms. (c) The mortar continues to move through the coarse aggregate toward the face of the form. (d) The movement of mortar toward the face is complete. As the operator moves the vibrator down and up, air bubbles move upward along the form face and out of the concrete. (From *Concrete Construction*, Vol. 17, No. 11, 1972, pp. 536−538, Concrete Construction Publications, Inc., 329 Interstate Road, Addison, IL 60101.)

particles are propelled from the vibrator head preferentially because of their greater mass. Momentum is transferred through particle collisions. Mortar then begins to flow between the coarse aggregate. The vibrator head must be moved up and down in the concrete to facilitate mortar flow and homogeneous mixing. A stationary head does not compact the concrete to any great degree. Undervibration is a frequent cause of honeycombing, since the coarse aggregate has been excited, but the mortar has not had sufficient time to flow to the same extent.

The other effect of the vibrator is to force entrapped air out of the concrete. To remove air most efficiently, the vibrator should be plunged rapidly into the concrete and removed slowly with an up-and-down (jigging) motion. The rapid penetration tends to force the concrete upward and outward, thereby helping air to escape initially. As the vibrator is slowly removed, the air is forced upward ahead of the vibrator. Air is also propelled to the formwork surfaces, where it is forced up and out.

The vibrator has only a limited "sphere of influence" so that for uniform compaction the vibrator must be inserted into the concrete at close intervals to ensure that the spheres of influence overlap; typically, these should be about 500 mm (18 in.) apart. A vibrator has no influence below the head, so that when successive lifts are to be vibrated together, the head should be completely immersed into the lower lift (if final set has not occurred; see below) and withdrawn slowly through the upper lift (see Figure 10.5).

The time of vibration for proper consolidation is about 10 to 20 s; vibration should stop when paste first begins to appear around the vibrator. Overvibration can cause undesirable segregation in high-slump mixes with a high mortar content. Vibration is best with low-slump mixes, and when these are properly designed, the effect of overvibration is relatively small. It is more likely that undervibration will occur, resulting in excessive amounts of entrapped air and inhomogeneous concrete, leading in extreme cases to honeycombing. Vibrators should not be used to move concrete laterally in forms, as this causes segregation.

Revibration of concrete an hour or two after initial consolidation may be needed in order to weld successive lifts together. This process can be beneficial and can be used to improve concrete consolidation by eliminating any cracking, voids, or weak areas created by subsequent settlement or bleeding, particularly around reinforcing steel or other embedded materials, or loss of moisture by defects in formwork. Thus, revibration can produce better concrete even after initial set has occurred, but if delayed until final set it will disrupt the developing microstructure and reduce the strength of the concrete.

Vacuum Dewatering

A useful method for consolidation of horizontal surfaces that will be subject to considerable wear is the vacuum dewatering process. In this process a rubber mat is laid on the surface (Figure 10.12) to provide a seal and a vacuum applied inside the mat sucks out water from the concrete. Filter pads prevent fine particles from being removed with the

Figure 10.12 Schematic representation of vacuum dewatering: (a) side view; (b) top view. (From H. Wenander, *Concrete Construction*, Vol. 20, No. 2, 1975, pp. 40–42, Concrete Construction Publications, Inc., 329 Interstate Road, Addison, IL 60101.)

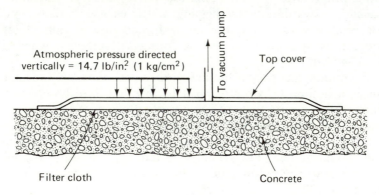

Atmospheric pressure directed vertically = 14.7 lb/in.² (1 kg/cm²)

To vacuum pump

Top cover

Filter cloth

Concrete

(a) Side view

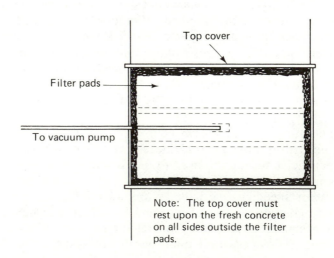

Top cover

Filter pads

To vacuum pump

Note: The top cover must rest upon the fresh concrete on all sides outside the filter pads.

(b) Top view

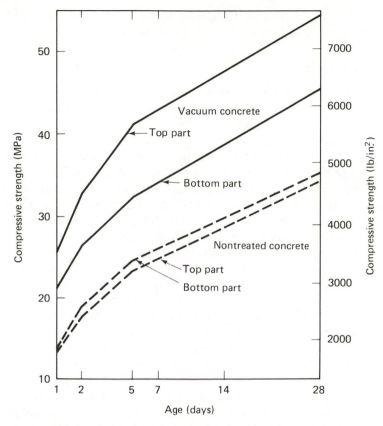

Figure 10.13 Effect of vacuum dewatering on the strength of concrete. (From H. Wenander, *Concrete Construction,* Vol. 20, No. 2, 1975, pp. 40–42, Concrete Construction Publications, Inc., 329 Interstate Road, Addison, IL 60101.)

water. In this way about 15 to 25% of the original water can be removed from the upper 150 to 300 mm of the slab (depending on the length of the treatment). The lower *w/c* ratio and the effective compaction of the concrete give dramatic improvements in strength and durability (see Figure 10.13). Improvements in performance are greater than if calculated simply on the basis of changes in overall *w/c* ratio because of the tendency for the *w/c* ratio to be higher at the surface of a slab due to bleeding.

Vacuum dewatering is started after the concrete has been vibrated and screeded. Subsequently, regular finishing can begin immediately so that dewatering can actually speed up the finishing process. Power finishing is needed because of the hardness of the surface. Initially, a planing disc should be used to remove slight irregularities caused by the dewatering process, followed by final trowelling.

Casting in Lifts

It is desirable to cast concrete in monolithic units between construction joints where possible. Deep placements should be cast in successive horizontal layers, or lifts, avoiding sloping surfaces as much as possible. Adjacent lifts should be "knitted" together by ensuring that the internal vibrator passes an inch or two into the lower lift. Successive lifts should not be more than about 2 ft thick, and should only be a few inches thick if external vibration is used. If excessive vibration of the lower lift has not occurred, there should be no discontinuity between lifts. The slump of concrete should be less in the upper lifts, for the tendency of even well-designed concrete to bleed slightly through settlement will increase the water content in the upper levels.

When the concrete in a lift hardens before the next lift can be placed, a *cold joint* or *construction joint* will result, since the two layers can no longer be consolidated together. The inherent weaknesses of such a joint can be minimized by proper procedures. The surface of the last lift should be left in a roughened state to provide a good mechanical bond. Before placement of the next lift, the surface should be scarified to remove any laitance (see Section 8.1). Air–water jets, wire brooming, or even sandblasting are useful techniques. The concrete should be dampened and a layer of mortar well worked into the surface. This ensures a good bond between fresh and hardened concrete and reduces the possibility of segregation at the bottom of the upper lift.

10.4 FINISHING CONCRETE

Finishing Flatwork

Special techniques have been developed for the finishing of surfaces of concrete slabs for floors and pavements. The object is to produce a dense, compacted, well-graded surface suitable for the conditions of service. Proper finishing of good-quality concrete can ensure a maintenance-free surface but cannot offset the inherent deficiencies of poorly designed concrete. The sequence of finishing operations is as follows.

Screeding

Excess concrete is struck off to bring the surface to the proper grade. This can be done mechanically or by hand. The screed is moved back and forth across the concrete with a forward motion, and surplus concrete is left in front to fill in any low spots. Vibration can accompany screeding. A *darby* or *bullfloat* is used immediately after screeding to firmly embed large aggregate particles and to remove any remaining bumps and hollows.

Floating

The concrete is allowed to harden until all bleed water has disappeared and the concrete can be walked on. The surface is then floated with flat wood or metal blades. This process firmly embeds the aggregate, compacts the surface, and removes any slight imperfections that may remain. Floating tends to bring cement and water to the surface, thus floating too early, or for too long, can be damaging and weaken the surface by forming a high *w/c* ratio layer of paste.

Trowelling

When floating is finished, the surface may be steel-trowelled if a really smooth, dense, wear-resistant surface is required. At this stage the object is compaction and surface defects cannot ᴏe repaired. Again, trowelling too early can be damaging. More than one trowelling can be done if desired.

Texturing

A well-trowelled surface will be smooth and shiny and prone to slipping or skidding, especially when wet. A textured surface may be needed for skid resistance. Scoring the surface with a wire broom or coarse fiber broom is common. After darbying, the surface can be texturized more readily, since it is less compacted; drawing burlap across the surface will provide texture.

Special wear-resistant aggregate toppings are often applied by distributing aggregate evenly over the top of freshly cast concrete, firmly embedding it with a darby, followed by floating and trowelling. Decorative aggregates for exposed finishes may also be applied in this fashion. After floating, excess paste at the surface can be removed by gentle washing and brushing to expose the aggregate.

Hardening

For additional durability and wear resistance, the concrete surface can be treated with certain chemicals. The treatment will cause the precipitation of insoluble compounds in the pores close to the surface, thereby making a stronger, denser layer. Formulations based on sodium silicate or on fluorosilicates are the most common. Reaction with the calcium hydroxide in the paste forms additional C–S–H.

10.5 SPECIAL ASPECTS OF HANDLING CONCRETE

Seasonal Concreting

When placing concrete under the extreme weather conditions that may occur in summer or winter, certain aspects of concreting can be adversely affected and special precautions are needed to ensure quality concrete. Here we discuss problems of concrete mixing and placing, while special curing procedures are discussed in Chapter 11.

Hot-Weather Concreting

The main problems associated with concreting in hot weather are high concrete temperatures and loss of moisture during placement and finishing. High concrete temperatures increase the water requirements to maintain a given slump; decrease setting times, and hence the time available for placement, consolidation, and finishing; increase the danger of plastic shrinkage; and lower the ultimate strength. The optimum concrete temperature should be in the range 10 to 15°C (50 to 60°F), or lower for mass concrete where there is an increased risk of thermal cracking. This can seldom be achieved in summer without artificial cooling. Concrete temperatures should not exceed 30 to 33°C (85 to 90°F) and should be kept as low as is practical.

Concrete temperatures can be regulated by controlling the temperature of the ingredients. The contribution of each constituent is determined by its temperature, specific heat, and weight fraction. This is the basis of Eq. (10.1), which can be used to calculate the concrete temperature, in either °C or °F:

$$T_{concrete} = \frac{H(T_a W_a + T_c W_c) + T_a W_{wa} + T_w W_w}{H(W_a + W_c) + W_{wa} + W_w} \qquad (10.1)$$

where H is the approximate specific heat of cement and aggregate (925 J/kg·°C or 0.22 Btu/lb·°F), W_a, W_c, W_{wa}, and W_w are the weights (kg or lb) of aggregate, cement, aggregate moisture, and mixing water; and T_a, T_c, and T_w are the temperatures of aggregate (including its moisture), cement, and mixing water, respectively. Using this equation it can be seen that cooling the water is the most effective means of lowering the concrete temperature and is easier than cooling the aggregate. The high specific heat of water offsets its small weight fraction so that its contribution is about the same as that of the aggregate. Using ice is even more effective, since heat is absorbed in melting the ice. Equation (10.1) is then modified as follows:

$$T_{\text{concrete}} = \frac{H(T_a W_a + T_c W_c) + T_a W_{wa} + T_w W_w - F_i W_i}{H(W_a + W_c) + W_{wa} + W_w + W_i} \qquad (10.2)$$

where W_i is the weight of ice and F_i its latent heat of fusion (335 kJ/kg or 145 Btu/lb). Ice must be completely melted before mixing is complete. Other simple precautions can be used to keep the temperatures of all materials as low as possible, such as storing materials in the shade or in cooled bins; spraying aggregates with water or covering with white reflective sheeting; and painting water tanks and lines white. The temperature of the cement is seldom a bother, since both its specific heat and weight fractions are low. A change in cement temperature of 9°C is needed to change the concrete temperature 1°C, and only if cement is delivered from the plant still warm are problems likely to occur. Normally, cement has cooled before being shipped, but sometimes in the summer, when the demand for cement is at its greatest, it may be shipped almost as soon as it is made.

Other precautions in the field can help keep the concrete cool. Concrete should be kept shaded from direct sun as much as possible while being transported and placed. It is better to confine actual concreting to early or late in the day. Forms, steel, and subgrade should be wetted down thoroughly just before placement. This not only helps keep the area cool, but it minimizes loss of moisture from the concrete by absorption or evaporation, which can contribute to plastic shrinkage. The concrete should also be protected from drying winds, which is another cause of plastic shrinkage cracking. (Plastic shrinkage is discussed in more detail in Chapter 18; Figure 18.7 gives charts that can be used to determine safe concrete temperatures and wind velocities.) The more rapid hydration of warm concrete hastens setting times, and this may be aggravated if rich mixes are used to offset the increased water requirements. Thus, there is a chance of developing cold joints between successive lifts, and the use of water-reducing, set-retarding admixtures can be used to advantage in such situations.

Cold-Weather Concreting

In the winter months the concern is the exact opposite: the need to maintain concrete at an adequate temperature and to prevent the concrete from freezing early in its life. Desirable concrete temperatures are given in Table 11.1; these make allowances for the cooling effect of the environment. The same considerations apply to warming concrete as to cooling it, but in reverse. The use of hot water is the easiest way of raising concrete temperatures. Warming aggregates also helps, as does storing them in heated enclosures. The use of frozen aggregates is not recommended because of difficulties in batching and increased variations in moisture contents.

Concrete should never be placed on frozen subgrade. Heat loss from the concrete may cause the lower part of a slab to freeze, while uneven settlement or subsequent thawing may cause cracking. The concrete should be protected from freezing for the first day or two. If fresh concrete is frozen, it is permanently damaged, and subsequent normal curing will only partially restore its properties. Hardened concrete should be allowed to partially dry for maximum protection before being exposed to freezing.

Formwork

Formwork is a very important aspect of concrete construction. A detailed discussion is not presented here simply because there is not space to deal adequately with the subject. Whole books have been devoted to formwork, and those recommended in the bibliography at the end of the chapter should be consulted. Nevertheless, the design and construction of formwork can be crucial to the successful implementation of a design and the economics of construction. Formwork is the sole support of fresh concrete and must be designed to withstand some structural loads during construction, including the effects of materials and equipment as well as the lateral pressures induced by fresh concretes. Structural failure of concrete formwork is the largest single cause of construction failures, and a number of states are now requiring formwork to be designed by a registered engineer. Formwork should be watertight, rigid, and properly aligned.

The choice of formwork material is governed solely by questions of cost and desired surface finishes. The use of prefabricated reusable forms is an increasingly cost-effective strategy, while on the other hand permanent forms that become the exterior surface can be used to advantage.

Inspection of Concrete

Frequent and thorough inspection of all concrete operations is also an important part of the whole operation. The good inspector has a good background of concrete and concrete operations, is conversant with the job specifications, has plenty of common sense, and gets on well with people. Adversary relationships between contractor and engineer will not produce quality concrete; problems must be solved with mutual cooperation. The inspector should exercise good judgment when interpreting test results and checking compliances with specifications.

Inspection should cover all aspects of concrete construction and starts with an inspection of the batching plant and its operation. Upon delivery of concrete to the job site, the concrete should be tested to ensure that it is of uniform quality and conforms to specifications. Standard ASTM Test Methods should be used unless special tests are specified. Before the concrete is placed, the inspector will ensure that formwork is constructed according to the plans, is properly aligned, is sturdy and watertight to resist concrete pressures, and is built of clean, sound materials. The inspector has also checked that the reinforcement is laid in correct location and spacing and with the required splicing of overlapping bars, and is free of rust. During placement the various operations should be regularly monitored to ensure that the concrete is not tending to segregate, is properly consolidated, and that special precautions appropriate to the conditions are followed. Finally, the inspector should ensure that the concrete is adequately cured.

Bibliography

General

ACI COMMITTEE 302, "Recommended Practice for Consolidation of Concrete," *Journal of the American Concrete Institute,* Vol. 68, No. 12, pp. 893–932 (1971).

ACI COMMITTEE 304, "Recommended Practice for Measuring, Mixing, Transporting and Placing Concrete," *Journal of the American Concrete Institute,* Vol. 69, No. 7, pp. 374–414 (1972).

ACI COMMITTEE 305, "Hot-Weather Concreting," *Journal of the American Concrete Institute,* Vol. 74, No. 8, pp. 317–332 (1977).

ACI COMMITTEE 306, "Cold-Weather Concreting," *Journal of the American Concrete Institute,* Vol. 75, No. 5, pp. 161–183 (1978).

"*Concrete Manual,*" 8th ed. U.S. Bureau of Reclamation, Denver, Colo., 1975.

Design and Control of Concrete Mixes, Metric ed., Engineering Bulletin. Canadian Portland Cement Association, Toronto, Ontario, 1978.

ILLINGWORTH, J. R., *Movement and Distribution of Concrete.* McGraw-Hill Co. Ltd., Maidenhead, England, 1972.

MURDOCK, L. J. AND K. M. BROOK, "*Concrete Materials and Practice,*" 5th ed. Edward Arnold (Publishers) Ltd., London, 1979.

ORCHARD, D. F., *Concrete Technology,* Vol. 2, *Practice,* 3rd ed. John Wiley & Sons, Inc., New York, 1973.

SCANLON, J. M., "Quality Control During Hot and Cold Weather," *Concrete International. Design and Construction,* Vol. 1, No. 9, pp. 58–65, 1979.

Specific Placements

ACI COMMITTEE 304, "Placing Concrete by Pumping Methods," *Journal of the American Concrete Institute,* Vol. 68, No. 5, pp. 327–344 (1971).

ACI COMMITTEE 304, "Placing Concrete with Belt Conveyors," *Journal of the American Concrete Institute,* Vol. 72, No. 9, pp. 474–490 (1975).

ACI COMMITTEE 304, "Preplaced Aggregate Concrete for Structural and Mass Concrete," *Journal of the American Concrete Institute,* Vol. 66, No. 10, pp. 785–797 (1969).

ACI COMMITTEE 506, "Recommended Practice for Shotcreteing," *Journal of the American Concrete Institute,* Vol. 63, No. 2, pp. 219–246 (1966).

ACI COMMITTEE 506, "Specifications for Materials, Proportioning and Application of Shotcrete," *Journal of the American Concrete Institute,* Vol. 73, No. 12, pp. 679–685 (1976).

Shotcrete, SP-14. American Concrete Institute, Detroit, Mich., 1966.

TOBIN, R. E., "Hydraulic Theory of Concrete Pumping," *Journal of the American Concrete Institute,* Vol. 69, No. 8, pp. 505–510 (1972).

WENANDER, H., "Vacuum Dewatering Is Back," *Concrete Construction,* Vol. 20, pp. 40–46 (February 1975).

Finishing

Concrete Construction, Vol. 24 (April 1979). Special issue on concrete finishing.

HERSEY, A. T., "Causes of Floor Failures," *Journal of the American Concrete Institute,* Vol. 70, No. 6, pp. 426–429 (1973).

YEAGER, J. C., "Concrete Finishing Practices," *Journal of the American Concrete Institute,* Vol. 70, No. 6, pp. 420–425 (1973).

Formwork

ACI COMMITTEE 347, "Recommended Practice for Concrete Formwork," *Journal of the American Concrete Institute,* Vol. 64, No. 7, pp. 337–373 (1967).

HURD, M. K., *Formwork for Concrete,* 3rd ed., SP-4. American Concrete Institute, Detroit, Mich., 1977.

WYNN, A. E., and G. P. MANNING, *Design and Construction of Formwork for Concrete,* 6th ed. Cement and Concrete Association, Slough, U.K., 1974.

Inspection

ACI COMMITTEE 311, "Recommended Practice for Concrete Inspection," *Journal of the American Concrete Institute,* Vol. 71, No. 7, pp. 347–352 (1974).

"Inspection of Concrete," *Journal of the American Concrete Institute,* Vol. 72, No. 6, pp. 269–290 (1975).

Manual of Concrete Inspection, 6th ed., SP-2. American Concrete Institute, Detroit, Mich., 1975.

"Responsibility for Inspection," *Journal of the American Concrete Institute,* Vol. 69, No. 6, pp. 320–333 (1972).

Problems

10.1 Why is it important that concrete be efficiently mixed?

10.2 Compare the advantages and disadvantages of central-mixed and truck-mixed concrete.

10.3 What precautions should be taken when transporting concrete if the job site is a long way from the batching plant?

10.4 Under what circumstances should it be permissible to add material to the concrete mix on the job site?

10.5 What precautions should be taken when belt conveyors are used to transport concrete at a construction site?

10.6 What are the special requirements of mix design when concrete is to be pumped?

10.7 How would you guard against segregation on the job site?

10.8 What are the advantages and disadvantages of slip forming?

10.9 Describe a method for placing concrete underwater, citing any special requirements and precautions.

10.10 What are the effects of (a) undervibration; (b) overvibration?

10.11 How should the method of compaction change as the workability of concrete decreases?

10.12 What is the difference between a cold joint and a construction joint?

10.13 Describe the purposes of the procedures used in finishing concrete flatwork.

10.14 What is honeycombing, and what causes it?

10.15 What is the purpose of controlling the temperature of the concrete ingredients (a) in summer; (b) in winter?

10.16 What is the function of an inspector on the job site?

11

curing

Concrete must be properly cured if its optimum properties are to be developed. An adequate supply of moisture is necessary to ensure that hydration is sufficient to reduce the porosity to a level such that the desired strength and durability can be attained. Concrete structures rarely fail because the specified design strength is not attained, but inadequate strength at the time the forms are stripped may cause problems. Premature cessation of moist curing and removal of formwork during winter construction have more than once been the cause of costly construction failures due to the loading of under-strength concrete. Concrete needs time to gain strength even when good curing methods are being used, and the strength should be checked prior to form removal. The loss of potential durability in the long term is a more widespread and insidious problem, since the maintenance-free service life is reduced.

11.1 CURING AT AMBIENT TEMPERATURES

Parameters Affecting Curing

As mentioned above, water must be supplied for the hydration of cement. Although cement paste will in practice never completely hydrate, because the largest grains become covered with a thick layer of C–S–H which inhibits reaction, the aim of curing is to ensure as much hydration as possible at reasonable cost. Also, in pastes with lower w/c ratios, self-desiccation can occur during hydration and thus prevent further hydration unless water is supplied externally (see Chapter 4). Theoretically, there is enough water in concrete to ensure complete hydration without additional water being supplied if the *w/c* ratio is 0.42 or greater. However, in practice, water is lost from the paste by evaporation, or by absorption of water by aggregates, formwork, or subgrade. To reduce this absorption, the formwork and subgrade should be dampened prior to placement. (The question of absorption by the aggregates is discussed in Chapter 9.) Once enough moisture is lost from the concrete, so that the internal relative humidity drops below about 80% either by evaporation or self-desiccation, hydration will stop and strength development will be arrested. The strength of the concrete will thus be reduced below its potential, and this reduction will be greater in the case of high-strength concrete (low *w/c* ratio) than it is in the case of low-strength concrete (high *w/c* ratio). Therefore, it is desirable to provide additional moisture during curing to ensure maximum hydration.

Figure 11.1 shows the effect of limited moist curing on the development of the compressive strength of concrete. On the cessation of moist curing, the rate of strength gain slows down as water is lost from the concrete, and further strength gain soon ceases. A 3-day period of moist curing will only allow the concrete to reach 75 to 80% of the potential 28-day strength which can be achieved with continuous moist curing. Furthermore, none of the additional 25 to 30% of strength gain that can accrue beyond 28 days will be realized.

Figure 11.1 Compressive strength of concrete dried in laboratory air after preliminary moist curing. (Adapted from *Concrete Manual,* 8th ed., U.S. Bureau of Reclamation, Denver, Colo., 1975; R. H. Price, *Journal of the American Concrete Institute,* Vol. 47, No. 6, 1951, pp. 417–432.)

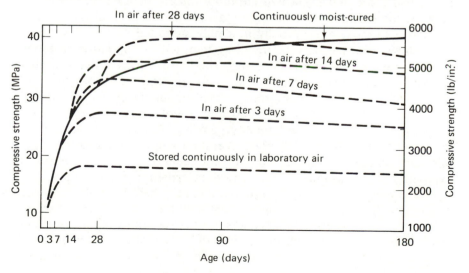

Interrupted Curing

Exterior concrete will be exposed to wetting and drying repeatedly during its lifetime, so that the opportunity for continued hydration still exists. Although it is true that resaturated concrete will resume its interrupted hydration, the amount of strength developed is not as high as it would have been if moist curing had not been interrupted. Moreover, it is not easy to fully resaturate concrete except by means of prolonged immersion so that the amount of additional hydration will be variable and unpredictable, and may be quite small. Interrupted moist curing of young concrete is particularly undesirable. In addition to the reasons discussed above, intermittent moist curing will also subject the concrete to wetting and drying cycles at a time when the concrete is weak enough to be susceptible to tensile stresses that may develop during drying. In the summer months, interrupted moist curing may subject the concrete to thermal length changes, which could also cause cracking.

Effect of Relative Humidity

Concrete will hydrate even if it is not in a fully saturated condition. This is because water is held in larger capillary pores by surface tension forces below 100% RH. The cement can draw on these water reservoirs for further hydration, but the rate will become slower as the relative humidity maintained within the paste is lowered. The partial emptying of the capillary pores interferes with the flow of water in the pore system. Water is used first for hydration of cement within its localized area of the paste, and those areas that hydrate more rapidly (fine cement particles present, for example) will become starved of water. Additional water will move to those areas rapidly in a fully saturated system, but much more slowly in a partially saturated system, and in the latter case this will become the rate-determining step. Therefore, concrete that is sealed against moisture loss hydrates and gains strength more slowly than concrete that is continuously moist cured under water. Water cannot enter the sealed concrete and the consumption of water during hydration lowers the internal relative humidity, slowing the rate of hydration. If the humidity falls below 80% RH, hydration will cease altogether, and this can happen in sealed low-w/c-ratio concretes, or in concretes exposed to an environment with a lower external relative humidity.

The level of the external relative humidity has the most influence on strength development through its control in the drying out of concrete in its early life. The effect of drying is most severe in fresh concrete and those factors are discussed in Section 18.2. The rate of drying of the concrete will have some influence on the residual strength gain after moist curing has been stopped. The surface temperature, wind velocity, and relative humidity are all important factors, as is the thickness of the concrete. The importance of drying rates is only of serious concern if short curing periods are contemplated.

Effect of Temperature

Figure 11.1 refers to curing at 21°C, but temperature affects the rate of strength development, and therefore the length of curing that is necessary. Figure 11.2a shows the effect of curing temperature on strength development after 1 day and 28 days. The increased early strength at higher temperatures is due to the fact that the cement simply hydrates more rapidly, but the decrease in later strength is not so easily explained. It is believed that neither the chemical nor the physical structure of the hydration products is radically changed by the hydration temperature up to about 45°C (115°F). The detrimental effect appears to be due to the fact that it produces a nonuniform distribution of hydration products, leaving weak zones in the cement that govern the strength. Figure 11.2b shows the effect of curing temperature on strength development to times of 90 days and indicates the relative length of curing times required to reach a specified strength at different temperatures. The effect of placing concrete at different temperatures and then holding it at 21°C during the curing period is shown in Figure 11.2c, where the same detrimental effect of higher temperatures is readily seen. Hydration occurs down to about −10°C. We should note that curing at low temperatures can result in a higher ultimate strength, even though the initial rate of strength development is low. Thus, as a general rule, the higher the initial temperature of the concrete, the lower will be its later strength. For 28-day strengths the optimum initial temperature is about 4°C for a Type I cement, as seen in Figure 11.2c. In general, the tensile strength is similarly affected by the curing parameters.

Time of Moist Curing

The most desirable objective is to continuously moist cure the concrete as long as possible—ideally, until the concrete has attained its specified strength. Obviously, this is seldom a practical proposition, and some compromise must be accepted to allow acceptable construction schedules and to minimize costs. It can be seen from Figure 11.1 that moist curing for about 7 days will ensure that the 28-day moist-cured strength will eventually be reached. The ACI Recommended Practice for Curing Concrete (ACI 308) suggests 7 days of moist curing for most

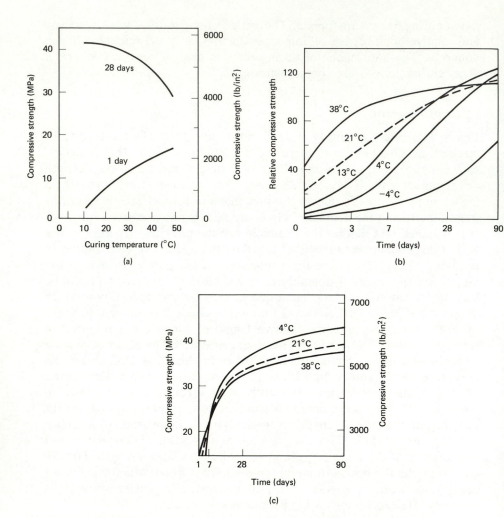

Figure 11.2 Effect of temperature on compressive strength development: (a) Comparison of 1-day and 28-day strengths (constant curing temperature). (Adapted from G. J. Verbeck and R. A. Helmuth, *Proceedings, Fifth International Symposium on the Chemistry of Cement, Tokyo, 1968*, Vol. 3, pp. 1–32.) (b) Curing temperature maintained continuously. (c) Initial concrete temperature, curing temperature maintained at 21°C. (Adapted from W. H. Price, *Journal of the American Concrete Institute*, Vol. 47, No. 6, 1951, pp. 417–432.)

structural concrete, or the time necessary to attain 70% of the specified compressive or flexural strength, whichever is less. For unreinforced mass concrete, minimum curing times should be longer: 2 weeks, or 3 weeks if a pozzolan is used. These longer times allow for the slower strength development of a low-heat cement and of the cement–pozzolan reaction.

These recommendations are for concretes placed and cured at temperatures above 4°C. At lower temperatures the slow rate of hydration will mean that strengths may well be too low even after 7 days of curing. The combined effects of time and temperature on the rate of hydration, and hence strength development, can be estimated using the maturity concept (see Chapter 15).

The danger of freezing is always present at low temperatures. Concrete should not be allowed to freeze until it has developed some strength (about 3.5 MPa or 500 lb/in.²). Generally, 24 h of moist curing at 4°C will not provide this protection without the use of insulation during curing to conserve the initial heat of the concrete and its heat of hydration and thereby keep it out of danger. Additional heat will also be required in more severe cold.

Effect of Carbon Dioxide

In the manufacture of some precast components (e.g., concrete block), the units are exposed to high levels of carbon dioxide, and this results in a more dimensionally stable product, as well as an increase in strength. This kind of curing is only effective with relatively porous concrete with a thin cross section to allow reasonable penetration of carbon dioxide throughout the concrete (see pp. 496–98).

Methods of Curing

There are many methods and materials that can be used for moist curing of concrete. These can be divided into two groups:

1. Water curing—those that supply additional moisture as well as prevent moisture loss.

2. Sealed curing—those that prevent loss of moisture only.

Water Curing

Supply of water to the concrete can be accomplished by ponding, spraying, or sprinkling, or by use of saturated coverings.

Ponding is a thorough method of curing horizontal surfaces and involves maintaining a layer of water on the surface by means of earth or sand dikes. This method is now seldom used, since it is labor-intensive and requires considerable supervision. Keeping concrete moist by means of a fine spray is another excellent method when water is plentiful and runoff is no problem. It can be used on both vertical and horizontal surfaces. A fine spray should be used, such as is provided by most lawn sprinklers or a soil-soaker hose. Ideally, spraying should be continuous, but if it is intermittent, care must be taken to see that the concrete does not dry out between applications. Spraying should not be too vigorous or erosion of the concrete surface may occur. Thus, curing by spraying also requires careful supervision.

The use of coverings that can hold quantities of water in addition to preventing evaporation is another means of water curing. Burlap, or other absorbent materials, are widely used and can be applied to both horizontal and vertical surfaces. Damp earth, sand, straw, or sawdust have been used on horizontal surfaces, but these materials have been largely supplanted by labor-saving curing procedures (see the next section). These materials can double as insulation during the winter by topping off with a dry layer. Coverings of this type require periodic moistening, as they will tend to dry out, so regular supervision is still required. Staining of concrete surfaces sometimes occurs by contact with soluble organic chemicals that may leach from the material, and these may occasionally retard setting or hardening of the surface. The leaching of tannic acid from sawdust with a high tannin content is an example.

Sealed Curing

Waterproof paper, plastic sheeting, and curing membranes are the most widely used materials for sealed curing. Their convenience and lower labor requirements have led them to displace the more traditional water-curing methods in many instances. Formwork can also act as an evaporation barrier, but wooden formwork will absorb moisture from the concrete if not kept damp during the curing period.

Waterproofed paper or plastic sheeting should be applied as soon as the surface has hardened sufficiently to prevent surface damage and after the concrete has been thoroughly wetted. Plastic sheeting is more versatile in that it is more flexible and can be used to cover more complex shapes. Plastic films can be bonded to absorbent materials, which help to retain and redistribute moisture that evaporates from the concrete and condenses on the cover. In this way the moisture can be returned to the concrete and improve its curing. Both paper and plastic can be made white to reflect sunlight and reduce absorption of heat in summer, or can be colored black to increase absorption of heat in winter. Specifications for sheet materials are given in ASTM C171.

Liquid membrane-forming curing compounds have become very popular in the curing of concrete pavements and floors, and can also be used on vertical surfaces. From the concrete's point of view, these compounds provide the least effective method of curing, since they do not entirely prevent evaporation from the concrete, and it is best if they are applied after some initial period of moist curing. In modern pavement laying, however, the application of a curing compound immediately follows the slip-form paver.

Membrane-forming compounds are formulated from resins, waxes, or synthetic rubbers dissolved in a volatile solvent or emulsified in water. Upon removal of the solvent by evaporation, an almost impermeable membrane forms on the surface and seals the concrete against moisture loss. Pigments can be added to the formulation: a white pigment in hot weather to reduce absorption of heat, and gray or black for cold weather. The use of a pigment is advisable so that it can be readily seen whether a complete covering has been applied. Curing compounds should not be used when a topping or overlay will subsequently be laid, or if the surface is to be painted, since the membrane will interfere with the bonding of these materials to the surface. A curing compound should not be used during construction of pavements in the fall which will be exposed to de-icing salts, since the membranes retard the air drying that is needed to improve the salt-scaling resistance of the surface. Specifications for membrane curing compounds are given in ASTM C309.

Comparison of Curing Methods

As mentioned above, water curing is better than sealed curing since it protects the concrete against self-desiccation. This is particularly important when using low-w/c-ratio concretes ($w/c < 0.40$), where self-desiccation can occur rapidly. The efficiency of sealed curing depends very much on the thickness of the covering and the integrity of the covering; care must be taken to ensure that separate sheets are properly overlapped and the joints sealed against moisture loss. Materials for sealed curing can be evaluated by ASTM C156.

The selection of a curing method should be carefully considered where self-desiccation can readily occur. This is strikingly illustrated in the case of curing expansive cements, where the formation of ettringite creates a high demand for water during hydration. As can be seen in Figure 3.10, the choice of an inadequate method will reduce the potential expansion and hence the protection against shrinkage cracking. This illustrates that curing compounds form a somewhat less effective vapor barrier than plastic sheeting, although a good membrane can provide an effective seal.

Curing in Special Situations

Mass Concrete

In the curing of mass concrete, temperature control becomes as important as moisture control. It is important to achieve a constant and uniform temperature throughout the mass as soon as possible after placement. The internal temperature should not rise more than 11°C above the mean annual ambient temperature. An internal cooling system may be needed to ensure this. Since low-heat cements are generally used, the period of moist curing must be extended. The use of a water spray will help to cool the surface and remove heat from the mass (see Chapter 21).

Hot-Weather Concreting

Hot, dry conditions that may prevail are very damaging to freshly placed concrete, and particular attention must be paid to curing procedures under such conditions. The avoidance of excessive plastic shrinkage is imperative. Concrete should be protected from drying conditions, direct sun or wind, and should be kept covered temporarily after finishing and before moist curing is commenced. Curing materials should be used that will reflect sunlight so that concrete temperatures will not become too high. Water curing is most desirable, and care should be taken to keep the concrete moist at all times because of possible stresses imposed on the concrete by alternate wetting and drying or by contact of cold water on a warm concrete surface. ACI Recommended Practice 305 addresses the special problems of hot-weather concreting.

Cold-Weather Concreting

During cold weather special precautions should be taken in curing, as outlined in the ACI Recommended Practice 306. The problems associated with temperatures below 4°C are:

1. Freezing of concrete while saturated and of low strength.

2. Slow development of strength.

3. Thermal stresses on cooling to ambient temperatures (if the fresh concrete is heated during curing).

Table 11.1

Recommended Concrete Temperatures (°C) for Cold-Weather Construction[a]

	Size of Section		
Condition of Placement and Curing	*Thin*	*Moderate*	*Mass*
Minimum temperature of fresh concrete, *as mixed*			
Above 0°C ambient	16	13	10
−18 to 0°C ambient	18	16	13
Below −18°C ambient	21	18	16
Minimum temperature, *as placed*	13	10	7
Maximum allowable drop in concrete temperature during first 24 h after end of protection	28	22	17

[a]$°F = (^9/_5 \times °C) + 32.$

To help offset the problems of 1 and 2, the concrete temperature should not be too low when placed (see Table 11.1). If the ambient temperature is not too low, the heat of hydration, together with adequate insulation of formwork and exposed surfaces, should protect the concrete from freezing in its early life. Lower concrete temperatures can be specified for mass concrete because less heat is dissipated during the curing period. Concrete temperatures above 21°C do not provide more protection from freezing because the loss of heat is greater during transporting and placing. Also, higher temperatures will require more mixing water and shorten the working time. Two days' protection is recommended when Type I cement is used and 1 day when Type III cement, or an accelerating admixture, is used. Batt insulation or special insulation blankets can be used; dry straw is also effective.

When temperatures are too low, insulation may not be sufficient to maintain a temperature that will give adequate strength development and prevent freezing. In this case, the concrete needs to be heated by an external heat source. This can be done by heating the formwork directly or by building an enclosure around the concrete and its formwork and heating the enclosure. Live steam is the best form of heating since moisture is also supplied, but "dry heat" is more commonly supplied through the use of oil-fired heaters, which should be vented outside. When these are used, special care must be taken to ensure that the concrete does not dry out. When concrete is externally heated during the curing period, it should not then be exposed directly to a low ambient temperature because of the danger of thermal shock. The concrete should be allowed to cool gradually; the maximum temperature drop during the first 24 h should not exceed that given in Table 11.1.

11.2 CURING AT ELEVATED TEMPERATURES

Low-Pressure Steam Curing

Curing in live steam at atmospheric pressure dramatically increases the rate of strength development of concrete. Steam curing is used primarily for precast concrete products such as masonary block, pipe, prestressed beams and wall panels, but can also be used for enclosed cast-in-place structures. In the precast concrete industry, steam curing allows increased production by a more rapid turnover of molds and formwork, shorter curing periods before shipment or prestressing, and less damage to the product during handling. The ACI Recommended Practice 517 outlines atmospheric pressure steam curing of concrete.

Maximum curing temperatures may be anywhere in the range 40 to 100°C (104 to 212°F), although the optimum temperature is in the range 65 to 80°C (150 to 175°F). The temperature will be a compromise between rate of strength gain and ultimate strength (see Figure 11.3), since, as we have already seen, the higher the initial temperature, the lower the ultimate strength. The optimum temperature depends on the application; the use of a lower temperature requires a longer curing period but gives a better ultimate strength.

The Curing Cycle

The length of the total curing period must allow for controlled heating and cooling of the concrete. A typical curing sequence is given schematically in Figure 11.4. The concrete should be allowed to remain at room temperature for some time after molding before being exposed to steam, to allow the product to undergo some initial hydration and improve its stability; this will improve later strength. The optimum length of this presteaming ("holding" or "delay") period is generally 2 to 6 h; it depends on the type of cement, maximum curing temperature, and so on, and should be determined for each application.

The rate of temperature rise and duration of the presteaming period are apparently interrelated. If a "presteaming" period is used, quite high rates of heating can be used (up to 33°C/h or 60°F/h), but in the absence of presteaming, the rate should not exceed 11°C/h (20°F/h). Generally, it is desirable to use as low a rate of heating as possible consistent with a convenient and economical curing schedule.

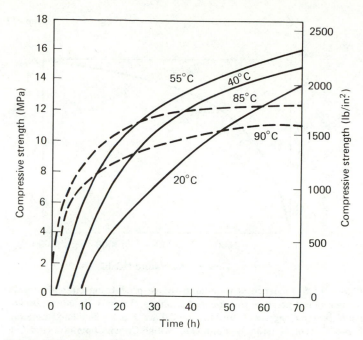

Figure 11.3 Effect of low-pressure steam curing on early-strength development. This data was obtained on concrete steam cured immediately after casting, but it is preferable to delay steam curing a few hours; see text. (Adapted from *Concrete Manual,* 8th ed., U.S. Bureau of Reclamation, Denver, Colo., 1975.)

Figure 11.4 Typical sequence for low-pressure steam curing.

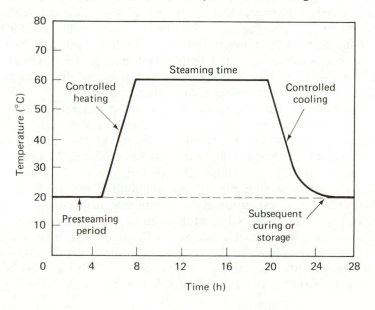

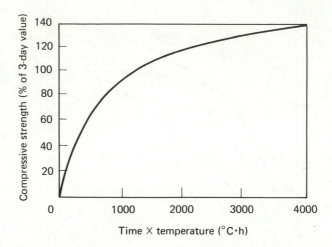

Figure 11.5 Influence of time-temperature product on strength development of low-pressure steam-cured products. (From R. W. Nurse, *Building Research Congress*, 1951, Division 2, Part D, p. 89, Building Research Establishment, Garston, Watford, U.K. British Crown Copyright, HMSO.)

The time the concrete spends at the maximum temperature (the "soaking" time) determines the amount of strength gain during the curing period. The strength gain can be related to the time–temperature product (Figure 11.5) or to the "maturity" (see Chapter 15); thus, the higher the maximum curing temperature, the shorter the curing time. Generally, as we have already observed, this higher rate of hydration will lead to a lower ultimate strength. Figure 11.3 shows that above 80°C (175°F) the effect of the high curing temperature on strength is noticeable before 3 days; but it has been suggested that by optimizing the "presteaming period" and avoiding high heating rates, the strength after 3 days can exceed the 28-day strength of normally cured concrete.

At the end of the curing period, the hardened concrete is much less susceptible to thermal shock than the fresh concrete was, and control of cooling rates is thus not as critical as control of the heating rates. Nevertheless, it is advisable to hold the rate of cooling in the range 22 to 33°C/h (40 to 60°F/h). Sometimes the steaming cycle is designed to provide only sufficient strength for safe handling of the product, and a period of additional moist curing at room temperature *(secondary curing)* is needed. Units can be placed in a fog room or merely stored in a moist condition under a waterproof covering for the additional time required. Products that have gained the desired strengths are generally allowed to dry before shipment; either air drying or accelerated drying using hot air can be used. Masonry block is often dried in this way; lowering the moisture content improves its dimensional stability and raises the compressive strength somewhat.

Properties of Concrete

The characteristics of low-pressure steam-cured concretes do not differ markedly from those of concretes cured under ambient conditions. The hydration of the cement compounds proceed more rapidly (C_3S is completely hydrated in 3 days at 100°C), but the hydration reactions are basically the same. C–S–H remains an amorphous material, although its composition changes with curing temperature. At higher temperatures there is an increasing tendency for the calcium sulfoaluminates to become unstable and much more sulfate and alumina are incorporated into the C–S–H. All ASTM types of portland cement can be used in steam curing. The reactions of pozzolans and slag additions are also accelerated by the higher curing temperatures. Air-entraining admixtures, water reducers, and set-controlling admixtures can be used with steam curing.

The decrease in ultimate strength is believed to be a result of a less uniform distribution of hydration products in the paste because of the rapid initial hydration. This is reflected in changes in pore-size distribution, particularly of the large capillary pores greater than 0.01 μm in diameter. This also results in a somewhat higher permeability, although the change is not great and durability is little affected by steam curing. Low-pressure steam-cured concretes generally show lower creep and shrinkage strains (about one-third less).

High-Pressure Steam Curing

If curing temperatures in excess of 100°C are desired, then saturated steam pressures must be allowed to develop, and a sealed enclosure must be used. The pressure vessel is generally known as an *autoclave,* and the term *autoclaving* is synonymous with high-pressure steam curing. The range of curing temperature used in autoclaving is 160 to 210°C (320 to 410°F) at steam pressures of 6 to 20 atm. The chemistry of hydration changes under these conditions and results in a product that has substantially different properties from products cured below 100°C. The most important improvements are:

1. Products are ready for use within 24 h; the strength is generally equivalent to 28-day strength under ambient curing.

2. Substantially less creep and shrinkage.

3. Better sulfate resistance.

4. Elimination of efflorescence (see Chapter 20).

5. Lower moisture content after curing.

High-pressure steam curing can only be used for precast concrete products. It can be used to advantage in the manufacture of specialty products, such as asbestos–cement composites, lightweight cellular concrete, and calcium silicate (sand–lime) bricks.

The Curing Cycle

The curing cycle is similar to that used in low-pressure steam curing and consists of a "presteaming" period, a "soaking" period, and controlled rates of heating and cooling. The rate of pressure release at the end of the soaking period should also be controlled. The length of the presteaming period depends on mix design, curing temperature, and other factors. The rate of temperature and pressure rise should be such that maximum temperature is reached in about 3 h. Sometimes a period of low-pressure steam curing is used before autoclaving (the two-stage process), in which case pressure buildup can be more rapid. Again, the length of soaking depends on the maximum curing temperature and the desired strength: 8 h at 175°C (350°F) is a typical period. Pressure release at the end of the soaking period should be quite rapid and be complete in 20 to 30 min. This allows flash evaporation of moisture from the product, which helps to cool it. As in low-pressure steam curing, the exact curing cycle will be determined by a compromise between satisfactory development of properties and economic considerations.

Chemistry of Autoclaving

If products containing only portland cement as the binder are autoclaved, very low strengths are obtained. Satisfactory strengths are achieved only when reactive siliceous material is added to the product, as can be seen in Figure 11.6. This occurs because under the conditions of high temperature and pressure, the chemistry of hydration is substantially altered. C–S–H forms initially, as in normal hydration, but is rapidly converted to a crystalline product, α-dicalcium silicate hydrate, according to Eq. (11.1):

$$(C_3S + C_2S) \xrightarrow{\text{H}} C_3S_2H_3 + CH \longrightarrow \alpha\text{-}C_2SH \quad (11.1)$$

$$\text{portland cement} \qquad \underset{\substack{\text{calcium} \\ \text{hydroxide}}}{C\text{–}S\text{–}H}$$

The conversion of C–S–H to α-C_2SH is accompanied by an increase in density and smaller volume of the solid phase, and therefore an increase

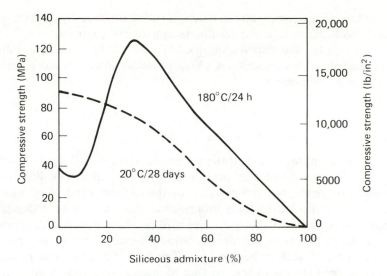

Figure 11.6 Effect of siliceous additions on the strength development of autoclaved concrete. (From C. A. Menzel, *Journal of the American Concrete Institute,* Vol. 31, No. 3, 1934, pp. 125–148.)

in porosity. This change is analogous to the conversion reaction that causes eventual strength regression in concretes made with calcium aluminate cements.

The addition of reactive silica modifies the hydration sequence as shown in Eq. (11.2):

$$(C_3S + C_2S) + S \xrightarrow{H} C_3S_2H_3 + CH + S \quad \rightarrow \quad C-S-H \rightarrow C_5S_6H_5 \quad (11.2)$$

portland cement   silica   tobermorite

The presence of silica promotes the formation of poorly crystallized C–S–H, but one which has less lime than that formed in normal hydration. (Hence it is not written as $C_3S_2H_3$.) This low-lime gel converts to another crystalline hydrate, called *tobermorite,* on continued heating. This change is not accompanied by a very large change in density, and hence the initial strength is maintained during the complete autoclave cycle. Equation (11.2) involves the pozzolanic reaction between reactive silica and lime, so pozzolanic materials such as fly ash or natural pozzolans can be used as the source of silica. However, at these elevated temperatures less-reactive forms of silica will also act as pozzolans, and ground quartz (silica flour) can be used as the source of silica. The fine aggregate may also provide reactive silica, as in the case of expanded shale block; lower quantities of added silica are then needed. Hydrated lime may be used in place of portland cement, as in the case of sand–lime bricks.

At these high temperatures calcium sulfoaluminate hydrates are not formed. Both the sulfate and aluminate apparently enter into the C–S–H and tobermorite structures. Alumina is said to increase the rate of crystallization of tobermorite. C_3AH_6 is occasionally observed but is only a minor component.

Properties of Autoclaved Products

It will now be appreciated that the optimum strength will occur when no α-C_2SH is formed, and this occurs at additions of 30 to 40% silica, depending on the exact composition of the cement used. Any silica in excess of that needed to form tobermorite does not react under these conditions but becomes merely an inert filler which causes a decrease in strength proportional to its concentration. Hence, there is an optimum silica content, as seen in Figure 11.6. The pozzolanic reaction proceeds faster at higher temperatures, so that shorter curing times are needed. Prolonged autoclaving at 180°C or higher may cause the formation of other crystalline calcium silicate hydrates with a concomitant strength reduction. It is believed that even the complete conversion to tobermorite is not desirable and that there is an optimum ratio of amorphous to crystalline material for maximum strength.

The drying shrinkage of autoclaved concrete is only about one-third that of concrete cured under ambient conditions. This is due to the preponderance of crystalline tobermorite, which does not undergo the chemical and physical changes that occur in amorphous C–S–H as moisture is removed (see the discussion in Chapter 18). Since there is still a capillary pore network, surface tension forces operate on drying and some shrinkage is observed; this is mostly reversible, however. The amount of shrinkage that occurs will depend on the proportion of amorphous C–S–H and the pore-size distribution; the presence of alumina will reduce shrinkage and this is probably a result of its influence on tobermorite formation. Creep of concrete is similarly reduced.

Since no calcium sulfoaluminate hydrates are present in the matrix, resistance to sulfate attack is markedly improved. The absence of free lime must also play a role and is also the reason that little efflorescence occurs in autoclaved products.

When properly cured, autoclaved products have a uniform light color and thus are very suitable for use with pigments. The color is more stable than in products cured by low-pressure steam or under ambient conditions. Defects caused by unsound materials or deleterious impurities also occur during autoclaving, rather than during service, and the defective components may be rejected before use.

There are few disadvantages of autoclaved products apart from the inherent technical limitations as to the kinds of products that can be so treated and the high capital cost of the plant. The bond strength between the concrete and the reinforcement is usually much lower (by about 50%) and the material tends to be more brittle than ordinary concrete.

Bibliography

ACI COMMITTEE 308, "Recommended Practice for Curing Concrete," *Journal of the American Concrete Institute,* Vol. 68, No. 4, pp. 233–243 (1971).

ACI COMMITTEE 516, "High Pressure Steam Curing: Modern Practice and Properties of Autoclaved Products," *Journal of the American Concrete Institute,* Vol. 62, No. 8, pp. 869–908 (1965).

ACI COMMITTEE 517, "Recommended Practice for Atmospheric Pressure Steam Curing of Concrete," *Journal of the American Concrete Institute,* Vol. 66, No. 8, pp. 629–646 (1969).

Design and Control of Concrete Mixes, Metric ed., Engineering Bulletin. Canadian Portland Cement Association, Toronto, Ontario, 1978.

Menzel Symposium on High Pressure Steam Curing, SP-32. American Concrete Institute, Detroit, Mich., 1972.

ORCHARD, D. F., *Concrete Technology,* Vol. 2, *Practice,* 3rd ed. John Wiley & Sons, Inc., New York, 1973.

TAYLOR, H. F. W., "Steam Curing of Portland Cement Products," in *The Chemistry of Cements,* ed. H. F. W. Taylor, Vol. 1, pp. 417–432. Academic Press, Inc. (London) Ltd., London, 1964.

Problems

11.1   What are the parameters that affect curing?

11.2   Is interrupted curing detrimental to concrete quality?

11.3   Is it absolutely necessary to cure concrete at 100% RH?

11.4   What are optimum curing conditions for strength and durability? Are these the most economical conditions?

11.5   What are the advantages and disadvantages of curing compounds?

11.6   What special precautions must be taken when placing concrete in the summer?

11.7   What is the effect of freezing on very young concrete? How can freezing be avoided?

11.8   What are the advantages and disadvantages of (a) low-pressure steam curing; (b) high-pressure steam curing?

11.9   Why is it necessary to add silica to concrete products that will be autoclaved?

11.10 Why must both the rate of heating and the rate of cooling be controlled in the high-pressure steam curing cycle?

12

architectural concrete

One of the advantages of concrete is its versatility not only as a building material but also as an architectural material. Concrete can be cast into a variety of complex shapes; it can be given a variety of special surface finishes, textures, and colors; and it can be used in sculptures, murals, and other special aesthetic creations. Architectural effects can be obtained by a suitable choice of concrete materials, formwork materials, special casting techniques, or texturing of hardened concrete surfaces. Generally, architectural concrete will cost more than plain structural concrete, but may actually be cheaper if other architectural treatments, such as painting or brick cladding, are included in the comparison. Since careful control of color, texture, and design details is more important for the overall aesthetic effect, it is important that special attention be paid to uniformity of the component materials and of the concrete, and that strict control over fabrication processes be maintained. These special considerations mean that better overall quality control is necessary; this will result in a better overall product and improved properties for a maintenance-free structure.

It is not possible to do justice in the space available to the infinite variety of aesthetic effects that concrete can offer. This chapter will only survey the various materials and processes that have been used to produce special effects. Details of architectural design that may be important for the overall effect will not be discussed in this book.

12.1 COLORED CONCRETE

Manipulation of color is one strategy an architect can use to advantage. Color in concrete can be achieved through use of special cements and pigments or by the selection of colored aggregates. Construction costs of white or colored concretes will be higher than for conventional concrete because special attention must be paid to handling and consolidation in order to attain uniform colors. This is discussed in the next section.

Cements

The gray color of portland cement may not set off the color of aggregates to advantage and will modify the colors obtained by the coloring agents. Therefore, white cement (see Chapter 3) is commonly used, even though it is considerably more expensive than ordinary portland cement. White cements will produce colored concretes with brighter, truer colors, and can be used to produce pastel shades as well as white concrete. However, because of atmospheric pollution, white or colored concrete (particulary light colors) may soon become discolored and may need periodic sand blasting or other cleaning to restore the original color. Some cements may be buff, tan, or light gray in color or have a greenish hue, depending on the nature of the raw materials available.

Coloring Agents

The most popular means of obtaining integrally colored concrete is to add color pigments during mixing. However, special care must be taken to ensure good dispersion of the pigment within the concrete. Some cement companies provide a range of colored cements by inter-grinding a pigment with a white cement at the factory. Good control of color uniformity can thus be obtained. Even so, it is difficult to make successive batches of concrete over a period of time with exactly the same color. This is one of the reasons why colored concrete is not often used. Pigments should provide a permanent color that will not fade on normal exposure to the elements. They should be particularly resistant to the alkaline effects of the lime, which can cause some pigments to lose their color. Typical pigments are listed in Table 12.1.

Color pigments are finely divided solids which can be classified as mineral admixtures, although their levels of addition are much lower than those of other mineral admixtures. They can be used with other admixtures; indeed, the use of a water-reducing admixture is often advisable since it can improve the dispersion of the pigment and minimize its tendency to float to the surface, as well as reducing laitance and efflorescence, which will reduce color intensity. The addition of calcium

Table 12.1

Typical Pigments for Integrally Colored Concrete

Color	Pigment	Typical Dose[a] (by Weight of Cement)
Red	Iron oxide (hematite)	5%
Yellow	Iron oxide	5%
Green	Chromium oxide	6%
Blue	Phthalocyanine	0.7%
	Cobalt oxide	5%
Black	Carbon	2%
	Iron oxide (magnetite)	5%
Brown	Iron oxide	5%

[a]Amount above which no increase in intensity is observed.

chloride should be avoided since it can interfere with color uniformity. In some cases colored admixtures are available which are carefully controlled blends of pigments with compatible admixtures. Pigments for cement and cement products are described in BS 1014.

Pigments can also be incorporated in products that are applied directly to horizontal surfaces. They can be used during the finishing process as a dry shake application to provide color at the surface only, rather than through the complete section. They can also be formulated with surface hardeners, abrasion-resistant toppings, and so on, that are used to treat horizontal wearing surfaces. A problem with this approach to coloring is obtaining uniform color and properly finished concrete. However, this is a more economical use of pigments, and greater intensity of color can be more easily achieved.

Chemical stains have also been used to color concrete surfaces. These materials are aqueous, acidic solutions of metallic salts. The salts react with the alkaline concrete to deposit metal hydroxides (or hydroxyl salts) within the pores. The colors produced depend on the nature of the metal salt, but are usually black, green, or various shades of brown. The acid etches the concrete slightly to assist in penetration of the stain. Generally, only the cement paste is stained, the aggregate being unaffected. Thus, stains can be used to advantage with exposed aggregate concrete or other textured surfaces where integrally colored concrete is not desired.

Aggregates

There are a wide variety of naturally colored rocks that can be used as concrete aggregates to give pleasing color effects. The range of colors is much more varied than can be obtained by the use of pigments. The kinds of rock that can be obtained in colored forms are very widespread, and some variety of color can be obtained almost anywhere. Most common are the brown and ochre shades that can be obtained in many river gravels and the pure white of quartz. The latter is particularly useful as a fine aggregate (silica sand) when white or colored concrete is used. Marbles offer the widest range of color, while granite (which is an excellent concrete aggregate) is obtained in shades of pink, gray, black, and white. The range of natural colors can be extended by the use of colored glasses and ceramics; however, many glasses are susceptible to the alkali–aggregate reaction. Colored aggregates are best used to advantage in exposed aggregate finishes or in sawn and polished terrazzo slabs. They are an expensive form of aggregate since the material must be carefully processed to give a uniform color and texture.

12.2 MIX DESIGN AND HANDLING

Mix design, choice of formwork materials, handling and compaction, and curing methods can all lead to color and textural variations in form-finished surfaces. The major problems are summarized in Table 12.2.

Mix Design

The principles of mix design for architectural concrete are no different from ordinary structural concrete. When the surface effects call for intricate as-cast finishes, particular attention must be paid to workability under vibration to ensure the intimate contact with forms that will give sharp, clean details. The slump should be as low as possible, consistent with the materials and type of placement, to reduce the possibility of bleeding, which can lead to color variations within a lift. To obtain surfaces of uniform color, a high cement content and additional fine sand are desirable. Where exposed aggregate finishes will be used, a higher-than-normal proportion of coarse aggregate is desirable to give a good density of exposed aggregate. The use of gap-graded aggregate with a narrow size range can provide a more uniform distribution of exposed aggregate. Fine sands, rather than coarse sands, should be used to maintain workability.

Mixing, Transporting, Placing and Compacting

Proper handling of concrete is critical to ensure high-quality concrete finished free of defects and blemishes. This is particularly true when white cement or pastel-colored cements are used, because they will clearly reveal differences in texture, surface contamination, and the presence of impurities. Accurate batching is as important for aesthetic

Table 12.2

Surface Blemishes on Form-finished Concrete

Description	Most Probable Causes
Color variation	Variations in materials
	Incomplete mixing
	Segregation during placing
	Variations in batching
Hydration discoloration	Variable absorption of formwork
	Formwork leakage
	Uneven application of form release agent
	Uneven curing
Aggregate transparency	Low sand content
	Smooth, flexible formwork
	Excessive vibration
	External vibration
Dye discoloration	Stains or dyes on form face
	Reactions with form coatings
	Impure release agent
Brown discoloration	Impure release agent
	Excessive amounts of release agent
	Rust from forms
Honeycombing	Insufficient fines
	Compaction inadequate
	Leaking joints
Bugholes	Oversanded mix
	Nonabsorptive forms
	Poor wetting characteristics of form face
	Inadequate vibration
Spalling or chipping	Low-strength concrete
	Insufficient form release agent
	Forms removed too early
Crazing	Mix too rich; w/c ratio high
	Formwork with low absorbency
	Insufficient curing
	Forms removed too early

considerations as it is for structural purposes. Variations in water and cement content can have a considerable effect on the color of the concrete. A low w/c ratio gives a darker color, due in a large part to the fact that the ferrite phase hydrates to a lesser extent in these areas. Equipment and conveyances should be kept clean and well maintained to prevent contamination by oil and grease. Mixing should always be exactly the same for each batch when color is used to ensure the same dispersion of pigment. Cement and pigment should be preblended, and mixing times may be longer than usual to ensure uniform dispersion.

Concrete should be placed as close as possible to its final position to minimize potential segregation. It should not be allowed to splatter the formwork, since this can cause surface defects. When as-cast finishes are required, consolidation is an important consideration. There is more tolerance when surfaces will later be textured, but even in such cases gross variations in consolidation may not be masked. Proper vibration is critical for a uniform finish that is free of blemishes and has clean, sharp edges. Chapter 10 describes the process of vibration and the correct procedures to be used for good vibration. When a well-designed low-slump concrete is used, it is unlikely that ill effects will be caused by overvibration; undervibration is more likely to cause surface blemishes. If vibration is not carried out properly, air driven to the surface of the forms will not be fully expelled. The vibrator must not be allowed to touch the forms, however, as this will result in areas rich in paste, which will cause textural and color variations. When exposed aggregate finishes are to be used, the vibrator should be kept at least 75 mm (3 in.) from the form face to prevent pockets of fine materials from forming.

Formwork

If an as-cast finish is required, special care must be taken in choosing materials (see Table 12.3) and in the construction and preparation of formwork. Concrete reproduces the surface against which it is

Table 12.3

Formwork Materials for Architectural Concrete

Form Material	*Comments*	*Number of Times Can Be Reused*
Plywood	Variable absorption may cause staining: coat with wood sealer	Up to 5
Plastic-coated plywood	Joints must be sealed, may cause staining	10
Steel	May cause variations in color and textures; rust may cause stains: coat with epoxies	50–100
Fiberglass		20–30
Form liners:		
Lumber	Special textures; variable absorption, and may retard set: coat with wood sealer	1–20
Rubber, PVC	Special textures; flexible; uniform surfaces	100
Fiberglass	Special shapes and textures	20–30

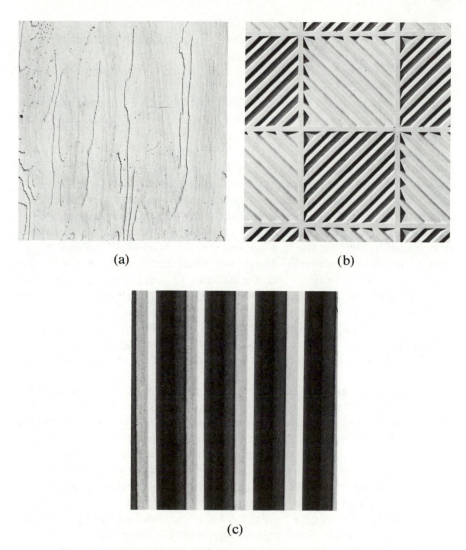

(a) (b)

(c)

Figure 12.1 Examples of surface texture provided by formwork: (a) wood grain; (b) rubber mat; (c) plastic liner. (Photograph courtesy of Portland Cement Association.)

cast very faithfully. This can be used to advantage to obtain wood-grain effects by casting directly against wood (Figure 12.1) or to obtain textures imprinted in rubber or plastic form liners. However, concrete will also faithfully reproduce blemishes and defects on the form surfaces, such as knots, patches, and marks stamped or drawn on the surface. Staining may be caused by untreated wood, by form oils, or by rusting steel.

Certain plywoods and fiberboards can cause severe staining, and protective coatings may be needed. Staining is particularly noticeable with white concrete, and special nonstaining release agents are available.

If the formwork is absorbent (e.g., wood), it will remove water from the concrete before setting. The resultant moisture movement brings cement particles to the surface. Thus, at areas of higher absorption there will be a low *w/c* ratio at the surface, which causes a darker coloration, as discussed above. This is known as *hydration discoloration* and can be eliminated by sealing the form surface, by using a material of uniform absorbency (not easy), or by choosing a nonabsorbent form material. Plastic sheeting can be used as a nonabsorptive form liner, but thin plastic film should not be used because wrinkles are not easily eliminated and will be reproduced on the surface. Concrete cast against nonabsorbent surfaces has a lighter color. Also, materials of low absorbency tend to increase the number and size of "bugholes" (called "blowholes" in some countries). These are small entrapped air voids that occur at the form surface and are difficult to remove completely even with adequate vibration. Bugholes are reduced by decreasing the amount of sand, particularly the coarser fractions, but low sand contents lead to greater color variations. Thus, some trade-off is needed between uniform color and the occurrence of bugholes. Form-release agents may encourage the formation of bugholes, and the incorporation of a surface-active agent is recommended.

Aggregate transparency occurs when concrete is cast against smooth nonabsorbent forms. A very thin layer of mortar always separates the coarse aggregate from the form. When in contact with a smooth surface, the calcium hydroxide crystals that form during hydration are oriented to form a transparent coating through which the aggregate is visible. This gives a mottled appearance. On rough surfaces or those in contact with absorptive forms, calcium hydroxide crystallizes randomly and forms a translucent layer. Aggregate transparency can be corrected by roughening the surface by light etching or abrading, or by using an absorptive form liner. Some release agents give a roughened surface.

Tightly constructed formwork that will not leak is important for good finishes. Loss of water at leaky joints will lead to hydration discoloration, while loss of paste gives changes in texture and color. Excessive loss of paste can lead to honeycombing and scabbing (Figure 12.2)—a situation that should be avoided even with structural concrete. Leakage at joints can also aggravate the bughole problem, as paste is expelled and air sucked in during the vibration cycle (see Figure 12.3). Whenever possible, joints should be sealed with tape, filled with a suitable caulking material, or protected by compressible gaskets.

Figure 12.2 Effect of leakage of formwork joints. (Photograph courtesy of Concrete Construction Publications, 329 Interstate Road, Addison, IL 60101.)

Figure 12.3 Effect of vibration on leaking formwork joints (movements and deflections exaggerated for emphasis). (From *Concrete Construction,* Vol. 17, No. 11, 1972, pp. 536 –538, Concrete Construction Publications, Inc., 329 Interstate Road, Addison, IL 60101.)

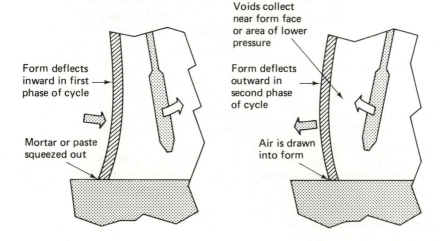

Voids collect
near form face
or area of lower
pressure

Form deflects
inward in first
phase of cycle

Form deflects
outward in
second phase
of cycle

Mortar or paste
squeezed out

Air is drawn
into form

Finishing and Flatwork

Trowelling of white or colored concrete flatwork improves color uniformity by smoothing out minor difference in texture and bringing more cement (and pigment) to the surface. This results in better color and a more reflective surface. However, if trowelling is delayed until the concrete is too stiff, "trowel burning" may occur. This is discoloration caused mostly by uneven compaction. To avoid trowel burning, concrete surfaces should be protected to eliminate excessive evaporation, the use of calcium chloride should be avoided, and plastic trowels should be used.

There are many other factors that can lead to mottling or discoloration of concrete flatwork. For example, the alkali content of the cement, or inappropriate curing procedures, can cause or aggravate discoloration. The mottled appearance of a surface can be caused by changes in the amount of hydration of the ferrite phase or by local deposition of salts within the pores. In some cases repeated washings may remove the discoloration but in other cases chemical treatment may be necessary.

Curing

Good curing practices are important for maintaining uniformity of concrete surfaces as well as for ensuring proper hydration and strength development. Concrete that is allowed to dry prematurely is lighter in color than concrete that is cured longer. This is the same effect as hydration discoloration, since insufficiently curing cement paste has the same effect as increasing w/c ratio. Thus, concrete should be uniformly cured or color variations will again result from hydration discoloration. If plastic sheeting is used to keep concrete moist, partial contact with the concrete should be avoided, since this can cause mottling. Waterproofed paper should not cause staining of surfaces. Concrete is best cured for several days before removing formwork to ensure uniform hydration and good strength development, to reduce possible damage during formwork removal, and to minimize cracking or crazing caused by rapid drying. Curing compounds are not recommended, since they can cause staining or discoloration.

Concrete sealers are sometimes used to protect white or colored concretes from contamination by atmospheric pollution. There is a wide variation in the performance of such materials: some can cause permanent discoloration of the concrete and they are generally avoided except in areas of heavy pollution.

General Cleanup and Patching

After removal of formwork, the exposed concrete should be cleaned and restored to a uniform, smooth surface. This requires packing bolt holes and tierod holes with mortar, repairing honeycombing, removing bulges and projections, removing stains, and filling in bugholes and air voids at the surface. These repairs should be made as soon as possible after form removal. If concrete placement and consolidation have been carefully controlled, there should be a minimum amount of cleanup needed, but in extreme cases considerable amounts of poor-quality concrete may need to be removed and replaced. Small holes and cavities should be packed with mortar, as dry as possible, and projections removed by chipping and tooling, and rubbed smooth. Small areas of honeycombing can be grouted, but large areas should be entirely removed and replaced. Staining generally occurs from form-release agents or rust marks which are most easily removed by light sandblasting, since they are only surface problems. The surface should then be wetted down and a grout of portland cement and fine sand should be applied over the whole surface by brushing, spraying, or rubbing. This fills up small air holes and covers up patches and other nonuniformities of the surface texture and appearance. The grout is well floated and the excess scraped off. After the surface has dried, it is rubbed with burlap to remove any remaining dried grout on the surface. If the dried grout is left too long (e.g., overnight), it is too difficult to remove. This treatment will result in a smooth, unblemished surface, uniform in color. It is important that exactly the same materials and proportions should be used throughout the patching operations. Trial mixes of the grout should be made up ahead of time to get the exact color desired. Similarly, any localized patching, such as repair of honeycombing, should use the same materials and proportions used initially in order to get a good color match.

12.3 TEXTURED CONCRETE SURFACES

Although color plays an important role in architectural concrete, the provision of texture and design is probably even more important. Textured surfaces are not only pleasing in themselves but also eliminate or mask small variations in surface texture and color on the form finish surface. Details of design are beyond the scope of this book, but it is appropriate to discuss ways of achieving textured finishes. There are three main approaches to this objective: textured form liners, exposed aggregate finishes, and mechanical finishes.

Textured Form Liners

The use of textured form liners can produce some interesting effects (Figure 12.4). These can range from ridging or dimpling, through simple repetitive geometrical patterns, to complex mural designs. Smooth, flexible liners can be molded to give a variety of textural patterns. They are generally made from rubbers, thermoplastics (such as PVC), or fiberglass; can be reused many times; and are suitable for both precast and cast-in-place concretes. Special designs for murals are sculptured either in sand beds or in expanded polystyrene blocks. Panels are generally cast face downward; this helps to obtain a clearer reproduction of the details of the design. Exposed aggregates or specially embedded materials can be incorporated into such designs, leading to an infinite variety of artistic effects. Mechanical texturing can also be used subsequently.

Exposed Aggregate Finishes

Exposing coarse aggregate at the concrete surfaces uses to advantage the variety of color, shape, and texture of the material. Exposure even of ordinary gravel or crushed stone can give pleasing effects. There are several methods that are used to obtain exposed aggregate finishes.

Figure 12.4 Example of concrete mural design. (Miami Beach Public Library Auditorium. Photograph courtesy of Portland Cement Association.)

The *face-up method* is widely used for patios and sidewalks and is a variation of the dry-shake method for producing special surfaces. A base of ordinary concrete is cast, screeded, and trowelled flat. The surface is covered uniformly with the desired aggregate, which is then floated into the surface and again trowelled flat. Just about the time of initial set, a thin layer of cement paste is carefully removed by washing with a light spray of water. High-pressure water jets have also been used, but require skill and experience for satisfactory results. Alternatively, a retarding admixture is applied to the surface of the concrete after trowelling. About 16 h later, when the rest of the concrete has set, the retarded surface paste can be brushed and washed off easily. Special toppings can be cast on top of plain concrete using colored aggregates and white or colored cements. Large pieces of aggregates can be hand-laid in special patterns.

A retarder can also be used as a form coating to retard the concrete in contact with the form. The retarder can be painted on directly, or cloth impregnated with retarder can be used to line the form. Special admixture formulations are available for exposed aggregate finishes. Both horizontal and vertical surfaces can be cast by this method. Horizontal casting is used in precast work and allows greater versatility. The panels are cast *face down* with the retarding form at the bottom. As in the face-up method, a special facing mix may be used which can consist of aggregates specially selected for color, shape, and texture. A greater concentration of coarse aggregate is possible, and white or colored cement can also be used. Plain concrete is then cast on top to form an integral panel. This approach is not possible in vertical casting: the aggregate to be exposed must be used throughout the concrete. The prepacked aggregate method can be used in vertical casting.

The kind and amount of retarder depends on the depth of exposure. A water-insoluble retarder is recommended for retardation to a depth of 1.5 to 3 mm ($^{1}/_{16}$ to $^{1}/_{8}$ in.), while water-soluble retarders penetrate more deeply into the concrete and will give greater exposure. The depth of exposure should not exceed one-third of the average diameter of particles. The choice of aggregate size is thus dependent in part on the depth of texture (see Figure 12.5), which in turn depends on the distance at which the texture should be visible (see Table 12.4) and the total area cast.

For only slight exposure (less than 3 mm or $^{1}/_{8}$ in.) sand blasting or acid etching can also be used. Sand blasting requires special equipment and is expensive. Also, uniformity of finish is difficult to achieve, while softer aggregates may be abraded sufficiently to lose their intensity of color. Light sand blasting can be used to minimize minor color variations, such as light mottling, or slight changes in color from lift to lift. Acid

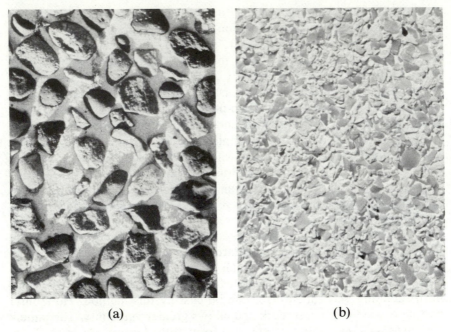

     (a)   (b)

Figure 12.5 Examples of exposed aggregate finishes. (Photographs courtesy of Portland Cement Association.)

etching is designed to dissolve away hardened paste at the surface, but is slow and difficult to control uniformly. It cannot be used with carbonate aggregates (limestones, dolomites, marbles) because the aggregates are attacked by the acid, which may result in discoloration or loss of integrity. A light acid etch is sometimes used to clean an exposed aggregate surface. The use of acid is hazardous and proper protective clothing should be worn.

  The *sand-bedding technique* is used to advantage when exposing

Table 12.4

Visibility Scale for Exposed Aggregate Concrete

Size of Exposed Aggregate	Distance at Which Texture Is Visible
6–12 mm (1/4–1/2 in.)	6–9 m (20–30 ft)
12–25 mm (1/2–1 in.)	12–22 m (40–75 ft)
25–50 mm (1–2 in.)	22–40 m (75–125 ft)
50–75 mm (2–3 in.)	40–55 m (125–175 ft)

large aggregates or when special effects are required, such as a flat-stone arrangement. A bed of sand, the same as used in the concrete, is laid down and the aggregate embedded in it to the required depth. The whole is then damped down for consolidation, and structural concrete is cast on top. After the concrete has hardened, the panel is removed and excess sand is washed off the face. This method allows a greater depth of exposure—12–50 mm—which is more than is possible with chemical retarders.

Mechanical Finishes

Mechanical finishes include abrasive action (sand blasting) and fracturing processes (tooling). We have already mentioned the use of sand blasting for exposed aggregate finishes due to the preferential abrading of the hardened paste. Light sand blasting may give a slightly roughened surface without exposing aggregate or may be used to expose aggregates to a depth of as much as 12 mm. The depth of abrasion depends on the size of grit, age of concrete, and time of treatment. The use of rubber templates to protect certain areas of the concrete can give pleasing designs (Figure 12.6).

Figure 12.6 Examples of decorative effects by selective sandblasting. (Photographs courtesy of Portland Cement Association.)

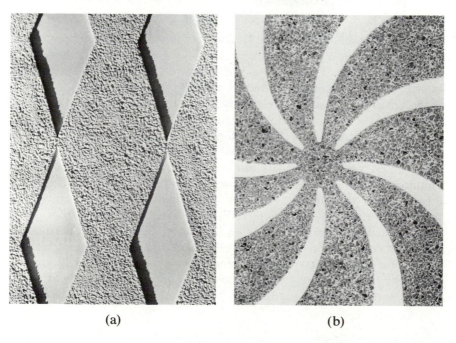

(a)   (b)

Figure 12.7 Texture produced by bush-hammering vertically cast ribs. (Endo Pharmaceutical Center, Garden City, N.Y. Photograph courtesy of Portland Cement Association.)

Various tools can be used to fracture the concrete at the surface. Scaling produces a fine texture at the surface. The scaler has three pneumatic chisels which rotate and fracture the concrete surface on impact. Bush hammering is often used to expose aggregate in the plane of the concrete surface, since mostly cement paste is removed. A bush hammer has a face, resembling a meat tenderizer, which strikes the surface repeatedly. Jack hammering or chiseling is done when the concrete is strong enough so that coarse aggregate is fractured as well as the paste. A variety of effects can be obtained by these methods. One common method is to cast ribs (or reeds) into the concrete surface using a form liner; the ribs are abraded, bush-hammered, or chiselled to produce a striated effect (Figure 12.7).

Surfaces can also be ground and polished. *Terrazzo* is a decorative finish that consists of a facing made of cement (usually white) and marble chips. When the terrazzo has hardened, the surface is coarsely ground to expose the aggregate. Any air voids are filled in with cement paste and the surface is finely ground and polished.

Bibliography

General

ACI COMMITTEE 303, "Guide to Cast-in-Place Architectural Concrete Practice," *Journal of the American Concrete Institute,* Vol. 71, No. 7, pp. 317–346 (1974).

Concrete Construction, Vol. 17 (November 1972). Special issue on architectural concrete.

"Finishing Architectural Concrete," *Concrete Construction,* Vol. 24, No. 5, pp. 319–321 (1979).

GENSERT, R. M., "Cost Factors for Cast-in-Place Architectural Concrete," *Journal of the American Concrete Institute,* Vol. 69, No. 1, pp. 36–45 (1972).

Color and Texture

Exploring Color and Texture, Concrete in Architecture 65. Portland Cement Association, Skokie, Ill., 1966.

Exposed Aggregate Concrete, Information Bulletin No. ISI71A. Portland Cement Association, Skokie, Ill., 1970.

Guide to Exposed Concrete Finishes. The Architectural Press, London, 1970. 161 pp.

"Ornamental Finishes," *Concrete Construction,* Vol. 24, No. 4, pp. 237–243 (1979).

White Concrete, Information Bulletin No. 15175.01A. Portland Cement Association, Skokie, Ill., 1971. (Originally published in *Modern Concrete,* December 1968 and January 1969.)

Special Problems

Concrete Construction, Vol. 23 (October 1978). Special issue on troubleshooting site-cast architectural concrete problems.

GREENING, N. R., AND R. LANDGREN, *Surface Discoloration of Concrete Flatwork,* Research and Development Bulletin No. 203. Portland Cement Association, Skokie, Ill., 1966.

READING, T. J., "The Bughole Problem," *Journal of the American Concrete Institute,* Vol. 69, No. 3, pp. 165–171 (1972).

Problems

12.1 Identify five different examples of architectural concrete in your community. Describe their appearance and discuss how the architectural effects were obtained. Is there any apparent deterioration of the concrete or of its appearance?

12.2 What special materials that can be used for architectural effects are available at the building supply stores in your locality?

13

concrete as a composite material

It has long been known that the properties of multiphase materials may well be superior to the properties of the individual phases taken separately, particularly when it comes to the weak or brittle phases. Perhaps the oldest written account of such a composite material, clay bricks reinforced with straw, occurs in *Exodus:*

> And Pharaoh commanded the same day the taskmasters of the people, and their officers, saying,
> Ye shall no more give the people straw to make bricks, as heretofore: let them go and gather straw for themselves.

Today, we know that neither rock nor pure cement paste make particularly useful building materials—the rock because it is too brittle, and the cement because it cracks on drying. However, together they combine to form perhaps the most versatile of all building materials.

We may define a *composite material* as a three-dimensional combination of at least two chemically and mechanically distinct materials with a definite interface separating the components. This multiphase material will have different properties from the original components. Clearly, concrete qualifies as such a multiphase material. At the macroscopic level, concrete consists of coarse aggregate embedded in a matrix of mortar; on a somewhat finer scale, the mortar itself consists of particles of sand embedded in a matrix of hydrated cement paste. On a microscopic scale, the hydrated cement paste consists of C–S–H and calcium hydroxide, containing an extensive network of capillary pores, which may be dry or filled with water, plus some grains of still unhydrated cement. On a still finer, submicroscopic scale, the C–S–H is a mixture of poorly crystallized particles of a variety of shapes and chemical compositions, surrounding a more-or-less continuous system of gel pores, which also may be dry, or partially or completely filled with water. Finally, the aggregates themselves are generally composite materials, consisting of a mixture of different minerals with a well-defined porosity.

The material structure of concrete is thus very complex. However, if we make some simplifying assumptions, we can then construct a mathematical model that will allow us to determine the relationship between the structure of concrete and its physical properties. In Chapter 4, in relating the mechanical properties of hydrated cement paste to the structure, we essentially modeled hardened cement paste as a composite consisting of homogeneous C–S–H and CH containing a continuous system of capillary pores; we were able to ignore the details of the C–S–H structure. Similarly, we are able to model concrete as a two-phase material, with aggregate particles embedded in a matrix of cement paste. That is, for the purposes of this model, we assume that the aggregate phase and the paste phase are each homogeneous and isotropic. Experience has shown that such a model can provide very good approximations to the mechanical behavior of concrete if the properties of the two components are known.

13.1 FACTORS AFFECTING COMPOSITE BEHAVIOR

Before applying this model to concrete, it is important first to consider the behavior of composites in general in more detail. In order to describe a system with one or more disperse phases (particles) embedded in a continuous matrix, the following parameters must be considered: (1) shape of particles, (2) size and size distribution of particles, (3) concentration and concentration distribution of particles, (4) orientation of particles, (5) spatial distribution (or topology) of particles, (6) composition of the disperse phase, (7) composition of the continuous phase, and (8) bond between the continuous and disperse phases.

Simply describing the geometry of the system in terms of these parameters would be very difficult. In addition, it may be difficult to determine if a phase is continuous or disperse. Therefore, it may be helpful to consider the two extreme cases of phase arrangement, the parallel and series systems shown in Figure 13.1 (a) and (b). These systems, and others like them, may be used to calculate the elastic parameters of the composite system if the elastic parameters of the individual phases are known. The parallel system, in which the two

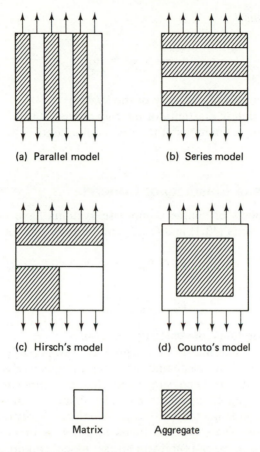

Figure 13.1 Models of the structure of concrete: (a) parallel (constant-strain) model; (b) series (constant-stress) model; (c) Hirsch's model (from T. J. Hirsch, *Journal of the American Concrete Institute,* Vol. 59, No. 3, 1962, p. 427); (d) Counto's model (from V. J. Counto, *Magazine of Concrete Research,* Vol. 16, No. 48, 1964, p. 129. Reproduced by permission of Cement and Concrete Association).

phases are subjected to uniform strains, provides the upper bound solution for the elastic parameter of interest. The series system, in which the phases are subjected to uniform stresses, provides the lower bound solution.

Using these two models to calculate the modulus of elasticity of the systems, it can be shown that for the *parallel model,*

$$E_s = E_1 V_1 + E_2 V_2 \qquad (13.1)$$

For the *series model,*

$$\frac{1}{E_s} = \frac{V_1}{E_1} + \frac{V_2}{E_2} \tag{13.2}$$

where E_s = modulus of elasticity of the system
E_1, E_2 = moduli of elasticity of the two components
V_1, V_2 = volume fractions of the two components

Modulus of Elasticity of Concrete

These models of simple composite systems have been applied to concrete. Equations (13.1) and (13.2) can be rewritten as

$$E_c = E_p V_p + E_a V_a \tag{13.3}$$

$$\frac{1}{E_c} = \frac{V_p}{E_p} + \frac{V_a}{E_a} \tag{13.4}$$

where E_c = modulus of elasticity of concrete, E_p, E_a = moduli of elasticity of the paste matrix and aggregate, respectively, and V_p, V_a = volume fractions of the paste and aggregate, respectively.

However, it has been suggested that a composite material like concrete, consisting of aggregate particles dispersed in a paste matrix, may have two fundamentally different structures. Concretes made with natural aggregates (i.e., hard particles in a soft matrix) conform more closely to the lower bound (uniform stress) model (Figure 13.1b). On the other hand, concretes made with lightweight aggregates (i.e., soft particles in a hard matrix) conform more closely to the upper bound (uniform strain) model (Figure 13.1a). In fact, neither model is quite correct, since concrete under load exhibits neither uniform stress nor uniform strain (see Chapter 14 and Figure 14.13). For concretes made with natural aggregates, the uniform stress model underestimates E_c by about 10%; the uniform strain model overestimates E_c by a greater amount.

Other models to represent concrete have also been proposed. For instance, a model that is the sum of Eq. (13.3) and (13.4) has been suggested by Hirsch, as represented in Figure 13.1c. This can be expressed as

$$\frac{1}{E_c} = x \left(\frac{1}{V_p E_p + V_a E_a} \right) + (1 - x) \left(\frac{V_p}{E_p} + \frac{V_a}{E_a} \right) \tag{13.5}$$

where x and $(1 - x)$ are the relative proportions of material conforming with the upper and lower bound solutions. It has been found that, at least for some concretes, x is approximately 0.5. It should be noted that when there is no bond between the matrix and the aggregate ($x = 0$), Eq. (13.5) reduces to Eq. (13.4), the uniform stress model. For perfect bond ($x = 1$), Eq. (13.5) reduces to Eq. (13.3), the uniform strain model.

One limitation of Eq. (13.4) and (13.5) is that for $E_a = 0$, they predict $E_c = 0$, which is clearly not true. Therefore, still another model, shown in Figure 13.1d, was proposed by Counto, giving the relationship

$$\frac{1}{E_c} = \frac{1 - \sqrt{V_a}}{E_p} + \frac{1}{[(1 - \sqrt{V_a})/\sqrt{V_a}] E_p + E_a} \tag{13.6}$$

These models are compared graphically in Figure 13.2.

Figure 13.2 Relations between the modulus of elasticity of concrete and the volume fraction of aggregate for the simplified structural models shown in Figure 13.1. (From K. Newman, in *Composite Materials*, L. Holliday, ed., Elsevier Publishing Company, Amsterdam, 1966, pp. 336–452.)

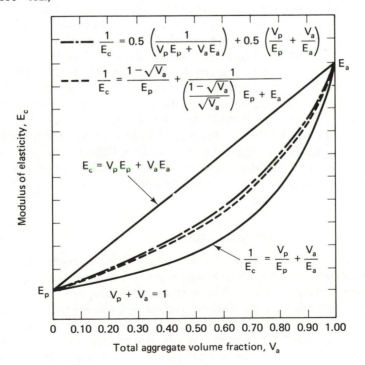

If we now consider, as a more realistic model for concrete, a two-phase material consisting of spherical particles in a continuous matrix, and assume (quite reasonably) that Poisson's ratio is the same for the composite material and each of the two phases, then E_c can be expressed as

$$E_c = \left[\frac{(1 - V_a)E_p + (1 + V_a)E_a}{(1 + V_a)E_p + (1 - V_a)E_a} \right] E_p \qquad (13.7)$$

This equation is valid only when (1) there is no interaction between aggregate particles, and (2) when there is perfect bond between the aggregate and the matrix. While (1) is a reasonably good assumption, (2) is not.

For normal-weight concrete, E_a is generally considerably higher than E_p. For this case of hard particles in a soft matrix, Eq. (13.4) still represents the theoretical lower limit for E_c, which can be reached only when there is no bond between the particles and the matrix. On the other hand, the theoretical upper limit is now represented by Eq. (13.7), [instead of Eq. (13.3)], which can be reached only when there is perfect bond. For the more realistic case of partial bonding, the value of E_c must lie somewhere between these (not very different) extremes.

For some lightweight concretes, $E_a \simeq E_p$. In this case, Eqs. (13.3), (13.4), and (13.7) all yield the result that E_c is independent of the relative amounts of cement and aggregate in the concrete; that is, $E_c = E_a = E_p$ for any volume fraction of aggregate. For other lightweight concretes, E_a is considerably lower that E_p. For this case of soft particles in a hard matrix, the elastic modulus of the concrete will be less affected by bond. For $E_p > E_a$, Eq. (13.3) represents the theoretical upper limit for E_c, which can be reached only if the material structure closely resembles the structure indicated in Figure 13.1a. For these lightweight concretes, Eq. (13.7) now represents the lower limit of E_c for spherical particles with a low elastic modulus evenly distributed in a matrix with a high elastic modulus.

Cement – Aggregate Bond

The strength of concrete depends on the strength of the paste, the strength of the coarse aggregate, and the strength of the paste–aggregate interface. There is considerable evidence to indicate that this interface is the weakest region of concrete; in general, bond failures occur before failure of either the paste or the aggregate. Bond forces are partly due to van der Waals' forces. However, the shape and surface texture of the coarse aggregate are important, since there may be a considerable amount of mechanical interlocking between the mortar and the coarse aggregate; flexural and tensile strengths of concretes made with rough aggregates may be up to 30% higher than those made with smooth aggregates. In addition, although we usually consider the aggregate to be chemically inert, there may be some chemical reactions between the cement and the aggregate which contribute to strength.

The bond region is weak because cracks invariably exist at the paste–coarse aggregate interface, even for continuously moist-cured concrete and before the application of any external load. These cracks are due to bleeding and segregation and to volume changes of the cement paste during setting and hydration. When ordinary curing takes place, which is accompanied by drying, the aggregate particles tend to restrain shrinkage because of their higher elastic modulus. This induces shear and tensile forces at the interface, which increase with increasing particle size, and which will cause addtional cracking if they exceed the bond strength. Under load, the difference in elastic moduli between the aggregate and the cement paste will lead to still more cracking.

Factors Affecting Bond Strength

Unfortunately, in spite of the importance of bond in determining the behavior of concrete, bond strength is very difficult to measure. There are still no standard tests for bond, but a number of the tests that have been used are shown in Figure 13.3. Nonetheless, we do have a reasonable understanding of the factors that affect bond strength.

Bond strength seems to depend on many of the same factors that control other types of strength. For example, it increases with decreasing *w/c* ratios. The bond strength also depends on the particular cement, varying as does the compressive strength of the cement, as shown in Figure 13.4. Curing temperature appears not to affect the bond. The effects of age on cement paste and bond strengths (both in flexure) are shown in Figure 13.5. Once these strengths become equal, the failure mode changes from bond failure to failure in the paste.

In order to have a good bond, the concrete must be properly vibrated. Since air voids tend to reduce the contact area between the mortar and the coarse aggregate, air entrainment also tends to reduce the bond strength. The absorption capacity of the aggregate may be important: porous aggregates (dried before use) can give excellent bond since the process of absorption will improve contact between the paste and the aggregate. Bond failures do not normally occur between the cement paste and the fine aggregate. Rather, bond failures seem to occur around the coarse aggregate particles. As the size of the aggregate

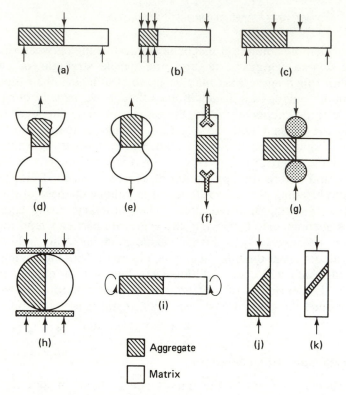

Figure 13.3 Techniques used for measuring aggregate-cement bond strength. (From K. M. Alexander, J. Wardlaw, and D. J. Gilbert, in *The Structure of Concrete*, ed. A. E. Brooks and K. Newman, Cement and Concrete Association, London, 1968, pp. 59–81.)

Figure 13.4 Relationship between bond strength and paste strength for eight different portland cements at a water/cement ratio of 0.35. (From K. M. Alexander, J. Wardlaw, and D. J. Gilbert, in *The Structure of Concrete*, ed. A. E. Brooks and K. Newman, Cement and Concrete Association, London, 1968, pp. 59–81.)

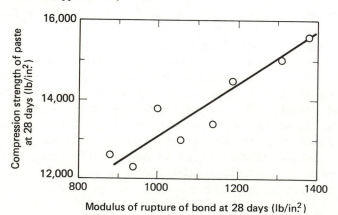

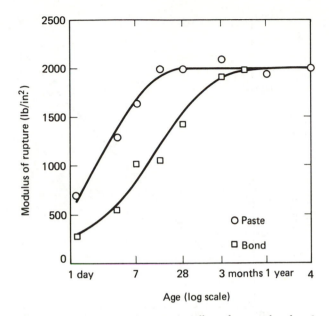

Figure 13.5 Long-term test showing the effect of age on bond and paste strength; water/cement ratio 0.35. (From K. M. Alexander, J. Wardlaw, and D. J. Gilbert, in *The Structure of Concrete*, ed. A. E. Brooks and K. Newman, Cement and Concrete Association, London, 1968, pp. 59–81.)

particles increases, the stresses at the paste–aggregate interface increase rapidly, leading to an apparent decrease in bond strength.

As stated earlier, many aggregates are not inert, and therefore bond is strongly dependent on the type of aggregate. In general, the more acid siliceous rocks develop the highest strengths. For extrusive rocks, the relationship between bond strength and silica content is shown in Figure 13.6. The "best" rocks may develop twice as much bond as the "worst" rocks. It appears that there is some pozzolanic reaction with the silica, leading to the formation of a reaction zone extending 25 to 200 μm into the paste and 0 to 100 μm into the aggregate, as may be seen from microhardness measurements (Figure 13.7). There also seems to be some reaction between the cement and limestone aggregates.

Not surprisingly, the surface texture of the aggregates is also very important in determining the bond behavior, the texture of the coarse aggregate particles in particular. An increasingly smooth surface texture leads to decreases in (1) strength, both compressive and tensile; (2) the stress at which cracking first becomes noticeable; (3) the total strain at failure; (4) the maximum volumetric strain; and (5) the modulus of elasticity (with this effect decreasing as the age of the concrete increases). Although not very much information is available, it would appear that a

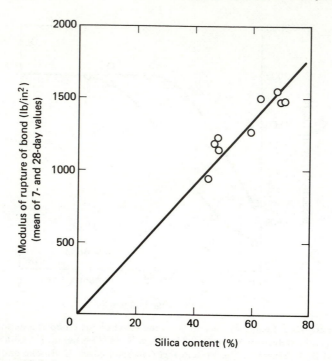

Figure 13.6 Dependence of bond strength on silica content of extrusive rocks. (From K. M. Alexander, J. Wardlaw, and D. J. Gilbert, in *The Structure of Concrete,* ed. A. E. Brooks and K. Newman, Cement and Concrete Association, London, 1968, pp. 59–81.)

coarser or rougher surface texture delays the onset of bond cracking. On the other hand, time-dependent deformations, such as creep and shrinkage, appear to be little affected by the surface texture.

The conditions of test have a large influence on the measured bond strength. In particular, drying greatly reduces the bond strength, due to the growth of shrinkage cracks at the interface. Bond failures may occur in both tension and compression induced shear. In tension, the bond strength may range from about 30 to 90% of the tensile strength of cement paste or of mortar, although occasionally (as shown in Figure 13.5) the bond strength may reach the tensile strength.

The intrinsic bond shear strengths are of the same order of magnitude as the tensile strengths. They can be described by Coulomb's law:

$$\tau_b = \tau_0 + \mu\sigma = \tau_0 + \sigma \tan \phi \tag{13.8}$$

where τ_b is the bond shear strength, τ_0 is the intrinsic shear strength, σ is the normal stress, and μ is the coefficient of friction, which is equal to

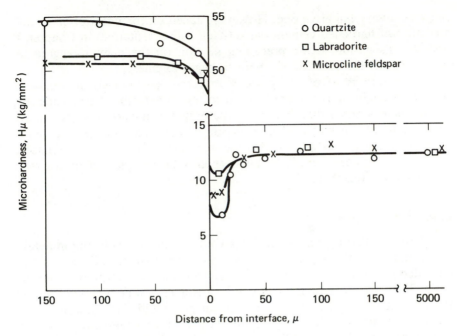

Figure 13.7 Variation in microhardness across the cement−aggregate interface. (From K. M. Alexander, J. Wardlaw, and D. J. Gilbert, in *The Structure of Concrete*, ed. A. E. Brooks and K. Newman, Cement and Concrete Association, London, 1968, pp. 59−81. Based on data from T. Yu. Lyubimova and E. R. Pinus, *Kolloidnyi Zhurnal* (U.S.S.R.), Vol. 24, No. 5, 1962, pp. 578−587.)

tan ϕ, where ϕ is the angle of internal friction. It has been found that ϕ is in the range 30 to 40° and is largely independent of the type of aggregate, the w/c ratio, or the strength of the matrix.

Of course, in addition to the bond between mortar and coarse aggregate, the bond between the concrete and reinforcing steel is also important. This type of bond is discussed in Chapter 16.

Nature of Strength of Concrete

Since concrete is an extremely complex multiphase material, its strength and its other mechanical properties must clearly derive, in some way, from the interactions between the various components. However, there is little agreement as to the relative magnitudes of the different forces involved. Indeed, the very nature of the forces between the different components, or even within a single phase, is only incompletely understood. Thus, any discussion of the nature of the strength of concrete must be a very tentative one. However, based on the description of the structure of hardened cement paste (itself rather tentative) in Chapter 4 and the discussion in the preceding section on the cement–aggregate bond, at least the outlines of a pattern begin to emerge.

Concrete is a relatively weak material, particularly in tension. Indeed, it might be considered useful to try to describe the reasons for this weakness rather than to discuss the nature of the strength. If we are to understand the mechanical behavior of concrete, we must explain both phenomena. The reasons for the apparent weakness are discussed in Chapter 14; here we will be concerned with the types of forces that hold concrete together.

Effect of Aggregate

The bonds within aggregate particles vary, of course, with the mineral-ogical nature of the aggregate. However, normal-weight aggregates are considerably stiffer and stronger than hardened cement paste or concrete itself. Thus, to understand the strength of concrete, it is the bonds within the cement paste, or between the cement paste and the aggregate, that must be examined. In general, the effect of the aggregate may be considered merely as a "dilution" of the cement paste. That is, the strength and deformation characteristics of concrete are governed primarily by the cement phase, although they will be modified by the inclusion of aggregate. As we have seen in the previous section, bond between the cement and the aggregate may be both mechanical and chemical. In either case, the strength of the cement–aggregate bond is less than that of either the cement or the aggregate. However, even this bonding must contribute some strength. It is known that where there is absolutely no bond between the cement and the aggregate, the aggregates have a negative effect on the properties of the composite. Usually, there is a situation of partial bonding, so the contribution of the aggregate can vary from a negative one to a positive one.

Bonding within Cement Paste

From our knowledge of the many phases that may be present in cement paste, and from scanning electron micrographs of the material, it is clear that the structure of cement is very complicated. The system will contain many boundaries: between C–S–H crystallites, between different phases within the C–S–H, at unhydrated cement grains, at pores, and so on. These boundaries themselves may be potential sources of weakness, even though there are in general forces acting across these boundaries. But, although no simple model can truly represent this relationship, we have no other way of dealing with the problem. On an intuitive level, therefore, the development of strength in cement can be described using Figure 4.7. The fresh paste has no strength; the system contains a substantial volume of liquid, and strength can only be provided by a solidification of material within the liquid phase. At 7 days (Figure 4.7b) there is considerable intergrowth, but the strength is still relatively low because the large water-filled spaces are still not filled by hydration products. By 28 days (Figure 4.7c), there is apparent fusing of cement grains through intergrowth of C–S–H, and much space is filled by calcium hydroxide. However, even after 90 days (Figure 4.7d) there are still considerable void spaces, although the microstructure is now very dense. The strength must arise from the intergrowth and the chemical or physical bonding of the constituents (as discussed in Chapter 4), but this strength will be considerably modified by the high porosity.

Bibliography

ALEXANDER, K. M., J. WARDLAW, AND D. J. GILBERT, "Aggregate–Cement Bond, Cement Paste Strength, and the Strength of Concrete," in *The Structure of Concrete,* ed. A. E. Brooks and K. Newman, pp. 59–81. Cement and Concrete Association, London, 1968.

COUNTO, V. J., "The Effect of the Elastic Modulus of the Aggregate on the Elastic Modulus, Creep, and Creep Recovery of Concrete," *Magazine of Concrete Research,* Vol. 16, No. 48, pp. 129–138 (1964).

HANSEN, T. C., "Influence of Aggregate and Voids on Modulus of Elasticity of Concrete, Cement Mortar, and Cement Paste," *Journal of the American Concrete Institute,* Vol. 62, No. 2, pp. 193–216 (1965).

HANSEN, T. C., "Theories of Multi-phase Materials Applied to Concrete, Cement Mortar, and Cement Paste," in *The Structure of Concrete,* ed. A. E. Brooks and K. Newman, pp. 16–23. Cement and Concrete Association, London, 1968.

HIRSCH, T. J., "Modulus of Elasticity of Concrete Affected by Elastic Moduli of Cement Paste Matrix and Aggregate," *Journal of the American Concrete Institute,* Vol. 59, No. 3, pp. 427–452 (1962).

HOLLIDAY, L., ed., *Composite Materials.* Elsevier Publishing Company, Amsterdam, 1966.

NICHOLLS, R., *Composite Construction Materials Handbook.* Prentice-Hall, Inc., Englewood Cliffs, N.J., 1976.

RAO, C. V. S. K., R. N. SWAMY, AND P. S. MANGAT, "Mechanical Behavior of Concrete as a Composite Material," *Materials and Structures (Paris),* Vol. 7, No. 40, pp. 265–272 (1974).

SHAH, S. P., AND F. O. SLATE, "Internal Microcracking, Mortar–Aggregate Bond, and the Stress–Strain Curve of Concrete," in *The Structure of Concrete,* ed. A. E. Brooks and K. Newman, pp. 82–92. Cement and Concrete Association, London, 1968.

STRUBLE, L., J. SKALNY, AND S. MINDESS, "A Review of the Cement-Aggregate Bond," *Cement and Concrete Research,* Vol. 10, No. 2, pp. 277–286 (1980).

Problems

13.1 Calculate the modulus of elasticity of concrete according to Eqs. (13.3), (13.4), (13.5), and (13.6) using the following material parameters: $E_p = 2 \times 10^6$ psi, $V_p = 0.25$, $V_a = 0.75$, and $E_a =$ (a) 8×10^6 psi; (b) 2×10^6 psi; (c) 0.5×10^6 psi.

13.2 Calculate the modulus of elasticity of concrete according to Eqs. (13.3), (13.4), (13.5), and (13.6) using the following material parameters: $E_p = 20$ GPa, $V_p = 0.20$, $V_a = 0.80$, and $E_a =$ (a) 100 GPa; (b) 20 GPa; and (c) 4 GPa.

13.3 For spherical particles in a continuous matrix, calculate the value of E_c from Eq. (13.7) using the conditions given in (a) Problem 3.1; (b) Problem 3.2.

13.4 Why is the cement–aggregate bond important?

13.5 Why is it difficult to measure the properties of the cement–aggregate bond?

13.6 What factors affect the cement–aggregate bond?

14

fracture and failure

Both hardened cement paste and concrete may be classed as brittle materials. Research in the last 20 years has shown very clearly that both the stress–strain behavior and the failure of these materials are governed by a process of microcracking, particularly associated with the interfacial region between the cement and the aggregate particles. In order to understand the nature of the strength of concrete, it is thus also necessary to understand the fracture mechanisms. However, before discussing the fracture mechanisms in detail, it is appropriate first to give a brief description of the science of fracture mechanics.

14.1 FRACTURE MECHANICS

Fracture mechanics may be defined as the study of the stress and displacement fields in the region of a crack tip in materials which are elastic, homogeneous, and isotropic, particularly at the onset of unstable crack growth (or fracture). This concept is applicable to brittle materials in which inelastic behavior is at a minimum. We will discuss classical fracture mechanics and then see to what extent these concepts apply to a composite such as concrete.

Theoretical Cohesive Stress

Ideally, the strength of a solid depends on the strength of its atomic bonds. Thus, to obtain an approximation to at least the order of magnitude of the fracture strength, we can consider the interaction

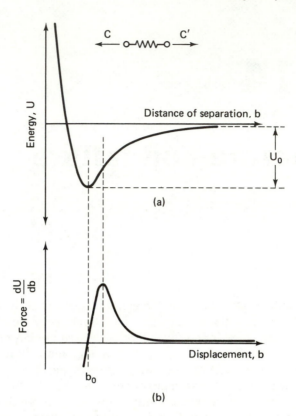

Figure 14.1 (a) Bonding energy as a function of distance of separation; (b) force–displacement curve.

between two atoms. Figure 14.1a shows a typical curve representing the energy of interaction of these two atoms as a function of their separation. This energy is a minimum at the equilibrium atomic spacing, b_0. The total energy required to separate these atoms completely is U_0. However, since one of the consequences of fracture is the creation of new surface areas along the fracture plane, this work to fracture may be equated to twice the surface energy, γ_s, of the new surface area created, as two new surfaces are created when a bond is broken.

If we differentiate the energy–distance curve, we can obtain the force–displacement curve shown in Figure 14.1b. The force is zero at the equilibrium atomic spacing; the initial slope of this curve is proportional to the modulus of elasticity, E. In practice, the force–displacement curve can be transformed to a stress–strain curve which can be approximated by a half sine wave, as shown in Figure 14.2. The area under this

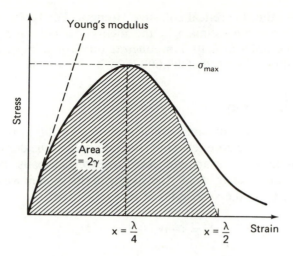

Figure 14.2 Atomic stress-strain curve.

curve represents the work to bring about fracture, $2\gamma_s$; the initial slope is E. The relationship between the stress, σ, and the displacement, x, is

$$\sigma = \sigma_{\max} \sin \frac{2\pi x}{\lambda} \qquad (14.1)$$

Thus, the work to fracture can be written as

$$2\gamma_s = \int_0^{\lambda/2} \sigma_{\max} \sin \frac{2\pi x}{\lambda} dx = \frac{\lambda \sigma_{\max}}{\pi} \qquad (14.2)$$

But for small displacements, Hooke's law is obeyed, and

$$\sigma = E\epsilon = E\frac{x}{b_0} \qquad (14.3)$$

From Eq. (14.1), since for small angles $\sin x \approx x$,

$$\sigma_{\max} = \frac{\lambda E}{2\pi b_0} \qquad (14.4)$$

So, equating Eqs. (14.2) and (14.4), we get

$$\sigma_{\max} = \left(\frac{E\gamma_s}{b_0}\right)^{\frac{1}{2}} \qquad (14.5)$$

where σ_{max} is the theoretical cohesive strength. For typical values of γ_s and b_0, a reasonable estimate of the theoretical cohesive strength for solids is of the order of $E/10$. For concrete this would imply a theoretical strength of about 2100 MPa (300,000 lb/in.^{2}).

Griffith Theory

Based on thermodynamic considerations, Griffith arrived at a similar solution of the theoretical cohesive strength. Considering an elastic body containing a crack and subjected to external loads, he calculated the condition at which the total free energy of the system was minimized. The total energy in the system is

$$U = (-W_L + U_E) + U_s \qquad (14.6)$$

where $-W_L$ = work due to the applied loads, U_E = strain energy stored in the system, and U_s = free surface energy in creating new crack surface. Then a crack would propagate when $dU/dc < 0$. Using this theory, we can derive the Griffith equation, which gives the theoretical fracture strength:

$$\sigma_f = \left(\frac{2E\gamma_s}{\pi C}\right)^{\frac{1}{2}} \qquad (14.7)$$

where C is one-half the crack length. At σ_f, $C \simeq b_0$. This is very similar to Eq. (14.5), even though it was derived in quite a different fashion.

The question then arises: Why is there this enormous discrepancy, perhaps two or three orders of magnitude, between theoretical values of strength [as predicted by Eq. (14.5) or (14.7)] and those actually measured? The answer to this question was first suggested by Griffith in 1920, who concluded that any real material must contain flaws, microcracks, or other defects that would have the effect of concentrating the stress sufficiently to reach the theoretical cohesive stress in highly localized regions of the specimen. Cracks would thus grow under an applied stress until failure occurred. It is easy enough to show that cracks can indeed concentrate the stress sufficiently to achieve these very high stresses locally. If we consider a plate in uniform tension, containing an elliptical hole (which in the limit might represent a crack), as shown in Figure 14.3, then the stress at the crack tip can be written as

$$\sigma_{max} = \sigma\left(1 + 2\sqrt{\frac{C}{\rho}}\right) \qquad (14.8a)$$

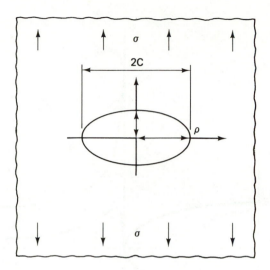

Figure 14.3 Elliptical hole in an infinite plate in tension.

where ρ is the radius of the crack tip, or

$$\frac{\sigma_{\max}}{\sigma} = 2\sqrt{\frac{C}{\rho}} \qquad \text{for } C \gg \rho \tag{14.8b}$$

Clearly, for $C \gg \rho$, that is, for a very "sharp" crack, this factor can become very large. The form of the stress distribution ahead of the crack is shown in Figure 14.4.

Stress-Intensity Factor

Although we can distinguish three modes of crack displacement, the crack-opening mode (Mode I) is the most important one to consider for brittle materials. If we consider the stress field near a sharp crack, as shown in Figure 14.5, the stresses can be written as

$$\left\{ \begin{array}{c} \sigma_x \\ \sigma_y \\ \tau_{xy} \end{array} \right\} = \frac{K_I}{(2\pi r)^{1/2}} \left\{ \begin{array}{c} \cos\left(\dfrac{\theta}{2}\right)[1 - \sin\left(\dfrac{\theta}{2}\right)\sin\left(\dfrac{3\theta}{2}\right)] \\[2mm] \cos\left(\dfrac{\theta}{2}\right)[1 + \sin\left(\dfrac{\theta}{2}\right)\sin\left(\dfrac{3\theta}{2}\right)] \\[2mm] \sin\left(\dfrac{\theta}{2}\right)\cos\left(\dfrac{\theta}{2}\right)\cos\left(\dfrac{3\theta}{2}\right) \end{array} \right\} \tag{14.9}$$

$$\sigma_z = \mu\,(\sigma_x + \sigma_y) \qquad\qquad \tau_{xz} = \tau_{yz} = 0$$

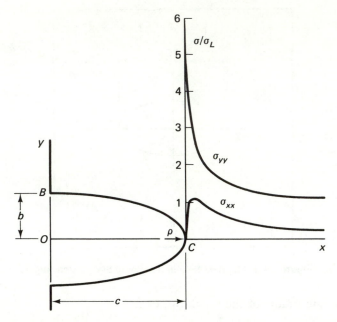

Figure 14.4 Stress concentration at elliptical hole, $c = 3b$. Note that concentration localized within $\approx c$ of the tip, with high stress gradients within $\approx p$ of the tip.

Figure 14.5 Coordinate system and stress components ahead of a crack tip (Mode I displacement).

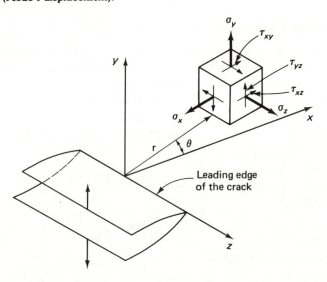

where μ is Poisson's ratio. The parameter K_I, the stress intensity factor, has the form $K_I = \sigma \sqrt{\pi C}$ for an infinite solid. For specimens with finite dimensions, this becomes $K_I = Y\sigma \sqrt{\pi C}$, where Y is a modification factor that takes into account the geometry of the specimen. K_I has the dimensions of stress $\times \sqrt{\text{length}}$, having units of $N \cdot m^{-3/2}$ (or $lb/in.^{-3/2}$), and may be considered to be a single-parameter description of the stress and displacement fields in the region of a crack tip. Its calculation assumes a linearly elastic material which is both homogeneous and isotropic. Although these assumptions are really incorrect for concrete, we generally assume that the approximations involved in applying linear–elastic fracture mechanics to concrete are reasonable. The underlying assumption is that when the stress intensity factor reaches some critical value, unstable fracture occurs. This *critical stress intensity factor* is designated K_{IC}, and is sometimes referred to as *fracture toughness*. It should be a fundamental material property, independent of how it is measured.

An alternative way of considering fracture involves not K_I, but the *strain energy release rate*, G_I. Unstable crack extension occurs when G_I reaches the critical strain energy release rate, G_{IC}. It can be shown that

$$G_{IC} = \frac{K_{IC}^2}{E} \qquad \text{plane stress}$$

or   (14.10)

$$G_{IC} = \frac{K_{IC}^2}{E(1 - \mu^2)} \qquad \text{plane strain}$$

so that these are equivalent ways of expressing the fracture criterion. It can also be shown that $G_{IC} = 2\gamma_s$.

Now if the fracture criterion described above is applied to concrete, it should be possible to calculate γ_s for concrete, or at least for hardened cement paste. However, when this calculation is performed, values of γ_s which are at least an order of magnitude larger than the accepted value for the surface energy of cement paste are obtained. This would imply that the simple fracture mechanics theory described above cannot be used on cement or concrete. It is now assumed that the discrepancy in γ_s values is due to the fact that in cement or concrete, the crack does not propagate in a straight line (as assumed by Griffith), but follows a very tortuous path, around aggregate particles or perhaps around different phases in the cement paste itself. There may also be multiple cracking, and some branch cracking at the crack tip. Nevertheless, it has been found that K_{IC} and G_{IC} can be useful in characterizing the behavior of concrete, although some nonlinear fracture parameter may indeed prove to give a better description of the fracture process.

Fracture in Compression

So far, we have only considered tensile stresses, while concrete is used primarily in compression. However, the Griffith analysis can be extended to include uniaxial compression, as well as biaxial stress states. Griffith showed that under biaxial compressive stresses, the presence of small cracks leads to tensile stresses at some points along the edge of the flaw, as long as the stress components are unequal. Using the normal convention that $\sigma_1 > \sigma_2 > \sigma_3$, with tension positive, the Griffith criterion becomes

$$\sigma_1 = \sigma_t \qquad \text{if } 3\sigma_1 + \sigma_3 > 0 \tag{14.11}$$

and

$$(\sigma_1 - \sigma_3)^2 + 8\sigma_t(\sigma_1 + \sigma_3) = 0 \qquad \text{if } 3\sigma_1 + \sigma_3 < 0 \tag{14.12}$$

where σ_t is the uniaxial tensile strength. This is shown graphically in Figure 14.6. This criterion predicts that the compressive strength of

Figure 14.6 Griffith's criteria of fracture under biaxial stress (tension positive).

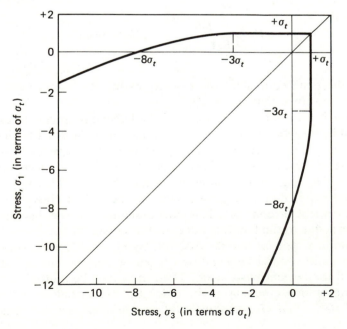

concrete is eight times the tensile strength, which is fairly close to the observed values. This failure criterion agrees well with experimental data in the tension–compression zone, but does not work too well in the compression–compression zone, where splitting will generally occur normal to the unloaded σ_2 direction. Nevertheless, there is now enough evidence to indicate that concrete failure is due to the propagation of microcracks under localized tensile stresses, regardless of the type of loading (with the exception of hydrostatic compression).

On the other hand, it has also been suggested that failure is controlled not by this limiting tensile stress, but rather by a limiting tensile strain, of the order of 1 to 2×10^{-4}. This might then explain the failure of a compression cylinder by splitting, when the lateral strain ($=$ $\mu \times$ longitudinal strain) reaches this limiting value. This mechanism is probably due to a bond failure at the cement–aggregate interface. It may well be that the failure criterion for concrete should specify both a limiting strain and a limiting stress.

Difficulties in Application

Fracture mechanics was first applied to concrete in 1959, and since then a number of studies have been carried out. However, as yet, no standard fracture tests or test specimens have been developed, so experimental results are often contradictory and are certainly difficult to interpret. The most important problems have to do with specimen size and with the heterogeneity introduced by the aggregate. A notch-sensitive material is one in which the presence of a sharp notch reduces the tensile or flexural strength beyond that caused by the mere reduction in cross-sectional area. In view of what we have discussed above, in order for fracture mechanics to apply, a material must be notch-sensitive. It is generally found that cement paste, mortar, and concrete are all notch-sensitive (although not all experimental studies are in agreement). One problem seems to be that for concrete, the zone of disturbance at the crack tip is large compared to the size of small specimens. It has been suggested that a specimen depth of at least 25mm (10 in.) is required if fracture studies on concrete are to be valid. The fracture process is further complicated by the aggregate, as the growth of a Griffith crack can be arrested by encountering an aggregate particle, or by crack-tip blunting due to debonding at the cement–aggregate interface.

In spite of these complications, at least an approximate estimate of the values of the fracture parameters has been obtained. For cement paste, K_{IC} values generally lie in the range 0.1 to 0.5 MN·m$^{-3/2}$ (100 to 450 lb-in.$^{-3/2}$), while the values for concrete range from about 0.45 to 1.40 MN·m$^{-3/2}$ (400 to 1300 lb-in.$^{-3/2}$). However, K_{IC} for the interfacial region seems to be much smaller, about 0.1 MN·m$^{-3/2}$, (100 lb-in.$^{-3/2}$), confirming the belief that this is the zone of weakness for concrete. K_{IC} seems to be fairly independent of specimen age after about 10 days of curing. It increases with decreasing w/c ratio and as the volume of coarse aggregate increases.

Whatever the details of the failure mechanism, it appears that the fracture-mechanics approach can be used to characterize the fracture of concrete. However, there is still considerable uncertainty not only of the values of parameters such as K_{IC} for different concretes, but even as to how these fracture parameters should be obtained. Indeed, the behavior of concrete cannot yet be completely described by a single fracture parameter. It has been suggested that a nonlinear elastic fracture criterion should be used for concrete, but the details of these arguments are beyond the scope of this book.

Coulomb–Mohr Theory

The fracture mechanics approach described above is an attempt to describe the failure mechanism in concrete from a microscopic point of view. However, it is also possible to describe failure criteria for concrete at the macroscopic, or phenomenological level. The most common such theory, which has been applied quite successfully, is the Coulomb–Mohr failure criterion. The Mohr theory can be expressed as

$$\tau = f(\sigma) \tag{14.13}$$

or

$$\sigma_1 - \sigma_3 = f(\sigma_1 + \sigma_3) \tag{14.14}$$

Using this theory, and plotting stresses in σ, τ coordinates, a Mohr's envelope is obtained, as shown in Figure 14.7; any combination of shear and normal stresses lying outside the shaded area represents failure. The Coulomb theory predicts that the Mohr envelope will be defined by a pair of straight lines such that

$$\tau = \pm(\tau_o - \mu\sigma) \tag{14.15}$$

where τ_o is the shear failure stress in the absence of normal stresses, and μ the coefficient of internal friction. In this case, the direction between the maximum principal stress and the normal to the fracture plane, ϕ, can be written as

$$\phi = \frac{\tan^{-1}(1/\mu)}{2} \tag{14.16}$$

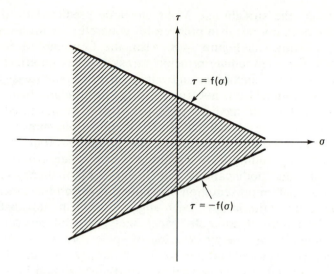

Figure 14.7 Mohr's envelope. The shaded area represents safe stress combinations.

Figure 14.8 Mohr envelopes for ultimate strengths of concrete under various combinations of stress, in terms of the uniaxial compressive stress.

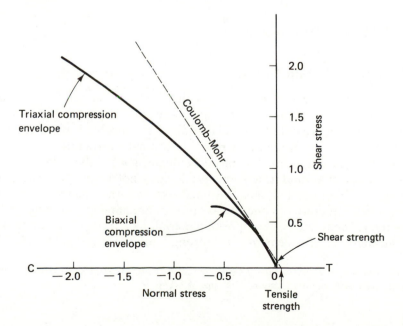

However, the straight-line Mohr envelope predicted by the Coulomb theory does not occur in practice for concrete; the Mohr envelope is curved, as shown in Figure 14.8. Also, the Coulomb–Mohr theory assumes that the intermediate principal stress, σ_2, has no effect; failure depends only on the maximum and minimum principal stresses. However, this assumption also is not true. Clearly, the ultimate shear strength is affected by the intermediate principal stress, as indicated by the different envelopes for biaxial and triaxial compression shown in Figure 14.8. To overcome these deficiencies, the Coulomb–Mohr theory has been modified, by adding to it a limiting tensile strain criterion. Even with this and other modifications, the Coulomb–Mohr theory is at best a very approximate representation of a failure criterion for concrete.

Other failure criteria have been suggested, but none satisfactorily describes the failure of concrete. Most likely, the failure criterion for concrete should be an energy criterion of some sort, since the energy involved in cracking should be capable of definition. However, at present, it would appear that at least two criteria for failure should be specified, as pointed out by Newman:

1. *A criterion for discontinuity* or the onset of failure when the concrete can no longer withstand the applied loading system without severe disruption of the structure;

2. *A criterion for ultimate strength* or the maximum loads that concrete can withstand under various stress combinations.[1]

14.2 MECHANISMS OF FAILURE

Stress – Strain Curve

Having looked in some detail at the fracture theories for concrete and at the nature of the cement–aggregate bond (Chapter 13), we can now turn to a more detailed examination of the mechanisms of fracture in terms of the structure of concrete. Unfortunately, fracture mechanics by itself deals only with the physical quantities involved in fracture: stress, crack size, and strain energy; it does not reveal the micromechanism of fracture. For this, we must examine the changes that take place in the structure of concrete as stresses are applied. As indicated in Chapter 13, concrete is a composite material; its strength is a function of the strength of the cement, the strength of the aggregate, and the

[1]K. Newman, in *The Structure of Concrete,* ed. A. E. Brooks and K. Newman, Cement and Concrete Association, London, 1968, pp. 255–274.

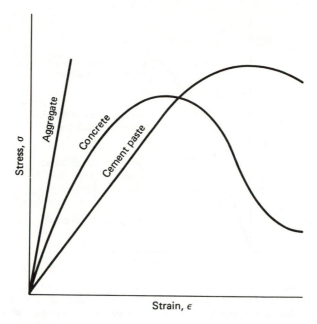

Figure 14.9 Typical stress–strain curves for aggregate, hardened cement paste, and concrete.

interaction between the components. As may be seen in Figure 14.9, the stress–strain curves of the two principal components are essentially linear, except perhaps at very high stress levels. However, their moduli of elasticity are quite different. Consequently, their differential response in concrete to an applied stress leads to inelastic behavior, and thus the stress–strain curve for concrete is highly nonlinear. This nonlinearity is due not only to the composite action of the material, but also to the nature of the cement–aggregate bond. It has been found that a reduction in bond strength leads to an increase in the nonlinearity of the stress–strain curves. Stronger concretes exhibit a more linear stress–strain curve, and the curve also becomes more linear when the aggregate stiffness approaches the stiffness of the matrix.

It is now well established that internal cracks and flaws in concrete exist even before any loads have been applied. Some of these are due to segregation and bleeding, particularly beneath large pieces of aggregate, or reinforcing steel. However, most of the cracks are bond cracks at the cement–aggregate interface. These are due not only to differences in the elastic moduli of the hardened cement and the aggregate, but also to their different coefficients of thermal expansion and their different response to changes in moisture content. The incompatibility in E values

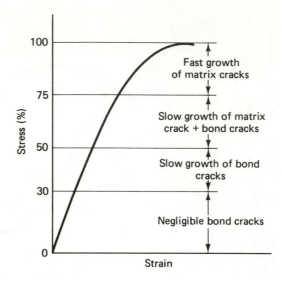

Figure 14.10 Diagrammatic stress–strain curve of concrete in compression. (From J. Glucklich, in *The Structure of Concrete*, ed. A. E. Brooks and K. Newman, Cement and Concrete Association, London, 1968, pp. 176–189.)

may lead to considerable stress concentrations under the differential volume changes due to continued hydration of the cement, temperature differentials, or to drying of the concrete. Since the bond strength is less than the strength of the matrix, which in turn is generally less than the strength of the aggregate, the interfacial region is the "weak link" in the structure, and cracks will tend to form in this region.

As in other brittle materials, the fracture process in concrete must pass through three stages: crack initiation, slow crack growth, and rapid crack growth. These three stages determine the stress–strain behavior and the failure of the concrete, as illustrated in Figure 14.10. Below about 30% of the ultimate load, the stress–strain curve is almost linear. Although bond cracks already exist, they are quite stable at low stresses and have little tendency to propagate. In addition to these preexisting cracks, there is probably some additional crack initiation at highly localized regions where the tensile strain concentrations are largest; these cracks, too, remain stable at low loads, but their formation probably accounts for the slight nonlinearity of the stress–strain curve even at low stresses. Between about 30 and 50% of the ultimate stress, the cracks begin to propagate, but very slowly. Most of the crack growth remains in the interfacial region, the system of bond cracks multiplies and grows in a stable fashion, and the stress–strain curve begins to show increasing curvature. At this stage, there is very little cracking in the cement matrix. However, once the stress exceeds about 50% of ultimate, cracks begin to extend into the matrix, and a much more extensive and continuous crack system begins to develop, with matrix cracks connecting the originally isolated bond cracks. (The development of a continuous crack system is related to the stress at which static fatigue occurs.) Finally, beyond about 75% of the ultimate stress, more rapid crack growth occurs in the matrix, and as these cracks extend, the crack system eventually becomes unstable and failure occurs.

It should be noted here that with a conventional, rather "soft" testing machine, concrete in compression will appear to fail "instantaneously" when the maximum load is reached. However, if a sufficiently stiff testing machine is used which maintains a constant rate of strain, the stress–strain curve will show a considerable descending branch, as shown in Figure 14.11. This indicates that even when the maximum load has been reached, the crack pattern is still not so extensive as to cause complete failure of the concrete in compression. Microcracked concrete maintains a redundant, stable structure until the peak of the σ–ϵ curve

Figure 14.11 Three branches of complete stress–strain curve for concrete. (From F. Watanabe, in *Mechanical Behavior of Materials*, Vol. 4, The Society of Materials Science, Japan, 1972, pp. 153–161.)

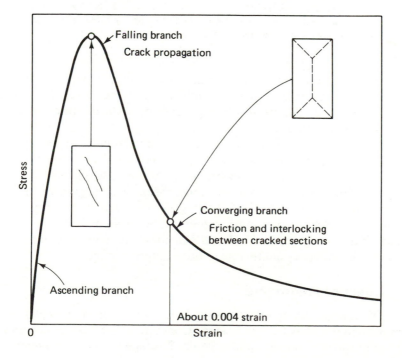

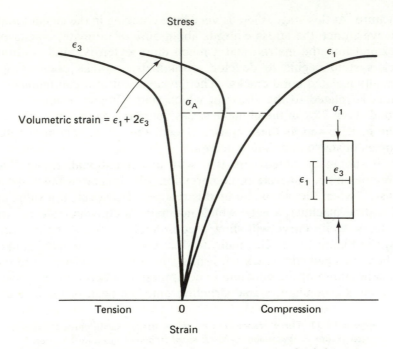

Figure 14.12 Stress-strain curves for a typical concrete loaded in uniaxial compression. (Adapted from K. Newman and J. B. Newman, in *Structure, Solid Mechanics and Engineering Design,* ed. M. Te'eni, John Wiley & Sons, Inc., New York, 1971, pp. 963–995.)

is reached. At this point, some additional strain capacity still exists before the cracks extend right through the concrete. Thus, the fracture of concrete is gradual, the result of progressive deterioration rather than the unstable growth of a single crack.

The beginning of matrix cracking corresponds to an apparent increase in Poisson's ratio, as shown in Figure 14.12 by the rapid increase in the ratio of ϵ_3 to ϵ_1 above the stress σ_A. At the onset of rapid cracking, the gross volume of concrete in compression also begins to increase. Clearly, these effects are due to the great increase in cracking that is taking place. Again, this must somehow be related to the composite nature of the concrete; for hardened cement paste alone in compression, the volume decreases continuously until failure occurs.

Effect of Aggregate

Of course, on a microscopic level, the picture is much more complicated than has been indicated. Concrete is heterogeneous, and the aggregate particles are not only irregular in shape but also are imperfectly bonded to the cement. Thus, highly localized stresses and strains may be quite different from the nominal applied stresses and strains. This is shown in Figure 14.13, based on studies using photoelastic coatings on the surface of a concrete specimen in compression. The localized strains may be as much as 4.5 times the average strains, and localized stresses may be more than twice as high as the average stresses. Since the largest strains occur in the contact zone, this accounts for the fact that the initial cracking occurs at the interface.

Figure 14.14 shows the idealized stresses around a single aggregate particle in concrete under compressive stress. For the normal case, where the particle is stiffer than the matrix, a stress analysis indicates the following order of failure: (1) tensile bond failure, (2) shear bond failure, (3) shear and tensile matrix failure, and (4) occasional aggregate failure. This small-scale anisotropy, brought on by the presence of the aggregate, may help to explain both the deformation and failure characteristics of concrete. Thus, even in compression, there will be some regions around aggregate particles undergoing a tensile stress. When the stress in this region reaches the theoretical tensile strength, a crack will start to grow. Compressive failure is thus preceded by progressive cracking, largely parallel to the axis of loading. (This is, of course, affected considerably by the end conditions of the particular compressive test—see Chapter 16.) Failure is eventually due primarily to the lateral tensile stresses produced by the microstructural failure of the material.

Figure 14.13 Distributions of strain and stress in the mortar and aggregate phases of a concrete specimen under uniaxial compression. Dashed lines represent the average values. Broken outlines represent aggregate particles. (From P. Dantu, *Annales de L'Institut Technique du Bâtiment et des Travaux Publiques,* Vol. 11, No. 121, pp. 55–77, January 1958.)

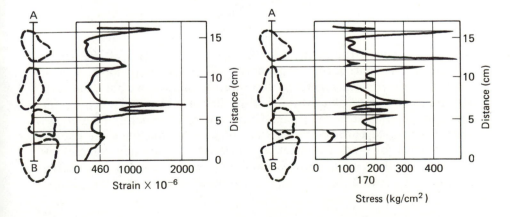

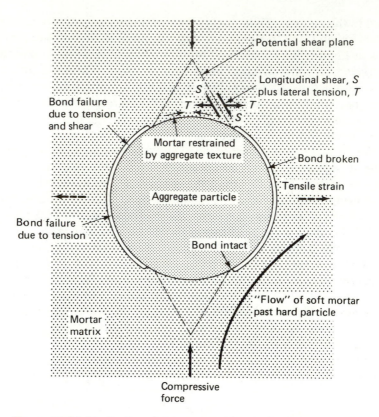

Figure 14.14 Idealization of stresses around a single aggregate particle at the discontinuity point under uniaxial compression. (From G. W. D. Vile, in *The Structure of Concrete,* ed. A. E. Brooks and K. Newman, Cement and Concrete Association, London 1968, pp. 275–288.)

Factors Stabilizing Crack Growth

Aggregates

While aggregate particles provide the source for many of the stress concentrations that occur in concrete, they also may have a very beneficial effect with regard to fracture. In normal-weight concrete, aggregates have a crack arresting action, because they are much harder and stronger than cement paste. Cracks will, in general, tend to go around aggregate particles rather than through them. Since the energy requirements for crack extension are increased by any new microcrack surfaces, as the cracks grow around aggregate particles the energy demand is increased. In this regard, it should be noted that unhydrated cement itself is very strong, and acts in the same way as aggregate in helping to arrest crack growth. Air voids or pores may also help to arrest cracks, by "blunting" the crack tip (i.e., increasing its radius of curvature).

Zone of Microcracking

The other effect that contributes to the strength of concrete is the occurrence of microcracking or "branch" cracking at the crack tip. Instead of a single crack, there is really a zone of microcracking with some remaining ligaments for stress transfer across the "crack." This, too, increases the amount of new crack surface, and the resulting energy requirement helps to stabilize crack growth. Thus, a crack that begins to grow will be arrested, or at least controlled, by the microcracking occurring at the crack tip. This will permit a further increase in load, until the next weakest crack starts to grow, and so on. Macroscopic crack growth is then due to the process of the formation of microcracks at a crack tip, followed by their coalescence. In this way, a crack eventually becomes dominant and extends through the concrete. The energy for the crack extension comes from either the work of external forces, or changes in the strain energy stored in the concrete. This effect, combined with the arresting effect of aggregate particles, helps to explain the extensive microcracking that takes place (rather than the growth of a single crack through the concrete), and also the nonlinearity of the stress–strain curve. The extent of branch cracking tends to be increased by more irregularly shaped aggregate particles.

Type of Loading

The behavior of concrete is greatly affected by the type of loading that occurs. For one thing, cracks tend to be more stable in compression than in tension. In addition, it has been found that the tensile failures that are produced by compressive loads occur at higher strains than those failures due to direct tensile stresses. Thus, as stated earlier, we probably need two criteria for failure: one for ultimate strength, the other for ultimate strain.

Static Fatigue

The failure of concrete through the growth of cracks can be used to explain one additional concrete phenomenon. It is well known that if concrete is loaded to about 75% or more of its short-term static strength, and if this load is sustained, the concrete will eventually fail (Figure

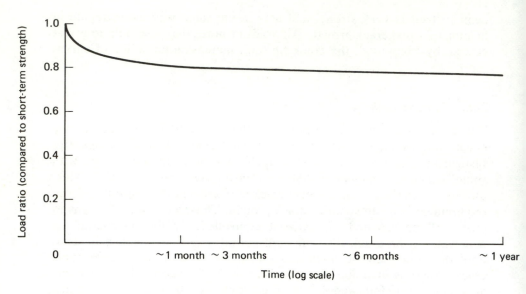

Figure 14.15 Static fatigue of concrete (schematic).

14.15). This is often referred to as delayed failure, or *static fatigue*. To explain this, it can be seen that the effect occurs only when the concrete is loaded in the range where unstable crack growth begins. Under such a sustained load, there will be slow crack growth both at the interface and in the matrix. Eventually, these cracks will grow until they reach some critical size, and fracture will then occur. Under this type of loading, the strain at failure will be higher than in short-term tests, because more cracking has time to take place. At lower sustained stresses, the cracking that does occur is compensated for by the consolidation of the concrete and its strengthening, probably due to an increase in secondary bonding.

The crack extension under sustained loading is sensitive to the presence of water. Dry concrete specimens take much longer to fail under sustained load than do saturated specimens. This suggests that a stress corrosion mechanism may be involved in this type of failure.

14.3 FATIGUE

Fatigue may be defined as the phenomenon by which a material is caused to fail by the repeated application of loads that are not large enough to cause failure in a single application. This implies that under repeated stresses, some internal progressive permanent structural change takes place in the concrete. This change, which may be referred to as fatigue damage, appears to involve primarily the growth of microcracks, although some viscous flow (or creep) may also take place. It has been found that fatigue occurs for compressive, tensile, and flexural loadings. In general, the factors that affect the static strength of concrete will have a similar effect on fatigue. However, since at failure the maximum strains are higher and the elastic moduli lower than for concrete under static loading, there must be some differences in the failure mechanism for fatigue loading. Since concrete may be subject to fatigue loadings in many applications, due to fluctuations in loading, temperature, or moisture content, it is important to examine this topic in more detail.

Fatigue Strength

Fatigue data are usually presented graphically by an $S-N$ diagram. This is a plot of the repeated stress, S (expressed as the stress ratio, or percentage of the static ultimate strength), versus the logarithm of the number of cycles of loading to cause failure, N. Unlike most metals, concrete does not appear to have a *fatigue limit* (i.e., the $S-N$ curve does not become horizontal) at least out to 10^7 cycles of loading. At 10^7 cycles the fatigue strength in compression, tension, or flexure is approximately 55% of the static strength. However, a single $S-N$ curve is inadequate to describe fatigue behavior; it is also necessary to take into account the large amount of scatter inherent in fatigue tests of concrete. Therefore, it is much more useful to present the fatigue data in terms of the probability of failure, as shown in Figure 14.16. The usual $S-N$ curve is that shown for $p = 0.5$.

Not only does the ratio of fatigue strength to static ultimate strength seem to be independent of the type of loading, it also seems to be independent of the details of the mix design. Cement, mortar, and concrete seem to have approximately the same $S-N$ diagrams, which also do not change significantly for different ages, w/c ratios, aggregates, air entrainment, and so on. However, concrete is very sensitive to the *range* of the cyclic stress (i.e., the difference between the highest and lowest stresses). It has been found that as the stress range decreases, the maximum stress that can be sustained for a given fatigue life increases.

The relationship between the stress range and the fatigue life can be expressed by the empirical *modified Goodman law*,

$$\sigma_{an} = \sigma_n(1 - \frac{\sigma_m}{\sigma_\mu})$$ (14.17)

where σ_{an} is the stress amplitude for a given fatigue life, σ_n the median fatigue strength for pure alternating stress, σ_m the mean stress, and σ_μ

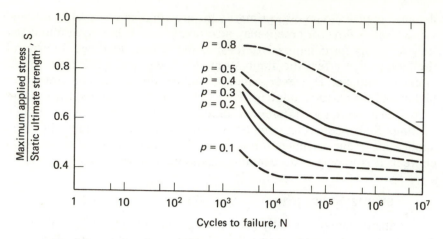

Figure 14.16 Typical set of fatigue curves for probabilities of failure for plain concrete subjected to reversed flexural loading. (From J. T. McCall, *Journal of the American Concrete Institute,* Vol. 55, No. 2, 1958, pp. 233–245.)

the ultimate strength. This "law" generally gives values that are conservative. Most commonly, this relationship is shown graphically on a modified Goodman diagram, as shown schematically in Figure 14.17a, (based on the assumption that the compressive strength of concrete is equal to 10 times the tensile strength). For design purposes, Figure 14.17b may be used, as recommended by ACI Committee 215, based on the assumption that the fatigue strength is the same for different modes of loading. This figure allows the determination of the maximum stress that concrete can withstand for 10^6 load cycles for a given minimum stress; for zero minimum stress, the conservative estimate is that the maximum load for 10^6 cycles is 50% of the static strength. The CEB also defines the fatigue strength as 50% of the static strength. On the other hand, the frequency of cyclic loading has no significant effect on fatigue strength as long as the maximum stress level is less than about 75% of the ultimate static strength. At higher loadings, creep becomes important, and the fatigue strength is reduced for low rates of cyclic loading.

Loading History

The effect of loading history on fatigue strength, particularly the effect of random varying loads, has not been studied in much detail. Although it is at best only an approximation, it has been found convenient

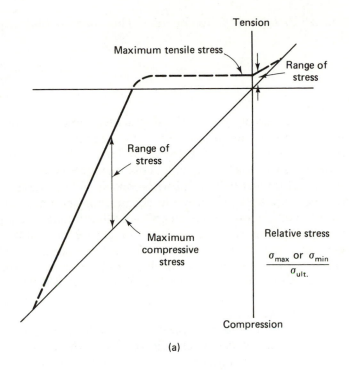

(a)

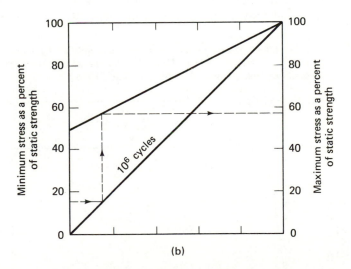

(b)

Figure 14.17 (a) Modified Goodman diagram for concrete (schematic). (From C. E. Kesler, Stanton Walker Lecture 1970, National Sand and Gravel Association, Washington, D.C.); (b) Fatigue strength of plain concrete in tension, compression, or flexure. (From ACI Committee 215, *Journal of the American Concrete Institute,* Vol. 71, No. 3, 1974, pp. 97–121.)

to apply the *Miner hypothesis* to concrete. This suggests that failure will occur when

$$\sum_i \frac{n_i}{N_i} = 1 \tag{14.18}$$

where n_i is the number of cycles of load at some stress condition and N_i the number of cycles required to cause failure at that condition. That is, the Miner hypothesis assumes that there will be a linear accumulation of damage due to each loading cycle. This hypothesis is not always conservative, and it must be applied with considerable caution.

The Miner hypothesis is unable to account for certain other effects of load history. For instance, it has been found that frequent rest periods during a fatigue test may increase the fatigue strength, although the improvement increases with the length of the rest period only up to about 5 min. This is shown in Figure 14.18, obtained from tests in which rest periods of varying length were observed after every 4500 load cycles. During these rest periods, the load was maintained at a load equal to the lower limit of the repeated load. This effect is probably due to some stress relaxation during the rest period. It would appear that a brief period at a high cyclic stress level followed by a lower stress level to failure results in a greater fatigue life than specimens subjected to only the lower stress continuously. Also, cyclic loading below the fatigue limit (i.e., below the load that would cause failure after 10^7 cycles) tends to

Figure 14.18 Effect of rest period on fatigue strength at 10 million cycles. (From H. K. Hilsdorf and C. E. Kesler, *Journal of the American Concrete Institute*, Vol. 63, No. 10, 1966, pp. 1059–1076.)

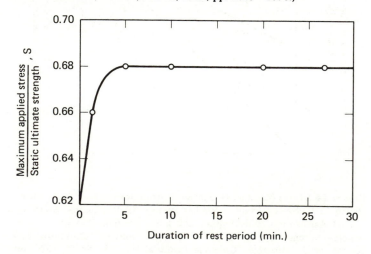

Duration of rest period (min.)

improve both the fatigue strength and the static strength by 5 to 15%, possibly due to some densification of the structure.

It has also been found that stress gradients may have a significant effect on the fatigue life. The fatigue strength of nonuniformly stressed specimens (simulating beams) was found to increase by about 15% over uniformly loaded specimens. A similar effect is found for static loading; it may be that eccentric loading somehow retards the growth of micro-cracks. Finally, it has also been shown that fatigue strength is improved by the application of lateral confining pressure.

Fatigue Strains

Even though fatigue strengths are lower than the static strength of concrete, the strains at failure are substantially larger in fatigue loadings than they are in simple static loadings. As yet, no limiting tensile strain has been found for fatigue loading, although, as we will see in Chapter 16, for static loading the limiting tensile strain is about 1 to 2×10^{-4}. Also, during fatigue loading, the slope of the stress–strain curve changes, from concave downward initially to concave upward, as shown in Figure 14.19. This is accompanied by a reduction in the secant modulus of elasticity (based on the maximum repeated stresses and strains) as shown in Figure 14.20, and this change in modulus can be used to estimate the fatigue life.

Figure 14.19 Variation of stress –strain curve with number of cycles. (From E. W. Bennett and N. K. Razu, in *Structure, Solid Mechanics and Engineering Design,* Part 2, ed. M. Te'eni, John Wiley & Sons, Inc., New York, 1971, pp. 1089 –1102.)

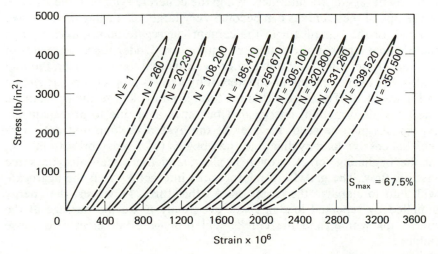

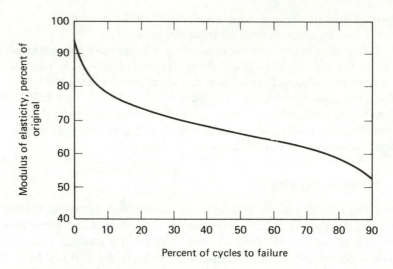

Figure 14.20 Reduction in secant modulus of elasticity as the result of repeated loads. From D. A. Linger and H. A. Gillespie, *Highway Research News,* No. 22, pp. 40–51, February 1966.

The fundamental mechanisms responsible for fatigue failure are only imperfectly understood. However, the main outlines of this type of failure can be explained using the concepts of stress concentration and fracture mechanics developed earlier in this chapter. We know that concrete contains a variety of internal flaws and cracks, which form during the hydration process and may become more extensive during periods when drying shrinkage occurs. These cracks are often found at the cement–aggregate interface. When the concrete is loaded, the stresses at the tips of these cracks can become very large, exceeding the cohesive strength of the cement (or of the cement–aggregate bond), even though the nominal stress in the concrete is quite low. Under repeated cycles of loading, these cracks will tend to grow. This may be seen from the fact that in the cyclic loading of concrete a hysteresis loop occurs, as shown in Figure 14.19. The area within the loop is equivalent to the irreversible energy of deformation, energy that becomes available to propagate the cracks. As may be seen, the area of the hysteresis loop first decreases with successive load cycling, but then begins to increase prior to failure. At the beginning, crack growth is slow, and there may also be some viscous flow. The growth of microcracks interacting with the aggregate will tend to stabilize the larger cracks that form. However, the energy supplied by repeated loadings will eventually make the damage at the crack tips worse, and the cracks will propagate sufficiently to cause failure.

Bibliography

ACI COMMITTEE 215, "Considerations for Design of Concrete Structures Subjected to Fatigue Loading," *Journal of the American Concrete Institute,* Vol. 71, No. 3, pp. 97–121 (1974).

Cement and Concrete Research, Vol. 3, No. 4 (1973). Special issue on fracture.

KAPLAN, M. F., "Crack Propagation and the Fracture of Concrete," *Journal of the American Concrete Institute,* Vol. 58, No. 5, pp. 591–610 (1961).

KESLER, C. E., "Fatigue and Fracture of Concrete," Stanton Walker Lecture 1970. National Sand and Gravel Association, National Ready-Mixed Concrete Association, Washington, D.C.

KROKOSKY, E. M., "Strength vs. Structure. A Study for Hydraulic Cements," *Materials and Structures (Paris),* Vol. 3, No. 7, pp. 313–323 (1970).

LAWN, B. R., AND T. R. WILSHAW, *Fracture of Brittle Solids.* Cambridge University Press, New York, 1975.

Mechanical Behavior of Materials, Vol. 4. "Concrete and Cement Paste. Glass and Ceramics." The Society of Materials Science, Japan. 1972.

NORDBY, G. M., "Fatigue of Concrete—A Review of Research," *Journal of the American Concrete Institute,* Vol 55, No. 2, pp. 191–219 (1958).

SHAH, S. P., AND S. CHANDRA, "Fracture of Concrete Subjected to Cyclic and Sustained Loading," *Journal of the American Concrete Institute,* Vol. 67, No. 10, pp. 816–825 (1970).

SWAMY, R. N., "Fracture Mechanics Applied to Concrete," pp. 221–281 in F. D. Lydon, ed., *Developments in Concrete Technology—1,* Applied Science Publishers Ltd., Essex, England, 1979.

Problems

14.1 Outline the basic assumptions of the Griffith theory of brittle fracture.

14.2 Describe the failure processes in concrete subjected to compressive loading.

14.3 Would this process be any different under pure tensile loading?

14.4 What is the significance of the parameter G_{IC}?

14.5 Why is the Mohr–Coulomb theory at best only an approximation for concrete?

14.6 Discuss the reasons for the nonlinearity of the stress–strain curve.

14.7 What is the effect of aggregate on the failure of concrete?

14.8   Discuss the statement: "The cement–aggregate interface is the weak point of concrete."

14.9   What is static fatigue, and why does it occur?

14.10 Why do you think concrete does not have a true fatigue limit?

14.11 Determine the maximum stress that can be applied to a concrete bridge deck which has a flexural strength of 15 MPa and a dead-load stress of 5 MPa, if it is to survive repeated flexural loads for 10^6 cycles.

14.12 What factors improve the fatigue life of concrete?

15

strength

The strength of hardened concrete is by no means its only important characteristic; durability, volume stability, and impermeability may be equally significant. It is, perhaps, unfortunate that it has become the custom to assess concrete quality only in terms of its compressive strength. The general assumption is that an improvement in concrete strength will improve its other properties as well. Although this is often the case, there are many important exceptions. For example, an increase in cement content that may be intended to increase the strength may also increase the amounts of shrinkage and creep. Thus, concrete strengths alone may be misleading in terms of choosing an appropriate mix design for a particular job. Nonetheless, since compressive strength is so universally used as an indication of concrete quality, this chapter will emphasize the factors affecting compressive strength, although strength under other kinds of loading will also be considered.

The factors that can affect the strength of concrete can be classified into four categories: constituent materials, methods of preparation, curing procedures, and test conditions. Methods of preparation were discussed in Chapter 10, curing procedures in Chapter 11, and test conditions will be discussed in detail in Chapter 16. Thus, we will here be concerned primarily with the effects of the concrete constituents — cement, aggregates, and water—on the strength and other mechanical properties of the concrete. However, we cannot consider these constituents in isolation from each other. Changes in one of the raw materials invariably lead to changes at least in the relative proportions of the others in order to maintain concrete quality.

In the discussion that follows, the effects of the factors governing strength will be considered primarily in terms of the compressive strength, since this parameter is of greatest importance for concrete and thus has received the most attention. However, other strength properties generally will be affected in a similar fashion.

15.1 EFFECT OF POROSITY ON STRENGTH

The primary factor that governs the strength of brittle materials is porosity. A number of models have been proposed to describe the basic strength–porosity relationship. The most common one, which seems to fit the experimental data well over a wide range of porosities in many systems, is the exponential expression:

$$S = S_0 e^{-kP} \tag{15.1}$$

where S is the strength, S_0 the strength at zero porosity ("intrinsic strength"), P the fractional porosity, and k a constant that depends on the system being studied. A logarithmic expression has also been proposed, based on a study of set plaster,

$$S = q \log \frac{P_r}{P} \tag{15.2}$$

where P_r is the porosity at $S = 0$ and q is a constant.

These equations and the many others that have been proposed represent, of course, a considerable simplification of the system. The nature of the material, the size and shape of the pores, and whether the pores are empty or are filled with a fluid all must have some effect. For instance, it appears that continuous porosity has a greater influence on strength than do isolated pores. Also, there is some evidence that large pores may be more effective than small pores in relieving stress concentrations at crack tips. Nevertheless, whatever the best form the strength–porosity relationship takes for a particular system, it is clear that strength depends mainly on the porosity. Indeed, it has been shown that for many materials, the ratio S/S_0 plotted against porosity follows the same curve. Figure 15.1 presents such a curve, with the plotted data representing a number of normally cured cements, more crystalline autoclaved cements, and a variety of aggregates.

Gel/Space Ratio

In their classic work published in 1946–1947, Powers and Brownyard showed that the increase in compressive strength of portland cement mortars "is directly proportional to the increase in [gel/space ratio]

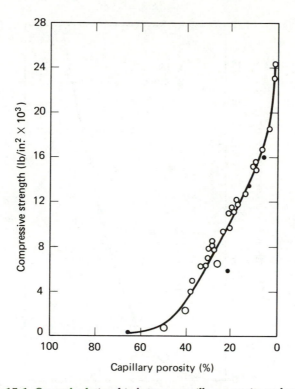

Figure 15.1 General relationship between capillary porosity and average strength of the various materials. (From G. J. Verbeck and R. A. Helmuth, *Proceedings, Fifth International Symposium on the Chemistry of Cement, Tokyo, 1968,* Vol. 3, pp. 1–32.)

regardless of age, original *w/c* ratio, or identity of cement." The gel/space ratio, *X*, is the ratio of the solid products of hydration to the space available for these hydration products. In other words, the gel/space ratio is a representation of the capillary porosity of the paste in terms of its measureable parameters. It was derived in Chapter 4 as[1]

$$X = \frac{0.68\alpha}{0.32\alpha + (w/c)} \qquad (4.15)$$

Before hydration begins, the available space is that occupied by the mixing water. After hydration has proceeded to a certain degree, the available space is then the sum of volumes of hydrated cement and of

---

[1]In mixes containing air, the *w/c* ratio should be replaced by $(w + A)/c$, where *A* is the volume of air.

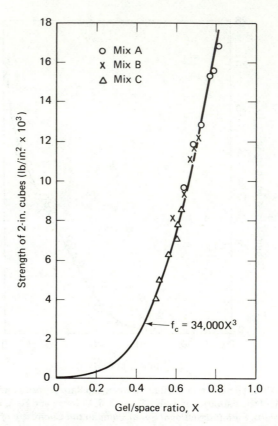

Figure 15.2 Compressive strength vs. gel/space ratio for cement – sand mortars: σ_c = compressive strength (lb/in.²); X = gel/space ratio. (From T. C. Powers, *Journal of the American Ceramic Society*, Vol. 41, No. 1, 1958, pp. 1–6.)

the remaining capillary pores. (Remember that 1 cm³ of cement will occupy about 2.1 cm³ of space when fully hydrated.)

It has been found that the relationship between compressive strength (σ_c) and the gel/space ratio can be written as

$$\sigma_c = AX^n \qquad (15.3)$$

where A is a constant representing the intrinsic strength of the cement gel (i.e., the strength at a gel/space ratio of 1.0) and n is a constant having values in the range 2.6 to 3.0, depending on the characteristics of the cement. The relationship found by Powers and Brownyard (Figure 15.2) can be written as

$$\sigma_c = \begin{cases} 235X^3 & \text{MPa} \\ 34{,}000X^3 & \text{lb/in.}^2 \end{cases} \tag{15.4}$$

This expression is independent of the mix design or age. It should be noted, however, that these data are based on strengths obtained from 2-in. mortar cubes. Tests on neat cement paste, or on specimens of different geometries, would presumably lead to different values of the constants, although the general trend would be the same. Indeed, Powers and Brownyard found that the gel/space ratio concept was also not entirely independent of the cement composition; higher C_3A contents (particularly above 7% C_3A) led to lower strengths at a given gel/space ratio.

It must be noted that in the discussion above, the gel porosity is included as a part of the solid C–S–H. This accounts for the relatively low intrinsic strength (230 MPa or 34,000 lb/in.²) of the paste, since C–S–H, which is a major component of the hardened paste, must itself be considered as a highly porous solid. It is possible to express the strength as a function of the *total* porosity (i.e., capillary + gel porosity), but there is no practical advantage in doing so, because the gel porosity is a constant fraction of the C–S–H volume.

15.2 FACTORS AFFECTING STRENGTH

Water/Cement Ratio

Even though the strength of concrete is dependent largely on the capillary porosity or gel/space ratio, these are not easy quantities to measure or predict. They are therefore not suitable for use in the mix design procedures outlined in Chapter 9. Fortunately, however, the capillary porosity of a properly compacted concrete at any degree of hydration is determined by the water/cement ratio. The paramount influence of the *w/c* ratio on the porosity of hardened cement paste at a given degree of hydration is shown in Figure 4.12b. Therefore, in practice, we can assure the strength of properly compacted concrete at a given age by specifying the *w/c* ratio. Even through Abrams' water/cement ratio law (Chapter 9) is not really a "law," in that it does not consider the degree of hydration, the air content of concrete, or the effects of the aggregate, it is sufficiently correct in the usual range of *w/c* ratios encountered in practice to be a very useful tool. However, the *w/c* ratio law as usually stated is a simplification of the true state of affairs.

One example of the limitation of the *w/c* ratio law is illustrated in

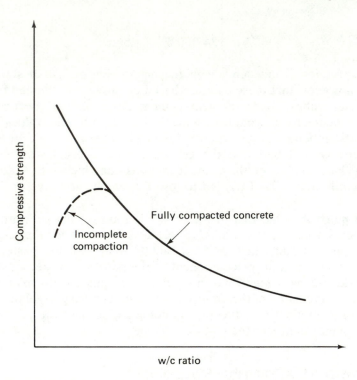

Figure 15.3 Relationship between compressive strength and *w/c* ratio.

Figure 15.3. Concrete that is not properly compacted will contain large voids, which contribute to its porosity. Thus, at lower *w/c* ratios, where full compaction is difficult to achieve, Abrams' law ceases to be valid. The point at which this occurs depends on the methods of compaction that are employed. In this regard it might be pointed out that the sophisticated methods of compaction being developed (primarily in precasting operations), and the use of superplasticizers, make possible the production of very-high-strength concretes using ordinary cement contents. That is, if we can fully compact the concrete, Abrams' law is obeyed even at very low *w/c* ratios.

A better description of the role of the *w/c* ratio in determining strength was given by Gilkey:

> For a given cement and acceptable aggregates, the strength that may be developed by a workable, properly placed mixture of cement, aggregate, and water (under the same mixing, curing, and testing conditions) is influenced by the: (a) ratio of cement to mixing water, (b) ratio of cement to aggregate, (c) grading, surface texture, shape,

strength, and stiffness of aggregate particles, (d) maximum size of the aggregate.

Thus, in fact, the *w/c* ratio "law" is really a family of relationships for different mixtures, as shown in Figure 15.4. Although there is general agreement that the *w/c* ratio law applies uniformly to ordinary, workable structural concrete (using graded aggregates up to a maximum size of 40 mm), this law may not necessarily apply to mixes in which the maximum aggregate size or aggregate amount is very different from "ordinary" concrete, as in neat pastes or mortars on the one hand, and mass concrete mixes on the other. Those involved in the design of "unusual" concrete mixes should recognize that a simple reliance on the *w/c* ratio dependency may lead to serious error.

Although we may specify a *w/c* ratio in our mix designs, there is often a great deal of uncertainty over what the true *w/c* ratio was when the concrete was placed. In normal practice, only the slump test is used to indicate whether there is additional water in the mix, either from the aggregates or because of a deliberate addition of water to make the

Figure 15.4 Water/cement ratio curves; approximate extreme workable ranges for a few mixtures of different proportions and/or coarse aggregate gradings. (From H. J. Gilkey, *Journal of the American Concrete Institute*, Vol. 57, No. 10, 1961, pp. 1287–1312.)

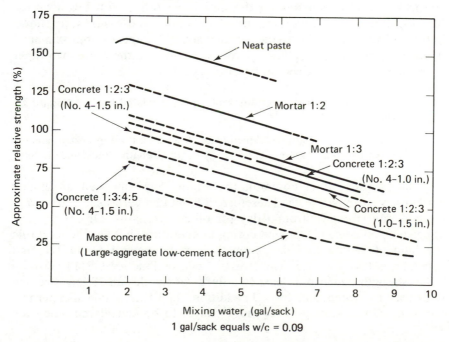

concrete easier to handle. It has thus been suggested that concrete be proportioned on the basis of performance standards, with no specification at all of *w/c* ratio. In practice, this would mean specifying concrete on the basis of strength, since durability tests are too expensive and time-consuming to be used routinely. Unfortunately, this would be a very retrograde step, since changes in strength may very well not correspond with changes in durability. In addition, the use of finer cements, certain admixtures, and special curing techniques could provide 7-day or 28-day compressive strengths that do not in any way represent the true quality of the concrete. Until some more reliable methods of specifying concrete performance in the field are developed, the *w/c* ratio remains the best tool for assuring satisfactory performance of concrete, in terms of both strength and durability. The development of methods for directly measuring the *w/c* ratio of fresh concrete, such as the Kelly–Vail technique (Chapter 8), may counter the criticism that we never really know the *w/c* ratio.

Time

The knowledge that strength is controlled by the gel/space ratio is not sufficient to permit a prediction of concrete strength at any given time, because the rate of hydration depends upon the particular cement and the curing conditions, as were discussed in Chapter 11. The rate of strength gain also depends on the initial *w/c* ratio, with low-*w/c*-ratio mixes gaining strength more rapidly than high-*w/c*-ratio mixes, as shown in Figure 15.5. This is due to the fact that strength depends approximately upon the cube of the capillary porosity. However, there are two other practical questions that should be discussed:

1. How can the strength at different times be related to the standard 28-day strength?

2. Should the strength developed after 28 days be used in design, since it is known that concrete continues to gain strength almost indefinitely?

In practice, it is common to obtain 7-day as well as 28-day compressive strength tests. Therefore, it would be very useful to extrapolate 28-day strengths from 7-day (or other) strengths. Of course, this depends upon cement type and curing temperature, but as a general rule the ratio of 28-day to 7-day strengths lies between 1.3 and 1.7, and generally is less than 1.5. The British Code of Practice CP114 allows a 7-day strength equal to at least two-thirds of the design 28-day strength to be used for acceptance tests. In addition, the British Code also permits the later gain in strength (beyond 28 days) to be considered when the

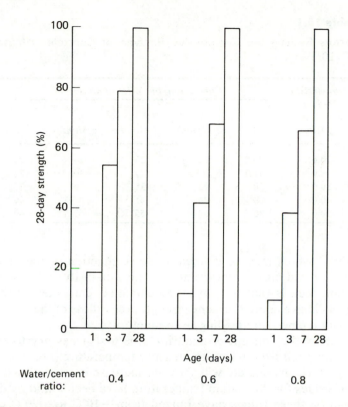

Figure 15.5 Relative gain of strength with time of concretes with different water/cement ratios (Type I cement). (From A. Meyer, *Betonstein Zeitung*, Vol. 29, No. 8, 1963, pp. 391–394.)

structure (or structural element) is to be loaded at a later age, using the factors shown in Table 15.1. These values do not, of course, apply when accelerators are used, or under extreme curing temperature conditions.

The Maturity Concept

The hydration of cement is greatly affected by both the time and the temperature of hydration, so the gain in strength of concrete is also largely controlled by these two factors. Thus, it was only natural that a considerable amount of research would be carried out to try to determine how the strength of concrete could be expressed as some function of the time and temperature of curing. This information could then be used to estimate the strength of concrete without the necessity of carrying out physical tests. Out of these studies came the concept of concrete

Table 15.1

Factors Relating the Compressive Strength of Concrete with Age,
CP110: Part 1

Characteristic (28-day) Strength f_{cu} MPa	Cube Strength (MPa) at an Age of:				
	7 Days	2 Months	3 Months	6 Months	1 Year
20.0	13.5	22	23	24	25
25.0	16.5	27.5	29	30	31
30.0	20	33	35	36	37
40.0	28	44	45.5	47.5	50
50.0	36	54	55.5	57.5	60

"maturity," which may be defined as some function of the product of curing time t and curing temperature T [i.e., maturity $= f(T \times t)$]. The assumption then is that for any particular mix, concretes of the same maturity will have about the same strength, regardless of the combination of time and temperature leading to the maturity.

To apply the concept of maturity, it is first necessary to establish some datum point from which to measure temperature (i.e., the temperature below which concrete will show *no* increase in strength with time). Different values for this datum temperature have been found by different investigators; these values have ranged from $-10°C$ to $-20°C$ (14°F to $-4°F$). The value $-10°C$ is most commonly used. A number of maturity functions have been proposed over the years.

The function that probably best correlates with the strength of concrete is the *Nurse–Saul expression*:

$$\text{maturity (°C} \times \text{days)} = \sum a_t(T + 10) \qquad (15.5)$$

where a_t is the time of curing in days and T is the temperature in °C. In some cases, this provides a very good correlation between maturity and compressive strength, as shown in Figure 15.6, as well as between maturity and elastic modulus.

A somewhat different maturity–strength function was proposed by Plowman, which can be stated in terms of the strength obtained in 28 days of curing at 64°F (using a datum temperature of 11°F):

percent of strength at a maturity of 35,600°F × hours

$$= A + B \log \left(\frac{\text{maturity in °F} \times \text{hours}}{1000} \right) \qquad (15.6)$$

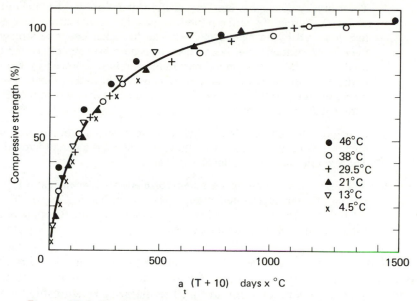

Figure 15.6 Percentage compressive strength as a function of the product of age and temperature (100% = 28 days at 23°C). (From S. G. Bergstrom, *Magazine of Concrete Research,* Vol. 4, No. 14, 1953, pp. 61 – 66. Reproduced by permission of Cement and Concrete Association.)

where A and B are constants related linearly to the strength at any age. Plowman stated that this function was valid for times up to at least 1 year and curing temperatures below 100°F (38°C). In addition, he provided values of the constants A and B for different strength classes of concrete. Using this function, a graph of strength versus $\log_{10}$ (maturity) would be linear, as shown in Figure 15.7.

Unfortunately, there are a number of limitations to the use of maturity for predicting concrete strengths.

1. The maturity functions that have been proposed do not take into consideration the effect of the humidity conditions during curing. However, it is clear from Figure 11.1 that this is a major consideration that cannot be ignored.

2. The various functions cannot be applied to mass concrete, because the rate of heat loss from such structures is much less than that from normal specimens (i.e., only the ambient temperature is considered; the contributions from the heat of hydration are ignored).

3. These functions are not very useful at low maturities, perhaps because the time should be calculated not from the time of mixing or casting, but from the later time at which the concrete actually begins to gain strength.

4. The maturity concept is not very useful if there are large temperature variations during the curing period, since a low initial curing temperature followed by normal curing will lead to higher concrete strengths than if the concrete had been cured at a normal temperature for the total time, while a high initial curing temperature followed by normal curing will have a detrimental effect. This is indicated in Figure 11.2b, where the temperatures shown on the various curves refer to the initial curing temperatures for the first 28 days.

5. Strength will also be affected by both the chemical composition and the fineness of the cement, since these both affect the rate of hydration. The *w/c* ratio may also influence the results.

6. Although there seems to be some correlation between maturity and strength for the accelerated strength tests described in Chapter 16, the relationship breaks down when the accelerated cure specimens are subsequently cooled and moist-cured.

Some investigators have found that maturity can be used to describe concrete strengths over a wide range of times and temperatures and concrete mixes. Others have found that the maturity relationships work only over very narrow ranges of time or temperature, or not at all. Nor

Figure 15.7 Strength vs. log maturity. (From J. M. Plowman, *Magazine of Concrete Research,* Vol. 8, No. 22, 1956, pp. 13–22. Reproduced by permission of Cement and Concrete Association.)

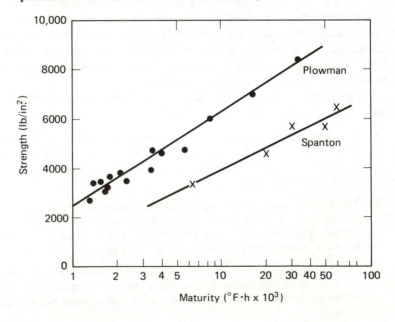

is there general agreement even among the proponents of the maturity concept about the best form of the maturity function to use. The contradictory findings are probably not due to experimental error. Rather, they are an indication that the process of strength gain in concrete is very complex, and cannot be generally modelled by the simple functions described above. In particular, these time–temperature functions should in no way be considered as physical "laws."

Nevertheless, in spite of its severe limitations, the maturity concept may be useful when trying to establish "after the fact" the strength of concrete in a structure at some previous time. This may be done by measuring core strengths at some later time and then using one of the maturity functions to estimate the strength at the earlier time. Also, the maturity concept may help to estimate the appropriate time for form removal when concreting at lower-than-normal temperatures. The maturity concept is not very widely used, probably because of the uncertainties described above, but concrete technologists should be aware of this method of estimating concrete strength, with all of its inherent limitations.

Cement

The effect of portland cement on concrete strength depends on the chemical composition and fineness of the cement. We have already seen (Chapter 3) that the strength of cement comes primarily from C_3S (early strength) and C_2S (later strength), and these effects carry through into concrete. Concretes made with cements of higher C_3S contents gain strength more rapidly, but may end up with slightly lower strengths at later ages. The effects of the five different types of cement on strength are shown in Figure 3.6. Although there are considerable differences in strength development up to about 1 month, at later ages the differences among the five types of cement become less important. Cements that hydrate more slowly, whether through changes in composition, curing conditions, or the use of admixtures, tend to develop higher ultimate strengths.

The effects of cement fineness on the strength of concrete are also considerable, since the rate of hydration increases with increasing fineness and leads to a higher rate of strength gain, as indicated in Figure 15.8. Typically, the maximum particle size is about 50 μm, with about 10 to 15% less than 5 μm, and perhaps 3% less than 1 μm. Apparently, the fraction of particles less than 3 μm in diameter has the greatest influence on 1-day strengths, while the 28-day strengths are most affected by the 3 to 30 μm fraction. However, even though more finely ground cements gain strength more rapidly, very fine grinding should be avoided. With

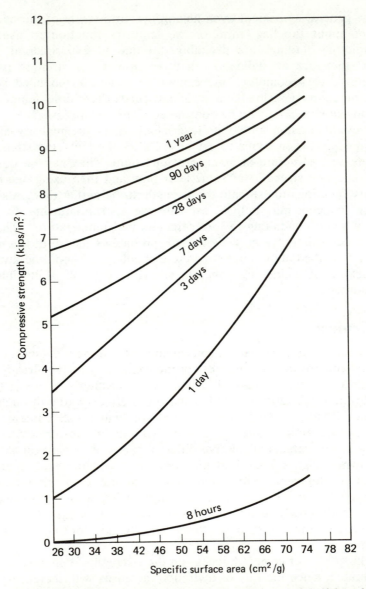

Figure 15.8 Effect of cement fineness on strength; $w/c = 0.4$. (Adapted from E. W. Bennett and B. C. Collings, *Proceedings of the Institution of Engineers*, Vol. 43, 1969, pp. 443–452. Reproduced by permission of the Institution of Civil Engineers.)

very fine particles, the interstitial water may lead to regions of high *w/c* ratio. On the other hand, there is evidence that particles having diameters greater than about 60 μm contribute very little to strength.

It must be recognized that the variability inherent in cement leads to a corresponding variability in concrete strength, and therefore requires a higher average concrete design strength (see Chapter 17). Not only do cements which are classified as the same ASTM type vary considerably from plant to plant, but within a given plant the cement characteristics vary over time, owing to changes in raw materials, burning conditions, and so on. It has been estimated that this variability in cement quality leads to a coefficient of variation in concrete strengths of the order of 5%.

Aggregate

Although the *w/c* ratio is the most important factor affecting strength, the properties of the aggregate cannot be ignored. The aggregate parameters that are most important are the shape and texture and the maximum size of the aggregate. The aggregate strength itself is of less importance, since aggregates are all generally much stronger than the cement paste, except in the case of lightweight aggregates or high-strength concrete.

The aggregate texture depends on whether the aggregates are naturally occurring gravels, which tend to be smooth, or crushed rocks, which tend to be rough and angular. As discussed in Chapter 13, surface texture affects both the bond and the stress level at which microcracking begins. The surface texture, therefore, may affect the shape of the $\sigma-\epsilon$ curve, but has little effect on the ultimate compressive strength of the concrete. On the other hand, since surface texture does affect the tensile or flexural strength of the concrete, the ratio between the flexural and the compressive strength depends on the type of aggregate, as shown in Figure 15.9. It has been found that at low *w/c* ratios, crushed rock will lead to higher concrete strengths because of the better mechanical bond, but this effect disappears as the *w/c* ratio increases. However, if mixes are considered on the basis of equal workabilities, this difference becomes unimportant because of the lower water requirement of smooth aggregates.

The use of a larger maximum size of aggregate affects the strength in several ways. Because the use of larger particles reduces the specific surface area of the aggregate, the bond strength is also less, and this tends to reduce the strength. Also, larger aggregate particles provide more restraint on volume changes in the paste, and thus may induce additional stresses in the paste, which tend to weaken the concrete.

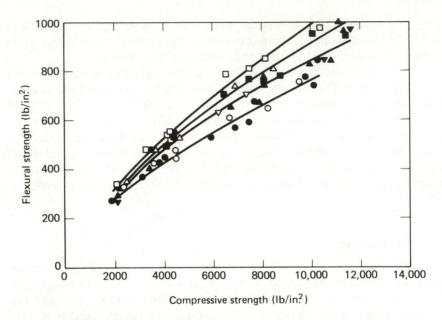

Figure 15.9 Relation between flexural strength of concrete beams and compressive strength of equivalent cubes, for different types of aggregates. (From R. Jones and M. F. Kaplan, *Magazine of Concrete Research,* Vol. 9, No. 26, 1957, pp. 89–94. Reproduced by permission of Cement and Concrete Association.)

These effects are offset, however, by the reduced water content necessary to achieve a suitable workability, so the net effect of using larger aggregate particles is slight. In very rich mixes there tends to be a reduction in strength due to the reduced bond, but in lean mixes there tends to be an increase in strength due to the reduced w/c ratio. Concretes made with larger aggregates tend to exhibit more variability, probably due to some tendency toward segregation.

It should be noted that the strength differences we are discussing here are not very large, and the indicated trends are true only for tests done with a given type of aggregate. The effects may well be obscured by the use of different types of aggregates.

In the range of aggregate contents normally encountered, the exact volume of aggregate is only of secondary importance in determining concrete strength. In general, at a constant w/c ratio, a higher strength can be achieved by using a leaner mix (Figure 15.10), probably because the total water content, and hence the paste content, are less and thus the total porosity is reduced compared to a richer mix. If a constant workability is maintained, the strength of concrete is dependent on the cement content (Figure 15.11), because of the reduced water requirement

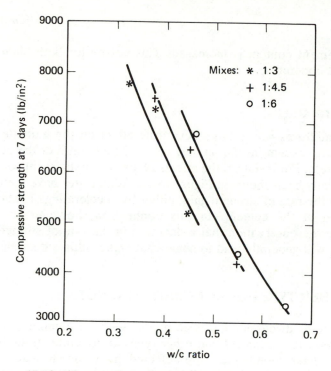

Figure 15.10 Influence of aggregate/cement ratio on the compressive strength of concrete. (From B. G. Singh, *Journal of the American Concrete Institute,* Vol. 54, No. 10, 1958, pp. 897–907.)

Figure 15.11 Strength in relation to cement content for air-entrained and non-air-entrained concrete of constant slump. (Adapted from *Concrete Manual,* 8th ed., U.S. Bureau of Reclamation, Denver, Colo., 1975.)

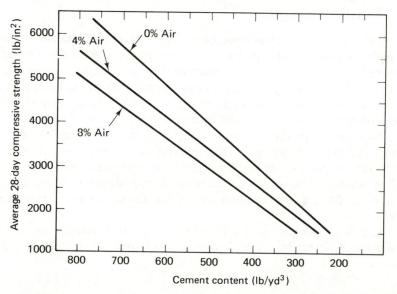

as the cement content is increased. This is true for both plain and air-entrained concretes.

Admixtures

Admixtures *per se* have very little effect on the ultimate strength of concrete, except insofar as they affect the *w/c* ratio or the porosity of the concrete. For instance, the effects of air-entraining agents on strength have already been shown in Figure 15.11. Admixtures have their greatest effect on the rate of strength gain, either by accelerating or retarding the hydration of the cement. In this connection, however, it is worth repeating the observation that a decrease in the rate of strength gain of concrete will generally lead to somewhat higher ultimate strengths.

15.3 OTHER TYPES OF CONCRETE STRENGTH

The foregoing discussion has dealt mainly with the compressive strength of concrete. It is assumed that other types of concrete strength (tensile, flexural, shear, bond, etc.) are affected in a similar manner by the variables discussed. Although this is generally true, there are no simple, unique relationships among the various measures of concrete strength. In this section we will look at the relationships that do appear to exist among the different types of concrete strength by trying to relate them to the compressive strength.

Tensile Strength

The tensile strength of concrete is much lower than the compressive strength, largely because of the ease with which cracks can propagate under tensile loads. Although tensile strengths are usually not considered directly in design (being assumed to equal zero), they are quite important, since cracking in concrete tends to be a tensile failure. However, the relationship between tensile and compressive strengths is not a simple one. It depends on the age and strength of the concrete, the type of curing, the type of aggregate, the amount of air entrainment, and the degree of compaction. In addition, the relationship also depends on the way in which the tensile strength is measured: direct tension, splitting tension, or flexure (see Chapter 16 for the details of these test procedures).

In general, as the age (or the strength level) increases, the ratio of tensile to compressive strength (σ_t/σ_c) decreases, as shown in Figure

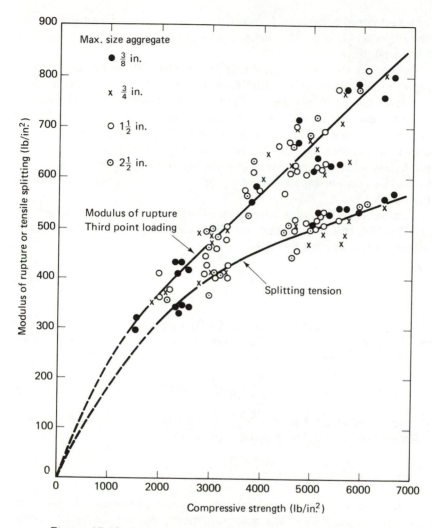

Figure 15.12 Relationships of compressive to flexural and tensile strengths of concrete. (From S. Walker and D. L. Bloem, *Journal of the American Concrete Institute,* Vol. 57, No. 3, 1960, pp. 283–298.)

15.12. Also, since crushed coarse aggregate seems to improve the tensile strength more than it does the compressive strength, the σ_t/σ_c ratio also depends on the type of aggregate. It has been found that, compared to moist curing, air curing reduces the tensile strength more than it does the compressive strength, probably because of the effect of drying shrinkage cracks. Therefore, the σ_t/σ_c ratio is less for air-cured than it is for moist-cured concretes. Incomplete compaction and air entrainment, on the other hand, affect the compressive strength more than they do the tensile strength.

In general, the ratio of the direct tensile strength to the compressive strength ranges from about 0.07 to 0.11. Since splitting tension (σ_{st}) values tend to be slightly higher than direct tension values, the σ_{st}/σ_c ratios are also slightly higher, from about 0.08 to 0.14. Flexural tests (which are the most common way of estimating tensile strengths) give results that are substantially higher than direct tension tests, and the ratio of flexural to compressive strength (σ_f/σ_c) ranges from about 0.11 to 0.23. The relationships between σ_{st} and σ_c, and σ_f and σ_c, are shown in Figure 15.12.

A number of formulas have been developed that relate σ_t to σ_c, but they all suffer from the limitations discussed above. A fairly common relationship that is used by ACI is

$$\sigma_f = \begin{cases} 0.62(\sigma_c)^{1/2} & \text{MPa} \\ 7.5(\sigma_c)^{1/2} & \text{psi} \end{cases} \tag{15.7}$$

The European Concrete Committee (CEB) suggests that the axial tensile strength can be related to the characteristic compressive strength (Chapter 17) by the expression

$$\sigma_t = 0.3(\sigma_c)^{2/3} \qquad \text{MPa} \tag{15.8}$$

with a range of variability of $\pm 30\%$. The CEB also suggests that σ_t can be assumed to be equal to only 60% of σ_{st}.

Shear Strength

Although pure shear is probably never encountered in concrete structures, failure of concrete quite commonly occurs through a combination of shear and normal stresses. Thus, a knowledge of the shear strength, τ, of concrete would be useful. However, it has not been possible to measure the strength of concrete in pure shear directly. The normal "pure shear" test, torsion of a hollow cylinder, produces a principal tensile stress equal to the shear stress but acting at 45° to the shear stress. Since concrete is weaker in tension than in shear, it is not possible to determine the shearing strength in this way.

To determine the shearing strength directly, tests have been carried out on beams of very short span, with the loads applied very close to the supports. In some tests, τ has been found to be only slightly larger than the tensile strength; in others τ has been found to be 50 to 90% of the

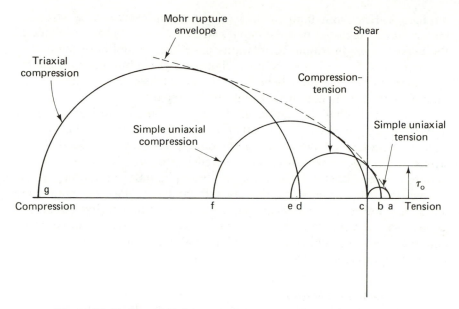

Figure 15.13 Typical Mohr rupture diagram for concrete.

compressive strength. This discrepancy may be due to the fact that either tensile or compressive stresses may also have been acting in these tests.

A reliable estimate of the shear strength can only be obtained from combined stress tests. This is probably best represented by a Mohr rupture diagram, as shown in Figure 15.13. The strength of concrete in pure shear is represented by the point at which the failure envelope intersects the vertical axis, τ_0. It has been found in this way that the shearing strength is approximately equal to 20% of the compressive strength of concrete. Of course, if normal stresses are also acting, the shearing strength can be made to exceed the uniaxial compressive strength.

Impact Strength

Even more than the other mechanical properties of concrete, the impact strength is strongly dependent on the method of testing, and this may account for the absence of any standard concrete impact tests. However, in spite of a number of investigations of impact strength, there is still no agreement on the relationship (if any) between impact and other strengths. Some investigators have found that impact is related to the compressive strength, and it has been suggested that the impact strength varies from 0.50 to 0.75 of the compressive cube strength. Others have suggested that the impact strength is more closely related to the tensile strength. Some correlations have been found between impact strength and flexural and compressive strengths, but only when the specimens were continuously moist-cured until tested. Others have found no correlations between impact and other strengths.

It is likely that the nature of the cement–aggregate bond is more important for impact than for other types of strength, as the type of aggregate seems to have a very great effect, as shown in Figure 15.14, where the impact strength is represented by the number of blows that concrete can withstand till there is "no rebound" of the impacting device. In general, coarse aggregates with greater surface roughness and angularity improve the impact strength. In addition, curing conditions appear to be very important, as dry and continuously moist-cured concretes seem to behave differently, probably because of the shrinkage cracks that develop when concrete is dried.

Figure 15.14 Relationship between compressive strength and number of blows to "no-rebound" for concretes made with different aggregates and ordinary portland cement. (From H. Green, *Proceedings of the Institution of Civil Engineers*, Vol. 28, Paper No. 6769, 1964, pp. 383–396. Reproduced by permission of the Institution of Civil Engineers.)

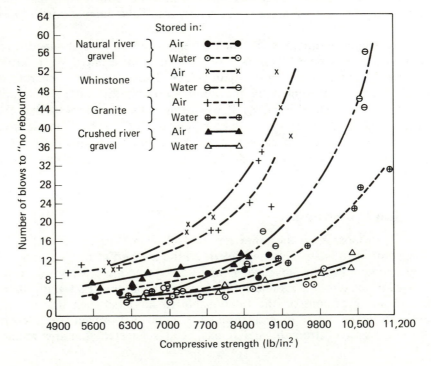

Multiaxial Stresses

In the discussion above, we have considered the different types of concrete strength separately. However, in practice, concrete is rarely subjected to only a single type of stress. In the general case, concrete might be subjected to a combination of compressive, tensile, and shearing stresses, all acting simultaneously. The behavior of concrete under these combined stresses is, unfortunately, still not completely understood. Not only is there no accepted standard test for concrete subjected to multiaxial stresses but, as was pointed out in Chapter 14, there is still no agreement on a general criterion of failure for concrete. In the discussion that follows, the effects of multiaxial stresses on concrete strength will be described from a phenomenological point of view; the theoretical considerations were discussed in Chapter 14.

It is probably easiest to describe the effects of combined stresses on strength by referring to a Mohr rupture diagram for concrete, shown schematically in Figure 15.13. Even though the Coulomb–Mohr theory may not apply exactly to concrete, this remains the most convenient way of representing failure under multiaxial stresses. In this figure, τ_0 represents the strength of concrete in pure shear, which was stated above to be about 20% of the compressive strength. The distance $c-f$ represents the uniaxial compressive strength (as might be found from the standard cylinder test). It may be seen, then, that an applied transverse compressive stress $(c-d)$ increases the compressive strength $(c-g)$. On the other hand, a transverse tensile stress $(c-b)$ will decrease the apparent compressive strength to $(c-e)$. The increase in compressive strength when a lateral confining pressure is applied can be very large, as shown in Figure 15.15. In addition, the application of a lateral confining pressure leads to a large increase in the compressive strain at failure. In general, the effect of a confining pressure on strength is more beneficial for weak than for strong concretes.

These effects depend to some degree on the details of the particular concrete mix being tested. For cement pastes, confining pressure leads to smaller strength increases than for concrete. Usually, for a given w/c ratio, the strength increases more under confining pressures as the aggregate content increases.

So far, we have looked only at triaxial compression. If we examine the case of tension plus biaxial compression, the tensile strength is reduced by the application of lateral compressive stresses. This effect, too, is larger for higher w/c ratio concretes, and is larger for concrete than for cement paste. In general, for this state of stress, strength increases with increasing aggregate content. For biaxial tension, the strength seems to be about the same as in uniaxial tension.

The various experimental investigations that have been carried out

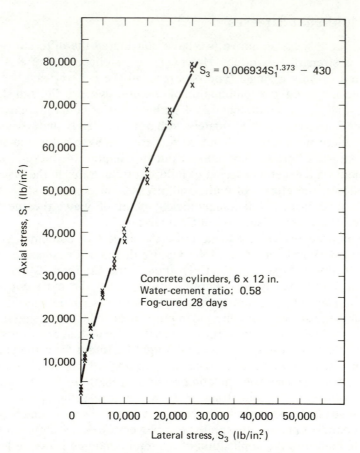

$$S_3 = 0.006934S_1^{1.373} - 430$$

Concrete cylinders, 6 x 12 in.
Water-cement ratio: 0.58
Fog-cured 28 days

Axial stress, S_1 (lb/in.²)

Lateral stress, S_3 (lb/in.²)

Figure 15.15 Relation of axial stress to lateral stress at failure in triaxial compression tests of concrete. (From W. H. Price, *Journal of the American Concrete Institute,* Vol. 47, No. 6, 1951, pp. 417–432.)

imply that compressive failure in concrete is governed by the strength of the paste; tensile failure seems to be governed by the cement–aggregate bond.

Most of the empirical relationships that have been developed for triaxial strength ignore σ_2, the intermediate principal stress. There is some evidence to show that σ_2 has only a small influence on behavior under triaxial loading. However, since concrete appears to have different Mohr failure envelopes for biaxial and triaxial compression, the effect of σ_2 cannot be ignored completely. (See the discussion in Chapter 14, and Figure 14.8.)

The effects of multiaxial stresses are quite complicated, and only a simplified picture has been presented above. In addition to the type of loading, one must also consider the *stress history*; that is, failure is determined not only by the ultimate stresses, but also by the rate of loading and the order in which these stresses were applied. In addition, as in triaxial tests on clay soils, the development of pore-water pressure during a triaxial test must be considered. If pore water is allowed to escape (i.e., a "drained" triaxial test), the compressive strength at failure will be higher than for the "undrained" case.

Bibliography

GILKEY, H. J., "Water–Cement Ratio versus Strength—Another Look," *Journal of the American Concrete Institute,* Vol. 57, No. 10, pp. 1287–1312 (1961); and discussions: Vol. 58, No. 6, pp. 1851–1878 (1961).

GREEN, H., "Impact Strength of Concrete," *Proceedings of the Institute of Civil Engineers,* Vol. 28, pp. 383–396 (1964).

KESLER, C. E., "Strength," in *Significance of Tests and Properties of Concrete and Concrete-making Materials,* ASTM STP 169A, pp. 144–159. American Society for Testing and Materials, Philadelphia, Pa., 1966.

MALHOTRA, V. M., *Maturity Concept and the Estimation of Concrete Strength— A Review,* Information Circular IC 277. Department of Energy, Mines and Resources, Mines Branch, Ottawa, 1971.

NURSE, R. W., "Steam Curing of Concrete," *Magazine of Concrete Research,* Vol. 1, No. 2, pp. 79–88 (1949).

PLOWMAN, J. M., "Maturity and Strength of Concrete," *Magazine of Concrete Research,* Vol. 8, No. 22, pp. 13–22 (1956).

POWERS, T. C., AND T. L. BROWNYARD, "Studies of the Physical Properties of Hardened Portland Cement Paste," *Journal of the American Concrete Institute,* Vol. 18, No. 2, pp. 101–132 (1946); Vol. 18, No. 3, pp. 249–336 (1946); Vol. 18, No. 4, pp. 469–504 (1946); Vol. 18, No. 5, pp. 549–604 (1947); Vol. 18, No. 6, pp. 669–712 (1947); Vol. 18, No. 7, pp. 845–880; Vol. 18, No. 8, pp. 933–992 (1947). Also published as Bulletin 22, Portland Cement Association, 1948.

PRICE, W. H., "Factors Influencing Concrete Strength," *Journal of the American Concrete Institute,* Vol. 47, No. 6, pp. 417–432 (1951).

PRICE, W. H., "The Practical Qualities of Cement," *Journal of the American Concrete Institute,* Vol. 71, No. 9, pp. 436–444 (1974).

SAUL, A. G. A., "Principles Underlying the Steam Curing of Concrete at Atmospheric Pressure," *Magazine of Concrete Research,* Vol. 3, No. 6, pp. 127–140 (1951).

VERBECK, G. J., AND R. A. HELMUTH, "Structures and Physical Properties of Cement Paste," *Proceedings, Fifth International Symposium on the Chemistry of Cement, Tokyo, 1968,* Vol. 3, pp. 1–32. Cement Association of Japan, Tokyo, 1969.

Problems

15.1 How, and why, does porosity affect the strength of concrete?

15.2 Compute a graph of S versus P using Eq. (15.1), where $S_0 = 34,000$ psi and $k = 2$.

15.3 Construct a graph using Eq. (15.4) of (a) strength versus w/c ratio at $\alpha = 0.75$; (b) strength versus α at $w/c = 0.50$. Compare the graphs using exponents of 2.6 and 3.0.

15.4 What factors other than w/c ratio affect the strength of concrete?

15.5 Why is it difficult to be sure of the exact w/c ratio in concrete?

15.6 Two concretes, A and B, made from the same mix design and materials, are cured in the field. It is determined that it is safe to strip the formwork from concrete A after 7 days moist curing at 20°C. If concrete B is first cured for 3 days at 10°C, how many additional days must it cure at 20°C before the formwork can be stripped?

15.7 Discuss the maturity concept.

15.8 Given a compressive strength of 35 MPa, estimate (a) the flexural strength; (b) the shear strength; and (c) the splitting tensile strength.

16

testing of hardened concrete

There are a number of reasons for carrying out tests on hardened concrete:

1. On a fundamental level, tests can be used to investigate the physical laws governing the mechanical behavior of concrete. Such tests would be used to help deduce and verify the relationships described in Chapters 13 and 14 between the physical and mechanical properties of the constituents of concrete, and the elastic properties and strength characteristics of the concrete made from those materials.

2. Often these physical laws may not be fully developed or understood. In this case a second type of testing is used, to determine the mechanical properties of a particular type of concrete in some specific application. This is done by simulating the expected conditions of use for a particular concrete as closely as possible, and observing the performance of the concrete. These results will not be fundamental in that it will usually not be possible to extrapolate them to predict the behavior of different concretes under different service conditions.

3. On a simpler level, a third type of test would be carried out when the physical laws governing the behavior of concrete are known but there is a need to evaluate the physical constants, such as E, which occur in these laws.

4. Finally, and most commonly, routine information on the quality of the concrete may be required. This type of testing usually goes under the name *quality control testing*. Here, extreme accuracy of measurement is less important; the speed and ease of carrying out the tests might well govern the choice of test procedures.

Although the tests to be described below are generally in the category of quality control tests, they are often used for the other three purposes mentioned. Before describing current concrete test methods in detail, however, two other very important concepts must be discussed.

16.1 NEED FOR "STANDARD" TESTS

The strength, durability, and other mechanical properties of concrete should not be considered in any sense as "fundamental" or "intrinsic" material properties. Such variables as specimen geometry, specimen preparation, moisture content, temperature, loading rate, and type of testing machine and loading fixture will all affect the observed mechanical behavior. Therefore, different test procedures would be expected to yield different results, and experience has shown this to be the case. Thus, when defining some mechanical property, such as compressive strength, it is also necessary to specify the test method used in determining the strength. To complicate the matter further, there are generally no unique relationships between values for mechanical properties measured in different ways. If, for instance, we measure the compressive strength of concretes using two specimen shapes, cylinders and cubes, the ratio (cylinder strength/cube strength) will not have a constant value, but will vary depending on the type of concrete.

Therefore, to try to minimize the confusion that would result if everyone were to use different test procedures, various "standard" test methods have been proposed. For these procedures, the test parameters are rather arbitrarily fixed, so that different people working in different laboratories in different areas can nevertheless carry out comparable tests, and so can generalize their test results. The standard test methods most commonly used in North America are those developed by ASTM, and these will be described in some detail below. However, some agencies, such as the U.S. Bureau of Reclamation, for example, may use somewhat different procedures, and most European countries have their own national standards. In addition, RILEM (The International Union of Testing and Research Laboratories for Materials and Structures) has published test methods which are widely used in Europe. The tests established by the British Standards Institute (BSI) will be described where they differ substantially from the ASTM tests because they are used in many parts of the world. The relevant test designations according to ASTM, BSI, and Canadian Standards Association (CSA) procedures are given in the Appendix.

It should be noted that as cement and concrete technologies change, the "standard" test methods also change; old test methods are abandoned

or modified, and new test methods developed. For instance, it is estimated that about one-third of the ASTM tests are revised annually.
Finally, there are properties of concrete (such as impact strength) for
which no standard test methods have yet been developed. As such
properties become important design characteristics, completely new tests
will have to be established for their determination.

16.2 SIGNIFICANCE OF TESTS

It must be emphasized that concrete tests are, in general, carried out *not*
on the concrete actually in the structure, but on small "companion"
samples of concrete that purport to represent the quality of the concrete
in the structure. Unfortunately, this can cause a great deal of confusion.
Usually, when we talk about the strength (or other mechanical properties)
of concrete, we are referring to the strength of the concrete in the
structure at some given age, but what is really measured is the strength
of small test specimens. Many studies have shown that there is *not* a
particularly good correlation between the strength of concrete defined by
ASTM standards and the strength of concrete actually in the structure.
This discrepancy arises from differences in the many variables that
determine the behavior mentioned above. Although standard tests may
well give some indication of concrete quality, many problems have arisen
where concrete samples gave acceptable test results while the concrete
in the structure was of lower quality and behaved quite differently.
Conversely, there have also been cases where perfectly good concrete
structures have been "repaired" on the basis of standard tests giving
unacceptable values.

As an example of how the discrepancy between real and measured
concrete properties arises, let us examine compressive strength. The
concrete in a standard cylinder is subjected to a different compactive
effort from the concrete in the structure; the size is different from the
size of structural members by several orders of magnitude; the temperature and relative humidity of curing are different; the cylinder remains
free of stress until tested, while structural members will be carrying at
least some load; under load, the concrete in the structure may be at least
partially restrained by steel reinforcement or by other structural elements. In view of this, it would be surprising if the cylinder strengths *did*
accurately reflect the strength of the concrete in the structure.

As a result, standard tests carried out on small, supposedly representative, samples do not in any way *guarantee* the quality of the
concrete. Rather, concrete design is based on the implicit assumption
that if the concrete has been placed, compacted, and cured properly in

the structure, and if the test results are satisfactory, there is a very high probability that the material in the structure will behave in an adequate manner. This is quite different from the common assumption that we actually know (or measure) the mechanical properties of the concrete in the structure. Moreover, material properties used in design are obtained from "standard" tests, so that the design procedures used in countries having different concrete test methods, must also be somewhat changed, or at least must use different material constants in the design equations.

At this point, one might well ask: If the properties we measure are not necessarily representative of the concrete in the structure, why bother to test the concrete at all? The answer to this question is that there *are* some very good reasons for so doing, but that an accurate measure of the properties of the concrete in the structure is not one of them. The reasons are:

1. Tests can help ensure that the concrete was batched properly, that is, that the proper ingredients, in the proper proportions, were used, and that the laboratory mix design was adequate.

2. They can indicate the statistical variability in the properties of the concrete being produced. High variability is one indication of poor concreting practice. (This aspect of quality control will be discussed in detail in Chapter 17.)

3. They may reveal problems arising due to inadvertent changes in materials or environmental conditions.

4. The simple existence of an extensive testing and inspection program will help ensure that the people involved in the production, delivery, and placement of the concrete do not become lax or careless in their operations.

5. If, in spite of all precautions, structural problems do arise, properly carried out and documented test reports will help to pinpoint the problem.

6. Test results may be needed before further construction operations can be carried out. For instance, strength tests may be used as a guide for the time of form removal.

16.3 TESTS FOR COMPRESSIVE STRENGTH

By far the most common test carried out on concrete is the compressive strength test. There are several reasons for this: (1) it is assumed that most of the important properties of concrete are directly related (at least qualitatively) to the compressive strength; (2) since concrete has very little tensile strength, it is used primarily in a compressive mode and therefore it is the compressive strength that is important in engineering practice; (3) the structural design codes are based mainly on the compressive strength of concrete; and (4) the test is easy and relatively inexpensive to carry out. The compressive strength test, and in particular the experimental variables that may affect the results, are described in great detail below. It should be remembered that the experimental variables discussed in terms of compressive strength will have similar effects on the other mechanical properties of concrete.

ASTM Cylinder Test

As we indicated in the previous section, the test for compressive strength is so sensitive to variations in procedure that it must be carried out strictly according to standard procedures so that results from different testing laboratories (or even within the same laboratory) are comparable. Therefore, the ASTM test is described in some detail. The normal compressive test specimen in North America is a cylinder with a length to diameter (*l/d*) ratio of 2:1, usually of dimensions 150 mm (6 in.) in diameter, 300 mm (12 in.) long. The specimens are prepared and tested as follows:

Molds

Molds may either be *reusable,* generally made of heavy-gauge metal, or *single-use,* generally made from formed sheet metal or waxed cardboard. The requirements for molds (ASTM C470) require that they be made of materials that are nonreactive with concrete, that they be watertight, and that they be sufficiently stiff so that they do not deform excessively on use. Unfortunately, it has been found that cardboard molds yield somewhat lower strengths (~3%) than do other types, possibly due to some water absorption by the cardboard or to expansion of the mold during setting.

Molding the Specimens (ASTM C31)

The mold is placed on a firm, level surface, free from vibration. From a representative sample of concrete (as discussed in Chapter 17), the mold is filled in layers, depending on the method of consolidation. If the slump is more than 75 mm (3 in.), the concrete is consolidated by rodding; if the slump is less than 25 mm (1 in.), the concrete is consolidated by vibration. For slumps between 25 and 75 mm, the specimens may be consolidated either way. (The specifications for a particular job may require a certain method of consolidation, which would supersede the foregoing guidelines.) The reason for these alternative methods is that a poorly compacted cylinder will have a lower strength than will a properly compacted one. For stiff mixes, which will always be compacted by vibration in the field, rodding simply will not compact the cylinders sufficiently, and vibration may cause segregation of wet mixes.

If the specimen is to be rodded, it should be filled in three equal layers. Each layer is rodded 25 times with a 16-mm (⅝-in.)-diameter steel rod with a rounded end. For the upper layers, the rod should penetrate about 25 mm (1 in.) into the underlying layer. Specimens to be vibrated are filled in two equal layers; either internal or external vibration may be used. After consolidation, the top surface is finished by striking off with the tamping rod or with a trowel.

Curing

Cylinders may be made for three reasons: (1) to check whether the laboratory designed mix achieves the specified design strength, as a basis for acceptance or for quality control; (2) to determine when forms can safely be removed; and (3) to determine when a structure may be put into service, or when prestressed concrete members may be tensioned. The curing conditions depend on the purpose for which the cylinders were made.

If the cylinders were made for reason 1, they must be stored at between 16 and 27°C (60 and 80°F) for the first 24 h, in such a way that moisture loss is prevented. The specimens are then removed from the molds and stored in a standard moist room or in saturated lime water at 23 ± 1.7°C (73 ± 3°F) until tested. If the cylinders are cast in a laboratory (ASTM C192), they must be maintained at 23 ± 1.7°C from the moment of casting. If the cylinders were made for the estimation of form-removal time, they are stored as near to the part of the structure in question as possible and should be protected from the elements in the same way as the structure. These specimens should be tested before the forms are removed. For specimens used to determine when a structure may be put into service, the cylinder molds should be removed at the same time that the forms are removed, so that the moisture condition of the specimen approximates the moisture condition of the structure. In some situations, in order to match the moisture conditions, only the specimen ends should be exposed.

Capping the Cylinders

Cylinders cast as described above will have end surfaces (particularly the top surface) that are rough and not plane or parallel. If tested in this condition, the apparent strength of the concrete would be considerably reduced as a result of stress concentrations that are introduced on loading. Convex ends will lead to lower measured strengths than concave ones, because the stress concentrations introduced are more severe. It is required that the specimen ends must be plane within 0.050 mm (0.002 in.). One way to achieve planeness is by grinding the ends; this is satisfactory but is expensive and time-consuming. The most common way of achieving this planeness requirment is to *cap* the ends of the cylinders (ASTM C617) with a suitable material. Three different capping materials are permitted. A thin layer of stiff portland cement paste may be used on freshly molded specimens. On hardened cylinders, either high-strength gypsum plaster or sulfur mortar may be used, the latter being much more common. The sulfur mortar is prepared by heating it to about 130°C (270°F) and then is poured into lightly oiled steel capping plates. The cylinders are set into the sulfur in such a way that the caps end up being about 3 mm (⅛ in.) thick and are aligned so that the deviation of each cap from perpendicularity with the axis of the specimen is less than 0.5°. After capping, the cylinders should be kept moist until tested; commonly, cylinders are tested soon after capping. (It should be noted that sulfur loses strength and pourability with use, and therefore it is required that the oldest material in a sulfur pot should not have been used more than five times.)

Determination of Compressive Strength (ASTM C39)

Once the specimens have been capped, they are ready for testing. The tests may be carried out in any suitable testing machine. Two hardened bearing blocks are used, a solid one that the specimen sits on, and a spherically seated one that will bear on the upper surface. These blocks must be plane to within 0.025 mm (0.001 in.). For 150-mm (6-in.)-diameter cylinders, the maximum diameter of the spherically seated block is 254 mm (10 in.). Because strength is dependent on the loading rate, the specimen is then loaded at a controlled rate of 0.15 to 0.34 MPa/s (20 to 50 lb/in.²/s) for hydraulic machines, or at a deformation rate of 1.3 mm/min (0.05 in./min) for mechanical machines, until failure, which is defined as the maximum load the specimen can carry. The maximum load and type of failure are then reported.

Other Compression Tests

The ASTM test described above is by no means the only compression test that is used; tests on cubes and prisms are also in use.

Cube Test

Cube tests are used as the normal compressive test in Great Britain, Germany, and in some other parts of Europe. The British Standard (BS 1881: Part 3) requires a 150-mm (6-in.) cubic mold, which is filled in three layers, each layer being rodded 35 times with a 25-mm square rod; alternatively, the cube may be compacted by vibration. It is then handled in much the same way as the cylinders described above, except that the storage temperature is 20 ± 1°C (68 ± 2°F). The cube is tested (BS 1881: Part 4) at right angles to the position at cast; this means that the faces of the cube in contact with the bearing platens were cast against the sides of a rigid steel mold. Therefore, the bearing faces are sufficiently plane as to require no capping or grinding. This is the chief advantage of the cube specimen. The specimen is then loaded at a rate of 0.25 MPa/s (33 lb/in.²/s) to failure.

Prism Test (ASTM C116)

This test was developed so that a value of compressive strength could be obtained from the broken portions of beams tested in flexure. This test is primarily a laboratory or research test; it is not an alternative to the cylinder test described above and will not give the same strength values, because of its different geometry. Further, the strength will not be the same as that of a standard cube because of the restraining effect of the overhanging parts of the specimen. However, it may be used to determine the relative compressive strengths of different concrete mixes. The specimen is tested as shown in Figure 16.1. It is oriented such that the width is less than, or equal to, the height. If the beam is of square cross section, the bearing surfaces may be the (flat) sides as originally cast; otherwise, the bearing faces must be ground or capped to conform to the planeness requirements described above. Again, a spherically seated testing head is used and the rates of loading are the same as for the cylinder test. The compressive strength is determined from the cross-sectional area of the bearing plates. Because this test was originally applied only to beams having a square cross section, it is often referred to as the "equivalent cube" test. A very similar test procedure is in BS 1881: Part 4.

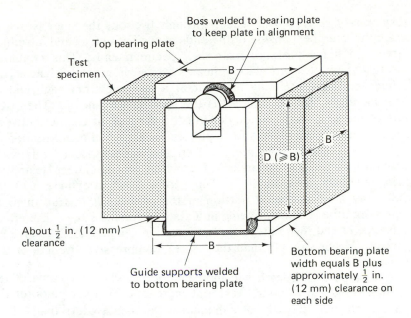

Figure 16.1 Device suitable for aligning bearing plates. (From ASTM C116. Reprinted, with permission, from the American Society for Testing and Materials, 1916 Race Street, Philadelphia, PA 19103. Copyright.)

Factors Affecting the Measured Compressive Strength

The compressive strength tests as described above appear to be perfectly straightforward. Unfortunately, the results obtained can be affected considerably by a number of factors. If we are to be able to interpret the strength values obtained by the standard (although arbitrary) procedures, it is necessary to examine in detail how the compressive strength is affected by the test parameters.

Stress Distribution in Specimens

The compression test assumes a state of pure, uniaxial compression. However, this is not really the case, because of friction between the ends of the specimens and the platens. This frictional force arises from the fact that, because of the differences in the modulus of elasticity and Poisson's ratio for steel and concrete, the lateral strain in the platens is considerably less than the lateral expansion of the ends of the specimen *if they were free to move.* Thus, through friction, the platens act to restrain the lateral expansion of the ends of the specimens and introduce a lateral confining pressure near the specimen ends. This confining pressure (which also introduces shear stresses) is greatest right at the specimen end and gradually dies out at a distance from each end of approximately $(\sqrt{3}/2)$ *d,* where *d* is the specimen diameter. (The exact nature of the stress distribution depends on the type of contact between the specimen and the platens for a particular test.) The manifestation of this lateral confining pressure is often the appearance of relatively undamaged cones (or pyramids) of concrete in specimens tested to failure, as shown in Figure 16.2a. Thus, for a standard cylinder with *l/d* = 2.0, only a small central portion of the cylinder is in true uniaxial compression, the remainder being in a state of triaxial stress. The effect of this type of end restraint, as we have seen in Chapter 15, is to give an apparently higher strength than the "true" compressive strength of the specimen.

The apparent strength will increase as the relative volume of the specimen subjected to lateral restraint increases. This accounts for the apparently higher strengths of cylinders with *l/d* ratios less than 2.0. It also accounts for the fact that, in general, cubes indicate higher strengths

Figure 16.2 Typical failure patterns for concrete cylinders in compression: (a) shear failure; (b) combined shear and splitting failure; (c) splitting failure.

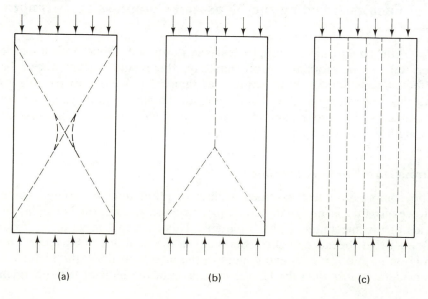

(a)   (b)   (c)

than cylinders; they simply are not long enough for the effects of the end restraint to die out. For a given specimen geometry, higher confining stresses will also result in higher apparent strengths (Chapter 15). Conversely, it should be noted that if a lubricant (or soft rubber) is placed between the ends of the specimen and the platens to eliminate friction, the outward flow of these materials under pressure may induce lateral *tensile* forces in the ends of the specimen. This will reduce the strength, and the specimens may fail by vertical splitting.

On a more fundamental level, to speak at all of a "compression" failure of concrete (or most other materials) is incorrect. Since compression tends to squeeze atoms and molecules closer together, it is hard to see how pure compression can lead to failure. In a compression test, however, there are also secondary tensile stresses induced in the specimen, at right angles to the axis of the specimen. Since concrete is relatively weak in tension, these stresses may cause cracking and failure. It has been suggested that the failure of concrete in tension will occur at limiting strains of about 0.0001 to 0.0002. For ordinary concrete, with Poisson's ratio approximately equal to 0.2, such lateral tensile strains will occur at fairly low compressive loads, and this could be the cause of failure, as shown in Figure 16.2c. This is probably the natural mode of failure in pure compression. The induced shearing stresses due to end restraint may cause an apparent shear failure of the specimen, as indicated in Figure 16.2a.

Since we cannot avoid some end restraint, it is likely that failure occurs through some combination of shear and tensile forces, as indicated schematically in Figure 16.2b. Tensile cracks may not be able to propagate through the portions of the specimen under a lateral confining stress.

Test Apparatus

As well as being plane, the ends of the specimen should also be perpendicular to the axis of the specimen. Since this is very difficult to achieve exactly, a spherically seated platen is used; a small deviation from parallelism will then not affect the strength. ASTM C39 specifies that the spherical head should be lubricated with conventional oil (rather than pressure-type grease), so that once the head is in full contact with the specimen, further tilting will not occur. Changing the degree of lubrication in the spherical seat can cause apparent changes in strength.

Different types of platens can also lead to different results. Figure 16.3 shows the effects of using "hard" or "soft" platens (i.e., platens of different stiffness) on the deformation of the specimen, since even steel platens will distort to some degree under load. For hard platens (Figure

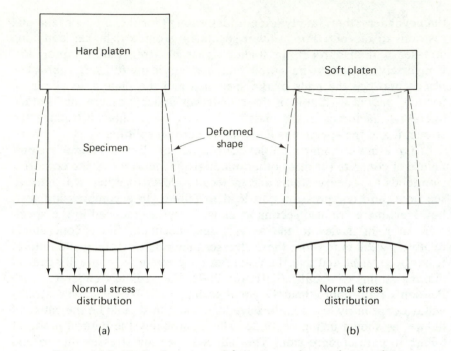

Figure 16.3 Idealized specimen deformation and normal stress distribution for (a) "hard" and (b) "soft" platens.

16.3a), the stress will be higher at the edges of the specimen; for soft platens (Figure 16.3b), the stress will be higher at the center of the specimen.

The type of testing head that best approximates true uniaxial compression is the so-called "brush platen." This consists of filaments about 5 × 3 mm in cross section, with gaps of about 0.2 mm between them. This allows the concrete to expand laterally with very little restraint. Such platens will, of course, give lower strengths than ordinary steel platens.

Finally, the type of testing machine itself can have a considerable effect on the strength measurements. Testing machines may be classified as *hard* (very rigid machines) or *soft* (less rigid machines). With a perfectly hard machine, the machine head will not follow rapid deformations of the specimen; a perfectly soft machine would follow these deformations exactly (as in a dead-weight test). While testing machines are, of course, neither completely hard nor completely soft, their relative rigidities can make quite a difference. In very soft machines, the energy

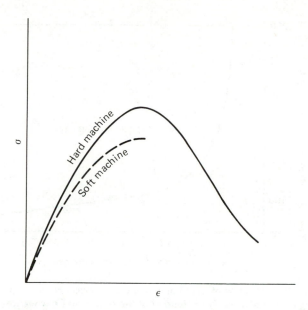

Figure 16.4 Effect of stiffness of testing machine on $\sigma-\epsilon$ curves of concrete.

stored in the machine is released as the specimen begins to fail; this additional energy will cause greater crack propagation and failure at lower loads than with very rigid machines, which cannot release their stored strain energy as easily. These effects are illustrated in Figure 16.4.

Effect of l/d Ratio

The standard cylinder has a length to diameter ratio of 2.0. It has been found that if other l/d ratios are used, the specimen strength changes due to the end effects described above. The general effect of the l/d ratio on strength is shown in Figure 16.5. However, the exact shape of the curve depends to some extent on the end conditions and on the type of concrete tested. It has been found that high-strength concrete is less affected by variations in specimen geometry. The effect of the l/d ratio is particularly significant when tests are carried out on drilled cores, where, out of necessity, the l/d ratio is often less than 2.0. In this case, correction factors (Figure 16.6) are provided (ASTM C42; BS 1881: Part 4) so that specimens with different l/d ratios can be compared on the same basis. If the l/d is greater than 2.0, the specimen can be cut to size before testing.

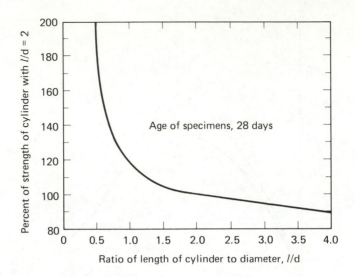

Figure 16.5 Relation of length and diameter of specimen to compressive strength. (From W. H. Price, *Journal of the American Concrete Institute*, Vol. 47, No. 6, 1951, pp. 417 –432.)

Specimen Geometry

Since different geometries of specimens —cylinders, cubes, or prisms — give different values of strength, which specimen shape, then, is "best"? Clearly, cubes are somewhat easier to test, since they require no capping. However, the tendency is to prefer cylinders to cubes, especially in research, as there is a feeling that they better represent the strength of the concrete. But as we have seen, small concrete specimens do not particularly represent the strength of the concrete in the structure; they give only comparative data. It thus makes little difference which specimen geometry is chosen, as long as it is used consistently. Although the ratio of cube strength to cylinder strength decreases as the strength of the concrete increases, it is often assumed that the strength of a cube is about 1.25 times the strength of a cylinder made of the same concrete.

Both the strength and the variability in strength of concrete decrease as the specimen size increases. This is generally explained by the "weakest link" theory. That is, we assume that the strength of a concrete specimen is governed by the weakest element ("link") within it; the larger the size of the specimen, the more likely it is to contain an element that will fail at a (given) low load. Such points of weakness are, in general, the types of flaws that were discussed in Chapter 14. However, as specimens become larger, the differences in the distributions of flaws

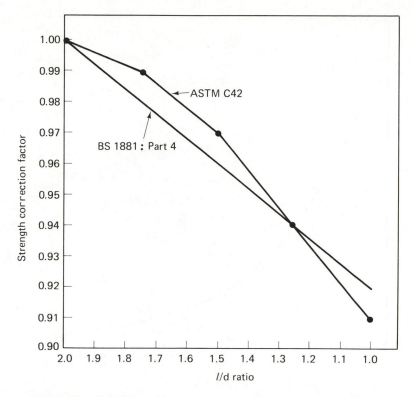

Figure 16.6 Correction factors for cylinders with l/d ratios less than 2.0. Applicable for normal concrete strengths of 2000 to 6000 lb/in.². Average values only given.

within each sample become less, and hence the variability decreases. Figure 16.7 shows the variation in strength for cylinders of $l/d = 2.0$, but with different diameters.

Rate of Loading

In general, the higher the rate of loading, the higher the measured strength. The reasons for this are not completely clear. It may be that slow rates of loading allow more subcritical crack growth to occur, thus leading to the formation of larger flaws and hence smaller fracture loads (Chapter 15). On the other hand, it may be that slower loading rates allow more creep to occur, which will increase the amount of strain at a given load. When the limiting value of strain is reached, failure will occur. More likely, the observed rate of loading effect is due to a

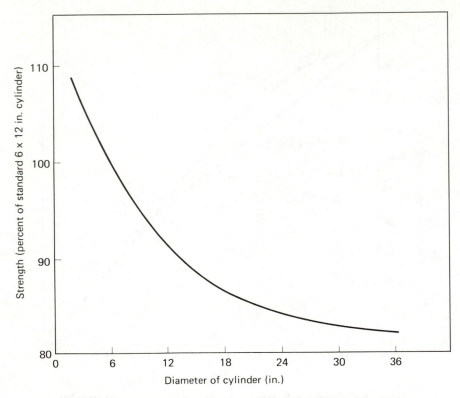

Figure 16.7 Effect of cylinder size on compressive strength of concrete, with *l/d* = 2.0. (From W. H. Price, *Journal of the American Concrete Institute,* Vol. 47, No. 6, 1951, pp. 417 –425.)

combination of these, and perhaps other factors as well. The extreme case of static fatigue has already been discussed in Chapter 14. The effects of varying the rate of loading are shown in Figure 16.8. Clearly, in order to obtain comparable results, the rate of loading in practice must be maintained within fairly narrow limits.

Moisture Content

Most concrete specifications, such as ASTM C39 described above, require that concrete be maintained and tested in a saturated state. It has been found that concrete that has been dried shows an increase in strength. The reasons for this are not completely understood; it may have something to do with the change in the structure of the C–S–H on drying, or it may simply represent a change in the internal friction and

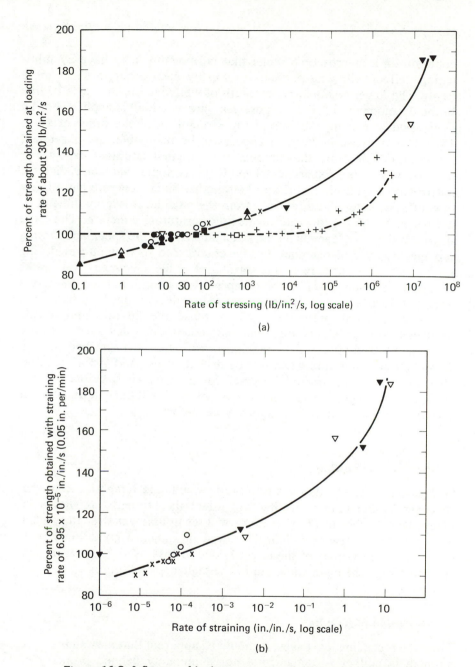

Figure 16.8 Influence of both stress rate and strain rate on the compressive strength of concrete. (a) Effect of rate of stressing on the compressive strength of concrete; (b) Effect of rate of straining on the compressive strength of concrete. (From D. McHenry and J. J. Shideler, in *Symposium on Speed of Testing of Nonmetallic Materials*, ASTM STP 185, 1955, pp. 72–82. Reprinted, with permission, from the American Society for Testing and Materials, 1916 Race Street, Philadelphia, PA 19103. Copyright.)

cohesion on a macroscopic scale; that is, moisture may have a "lubricating" effect, allowing particles to slip by each other in shear more easily. The lower compressive strength of wet concrete may also be due to the development of internal pore pressure as a load is applied. For an oven-dried specimen, the increase in strength is of the order of 10 to 15%. This increase in strength appears to be reversible, as subsequent resaturation will return the concrete to its original saturated strength.

However, the experimental evidence is somewhat contradictory, particularly for tensile and flexural strengths. Since concrete has a fairly low diffusion rate, and dries only from the outside, it is very difficult to get perfectly "dry" concrete, or to fully resaturate concrete. Thus, the degree of dryness depends on the size and shape of the specimen. Drying too quickly may also induce tensile cracks due to nonuniform drying (and hence differences in drying shrinkage) of the specimen. The cracks do not have much effect on the compressive strength but will lower the apparent flexural and tensile strengths. Slow drying, on the other hand, where cracking is prevented, will increase the flexural and tensile strengths. Similar effects are sometimes noted when dry specimens are resaturated. The effect of the moisture content on strength becomes an important consideration when testing drilled cores. ASTM C42 requires that such cores be soaked in water for at least 40 h before testing; however, the degree of saturation thus obtained will depend on both the specimen size and on its previous drying history.

Temperature at Testing

For reasons that are also not fully understood, the temperature of the specimen at the time of testing will affect the strength, as shown in Figure 16.9. Higher test temperatures will result in lower strengths, even for concretes that were identically cured in standard conditions. It is likely that at least part of the effect is due to loss of moisture from the specimen while being conditioned to the higher temperatures.

Interpretation of Results

From the preceding few pages, it should be apparent that the compressive strength is neither a fundamental nor a uniquely defined property of concrete. Its value depends very markedly on the method of measurement. Thus, since compressive strength results are of value primarily on

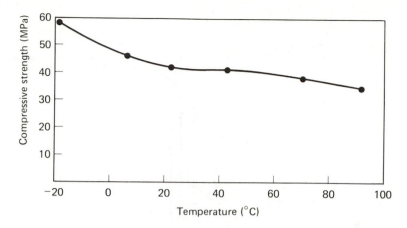

Figure 16.9 Compressive strength as a function of temperature at the time of testing. (From J. C. Saemann and G. W. Washa, *Journal of the American Concrete Institute,* Vol. 54, No. 5, 1957, pp. 385–389.)

a comparative basis, there must be strict adherence to a standardized test method.

16.4 OTHER CONCRETE TESTS
Tensile Strength

No standard tests have been adopted either by ASTM or BSI to provide a direct measurement of the tensile strength of concrete. The problem of secondary stresses induced through gripping makes the test results difficult either to interpret or to reproduce. Recently, however, RILEM has prepared a recommendation for a direct tension test of concrete, designed primarily for research rather than for routine control. This method involves applying direct tension to either cylindrical specimens or prismatic specimens (with a square cross section) through end plates glued to the concrete. The ends of the specimen must be sawed off to remove end effects due to casting or vibration; they must be perpendicular to the axis of the specimen within ¼°. Further, the ends must be carefully cleaned so that the glue (typically a polyepoxy resin) adheres uniformly to the entire surface. The load is applied at a rate of 0.05 MPa/s (7 lb/in.²/s) until failure occurs. As yet, there is not enough experience with this test method for a proper assessment of its usefulness to be made.

There are, however, two common methods for estimating the tensile

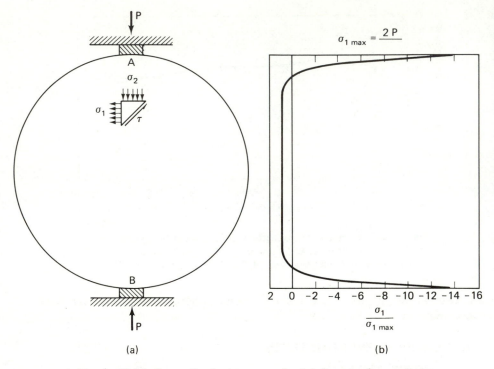

Figure 16.10 Stress distribution across loaded diameter for a cylinder compressed between two flat plates.

strength of concrete through indirect tension tests. These methods are the splitting tension test, which will be discussed in this section, and the flexure test, which will be described in the next section. The splitting tensile test is carried out on a standard cylinder, tested on its side in diametral compression, as shown in Figure 16.10a. The element on the vertical diameter shown in Figure 16.10a is subjected to the following stresses:

$$\text{vertical compression } \sigma_c = \frac{2P}{\pi LD}\left[\frac{D^2}{r(D-r)} - 1\right] \qquad (16.1)$$

$$\text{horizontal tension } \sigma_t = \frac{2P}{\pi LD} \qquad (16.2)$$

where P is the applied compressive load, L the cylinder length, D the cylinder diameter, and r the distance of the element from the top of the cylinder.

It is not practical to apply a true "line" load along the top and bottom of the specimen, partly because the sides are not smooth enough, and partly because this would induce extremely high compressive stresses near the points of load application. Therefore, the load is usually applied through a narrow bearing strip of relatively soft material. The tensile stress distribution along the vertical diameter of the specimen is then as shown in Figure 16.10b. That is, there are very high compressive stresses near the ends of the vertical diameter, and a nearly uniform tensile stress acting over about the middle two-thirds of the specimen. Since the concrete is much weaker in tension than in compression, failure will be in splitting tension at a much lower load than would be required to crush the specimen in compression, thus permitting an estimate to be made of the tensile strength of the concrete.

As specified in ASTM C496, the test is carried out on cylindrical specimens, either cast cylinders or drilled cores. The bearing strips are made from 3-mm plywood which is free of imperfections, and are about 25 mm wide. The specimen is aligned in the machine, and the load is applied at a rate of 690 to 1380 kPa/min (100 to 200 lb/in.²/min) splitting tensile stress until the specimen fails. The splitting tensile strength is then calculated from Eq. (16.2). BS 1881: Part 4, describes an essentially similar test, except that bearing strips only 12 mm wide are specified. It should be noted that it is also possible to carry out splitting tension tests on cubes. This is done by loading through two hemispherical bars along the center lines of two opposite faces. This gives much the same results as testing a cylinder; the horizontal tensile stress is

$$\sigma_t = \frac{2P}{\pi a^2} \tag{16.3}$$

where a is the side of the cube. There is a RILEM recommendation for such a test (although the recommendation states that tests on cylinders are preferable). In general, it has been found that the values of tensile strength obtained from splitting tests tend to be about 15% higher than values obtained from direct tension tests.

Flexural Strength

The other way of estimating the tensile strength of concrete is by the flexural test (ASTM C78). The specimen is a beam 152 × 152 × 508 mm (6 × 6 × 20 in.). The mold is filled in two equal layers, each layer being rodded 60 times, once for each 13 cm² (2 in.²) of top surface area. Stiffer mixes may be consolidated by vibration, as specified for compression cylinders. The beams are cured in the standard manner and are then

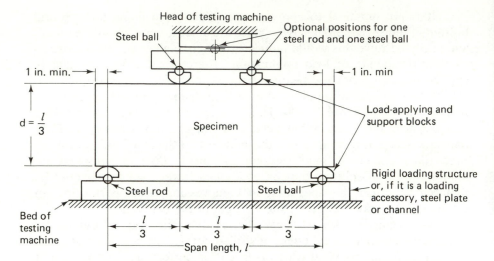

Figure 16.11 Diagrammatic view of a suitable apparatus for flexure test of concrete by third-point-loading method. This apparatus may be used inverted. If the testing machine applies force through a spherically seated head, the center pivot may be omitted, provided that one load-applying block pivots on a rod and the other on a ball. (From ASTM C78. Reprinted, with permission, from the American Society for Testing and Materials, 1916 Race Street, Philadelphia, PA 19103. Copyright.)

tested in flexure in third-point loading, as shown in Figure 16.11. Tests may also be carried out on beams sawn from hardened concrete. The cast specimens are tested turned on their sides with respect to their position as molded. This should provide smooth, plane, and parallel faces for loading. If for some reason this is not the case, the specimens must be ground, capped, or shimmed. Specimens are loaded at a rate of 860 to 1200 kPa/min (125 to 175 lb/in.²/min). BS 1881: Part 4, provides for a similar test procedure, except that the standard specimen dimensions are $150 \times 150 \times 750$ mm ($6 \times 6 \times 30$ in.). The theoretical maximum tensile strength, or *modulus of rupture, R,* is then calculated from the simple beam bending formula for third-point loading:

$$R = \frac{PL}{bd^2} \tag{16.4}$$

where P is the maximum total load indicated, L the span length, b the specimen width, and d the specimen depth. Equation (16.4) holds only if the beam breaks between the two interior loading points (i.e., in the middle third of the beam). If the beam breaks outside these points by not more than 5% of the span length, Eq. (16.4) is replaced by

$$R = \frac{3Pa}{bd^2} \tag{16.5}$$

where a is the average distance between the point of fracture and the nearest support. The results of tests where failure occurs even closer to the supports are discarded.

This test tends to overestimate the "true" tensile strength by about 50%, largely due to the fact that the simple flexure formula [Eq. (16.4)] assumes that the stress varies linearly across the cross-section of the beam. But because concrete has a nonlinear stress–strain curve, this assumption is not true. Near failure, particularly, the stress block is more nearly parabolic than triangular, as indicated in Figure 16.12. However, the test remains very useful, since concrete members tend to be loaded in bending rather than in axial tension, and thus the values obtained from a flexure test are a better representation of the concrete property that is of interest. This test is most widely used for quality control of highways

Figure 16.12 Stress distribution across the depth of a concrete specimen in flexure.

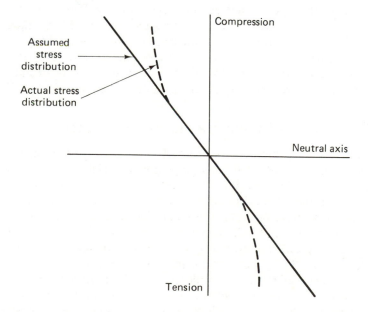

and airport runways, where it gives more useful information than do compression tests.

For smaller specimens than the ones described above, it is also possible to carry out flexure tests in center-point loading as described in ASTM C293. This test method is not as good as third-point loading and is not a substitute for it. The difficulty is that, ideally, in the third-point loading test the specimen, in the middle third of the span, is subjected to a pure moment, with zero shear. In the center-point test, there are substantial shear forces as well as unknown stress concentrations at the point of load application, which act along the line on which the specimen generally fails.

As in the case of the compression test, the test parameters can greatly affect the observed strengths. In particular, the size effect is very important. One additional reason for the fact that flexure tests give higher values than a direct tension test is that in direct tension, the total volume of the specimen is stressed, while in flexure, only a relatively small volume of material near the bottom of the beam is subjected to high stresses. Thus, if we believe the "weakest link" theory, the likelihood of finding a sufficiently weak element of concrete is reduced in flexure. This also contributes to the observation that center-point loading yields higher strength values than does third-point loading. If the load points are moved still farther apart, the strength continues to decrease. The effects of specimen size are shown in Figure 16.13. However, as the specimen size increases, the coefficient of variation decreases, which is why flexural strengths determined by third-point loading show less scatter than when center-point loading is used. Similarly, the apparent flexural strength increases as the rate of loading increases (Figure 16.14), as found in compression tests. Temperature effects in flexure are also similar to those described above.

Bond between Concrete and Reinforcement

Concrete is commonly reinforced with steel, either bars or prestressing cables. In order for a composite material such as reinforced concrete to be effective, there must be adequate bond between the concrete and the steel. Presumably, bond strength develops due to adhesion and friction between the steel and concrete, since there is no evidence of any beneficial chemical reaction. However, the bond strength is a particularly difficult property to measure, owing to the number of variables involved. In addition, as the concrete cures and dries, there may be volume changes (i.e., shrinkage), which act to reduce the amount of bond. If the concrete is cracked, or is very permeable, some corrosion of the steel may take place, which will also change the bond strength.

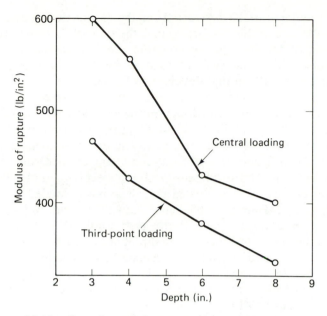

Figure 16.13 Effect of depth of specimen on the modulus of rupture of concrete. (From P. J. F. Wright, *Magazine of Concrete Research*, Vol. 4, No. 11, 1952, pp. 67–76. Reproduced by permission of the Director of Transport and Road Research Laboratory.)

Figure 16.14 Effect of rate of loading on modulus rupture. [From D. J. McNeely and S. D. Lash, *Journal of the American Concrete Institute*, Vol. 60, No. 6, 1963, pp. 751–761.)

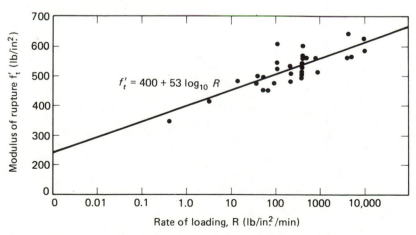

The other major variables (apart from the properties of the concrete itself) are the size of the reinforcing bar, the type of deformed bars used, the position of the bars in the concrete (i.e., compression or tension reinforcement), and the type of loading (pull-out or flexure), which affects the crack pattern that develops in the concrete.

There is no standard test available that can be used to establish the bond strength of concrete for design purposes. However, a pull-out test has been developed (ASTM C234) for the *comparison* of different concretes based on their bond strength. The test specimens are 150-mm (6-in.) cubes with standard No. 6 (19-mm-diameter) deformed steel bars embedded in them. The specimens (with the steel bars in place) are cast, compacted, and cured in the usual way. The bar is loaded at a rate not greater than 34 MPa/min (5000 lb/in.²/min). The load and the slip are recorded at intervals throughout the test, until (1) yield of the steel occurs; (2) the concrete splits; or (3) slip of at least 2.5 mm occurs at the loaded end. The bond strength is calculated as the load divided by the nominal embedded surface area of the bar. In practice, it is assumed that the bond strength is related to the compressive strength of the concrete, as shown in Figure 16.15.

Currently, ACI 318 does not require direct calculations of bond stress because it cannot be accurately measured. It uses the concept of calculating the required length (l_d) of bar needed to develop sufficient

Figure 16.15 Variation of bond with strength of concrete. (From W. H. Price, *Journal of American Concrete Institute,* Vol. 47, No. 6, 1951, pp. 417–432.)

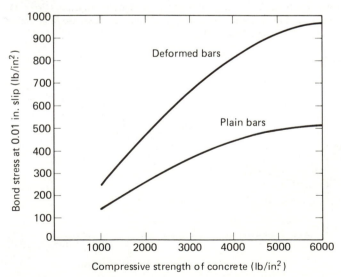

bond. The expression currently used is

$$l_d = \frac{KA_bF_y}{\sqrt{f_c'}} \tag{16.6}$$

where K is a constant depending on the size of bar (the bond stress is implicit in this constant), A_b the cross-sectional area of a bar, F_y the specified yield strength of the reinforcement, and f_c' is the specified concrete compressive strength.

The bond strength decreases markedly at higher temperatures, probably due to the difference in thermal coefficients of expansion between steel and concrete. Another factor that is very important in determining the bond strength is how well the concrete is vibrated around the steel, and whether any bleeding occurs that might trap water under the steel, as shown in Figure 8.5. Bond strength will be reduced by freeze–thaw cycling, by fatigue loading, and by alternate wetting and drying, probably due to progressive cracking around the reinforcement.

Modulus of Elasticity

As we have seen in detail in Chapter 14, the stress–strain curve for concrete is not linear. However, to calculate the stiffness or expected deflection of structural members, it is necessary that some estimation of the modulus of elasticity, E, be obtained. Although it is possible to define E in a variety of ways for a material with a nonlinear stress–strain curve (see Chapter 18), the most common way for concrete is to measure the *chord modulus of elasticity, E_c*, following the method outlined in ASTM C469. A standard cylindrical specimen to which strain gauges have been attached is first loaded and unloaded, primarily to properly seat the strain gauges. It is then loaded slowly (at a stress rate of 241 ± 34 kPa/s or 35 ± 5 lb/in.²/s) in compression, and a stress–strain curve obtained. The chord modulus of elasticity is calculated using the expression

$$E_c = \frac{S_2 - S_1}{\epsilon_2 - 0.000050} \tag{16.7}$$

where S_2 is the stress corresponding to 40% of the ultimate load; S_1 the stress corresponding to a longitudinal strain, ϵ_1, of 50×10^{-6}; and ϵ_2 the longitudinal strain produced by stress S_2. From the same test, Poisson's ratio, μ, can be calculated as

$$\mu = \frac{\epsilon_{t_2} - \epsilon_{t_1}}{\epsilon_2 - 0.000050} \tag{16.8}$$

where ϵ_{t_2} and ϵ_{t_1} are the transverse strains at midheight of the specimen produced by stresses S_2 and S_1, respectively.

The modulus of elasticity in shear, G, is not obtained by direct measurement. Rather, it is calculated from the elastic relationship

$$G = \frac{E}{2(1 + \mu)} \tag{16.9}$$

A similar procedure is followed in BS 1881: Part 5, except that E is defined as the slope of the stress–strain curve on loading between 1MPa (145 lb/in.²) and approximately one-third of the compressive strength of the concrete.

Dynamic Modulus of Elasticity

It is often desirable to have some nondestructive test for the stiffness of a concrete specimen so that progressive changes in the material due to sustained chemical attack, repeated freeze–thaw cycles, aging, or other factors can be measured. For this purpose, dynamic modulus of elasticity measurements are very convenient, since they are nondestructive and since there is also a general (if empirical) relationship between the modulus of elasticity and the compressive strength.

The standard way of carrying out these tests is to measure the dynamic modulus of elasticity, E_d, by vibration of concrete specimens at their natural frequency. The method is described in detail in ASTM C215, using the test setup shown schematically in Figure 16.16. Basically, a test specimen (either a cylinder or a prism) is supported at its nodal point (or points) so that it may undergo free-free vibration without significant restriction (i.e., for transverse vibration it is supported at 0.224 of the length from each end of the specimen; for longitudinal or torsional vibration it is supported at the center). The driver and pickup unit are placed as shown in Figure 16.16. The specimen is then forced to vibrate at various frequencies; the frequency of vibration that gives the maximum output indicates the fundamental frequency. E_d can be calculated from either the transverse or longitudinal frequency, using the general expression

$$E_d = KWm^2 \qquad \text{MPA (lb/in.²)} \tag{16.10}$$

where W is the weight of specimen (kg or lb), m the fundamental frequency of vibration (Hz or cycles/s), and K a constant which depends on the type of vibration (transverse or longitudinal), specimen dimensions, and specimen shape.

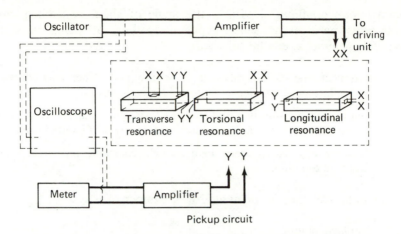

Figure 16.16 Schematic diagram of typical apparatus showing driver and pickup positions for the three types of vibration. (From ASTM C215. Reprinted, with permission, from the American Society for Testing and Materials, 1916 Race Street, Philadelphia, PA 19103. Copyright.)

Similarly, the dynamic shear modulus, G_d, is obtained from the expression

$$G_d = BWm^2 \qquad \text{MPa (lb/in.}^2) \qquad (16.11)$$

where B is a constant depending on the shape and size of the specimen. ASTM C215 gives formulas for the calculation of K and B for various specimen geometries.

Finally, the dynamic Poisson's ratio, μ_d, can be obtained from

$$\mu_d = \frac{E_d}{2G_d} - 1 \qquad (16.12)$$

In general, the dynamic elastic modulus determined in this way will be considerably higher than the static modulus of elasticity, since the dynamic modulus is approximately equal to the initial tangent modulus (see Chapter 18) rather than the chord modulus defined by ASTM C469. BS 1881: Part 5, which is very similar, specifies only the longitudinal vibrations.

Triaxial Strength

ASTM C801 provides a standard test for triaxial loading of concrete, in which two of the three principal stresses are always equal. The test permits determination of the compressive and shear strengths of concrete under a lateral confining pressure, the angle of shear resistance, and several other parameters. Three types of loading, all using standard cylindrical specimens, can be selected:

1. Hydrostatic pressure is raised to some level and is then held constant until failure occurs under an increasing axial load.

2. Hydrostatic pressure is raised to some level, and then the axial stress is held constant while failure occurs under increased lateral loads.

3. The ratio of axial to lateral loads is held constant, and stresses are increased to failure.

The data may then be presented in a number of ways:

1. Graphical plot of

$$\sigma_1 = \sigma_c' + K(\sigma_3)^a \tag{16.13}$$

where σ_1 is the maximum principal stress, σ_3 the minimum principal stress, σ_c' the unconfined compressive strength, and K and a are empirical constants.

2. Graphical plot of the stress difference $(\sigma_1 - \sigma_3)$ versus axial strain.

3. Graphical plot of axial stress vs. axial strain for different confining pressures.

4. Mohr stress circles, with shear stresses as ordinates and normal stresses as abscissas (see Chapter 15), so that the shear strength parameters can be determined.

This test procedure can be modified to measure creep under triaxial loads as well.

Accelerated Tests

The standard compression test was first codified in 1921, when construction proceeded at a more leisurely pace than it does today. At that time, when rather coarsely ground cements that gained strength relatively slowly were common, acceptance tests based on the 28-day strength of the concrete were not unreasonable. However, modern developments in construction procedures, such as slip forming, have accelerated the rate at which concrete is placed, and the more finely ground cements used today gain strength more rapidly. In addition, we have come to realize that the strength of standard test cylinders is, in any event, not a true representation of the strength of the concrete in the structure. Therefore, a considerable amount of work has been done to enable engineers to predict the potential 28-day strength (and hence the quality) of concrete within a few hours after casting.

To do this, it is necessary in some way to accelerate the rate of curing of the concrete, generally by the application of heat. Since the strength gain of concrete is a function of the time and temperature of curing, it should be possible to correlate the strength of concrete cured at a high temperature for a short time with the ''standard'' 28-day strength. It should be noted that the normal 1-day strength of concrete (or even the 3-day strength) cannot be used to predict the 28-day strength, because these early strengths are very sensitive to the fineness of the cement, the curing temperature in the first few hours after casting, and the presence of admixtures. In addition to the application of heat, some of these test methods also involve the use of accelerating admixtures or the application of pressure. However, only three of the many procedures have been developed into standards, as described in ASTM C684.

1. *Warm water method.* Concrete cylinders are made in the normal way and then are immersed in water at 35 ± 3°C (95 ± 5°F). After 23½ h, the cylinder is demolded, capped, and tested at an age of 24 h ± 15 min.

2. *Boiling water method.* Here, the standard cylinder is first cured in a moist environment at 21 ± 6°C (90 ± 17°F) for 23 h. It is then placed in boiling water. After 3½ h, it is removed, cooled for 1 h, capped, and tested at an age of 28½ h ± 15 min.

3. *Autogenous method.* This method uses the heat of hydration of the specimen to accelerate the curing. The standard cylindrical specimen is placed in a heavily insulated container and is held there for 48 h. The maximum and minimum concrete temperatures during this period are also recorded (which may indicate if anything abnormal has occurred). The specimen is then capped and tested at an age of 49 h ± 15 min.

Interpretation of Accelerated Curing Results

The purpose of these tests is to provide a very early indication of the potential strength of the concrete. These tests are generally used to predict the 28-day strengths, on which most design procedures are based. They have about the same variability as conventional tests and there is thus no reason why the results cannot be used directly for design purposes, although this would require suitable changes in the strength values now used for design. It must be emphasized that the values obtained from these three accelerated tests are not equal to each other. Neither are they equal to the standard 28-day strength, being generally lower. They are merely different numbers which may be used to evaluate the concrete quality. It should be noted also that there is really no "universal" curve which can be used to obtain 28-day strengths from any one of these accelerated methods. If accelerated tests are to be used as predictors of the 28-day strength, they must be carried out on the same materials. It has been found that the ratio of accelerated strength to 28-day strength increases as the cement content increases and as the initial mixing temperature increases.

These tests are becoming increasingly common as a quality control measure. Since there is nothing fundamental about the 28-day tests, it is likely that in the future, quality control will become based on these accelerated tests, simply so that errors can be corrected before too much additional concrete is placed.

16.5 ASSESSMENT OF CONCRETE QUALITY

Core Tests

So far, we have discussed only tests carried out on companion samples of concrete which purport to represent the concrete in the structure, even though we have seen that this relationship can be a tenuous one. However, situations arise where it is desirable to have some measure of the strength of the concrete actually in the structure. This is the case particularly when it is suspected that low cylinder strengths are due to improper specimen preparation. The appearance of cracking or other signs of distress may also warrant an investigation of concrete strength. In addition, if it is desired to use a concrete structure for a higher-stress situation than the one it was originally designed for (i.e., trying to increase allowable loads on the structure), a study of both the concrete strength and the position and size of the reinforcing steel may be necessary.

The common way of measuring the strength of the concrete in the structure is to cut cores using a rotary diamond drill (ASTM C42; BS 1881: Part 4). These cores (which may contain some embedded steel) are then soaked in water, capped, and tested in the usual way. As mentioned earlier, if the l/d ratio of the cores is less than 2.0, the core strengths must be corrected by the appropriate factor. Although this seems like a perfectly straightforward way of assessing concrete quality, there are a number of problems in interpreting the strength values obtained.

1. The strengths of cores are generally lower than those of standard cylinders, because the curing of concrete on site (where it is allowed to dry out and is subject to temperature variations) is not as favorable to strength development as curing in a moist room. Also, it is possible that

some damage to the concrete occurs due to vibration of the core drill. However, the design considerations are based on standard cylinders rather than on the true strength that the concrete develops in the structure, so it is far from clear what the significance of low core strength is. (Of course, there is no problem if core strengths are higher than the specified concrete strength.)

2. The ratio of core strength to cylinder strength is not constant. It decreases as the strength of the concrete increases, from a ratio of about 1.0 for cylinder strengths of 20 MPa (3000 lb/in.²) to 0.7 for cylinder strengths of 60 MPa (9000 lb/in.²).

3. The strength of the core will depend on its position in the structure. Generally, cores taken near the top surface of a structural element (or near the top of a lift) are weaker than those at the bottom, simply because of the effects of bleeding and of the settlement of the coarse aggregate.

4. Concrete as cast in the field is anisotropic, since bleeding can cause the creation of a weak cement–aggregate bond under aggregate particles (Figure 8.5). As may be seen from Figure 16.17, these planes of weakness are always horizontal in the concrete as cast, because of the influence of gravity. Therefore, they will tend to be perpendicular to the applied load (case I) for specimens cast with the axis of loading vertical, and parallel to the applied load (case II) for specimens cast with the axis

Figure 16.17 Planes of weakness due to bleeding: (a) axis of specimen vertical; (b) axis of specimen horizontal.

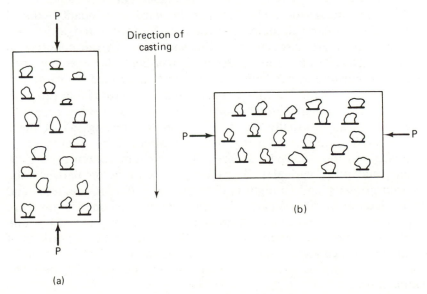

of loading horizontal. Cracks perpendicular to the applied load affect tensile loads much more than compressive loads; the opposite is true for cracks parallel to the direction of loading. Therefore, planes of weakness due to bleeding will lead to a greater reduction in ultimate tensile strength for case I, and a greater reduction in ultimate compressive strength for case II. It has been found experimentally that the strength of concrete cast with the axis of loading vertical is about 8% higher in compression, and about 8% lower in tension, than the strength of concrete cast with the axis of loading horizontal. These values seem to be independent of the mix parameters. Therefore, when trying to evaluate the concrete strength from drilled cores, this effect should be considered when cores are drilled horizontally, as in walls or columns.

In order to get a proper statistical sample of concrete from the structure under consideration, many cores may have to be obtained. Thus, in addition to the problems of interpretation, cores are expensive and time-consuming, besides leaving holes in the structure that must be repaired.

Nondestructive Tests

A great deal of work has been done since the late 1940s to try to develop rapid, nondestructive tests that would provide a reproducible measure of the quality of concrete in a structure. Many such tests have been proposed, but they all still lead to difficulties in interpretation of the results. While the development of a highly reliable *in situ* test to monitor concrete quality remains an elusive goal, some of the tests are useful for several reasons: (1) they can be used as a quality control measure, (2) they can help determine the time for form removal, and (3) they can help in the assessment of the soundness of the concrete in existing structures (e.g., after a structure has been damaged by fire). However, it must be remembered that the tests described below do *not* measure concrete strength; rather, they attempt to provide an *estimate* of the concrete strength through correlation with some other property. Unfortunately, as is usually the case in concrete testing, all of these nondestructive tests give results that are affected by a number of parameters—aggregate type and size, age, moisture content, mix proportions, and other variables. Therefore, the correlation between the measured property and strength is different for different concretes and must be determined for the particular concrete in question. The tests are useful primarily for indicating differences in concrete quality from one part of a structure to another. They may thus indicate those portions of the structure which require much closer examination, which will usually involve the drilling of some cores and possibly petrographic or other examinations.

The available nondestructive *in situ* tests can be classified in a number of different categories. Only a few tests have been adopted as ASTM standards (and a larger number are described in BS 4408: Parts 1–5), but this does not imply that the tests listed in ASTM are in any way better or more accurate than the other tests available; it simply indicates the new tests are being developed more rapidly than the rate at which standards can be proposed, written, and adopted.

Surface Hardness Methods

These are probably the oldest nondestructive tests, having been developed in Germany in the 1930s. Basically, the surface of the concrete is impacted in some standard way (using a given mass and energy of impact), and the size of the resulting indentation is measured. Although there is no theoretical relationship between surface hardness and strength, it is possible to develop empirical correlations that can be used for a particular concrete. Several surface hardness tests exist, but the most common one is the Frank spring hammer, which employs a spring-loaded mechanism to impact a ball against the concrete surface. The diameter (or depth) of the indentation is then correlated with concrete strength. To interpret these tests, it is necessary to have information about the mix proportions, type of cement and aggregate, moisture content, and age of the concrete. Even then, the accuracy of these methods is only within about 20 to 30%.

Rebound Hardness

Probably the most common nondestructive test is the rebound test, using a Schmidt rebound hammer. This device (Figure 16.18a) was developed in 1948, and is universally used because of its simplicity. The test measures the *rebound* of a hardened steel hammer impacted on the concrete by a spring. Again, although there is no theoretical relationship, empirical correlations between rebound hardness and strength can be obtained (Figure 16.18b). This method is described in detail in ASTM C805 (BS 4408: Part 4) and suffers from the same limitations as the surface hardness method, that is, the results will be affected by:

1. The surface finish of the concrete being tested: trowelled surfaces give higher values than formed surfaces, and ground and unground surfaces cannot be compared.

2. The moisture content of the concrete: dry concrete gives higher values than does wet concrete.

3. Temperature: frozen concrete will give very high values, and must be thawed before testing; the temperature of the hammer will also affect the rebound number.

4. The rigidity of the member being tested.

5. The carbonation of the surface, which can increase the hardness values by as much as 50%.

6. The direction of impact (upward, downward, horizontal).

The general view held by many users of the Schmidt rebound hammer is that it is useful in checking the uniformity of concrete and in comparing one concrete against another, but that it can only be used as a rough indication of the concrete strength in absolute terms.

Penetration Resistance

This type of test involves measurement of the resistance of concrete to penetration by a steel probe driven by a given amount of energy, as described in ASTM C803. The most common device of this type is the Windsor probe (Figure 16.19). This consists of a powder-activated driving unit which "fires" a probe into the concrete; the depth of penetration (or operationally, the exposed probe length) is measured, and this can be correlated with strength. Since in this technique there is considerable penetration into the concrete, surface texture and carbonation have less effect than in the tests previously described. However, mix proportions and material properties are still important, and the device must be calibrated for the material in question. Harder aggregates tend to give higher apparent compressive strengths.

Pull-out Tests

Pull-out tests involve the determination of the force required to pull a steel insert out of the concrete in which it was embedded *during casting*. A suitable apparatus for this purpose is described in ASTM C900-78T (Figure 16.20a). Assuming that the failure surface is a frustum, the pull-out strength, f_p, can be calculated by

$$f_p = \frac{F}{A} \qquad (16.14a)$$

where F is the force on the ram at failure and the area of the frustum,

$$A = \frac{\pi}{4}(d_3 + d_2)\,[4h^2 + (d_3 - d_2)^2]^{1/2} \qquad (16.14b)$$

Essentially, this provides a measure of the *shear strength* of the concrete which can then be correlated with the compressive strength (Figure

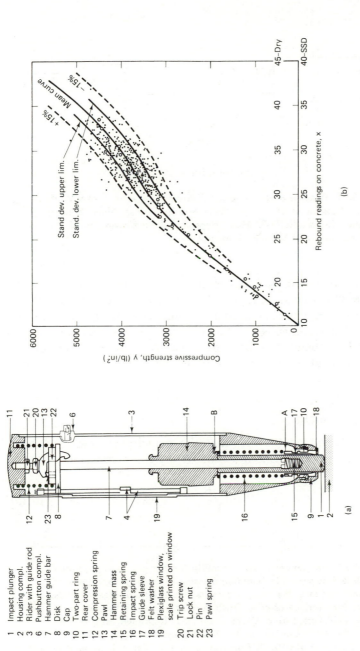

Figure 16.18 Schmidt rebound hammer: (a) Longitudinal section of the Type N concrete test hammer (conditon on impact) (reproduced courtesy of Proceq. S.A.); (b) Calibration chart for concrete made with crushed limestone and natural sand aggregates. Five hundred standard 6 × 12 in. cylinders tested SSD at 28 days. Test hammer No. 3080 was calibrated in the horizontal position. Add 5 points for downward direction; deduct 5 points for upward direction. (From N. Zoldners, *Journal of the American Concrete Institute*, Vol. 54, No. 2, 1957, pp. 161 – 165.)

1   Impact plunger
2   Housing compl.
3   Rider with guide rod
6   Pushbutton compl.
7   Hammer guide bar
8   Disk
9   Cap
10 Two-part ring
11 Rear cover
12 Compression spring
13 Pawl
14 Hammer mass
15 Retaining spring
16 Impact spring
17 Guide sleeve
18 Felt washer
19 Plexiglass window,
    scale printed on window
20 Trip screw
21 Lock nut
22 Pin
23 Pawl spring

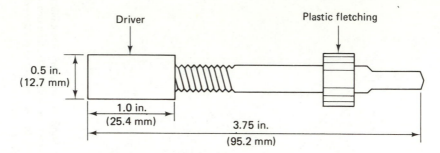

Assembled driver and probe

Figure 16.19 Windsor probe: assembled driver and probe. (From V. M. Malhotra, Department of Energy, Mines and Resources, Mines Branch Investigative Report IR-71-3, 1971. Reproduced by permission of the Minister of Supply and Services, Canada.)

16.20b). The test is economical and rapid, although it does leave a hole in the concrete which must be repaired. It is probably better than those discussed above, because a greater depth and volume of concrete are tested. The big disadvantage is that this test must be planned in advance, and the assembly embedded in the concrete during casting. Therefore, it cannot be used to evaluate existing structures where the concrete quality is suspect.

Ultrasonic Pulse Velocity

This method is based on the fact that the velocity of sound, *V*, in a material is related to the elastic modulus, *E*, by the expression

$$V = \sqrt{\frac{E}{\rho}} \tag{16.15}$$

Figure 16.20 Pull-out test: (a) Assembly for pull-out test. Hydraulic center-pull jack is seated in centering opening of bearing ring. Assumed fracture surface (for stress computation) is shown by light dashed lines. Typical fracture surface of conic frustum is shown in section by heavy dashed lines. (Adapted from ASTM C900-78-T. Reprinted, with permission, from the American Society for Testing and Materials, 1916 Race Street, Philadelphia, PA 19103. Copyright.); (b) Correlation between pull-out strength and compressive strength (91-day test results). (From V. M. Malhotra, Department of Energy, Report IR-72-56, 1972. Reproduced by permission of the Minister of Supply and Services, Canada.)

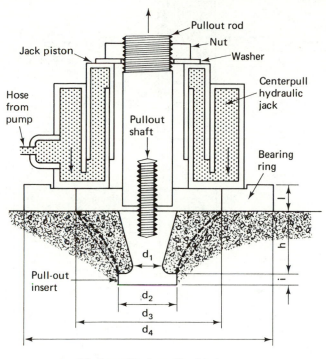

(a) Assembly for pull-out test

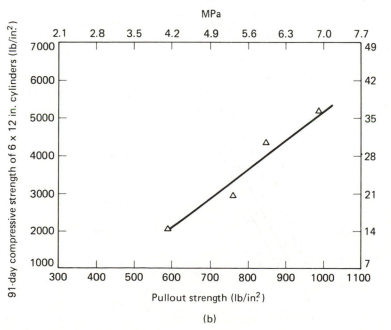

(b)

445

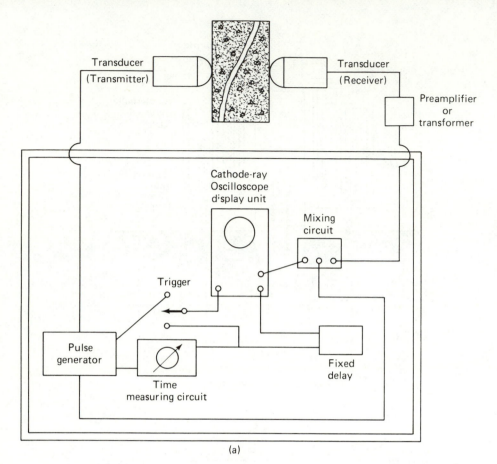

(a)

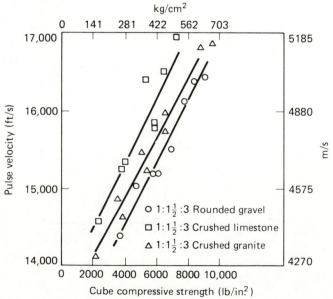

(b)

where ρ is the density of the material. Since the pulse velocity depends only on the elastic properties of the material and not on the geometry, this is a very convenient technique for evaluating concrete quality. In essence, an apparatus such as that shown schematically in Figure 16.21a is used to determine the pulse velocity through a known thickness of concrete, using the procedures outlined in ASTM C597 or BS 4408: Part 5. A number of commercial devices are available which meet these requirements, with an accuracy of measurement of about $\pm 1\%$.

In fact, mechanical impulses applied to a material generate three types of waves: longitudinal (compressional), shear (transverse), and surface (Rayleigh). The longitudinal waves are the fastest, and they are the ones that are most useful for testing purposes. The pulse velocity may be used directly as a quality control measure, but more commonly it is correlated with strength, as shown in Figure 16.21b. However, more recent work shows this correlation to be a poor one. Pulse velocity can be measured in several configurations, as shown in Figure 16.22, although method A seems to give better results. The pulse velocity is affected by a number of factors:

1. Smoothness of the contact surfaces: if the surfaces are not reasonably smooth, they should be ground smooth; a coupling medium such as grease must also be used to ensure good contact between the trans-ducers and the concrete.

2. The pulse velocity seems to depend on the path length, decreasing somewhat as the path length is increased.

3. The pulse velocity is not sensitive to temperature in the range 5 to 30°C. At higher temperatures, the pulse velocity is decreased, and at temperatures below freezing, it is increased.

4. Pulse velocity increases with increased moisture content.

5. The presence of steel bars will tend to increase the pulse velocity.

6. For a given pulse velocity, the compressive strength is higher for older specimens (Figure 16.23).

In spite of these difficulties, pulse velocity measurements are becoming increasingly common as a means of assessing concrete quality.

Figure 16.21 (a) Schematic diagram of pulse velocity testing circuit. (From ASTM C597. Reprinted, with permission, from the American Society for Testing and Materials, 1916 Race Street, Philadelphia, PA 19103. Copyright.); (b) Correlation of pulse velocity with compressive strength. (From R. Jones, *Non-destructive Testing of Concrete*, Cambridge University Press, Cambridge, England, 1962.)

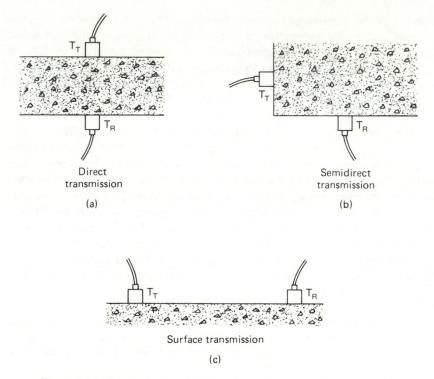

Figure 16.22 Methods of measuring pulse velocity through concrete: (a) direct transmission method; (b) semidirect transmission method; (c) surface transmission method.

They are used to study uniformity, for quality control, estimations of strength, setting characteristics, durability and the extent of cracking in a member, as well as to determine the modulus of elasticity.

Other Techniques

A number of other techniques have been studied that can provide some information about the concrete. These will be mentioned very briefly, as their detailed description is beyond the scope of this book. Both X-rays and gamma rays will penetrate concrete to some degree. X-rays are used primarily as a laboratory technique to examine the internal structure (i.e., cracking, aggregate distribution, etc.) of concrete. Gamma radiography has been used to locate reinforcing bars and to measure the density and thickness of concrete, as described in BS 4408: Part 3. Neutron back-scattering has been used to measure moisture content, and neutron activation analysis has been used to determine cement content. These studies, however, are still of a preliminary nature. A number of magnetic

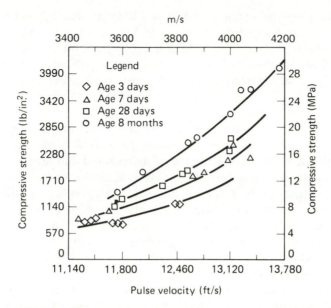

Figure 16.23 Influence of age of concrete on the correlation between pulse velocity and strength. (From I. Facaoaru, *Proceedings, Symposium on Nondestructive Testing of Concrete and Timber,* Institution of Civil Engineers, London, June 1964, pp. 23–33. Reproduced by permission of the Institution of Civil Engineers.)

devices are available that will measure the depth of cover of reinforcement as well as the position of the reinforcing bars, as described in BS 4408: Part 1. Electrical methods are being developed that can be used to determine the moisture content and thickness of concrete.

Analysis of Hardened Concrete

Occasionally, the question arises as to whether the concrete was cast at the specified mix composition. It is then desirable to have some method of determining the composition of the hardened concrete. This is a very difficult proposition because of the variability of the materials involved. In principle, it would be desirable to be able to determine (1) the original cement content and type of cement, (2) the w/c ratio, (3) the aggregate content and grading, (4) the amount and type of admixtures used, and (5) the nature of the air-void system.

In Chapter 8 it was suggested that it was possible to determine the nature of the air-void system by examining a polished section of the concrete under a microscope, using a "point-count" method to determine the volume and spacing of air voids in hardened concrete (ASTM C457).

This method can be extended (ASTM C856) to give a good indication also of the relative amounts of coarse and fine aggregate and the amount of cementitious matrix. The volumetric composition of the material is obtained from an observation of the frequency with which each component coincides with a regular array of points on a plane section through the material. Thus, it is possible to make at least some estimates of (3), (4) and part of (5) above.

In addition, there is a chemical method of determining the cement content (but not the type), according to ASTM C85. This method involves the dissolution of a sample of dehydrated and crushed concrete in hydrochloric acid and the determination of either, or both, of the SiO_2 and CaO contents of the resulting solution. This permits an estimation of the cement content, assuming that only the cement dissolves. The method is not reliable when the aggregates used yield significant amounts of CaO and SiO_2 under the test conditions. Chemical tests are very expensive and are subject to considerable error, particularly for concretes with low cement contents.

The determination of the original w/c ratio is also subject to considerable error, and no ASTM standard exists, although there is a British Standard (BS 1881: Part 6). This method involves drying the concrete at 105°C and then saturating in vacuum with CCl_4. The amount absorbed is determined and the equivalent amount of water is calculated. This gives, in essence, a measure of the volume of capillary pores (i.e., those pores originally filled with water which remain unfilled after the cement hydration). The sample is then redried to drive off the CCl_4, and the combined water is determined by *igniting* the sample (i.e., measuring the weight loss on heating to 1000°C and correcting for the CO_2 content). The original water content is then the sum of these two. If the cement content is determined by the method described above, the original w/c ratio can be calculated. It is claimed that this estimate is within ±0.02 of the true w/c ratio in non-air-entrained, normal-weight concretes. BS 1881: Part 6 also describes in detail methods for determining the cement content, type of cement (in some circumstances), aggregate content and grading, type of aggregate, bulk density, and the chloride, sulfate, and sulfoaluminate contents.

Unfortunately, there is as yet no reliable method for determining either the type or amount of chemical admixtures that may have been used. Thus, the determination of the original mix composition is both an expensive and a difficult problem, made worse by the fact that in the interval between casting and testing, the concrete may have been subject to leaching by chemical attack and to carbonation.

In general, no one method is sufficient to assess the quality of concrete in an existing structure. Rather, depending on the particular circumstances, a number of different tests might be used, such as hardness tests, core tests, and petrographic examination of the concrete. The object is to get the required information as economically as possible, with due consideration to the safety of the structure and the consequences of incorrect analysis. This area of concrete technology is still fraught with uncertainty, and should not be lightly undertaken by inexperienced investigators.

Bibliography

Accelerated Strength Testing, SP-56. American Concrete Institute, Detroit, Mich., 1978.

Concrete Manual, 8th ed. U.S. Bureau of Reclamation, Denver, Colo., 1975.

MALHOTRA, V. M., *Testing Hardened Concrete: Non-destructive Methods,* Monograph no. 9. American Concrete Institute, Detroit, Mich., 1976.

NEVILLE, A. M., *Hardened Concrete: Physical and Mechanical Aspects,* Monograph No. 6. American Concrete Institute, Detroit, Mich., 1971.

PRICE, W. H., "Factors Influencing Concrete Strength," *Journal of the American Concrete Institute,* Vol. 47, No. 6, pp. 417–432 (1951).

Significance of Tests and Properties of Concrete and Concrete-making Materials, ASTM STP 169B. American Society for Testing and Materials, Philadelphia, Pa., 1978.

Problems

16.1   Why are tests carried out on hardened concrete?

16.2   What is the purpose of having detailed specifications in standard tests?

16.3   Compare cylinder and cube strengths for compressive strength.

16.4   Why does increasing the rate of loading increase the apparent strength?

16.5   Why is it necessary to cap concrete cylinders?

16.6   (a) How would cracks oriented as in (1), (2), or (3) affect the measured compressive strength?
(b) How would these cracks affect the tensile strength?

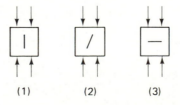

(1)   (2)   (3)

16.7   Why is the l/d ratio a sensitive parameter in the compressive strength test?

16.8   Why does the moisture content of a concrete specimen affect its strength?

16.9   Compare the methods of measuring tensile strength by (a) direct tension; (b) splitting tension; (c) flexure.

16.10 Why are accelerated tests of concrete strength useful?

16.11 What are the problems involved in interpreting the compressive strength of cores taken from a concrete structure?

16.12 Why are nondestructive tests used?

16.13 Can nondestructive tests give an absolute measure of strength? Discuss.

16.14 What test methods can be used to detect internal cracking in concrete?

17

quality control

By now, we have discussed the many factors that are involved in the production of high-quality concrete: materials, proportioning, handling and placing, curing, and testing. It should, therefore, come as no surprise that concrete, in common with other engineering materials, is inherently a *variable* material. That is, tests on nominally identical samples of concrete will show some variation in mechanical properties between samples. Clearly, this variability in properties must be considered when writing concrete specifications.

In general, the factors that contribute to this variability may be grouped as follows:

1. *Materials*. This includes variability in the cement itself; in the grading, moisture content, mineral composition, physical properties, and particle shape of the aggregates; and in the admixtures used.

2. *Production*. This involves the type of batching plant and equipment, the method of transporting the concrete to the site, and the procedures and workmanship used to produce and place the concrete.

3. *Testing*. This includes the sampling procedures, the making and curing of test specimens, and the test procedures used.

It is, of course, very difficult to assess the relative importance of these three groups of factors; in any event, their importance will vary for different regions and different construction projects. Since the variability in concrete quality is some function of the variabilities of each of these three factors, no one of these can be ignored in concrete production.

The term "quality assurance" is frequently used in the construction industry. Quality assurance refers collectively to all of the steps taken to ensure adequate confidence that the concrete will perform satisfactorily in service. Quality control applies to each action used to measure the properties of the concrete, or its components, and to control them within the established specifications.

17.1 MEASUREMENT OF VARIABILITY

Before discussing quality control, however, it is necessary to define the concept of variability more precisely. It has been found that the *distribution* of concrete strengths can best be approximated by the *normal (Gaussian)* distribution. Such a distribution is completely defined by two parameters: the mean, μ, and the standard deviation, s. The equation of the normal distribution curve is

$$y = \frac{1}{s\sqrt{2\pi}} \exp\left[\frac{-(x - \mu)^2}{2s^2}\right] \tag{17.1}$$

The mean is simply the arithmetic mean of all the values. If we represent the values of strength by x, then

$$\mu = \frac{\Sigma x}{n} \tag{17.2}$$

The standard deviation or the *root-mean-square* deviation is a measure of the *dispersion,* or variability, of the values. It can be calculated by

$$s = \sqrt{\frac{\Sigma(x - \mu)^2}{n - 1}} \tag{17.3}$$

The value of s is more easily obtained with a calculator if Eq. (17.3) is rewritten in the form

$$s = \sqrt{\frac{\Sigma x^2 - (\Sigma x)^2/n}{n - 1}} \tag{17.4}$$

The *variance* is defined as s^2, and the *coefficient of variation, V,* as

$$V = \frac{s}{\mu} \times 100\% \tag{17.5}$$

It is possible to "normalize" Eq. (17.1) (i.e., make it free from physical units) by making the substitution

$$z = \frac{x - \mu}{s}$$ (17.6)

The value z simply represents the number of standard deviations that x is away from the mean. Equation (17.1) can then be written

$$y = \frac{1}{s\sqrt{2\pi}} e^{-z^2/2}$$ (17.7)

When the function described by Eq. (17.1) or (17.7) is graphed, it gives the familiar bell-shaped curve shown in Figure 17.1 in terms of both x and z units. This curve has the following properties, the proofs of which can be found in any elementary text on probability theory:

1. The maximum ordinate is at $z = 0$ (or $x = \mu$).

2. The curve is symmetrical about $z = 0$.

3. Points of inflection exist at ± 1 standard deviation from the mean, that is, at $z = \pm 1$ (or $x = \mu \pm 1s$).

4. The total area under the normalized curve is 1.

5. The area under the curve between any two points is the probability of occurrence of a value in that interval. That is, the probability of a value lying within 1 standard deviation of the mean is 68.27%; within 2 standard deviations from the mean, 95.45%; within 3 standard deviations from the mean, 99.73%.

6. If we know the standard deviations $s_1, s_2, \ldots, s_n$ of all the components that contribute to the total standard deviation, the total standard deviation is given by

$$s^2 = s_1^2 + s_2^2 + \cdots + s_n^2$$ (17.8)

It may be seen, of course, that such a curve cannot be a perfect representation of the variability in concrete properties, since it extends to $\pm$ infinity. However, if we assume that the only area of interest is the region within 3 standard deviations from the mean, which contains 99.73% of all expected values, this curve is a very useful means of characterizing concrete properties.

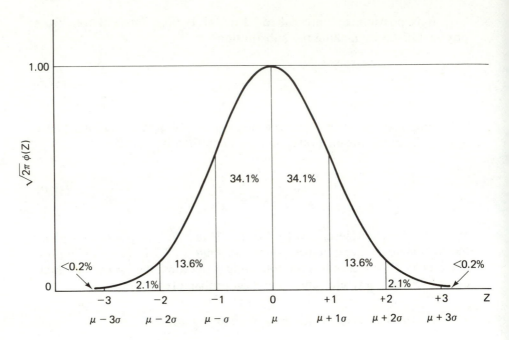

Figure 17.1 Areas under the normal curve.

Applications to Concrete

Now, assuming that the variation in concrete properties is to be described by the normal distribution, let us see how these concepts are applied in practice. In the discussion that follows, reference will be to the compressive strength, because it is here that the use of the variability concept is most common. However, it must be remembered that this concept is a general one and may be applied equally to any other quantifiable concrete properties, such as *w/c* ratio, flexural strength, and elastic modulus.

Once we accept the concept that there is a statistical distribution of concrete strengths, several important implications emerge:

1. We cannot design concrete structures on the basis of mean strength (or average strength) as defined by Eq. (17.2). If we did, this would mean that about one-half of the concrete placed would have strengths that fall *below* the design value, which would be unacceptable. On the other hand, we cannot insist that *all* concrete strengths be above the design value; since concrete strengths are approximately normally distributed, this is an impossibility. Therefore, we must arbitrarily decide what constitutes an acceptable percentage of specimens falling below the

"minimum" design values. Using this percentage, and knowing (or assuming) the standard deviation in strength that can be expected, we can then determine the required mean strength for which to design the concrete mix.

2. When carrying out tests on concrete, we are trying to evaluate the distribution in strength of all of the concrete in the structure, based upon a limited sample size. Clearly, enough test data must be collected so that the tests are truly representative of the concrete in the structure. However, because of the rather tenuous relationship between the strengths of test specimens and the quality of concrete in the structure (Chapter 16), we can at best only *estimate* the strength of the concrete in the structure.

3. Because variations in concrete strengths are due not only to mix variations, but also to sampling variations, there are two risks that must be balanced: the "producer's risk" that satisfactory concrete will be rejected, and the "consumer's risk" that bad concrete will be accepted. This consumer's risk can be large indeed if insufficient testing is carried out.

4. There must be some plan of action that can be followed if the concrete is considered not to have complied with the specifications.

It should be remembered, however, that the acceptance of the fact that a certain proportion of concrete test cylinders will be below the design strength of the concrete is much less "risky" than might appear at first glance, for several reasons:

1. Concrete batches tend to get intermixed as the concrete is placed in the forms, and this tends to "average out" the strengths.

2. As we have seen, most acceptance tests are based on 28-day strengths; however, concrete continues to gain strength beyond this time, and this will tend to compensate for low strengths by the time the concrete is actually in service.

3. In part because of the steel reinforcement of concrete, there can be a considerable redistribution of stresses in a structure.

4. Perhaps most important, many years of experience have shown that structures built in accordance with these concepts behave satisfactorily.

ACI Approach to Variability

The approach of ACI Committee 214, Evaluations of Results of Compression Tests of Field Concrete, to the variability inherent in concrete strength is to require an average concrete strength sufficiently in excess of the specified design strength so that only an allowable proportion of low strengths will occur. The required average strength is computed from the expression

$$\sigma_{cr} = \frac{\sigma'_c}{1 - tV} \qquad (17.9a)$$

or

$$\sigma_{cr} = \sigma'_c + ts \qquad (17.9b)$$

where σ_{cr} is the required average strength, σ'_c the specified design strength, t a constant depending on the proportion of tests allowed below σ'_c (Table 17.1), V the predicted value of the coefficient of variation, and s the predicted value of the standard deviation. The reliability of this expression depends on the number of samples used to establish V; preferably at least 30 samples should be used. It is also possible to estimate the desired average strength from Figure 17.2a or b, depending on whether Eq. (17.9a) or (17.9b) is used.

It should be noted that this approach provides a great deal of flexibility. Different classes of concrete could be permitted different probabilities of low-strength concrete, depending on the type of structure and on the economic and other consequences of failure. It is the opinion of Committee 214 that reasonable control of structural concrete would be obtained if the probability of a test below σ'_c is no greater than 1 in 10.

Table 17.1

Values of t Used in Eq. $(17.9)^a$

Percentages of Tests Falling Within the Limits $\mu \pm ts$	Chances of Falling below Lower Limit	t
40	3 in 10	0.52
50	2.5 in 10	0.67
60	2 in 10	0.84
68.27	1 in 6.3	1.00
70	1.5 in 10	1.04
80	1 in 10	1.28
90	1 in 20	1.65
95	1 in 40	1.96
95.45	1 in 44	2.00
98	1 in 100	2.33
99	1 in 200	2.58
99.73	1 in 741	3.00

^{a}From ACI 214. Reproduced with permission.

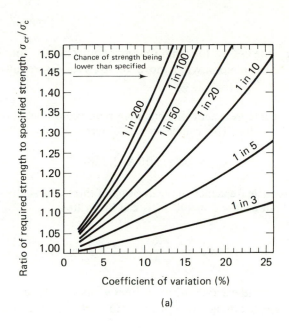

(a)

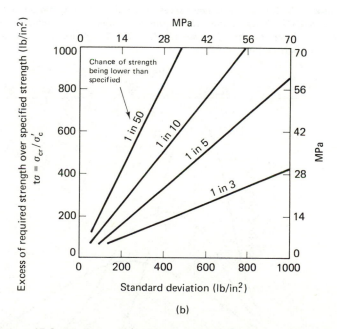

(b)

Figure 17.2 (a) Ratio of required average strength σ_{cr} to specified strength σ_c' for various coefficients of variation and chances of falling below specified strength; (b) Excess of required average strength σ_{cr} to specified strength σ_c' for various standard deviations and chances of falling below specified strength. (From *Recommended Practice for Evaluation of Strength Test Results of Concrete*, ACI 214-77, American Concrete Institute, Detroit, Mich., 1976.)

However, the ACI Building Code Requirements for Reinforced Concrete (ACI 318) is somewhat more restrictive. Its requirements are twofold:

1. The probable frequency of tests more than 500 psi below σ'_c should not exceed 1 in 100. This can be written in terms of s as

$$\sigma_{cr} = \sigma'_c - 500 + 2.326s \qquad (17.10)$$

2. The probable frequency of the *average* of three consecutive tests below σ'_c will not exceed 1 in 100. This can be written

$$\sigma_{cr} = \sigma'_c + \frac{2.326s}{\sqrt{3}} = \sigma'_c + 1.343s \qquad (17.11)$$

When the standard deviation is below about 500 psi, Eq. (17.10) governs; for higher standard deviations, Eq. (17.11) governs.

Figure 17.3 Normal frequency curves for coefficients of variation of 10, 15, and 20%. (Adapted from *Recommended Practice for Evaluation of Strength Tests Results of Concrete*, ACI 214-77, American Concrete Institute, Detroit, Mich., 1976.)

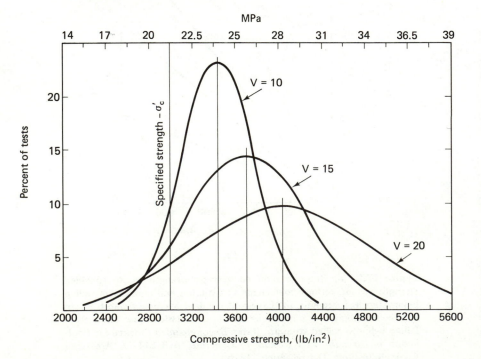

It should be clear that, whatever the exact acceptance criteria adopted, there is a considerable economic benefit in achieving the tightest quality control (i.e., the lowest coefficient of variation) possible. Figure 17.3 shows the normal distribution curves for concrete with different coefficients of variation, showing the average strength σ_{cr} based on a probability of 1 in 10 that a test will fall below $\sigma'_c = 21$ MPa (3000 lb/in.²). It can be seen that for $V = 10\%$, $\sigma_{cr} = 23.7$ MPa (3440 lb/in.²), while for $V = 20\%$, $\sigma_{cr} = 27.8$ MPa (4030 lb/in.²). Thus, as V increases, the average design strength must be increased. Since this is done principally by adding cement, it can be quite expensive to do so. The price of cement continues to rise: at the present cost of cement, it might require about \$1.50/yd³ extra to design for $V = 20\%$ rather than $V = 10\%$. As energy costs continue to rise, so will the cost of cement; poor quality control will become more and more expensive. Apart from producing generally inferior concrete due to the implied lack of quality control, this will also lead to a "waste" of strength. On the other hand, these costs must be balanced against the costs of providing better quality control.

Other Approaches to Variability

The ACI approach to quality control is not the only one. Concrete design procedures in the United Kingdom (CP110: 1972 *The Structural Use of Concrete*) explicitly recognize the variability of concrete. It is assumed that concrete strengths are normally distributed. The design criterion then becomes

$$\sigma_m = \sigma_c + ks \qquad (17.12)$$

where σ_m is the target mean strength (i.e., the average strength for which the mix is to be designed), σ_c the characteristic strength (i.e., the nominal design strength), s the standard deviation of the available strength data, and k is a constant, derived from the normal distributions, which depends on the proportion of test results that may be expected to fall below the level of σ_c. [This has the same form as Eq. (17.9b).] CP110 specifies a probability of failure of 1 in 20 (as opposed to the ACI Committee 214 recommendation of 1 in 10), for which $k = 1.64$. The product ks is known as the "margin"; the minimum margin that may be used, however, depends on the number of test data that are available. If s is calculated on the basis of at least 100 tests of similar concrete within 12 months, the margin must not be less than 3.75 MPa (550 lb/in.²); if s is calculated from at least 40 tests within 6 months, the margin must not be less than 7.5 MPa (1100 lb/in.²). If there is insufficient data to calculate s in this way, the margin should be at least 15 MPa (2200 lb/in.²) until sufficient

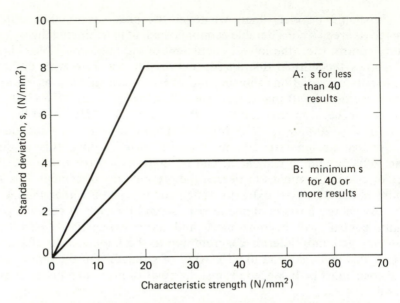

Figure 17.4 Relationship between standard deviation and characteristic strength. (From D. C. Teychenné, R. E. Franklin, and M. C. Erntroy, *Design of Normal Concrete Mixes,* Building Research Establishment, Transport and Road Research Laboratory, 1975. Reproduced with permission of the Controller of Her Brittanic Majesty's Stationery Office.)

test data become available. This concept is very similar to the RILEM recommendations.

If σ_c is that 28-day cube strength below which not more than 1 in 20 test results may be expected to fall, this requirement will be met if the average of any four consecutive test cubes exceeds the specified σ_c by at least 0.5 times the margin defined above. However, no individual test may be less than $0.85\sigma_c$.

On the other hand, the British "Design of Normal Concrete Mixes" (Chapter 9) relates the minimum standard deviation that may be used both to the amount of test data available and to the characteristic strength of the concrete, as shown in Figure 17.4.

The Canadian requirements (CAN3-A23.1) are similar to the ACI Building Code requirements. The average of any three consecutive tests must exceed σ_c', and no individual test may be more than 3.5 MPa below σ_c'. These requirements will be met 99 percent of the time if:

1. The average strength is $1.4s$ above σ_c' when $s < 3.5$ MPa.

2. The average strength is $2.4s - 3.5$ MPa above σ_c' when $s > 3.5$ MPa.

s must be calculated from at least 30 consecutive tests of similar concrete.

Other national standards impose different acceptance criteria; however, they all generally have two things in common: the assumption that concrete strengths are normally distributed, and an arbitrary choice of the percentage of test results allowed to fall below the design strength.

Choice of Design Strength

A problem that does remain in applying probabilistic principles to concrete quality control is the choice of the measure of variability. There is still no agreement on whether a certain degree of quality control should be achieved by using a constant coefficient of variation or a constant standard deviation. In fact, it seems, based on a great deal of test data, that up to some limiting value of strength (in the range 21 to 28 MPa or 3000 to 4000 lb/in.²), the coefficient of variation remains constant. Beyond that value, the standard deviation remains approximately constant. Therefore, considerable judgment and empirical observations at the job site must still be used to establish the appropriate relationship between the mean strength and the minimum strength. In fact, according to the ACI Buidling Code, where there is a record of strength tests using similar materials and conditions to those expected, the strength to be used as a basis for mix design should exceed the required σ'_c by the amounts shown:

2.8 MPa (400 lb/in.²) if the standard deviation is less than 2.1 MPa (300 lb/in.²)
3.8 MPa (550 lb/in.²) if the standard deviation is 2.1 to 2.8 MPa (300 to 400 lb/in.²)
4.8 MPa (700 lb/in.²) if the standard deviation is 2.8 to 3.4 MPa (400 to 500 lb/in.²)
6.2 MPa (900 lb/in.²) if the standard deviation is 3.4 to 4.1 MPa (500 to 600 lb/in.²)
8.3 MPa (1200 lb/in.²) if the standard deviation exceeds 4.1 MPa (600 lb/in.²) or if a suitable record of tests is not available.

The strengths indicated by Eqs. (17.10) and (17.11) may be used only after sufficient job data are available.

Clearly, different types of construction and equipment will lead to different levels of variability. It is very difficult to decide on an acceptable level. However, based on large numbers of strength tests, Table 17.2 gives the commonly accepted standards of quality control.

Number of Samples

In order to carry out the sort of statistical analysis described above, a sufficient number of samples must be obtained. In addition, sampling should also provide a continuing check on the concrete being provided.

Table 17.2

Standards of Concrete Control[a]

	Overall Variation				
	Standard Deviation for Different Control Standards, MPa (lb/in.²)				
Class of Operation	*Excellent*	*Very Good*	*Good*	*Fair*	*Poor*
General construction testing	Below 2.8 (400)	2.8–3.5 (400–500)	3.5–4.2 (500–600)	4.2–4.9 (600–700)	Above 4.9 (700)
Laboratory trial batches	Below 1.4 (200)	1.4–1.8 (200–250)	1.8–2.1 (250–300)	2.1–2.5 (300–350)	Above 2.5 (350)
	Within-Test Variation				
	Coefficient of Variation for Different Control Standards (%)				
Class of Operation	*Excellent*	*Very Good*	*Good*	*Fair*	*Poor*
Field control testing	Below 3.0	3.0–4.0	4.0–5.0	5.0–6.0	Above 6.0
Laboratory trial batches	Below 2.0	2.0–3.0	3.0–4.0	4.0–5.0	Above 5.0

[a]From ACI 214. Reproduced with permission.

A very common requirement is that adopted by the ACI, which requires that samples be taken at least once a day and at least once for every 110 m³ (150 yd³) of concrete or for each 450 m² (5000 ft²) of surface area placed. This procedure should be followed for each type of concrete being used, with the provision that at least five samples should be obtained for every type of concrete. If the total quantity of concrete of a given type is less than 40 m³ (50 yd³), tests may be waived. (The procedures for sampling fresh concrete are described in Chapter 8.)

In order to generate random samples, as required by the statistical considerations described above, a sampling plan should be established in advance of the construction project. Samples should be taken from *predetermined* batches in accordance with the sampling plan. The practice of testing only batches of high or low slump, or batches which appear to be "representative" to the inspector should be avoided, since this will not give an accurate representation of either concrete quality or uniformity.

17.2 QUALITY CONTROL CHARTS

Statistical principles alone are insufficient for quality control—we must be able to apply them efficiently so that we can determine as quickly as possible both the concrete quality and possible changes in quality. We may find, for instance, that the mean strength designed for is being achieved but that the standard deviation is larger than expected. This would require an increase in the mean strength, by adjusting the mix. On the other hand, a lower than expected standard deviation might permit a reduction in mean strength. Equally important, it might become apparent that *changes* in either the mean strength or standard deviation have occurred at some stage of the construction process. This might be due to changes in the raw materials (e.g., moisture content of the aggregate), the batching equipment, or a variety of other reasons. In any event, the sooner such changes are detected, the sooner we can take the appropriate remedial measures.

This implies that concrete strengths must be assessed on a continuous basis. This can most effectively be done using *quality control charts* of the type shown in Figure 17.5. Three types of charts are employed; in each case, the horizontal axis represents consecutive sample numbers.

1. Figure 17.5a is a chart of the individual strength values. The line for required average strength, σ_{cr}, is obtained using Eq. (17.9), based on the allowable number of tests below the specified strength, σ_c'. This chart simply shows the number of "low" tests and gives some indication of the scatter. However, this chart is relatively insensitive to changes in the concrete quality.

2. Figure 17.5b is a graph of the *moving* average for strength, where each point represents the average of the previous five sets of strength tests. (Each strength test will normally consist of breaking two or three cylinders.) This chart tends to "smooth out" chance variations and can be used to indicate trends that may be due to the influence of seasonal changes, changes in procedures, and so on. The number of tests to be averaged for each point can, of course, be varied to suit the particular job requirements. For instance, since we have stated previously that the ACI Building Code requires that the frequency of an average of three consecutive tests below σ_c' will not exceed 1 in 100, a chart where each point represents the average of the three previous tests would be appropriate. Table 17.3 provides a guide for concrete strength specifications, indicating the strength level below which individual tests or averages of a series of tests should not fall. (From the statistical theory that we have assumed, we would expect that a failure to meet these specifications would occur about 1 time in 50.) A chart such as the one in Figure 17.5b would therefore have σ_c' as its lower limit. A failure to meet the limitations of the chart (or Table 17.2) would indicate that the required average strength σ_{cr} is not being reached, either due to lower strengths, to more variability than anticipated, or to poor testing techniques.

3. Figure 17.5c represents the moving average of the *range* (the difference between the highest and lowest values in a set), where each

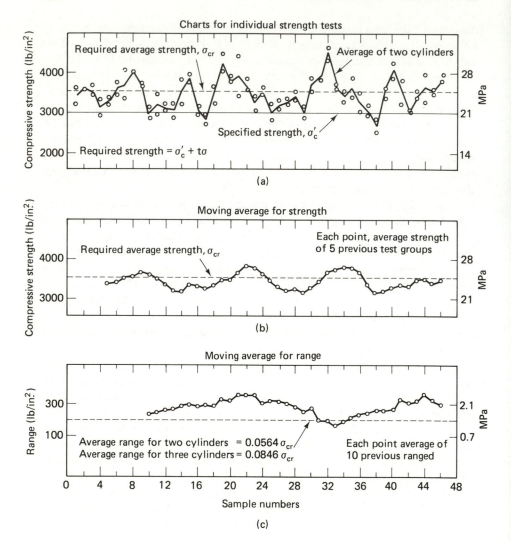

Figure 17.5 Quality control charts for concrete. (From *Recommended Practice for Evaluation of Strength Test Results of Concrete*, ACI 214 – 77, American Concrete Institute, Detroit, Mich., 1976.)

point represents the average of the ranges of the 10 previous sets of strength tests. The *maximum* average range allowable is also shown. This chart provides a check on the adequacy of the test procedures. If poor test procedures are indicated, they must be corrected. The adequacy of the test procedures, as indicated by the range between companion cylinders comprising a strength test, can be expressed by

$$\overline{R}_M = \sigma_{cr}\, v_1\, d_2 \qquad\qquad (17.13)$$

Table 17.3

Evaluation of Consecutive Low-Strength Test Results[a]

Number of Consecutive Tests Averaged	Averages Less Than Indicated Require Investigation[b]				Probability of Averages Less Than σ_c' (%)
			Criteria for Original Selection of σ_{cr}		
		1 Test in 10 Below σ_c'		*1 Test in 100 Less Than* $[\sigma_c' - 500 \; lb/in.^2$ $(3.5 \; MPa)]$	
	For V = 15 (%)	For Given s		For Given s	1 Test in 10 Below σ_c'
1	$0.86\sigma_c'$	$\sigma_c' - 0.77s$		$\sigma_c' - 500 + 0.76s$	10.0
2	$0.97\sigma_c'$	$\sigma_c' - 0.17s$		$\sigma_c' - 500 + 0.88s$	3.5
3	$1.02\sigma_c'$	$\sigma_c' + 0.10s$		$\sigma_c' - 500 + 1.14s$	1.3
4	$1.05\sigma_c'$	$\sigma_c' + 0.26s$		$\sigma_c' - 500 + 1.30s$	0.5
5	$1.07\sigma_c'$	$\sigma_c' + 0.36s$		$\sigma_c' - 500 + 1.41s$	0.2
6	$1.08\sigma_c'$	$\sigma_c' + 0.44s$		$\sigma_c' - 500 + 1.49s$	0.1

[a]From ACI 214. Reproduced with permission.

[b]The probability of averages less than the levels indicated is approximately 2% if the population average equals σ_{cr} and the standard deviation or coefficient of variation is at the level assumed.

where $\overline{R}_M$ is the average range, v_1 the within−test coefficient of variation, and d_2 is a constant depending on the number of cylinders tested. Since v_1 should not exceed 5% for good control, the average range should be

$$\overline{R}_M = (0.05 \times 1.128)\sigma_{cr} = 0.0564\sigma_c \text{ for two companion cylinders}$$

$$\overline{R}_M = (0.05 \times 1.693)\sigma_{cr} = 0.08465\sigma_{cr} \text{ for three companion cylinders}$$

The control charts described above are most commonly used in North America. In the United Kingdom, cumulative sum (cusum) control charts are now widely used. The feature which differentiates these charts from those described above is that the points plotted on the chart contain information from *all* of the observations up to and including the plotted point. It is claimed that these charts are more sensitive to changes in concrete quality. To construct a cusum chart, the cumulative sum of the algebraic differences between each result and the assumed (or design) value is plotted as successive results are obtained. The slope over any part of the cusum chart is then proportional to the difference between the actual and assumed values. A horizontal plot indicates that the actual results are the same as the design value. A rise in the plot indicates that the average value is greater than the design value, while a fall in the plot indicates that the average values are less than the design values. Small changes in the average appear as quite different slopes. Cusum charts may be prepared for the mean strength or the standard deviation.

A detailed account of this method is outside the scope of this book. In practice, in order to facilitate its use for concrete quality control, a series of charts, graphs and masks (developed by the British Ready Mixed Concrete Association) are used.

17.3 FAILURE TO MEET SPECIFIED REQUIREMENTS

The application of the statistical considerations described above is relatively straightforward, and if the concrete is, on the basis of these considerations, judged to have complied with the specifications, then everybody (the concrete supplier, the contractor, the testing agency, and the owner) is happy. However, if the concrete is judged *not* to have complied with the specifications, there are really no clear-cut procedures available to deal with the resulting situation.

Nevertheless, there are rational procedures that can be applied. They may be both expensive and time-consuming, but their application will generally result in enough information to make a sensible judgment on accepting or rejecting the structure. The procedures suggested below are discussed in order of increasingly severe consequences, based on the RILEM "Recommended Principles for the Control of Quality and the Judgment of Acceptability of Concrete" and shown graphically in the flowchart, Figure 17.6.

The first and simplest check is on the sampling and testing techniques, which are all too often at fault. It may be possible to determine that the test procedures described in Chapter 16 were not correctly applied or that the equipment used was not calibrated correctly or was otherwise unsuitable. If this is found to be the case, and if there is no other reason to suspect the concrete quality, little or no further action will be required beyond perhaps a few tests of the in-place concrete just to make sure of its quality.

Similarly, the materials and batching plant should be investigated, to make sure that the materials used were the ones specified, that the batching equipment was producing concrete with the desired mix proportions, and that the mix design itself was adequate. If the supplier is clearly at fault, the cause of the problem is then known, and the appropriate remedial measures can be taken, at least for the subsequent concrete production. These measures might involve redesign of the mix or more careful inspection and control.

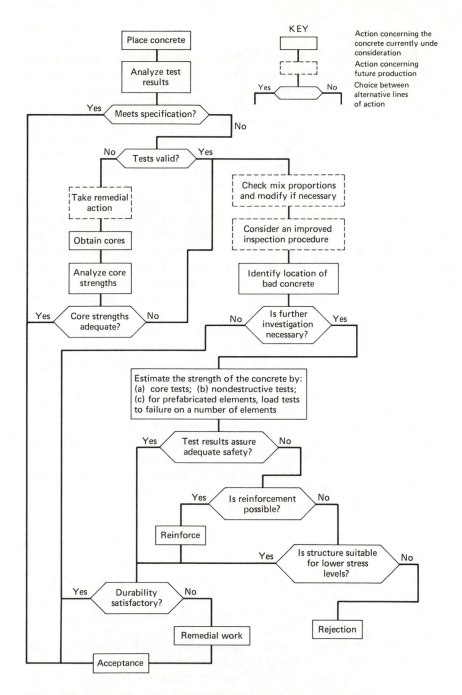

Figure 17.6 Flowchart for actions following judgment of compliance with specifications. (Adapted from *Materials and Structures (Paris)*, Vol. 8, No. 47, 1975, pp. 387–403.)

Often, however, no easily assignable reasons for the noncompliance of the concrete can be found, and even if they are found there is still the problem of assessing the quality of the noncomplying concrete already placed. Thus, at this stage it becomes necessary to try to evaluate the quality of the concrete in the structure. Most important, we want to know the load-carrying capacity of the structure. It would also be desirable to find out why the concrete did not comply with the specification (if this has not already been ascertained), so that the costs of repairing the damage can be properly assigned.

For these purposes, *in situ* tests as described in Chapter 16, such as coring, hardness tests, and so on, can be used. Almost always, the various building codes require that some cores be taken from the structure. Sometimes that portion of the concrete represented by the low test results can be located within the structure; otherwise, problem areas can often be found by using either hardness tests or ultrasonic tests. In any event, sufficient cores should be taken to give statistically valid results. The ACI and CSA requirement is that at least three cores should be taken for each case of a cylinder test more than 3.5 MPa (500 lb/in.²) below σ_c'. The concrete is considered to be adequate if the average of three cores is at least $0.85\,\sigma_c'$ and no single core is less than $0.75\sigma_c'$. Since core tests are not necessarily a reliable indicator of concrete quality, they should usually be supplemented with either hardness tests, ultrasonic tests, or both. If the strength tests on cores meet the requirements mentioned above, the concrete is considered to be acceptable. However, this specification is itself open to question. It is not clear that a correlation exists between structural behavior and core strengths of $0.85\sigma_c'$, particularly if we consider concrete durability as well as strength (see Section 16.5, Core Tests). So, marginally acceptable core strengths should be treated with considerable caution.

Structural Safety

At this stage, if the concrete is still judged to be inadequate, a further investigation is required to assess both the safety and the durability of the sturcture. This will generally involve a determination not only of the concrete strength (from core tests), but also of the location and size of the steel reinforcing. The latter information may involve removal of the concrete to expose the steel, or more generally will involve the use of magnetic indicators which are available for this purpose. Once the concrete strength and the exact details of the steel reinforcement are known, the load-carrying capacity of the structure can be determined analytically, with a much higher degree of certainty than was possible in the original design. This analysis can then be used to assess the structure; it may be suitable for the intended purpose, or it may be possible to reclassify the structure for a lower level of live loads. If neither of these alternatives ensures the safety of the structure, it must be rejected, unless satisfactory repairs and strengthening of the structural elements can be carried out. Even if structural safety is assured, it may be necessary to consider repairs to improve the durability of the concrete.

If the analytical assessment of the structure is found to be impractical, it is also possible to carry out a static load test on the structure or structural element in question. Since this is often a costly and hazardous procedure, which is also open to interpretation, it is recommended only as a last resort. As with the analytical method, load tests will generally not provide information on the durability of the concrete, which must be assessed separately.

The procedures discussed above are seldom followed in practice and are not properly codified in North America. Investigations tend to be costly and time-consuming, and because of their inherent uncertainties, one is still left in the end to rely on "engineering judgment." However, given the fact that the principles of producing quality concrete are by now well known, these cases do occur more frequently than one would like to admit. One unfortunate consequence is that there is a tendency for some testing agencies to try to rationalize the discarding of low test results by saying that those results are "obviously nonrepresentative" for one reason or another, and field inspectors will show a bias (often unconsciously) toward sampling only concrete that appears to be "good," to avoid the morass of claims and counterclaims resulting from the reporting of low test results.

Although we can never completely eliminate the production of poor concrete, it is nonetheless clear that the present acceptance procedures are inadequate, because they are based largely on 28-day strengths, as discussed in detail in Chapter 16. Much of the trauma of finding poor concrete would be eliminated if more rapid methods of assessing concrete quality were widely adopted. For this purpose, acceptance tests based on an analysis of the composition of fresh concrete are most promising, since they would provide an answer almost immediately. A wider use of accelerated tests for hardened concrete would also be helpful.

Bibliography

ACI COMMITTEE 217, "Practice for Evaluation of Concrete in Existing Massive Structures for Service Conditions," *Concrete International: Design and Construction,* Vol. 1, No. 3, pp. 47–61 (1979).

Building Code Requirements for Reinforced Concrete, ACI 318-71 (rev. 1978). American Concrete Institute, Detroit, Mich., 1978

Code of Practice for the Structural Use of Concrete CP110: Part 1. British Standards Institution, London (1972).

Code for Ready Mixed Concrete. British Ready Mixed Concrete Association, Shepperton (1975).

MALHOTRA, V. M., "Contract Strength Requirements—Cores versus *in situ* Evaluation," *Journal of the American Concrete Institute,* Vol. 74, No. 4, pp. 163–172 (1977).

Realism in the Application of ACI Standard 214-65, SP-37. American Concrete Institute, Detroit, Mich., 1973.

Recommended Practice for Evaluation of Strength Test Results for Concrete, ACI 214-77. American Concrete Institute, Detroit, Mich., 1976.

"Recommended Principles for the Control of Quality and the Judgment of Acceptability of Concrete," *Materials and Structures (Paris),* Vol. 8, No. 47, pp. 387–403 (1975).

TEYCHENNÉ, D. C., "Recommendations for the Treatment of the Variations of Concrete Strength in Codes of Practice," *Materials and Structures (Paris)* Vol. 6, No. 34, pp. 259–268 (1973).

Problems

17.1 Why is concrete an inherently variable material?

17.2 As the project engineer on a large construction site, what remedial action would you take if, on average, 1 in 8 test cylinders fail at less than the specified strength?

17.3 If the specified design strength is 4500 lb/in.^{2}, to what average strength should the concrete be designed in order that only 1 in 10 test cylinders will fall below the specified strength? The coefficient of variation is determined to be (a) 5%; (b) 10%; (c) 20%.

17.4 As an engineer responsible for the concrete specifications for a structure to be built in a remote area, where there is no experience with local variables, how would you estimate the expected variability of the concrete?

17.5 What strength should be specified in the mix design of a concrete that must attain 40 MPa in 95% of test cylinders given a known standard deviation of 4 MPa?

17.6 How is concrete durability taken into account in concrete quality control?

18

deformation

Deformations of concrete under service conditions arise from a number of different stimuli: applied stress, change of moisture content, and changes in temperature, for example. The response of the concrete to these stimuli is complex, resulting in reversible, irreversible, and time-dependent deformations. This chapter deals with three main types of deformations:

1. Instantaneous deformations that occur when an external stress is first applied.

2. Shrinkage, which occurs on loss of moisture from the concrete.

3. Creep, which is the time-dependent deformation that occurs on the prolonged application of stress.

In each case, strains are of the same order of magnitude, so each type of deformation must be taken into account. Strains due to thermal changes are considered in Chapter 19.

18.1 ELASTIC STRAINS

Concrete is not a truly elastic material. The nonlinearity of the σ–ϵ curve (Figure 18.1) is due to an irreversible response of the concrete even after short loading times, which can be measured when the concrete is unloaded. Thus, the term "instantaneous" strain rather than "elastic" strain is probably more appropriate.

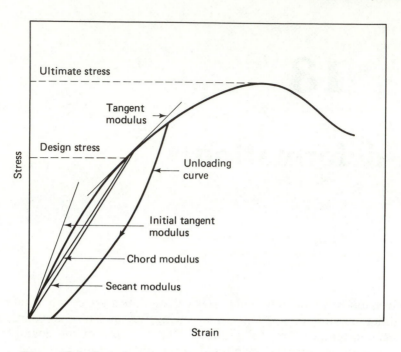

Figure 18.1 Typical stress–strain diagram for concrete, showing the different elastic moduli.

Determination of Modulus of Elasticity

Measurement of Modulus

There is an obvious difficulty in determining the modulus of elasticity (Young's modulus) from a nonlinear stress–strain diagram. The closest approximation to a modulus of elasticity derived from a truly elastic response is the *initial tangent modulus,* which is the slope of the tangent to the curve at the origin (Figure 18.1). This modulus is of little practical significance, since it applies only to small stresses and strains. It is likely that the dynamic modulus of elasticity, determined from ultrasonic measurements (Chapter 16), may be a reasonable estimate of the initial tangent modulus, since its determination involves only very small displacements of the material. A more practical measure of the modulus of elasticity is the *secant modulus.* All strain occurring at a given stress is regarded as elastic, and thus the slope of the secant between the origin and any point on the curve will give the secant modulus for the corresponding applied stress. The secant modulus thus includes an element of nonlinearity, and clearly its value depends on the level of applied stress chosen.

The initial tangent or the secant moduli are not always easily determined in compression. There is often observed a small element of strain on initial loading, caused by preexisting cracks closing up under the applied load, which gives rise to a slight concavity at the beginning of the stress–strain curve. In such a case the *chord* modulus (ASTM C469, Chapter 16) is used, which is the slope of the line drawn between two points on the stress–strain curve. The initial tangent modulus would correspond to the slope of the curve at a strain of 0.00005. The lower strain limit is chosen to avoid any error caused by the influence of cracks on the initial portion of the stress–strain curve. The chord modulus and the secant modulus determined at 40% of the ultimate strength will not be very different, although the former will be more precise. The chord modulus is a more conservative measure than the initial tangent modulus and more easily determined experimentally. On the other hand, the chord modulus will underestimate the additional strain that occurs when an additional stress is imposed on the concrete. The tangent modulus measured at the point of interest is a better measure of the concrete response to relatively small additional stresses.

Prediction of Modulus

The modulus of elasticity used in concrete design is seldom determined by direct test but is generally estimated from an empirical relationship between modulus and strength. In the ACI Building Code the relationship used is

$$E_c = \begin{cases} 33\rho^{3/2}(\sigma_c')^{1/2} & \text{lb/in.}^2 \\ 0.043\rho^{3/2}(\sigma_c')^{1/2} \text{ MPa} \end{cases} \tag{18.1a}$$

where E_c is the secant modulus of elasticity (at about 40% of ultimate strength), ρ the unit weight of the concrete in kg/m³ (lb/ft³), and σ_c' the compressive strength of a standard 150×300 mm (6×12 in.) cylinder in MPa (lb/in.²). Assuming a density for normal-weight concrete of 2320 kg/m³ (145 lb/ft³), this reduces to

$$E_c = \begin{cases} 57,000\,(\sigma_c')^{1/2} & \text{lb/in.}^2 \\ 4.730\,(\sigma_c')^{1/2} & \text{GPa} \end{cases} \tag{18.1b}$$

In the British Code of Practice (CP110, Part 1) both the static and dynamic moduli of elasticity are related to the compressive cube strength of the concrete as shown in Table 18.1 for concretes having a density of

Table 18.1

Relationship between Compressive Strength and Modulus of Elasticity

Compressive Strength (MPa)	Static Modulus (GPa)	Dynamic Modulus (GPa)	Static Modulus according to CEB (GPa)
20	25	35	29
25	26	36	30.5
30	28	38	32
40	31	40	35
50	34	42	37
60	36	44	39

2300 kg/m³ or more. These values will generally be correct to within ±4 GPa. The values for the static modulus of elasticity given in Table 18.1 may be very closely estimated from the relationship

$$E = 9.1\rho(\sigma_c)^{1/3} \qquad \text{GPa} \tag{18.2a}$$

For the lightweight aggregate concretes having densities between 1400 and 2300 kg/m³, the value of the *static modulus* given in Table 18.1 should be multiplied by $(\rho/2300)^2$.

For normal-weight concrete, CEB (the European Concrete Committee) uses the expression

$$E = 9.5(\sigma_c + 8)^{1/3} \qquad \text{GPa} \tag{18.2b}$$

where σ_c is the *characteristic strength* and the number "8" is introduced to allow for the (assumed) difference between the mean compressive strength of the concrete and the specified characteristic strength. The values predicted by this equation for different *characteristic* strengths are also shown in Table 18.1. These values are considered to be accurate to within ±30%. For lightweight concrete the CEB recommends multiplying the values in Table 18.1 by $(\rho/2400)^2$, corresponding to the equation

$$E = 9.5(\rho/2400)^2(\sigma_c + 8)^{1/3} \qquad \text{GPa} \tag{18.2c}$$

Different but essentially similar relationships are used in other countries.

Various attempts have been made to describe the shape of the stress–strain diagram analytically in terms of σ and ϵ. However, there is no standard curve, since its shape depends on other properties of concrete and the mode of testing, and therefore the usefulness of such expressions would seem to be limited, at best. A qualitative description of the stress–strain curve is given in Chapter 14.

Factors Affecting the Modulus of Elasticity

We have seen that the modulus of elasticity is related to compressive strength and density and can be calculated from the empirical equations (18.1) and (18.2). Thus, factors that affect strength should similarly influence modulus, and this is by and large true. The dominant parameter is, of course, the porosity, and the modulus will decrease markedly as the w/c ratio is increased. It should be remembered that since the porosity changes that occur with w/c are wholly within the paste, this change in stress–strain behavior is due to changes in paste response.

The modulus of elasticity of cement, E, has been found to be proportional to approximately the cube of the gel/space ratio. In terms of the capillary porosity, the relationship can be written as

$$E = E_g(1 - P_c)^3 \qquad (18.3)$$

where E_g is the modulus of elasticity of the hardened cement paste at zero porosity, and P_c is the capillary porosity expressed as a ratio. A similar relationship holds for the shear modulus, G (Figure 18.2). For ordinary portland cement, E_g appears to be about 32 GPa (4.5×10^6 lb/in.^{2}), while G_g is about 12 GPa (1.75×10^6 lb/in.^{2}).

One apparent inconsistency in the strength–modulus relationship is the moisture dependency. The strength of saturated concrete is lower than that of dry concrete, while for elastic modulus the reverse is true.

The modulus of elasticity is much more sensitive to the amount and nature of the aggregate (Figure 18.3). Since the value of the elastic modulus is partly dependent on progressive microcracking at the paste–aggregate interface, the shape, texture, and total amount of aggregate will influence its value. For example, concrete made with crushed stone will have a lower modulus than will concrete made with rounded gravel, as will concrete made with a lower maximum aggregate size. The modulus of elasticity of the aggregate affects the modulus of concrete as discussed in Chapter 13.

The much lower elastic modulus of porous, lightweight aggregates, such as expanded clay or shale, has a strong influence on the elastic modulus of concrete. The elastic modulus of lightweight concrete is 40 to 80% that of normal-weight concrete (see Table 18.2). The pattern of

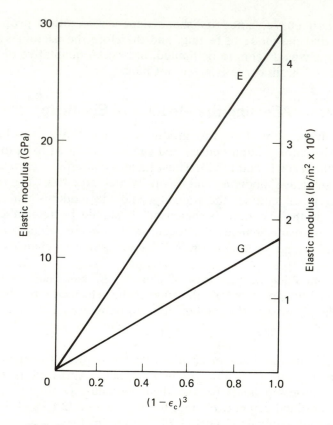

Figure 18.2 Dependency of elastic moduli on the porosity of cement paste. (From G. J. Verbeck and R. A. Helmuth, *Proceedings, Fifth International Symposium on the Chemistry of Cement, Tokyo, 1968,* Vol. 3, pp. 1–32.)

microcracking will be quite different and the shape of the stress–strain curve will change. Stress concentrations at the interface will now be much less, so that the σ–ϵ behavior is nearly linear over a much wider stress range. Thus, the influence of aggregate shape, texture, and amount is much less than for normal-weight concrete. Equation (18.1a) can also be used for lightweight concrete. High-strength concrete (see Chapter 21) also has a more linear σ–ϵ curve, but a higher modulus of elasticity (about 40 GPa, or 6×10^6 lb/in.²).

The stress–strain curve, and hence the modulus of elasticity, is dependent on the conditions of testing. An important parameter is the rate of application of load (Figure 18.4). If the stress is applied almost

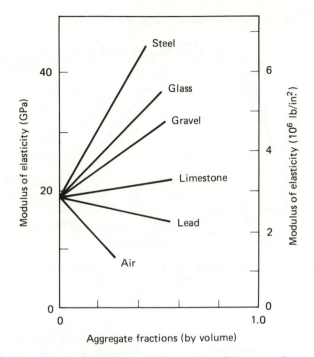

Figure 18.3 Effect of aggregates on the modulus of elasticity of concrete. (Adapted from O. Ishai, *Journal of the American Concrete Institute,* Vol. 59, No. 9, 1961, pp. 1365–1368.)

instantaneously (a few tenths of a second, say) then little microcracking can occur prior to failure, and the stress–strain curve is more nearly linear. Increasing the loading time up to 2 min will increase the strain by 15 to 20%. Fortunately, in the time range normally required to test a specimen (2 to 10 min), the effect of loading time is very small. Further

Table 18.2

Moduli of Elasticity for Concrete and Its Components, GPa (lb/in.^{2})

	Normal-Weight	*Lightweight*
Aggregate	70–140 (10–20 × 10^6)	14–35 (2–5 × 10^6)
Cement paste	7–28 (1–4 × 10^6)	7–28 (1–4 × 10^6)
Concrete	14–42 (2–6 × 10^6)	10–18 (1.5–2.5 × 10^6)

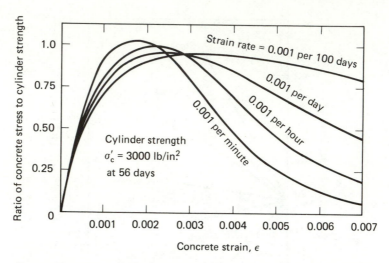

Figure 18.4 Stress–strain curves for various strain rates of concentric loading. (Adapted from H. Rusch, *Journal of the American Concrete Institute*, Vol. 57, No. 1, 1960, pp. 1–23.)

increases in strain occur when the loads are applied very slowly and can be treated as superimposed creep. Although there is strictly a creep contribution to "elastic strains" as measured in a compression test, it can be ignored for all practical purposes.

Dynamic Moduli

In determining the dynamic modulus of elasticity, the concrete is subjected to very small strains which will be almost entirely elastic. Therefore, the dynamic elastic modulus is more closely related to the initial tangent modulus than to the chord modulus, and is about 20 to 30% higher than its static counterpart, depending on the strength of the concrete. The dynamic elastic modulus is used primarily to evaluate soundness of concrete in durability tests or in *in situ* evaluations. However, it may be a more appropriate value to use when the concrete is to be used in structures that will be subjected to dynamic loadings (i.e., earthquake or impact).

The shear modulus (modulus of rigidity, G), which governs behavior under shear stresses, is not easily determined from direct stress–strain measurements. It is probably best determined by dynamic methods using the resonant frequency of torsional vibration of a concrete specimen (see Chapter 16).

Poisson's ratio, μ, can be conveniently determined by direct strain measurements in uniaxial compression (ASTM C469), or it can be determined dynamically using Eq. (16.12). Poisson's ratio for saturated concrete lies in the range 0.2 to 0.3, but decreases to about 0.18 on drying. This change is not completely reversible on resaturation. For different concretes, Poisson's ratio generally falls in the range 0.15 to 0.20, but the variation with different concrete properties is not known precisely. When determined dynamically, the value is somewhat higher, averaging about 0.24, and this is probably more representative of elastic behavior.

The bulk modulus (K) of cement has a range of values from about 3.5 GPa (0.5×10^6 lb/in.²) at high porosities to about 20 GPa (3×10^6 lb/in.²) for very low porosities.

18.2 SHRINKAGE

Volume changes accompany the loss of moisture from either fresh or hardened concrete. However, the term "drying shrinkage" is generally reserved for hardened concrete, while "plastic shrinkage" is used for fresh concrete, since its response to loss of moisture is quite different. "Carbonation shrinkage," which occurs when hydrated cement reacts with atmospheric carbon dioxide, can be regarded as a special case of drying shrinkage. "Autogenous shrinkage," which occurs when a concrete can self-desiccate during hydration, is also a special case of drying shrinkage. Shrinkage is a paste property; in concrete the aggregate has a restraining influence on the volume changes that will take place within the paste.

Plastic Shrinkage

Loss of water from fresh concrete, if not prevented, can cause cracking. The most common situation is surface cracking due to evaporation of water from the surface, but suction of water from the concrete by the sub-base or by formwork materials can cause cracking also or can aggravate the effects of surface evaporation. In fresh concrete the space between the particles is completely filled with water. When water is removed from the paste by exterior influences, such as evaporation at the surface, a complex series of menisci are formed. These, in turn, generate negative capillary pressures which will cause the volume of the paste to contract. Capillary pressures continue to rise within the paste until a critical "breakthrough" pressure (P_c) is reached, at which point the water is no longer evenly dispersed through the paste and rearranges to form discrete zones of water with voids between. The maximum rate of plastic shrinkage occurs just prior to the "breakthrough" pressure, and little shrinkage occurs afterward.

The effect of aggregate on plastic shrinkage can be seen in Figure 18.5. Plastic shrinkage can in theory be beneficial by causing compaction of the paste; however, in practice, the effects of plastic shrinkage are not uniform throughout the mass, and differential volume changes can cause cracking under induced tensile stresses. Plastic shrinkage cracking (Figure 18.6) is most common on horizontal surfaces of pavements and slabs where rapid evaporation is possible, and its occurrence will destroy the integrity of the surface and reduce its durability. It is aggravated by a combination of high wind velocity, low relative humidity, high air temperature, and high concrete temperatures. These conditions are most prevalent during the summer months, but can occur at any time. If the rate of surface evaporation exceeds 0.5 kg/m²/h (0.1 lb/ft²/h), loss of moisture may exceed the rate at which bleed water reaches the surfaces and creates the negative capillary pressures which cause plastic shrink-

Figure 18.5 Effect of aggregate on plastic shrinkage. (From R. G. L'Hermite, *Proceedings, Fourth International Symposium on the Chemistry of Cement, Washington, D.C., 1960,* Vol. 2, pp. 659–694.)

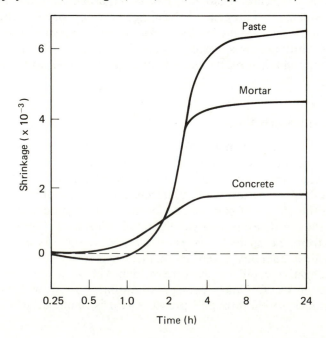

5 cm

Figure 18.6 Plastic shrinkage cracking. Core taken from pavement. (Photograph courtesy of Bryant Mather.)

age. Precautionary measures are thus desirable and should always be used if evaporation rates exceed 1.0 kg/m²/h (0.2 lb/ft²/h). Charts are available to estimate evaporation rates under given environmental conditions (Figure 18.7). Plastic shrinkage can be controlled by reducing wind velocities with windbreaks, by reducing the temperature of the concrete, or by increasing the rate of setting of the concrete (if construction practices allow it). By far the most effective method of control is to ensure that the concrete surface is kept wet until the surface has been finished and routine curing begun. Temporary wet coverings, waterproof sheeting, or a fog spray are appropriate methods.

Autogenous Shrinkage

If no additional water beyond that added during mixing is provided, it is possible that the concrete will begin to dry out even if no moisture is lost to the surroundings. This can happen in concretes with a low *w/c* ratio (theoretically below 0.42) and is due to the internal consumption of water during hydration. The phenomenon is known as *self-desiccation* and leads to autogenous shrinkage. This cause of shrinkage is relatively rare (except perhaps in mass concrete) and can be treated as a special case of drying shrinkage, since it is immaterial whether the water is removed by physical or chemical processes. Conversely, if the concrete

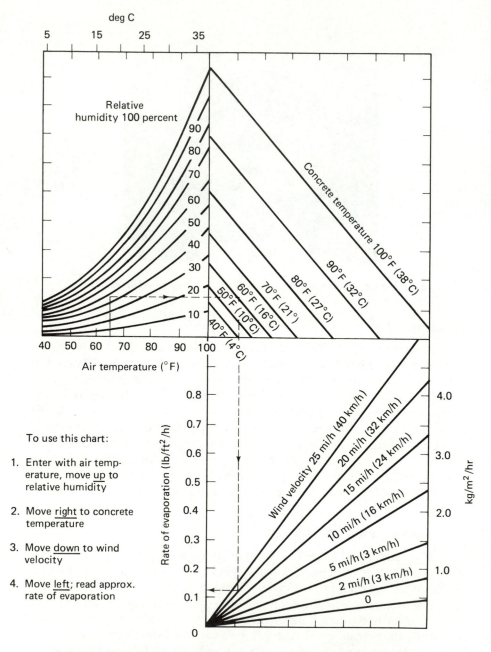

deg C

Relative humidity 100 percent

Concrete temperature 100°F (38°C)

90°F (32°C)

80°F (27°C)

70°F (21°)

60°F (16°C)

50°F (10°C)

40°F (4°C)

Air temperature (°F)

To use this chart:

1. Enter with air temperature, move <u>up</u> to relative humidity

2. Move <u>right</u> to concrete temperature

3. Move <u>down</u> to wind velocity

4. Move <u>left</u>; read approx. rate of evaporation

Rate of evaporation (lb/ft²/h)

kg/m²/hr

Wind velocity 25 mi/h (40 km/h)

20 mi/h (32 km/h)

15 mi/h (24 km/h)

10 mi/h (16 km/h)

5 mi/h (3 km/h)

2 mi/h (3 km/h)

0

Figure 18.7 Chart to calculate the rate of evaporation of water from freshly placed concrete. (From *Engineering Bulletin*, 11th ed., Portland Cement Association, Skokie, Ill., 1968.)

is continuously cured under water from the time of casting, a very slight expansion may be observed due to the formation of ettringite, or hydration of free MgO.

Drying Shrinkage

Drying shrinkage of hardened concrete is a much more important phenomenon than those already described. Inadequate allowance for the effects of drying shrinkage in concrete design and construction can lead to cracking or warping of elements of the structure due to restraints present during shrinkage. The most obvious example is the necessity of providing shrinkage control joints in pavements and slabs. These joints prevent random, irregular shrinkage cracking and confine it to a desired path in a form in which the crack can readily be filled with a sealant to prevent the entry of foreign materials. The response of the paste to moisture loss is modified by the presence of aggregate and the shape of the concrete member. Thus, in order to understand precisely how different experimental parameters affect the drying shrinkage of concrete a study of hardened cement paste itself is appropriate. The factors that affect shrinkage are listed in Table 18.3.

Table 18.3

Parameters Affecting Drying Shrinkage and Creep of Concrete

Paste parameters
    Porosity   } w/c ratio and degree of hydration
    Age of paste
    Curing temperature
    Cement composition
    Moisture content
    Admixtures
Concrete parameters
    Aggregate stiffness
    Aggregate content (cement content)
    Volume-to-surface ratio
    Thickness
Environmental parameters
    Applied stress   } affects only creep
    Duration of load
    Relative humidity
    Rate of drying
    Time of drying

Behavior of Cement Paste

The results discussed in this section are based on laboratory work using cement paste specimens with relatively small cross sections. If such specimens are not used, the rate of diffusion of water from the specimen to the surroundings makes the attainment of equilibrium states difficult to achieve and complicates the interpretation of data. It should be emphasized, however, that the fundamental processes underlying drying shrinkage are not yet fully understood.

One important aspect concerning the drying shrinkage of cement paste, and hence of concrete, is the fact that part of the total shrinkage that occurs on the first drying is irreversible (see Figure 18.8). Thus, the subsequent volume expansions that occur on rewetting and the volume contractions that occur on later drying are smaller. Experimental parameters generally have some effect on the amount of irreversible shrinkage as well as on total shrinkage. This improvement in dimensional stability after first drying can be used to advantage in precast concrete products.

Since moisture loss is the underlying cause of drying shrinkage, the relationship between the two should be of interest. Figure 18.9 shows a typical shrinkage–weight loss curve for cement paste; five approximately linear domains are observed. Domains (1) and (2) have been attributed to loss of water from capillary pores, domain (3) represents loss of adsorbed water from the surfaces of C–S–H particles, domain (4) results from loss of water that contributes to the structure of C–S–H, and domain (5) is due to decomposition of C–S–H. The closely related plot of shrinkage versus relative humidity (Figure 18.10) emphasizes the differences between the various domains. Since in most civil engineering

Figure 18.8 Typical behavior of concrete on drying and rewetting.

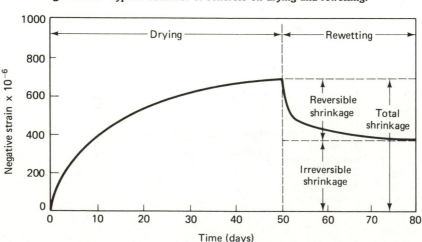

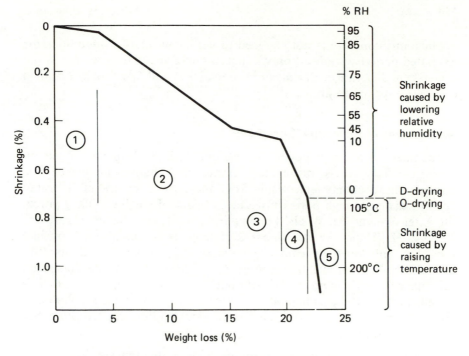

Figure 18.9 Shrinkage – moisture loss relationships in pure cement pastes during drying.

Figure 18.10 Shrinkage of cement paste as a function of relative humidity.

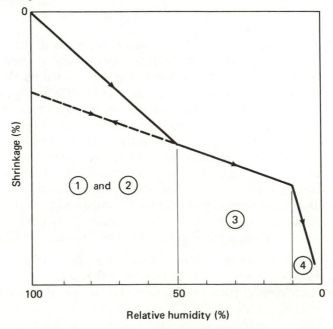

applications concrete is not exposed to very low relative humidities for extended periods, we need only concern ourselves with behavior in the upper humidity range. Shrinkage in domains 4 and 5 will only occur in concrete exposed to fire.

Mechanisms of Reversible Shrinkage

Three phenomena are believed to contribute to bulk shrinkage of cement paste: capillary stress, disjoining pressure, and changes in surface free energy. These phenomena result from the special nature of hydrated cement paste: its high porosity with a network of small capillary pores, the extensive van der Waals' bonding in C–S–H, and the small particles of C–S–H. Thus, the net linear shrinkage is a function of internal pressures developed by capillary stresses P_{cap}, disjoining effects P_{dis}, and changes in surface free energy P_{sfe}. These effects are illustrated schematically in Figure 18.11. Although it is linear shrinkage that is usually measured experimentally, we are actually explaining a volumetric contraction. P_{cap}, P_{dis}, and P_{sfe} are dependent on both relative humidity and temperature.

Capillary stress. The development of hydrostatic tension induces an isotropic compressive stress within the rigid solid skeleton. Hydrostatic tension is only developed when a meniscus is formed in a capillary, the stress being given by

$$P_{cap} = \frac{2\gamma}{r} \qquad (18.4)$$

where γ is the surface tension of water and r the radius of the meniscus. [Equation (18.4) is derived for a cylindrical capillary and is more complex for more complex shapes.] The value of r is determined by the relative humidity. The difference between domains (1) and (2) in Figure 18.9 is thus simply explained by the fact that the larger capillaries, which are emptied at relative humidities down to about 95%, have a relatively large volume-to-surface ratio and develop rather small stresses. At lower humidities, although the capillary volume becomes much smaller, stresses rise quite rapidly. Capillary stresses cannot exist below about 40 to 45% RH since menisci are no longer stable. The removal of hydrostatic stress at this relative humidity should be accompanied by a relaxation in the solid and an *increase* in length. This is occasionally observed, but is usually obscured by irreversible shrinkage that continues to increase.

Disjoining pressure. Water is adsorbed on the surfaces of C–S–H at all relative humidities, the thickness increasing with increasing humidity. An assembly of colloidal particles, as in C–S–H, has van der Waals' forces

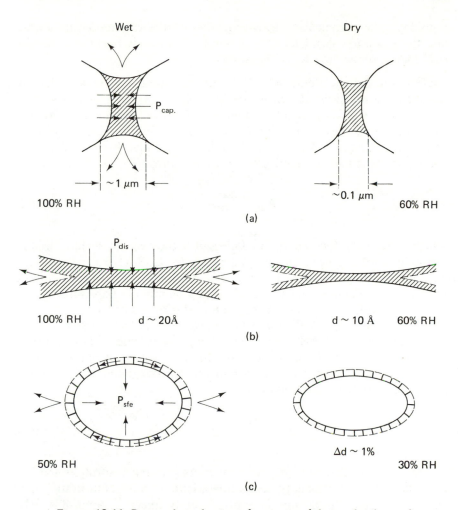

Figure 18.11 Proposed mechanisms for causes of drying shrinkage of cement paste: (a) capillary stress; (b) disjoining pressure; (c) surface tension.

attracting adjacent particles and bringing their adjacent surfaces in close contact. Adsorption of water on the C–S–H surfaces creates a disjoining pressure (which results from the orientation of water molecules in the adsorbed film). The disjoining pressure increases with the increasing thickness of the adsorbed water (i.e., increasing relative humidity), and when the disjoining pressure exceeds the van der Waals' attractions, the particles will be forced apart, creating a dilation. Conversely, under a decreasing disjoining pressure (which accompanies a lowered relative humidity), the particles are drawn together by the van der Waals' forces and there is a net shrinkage. Disjoining pressure only becomes a significant factor above 50% RH (Chapter 4).

Surface free energy. Below 40% RH when no capillary stress or disjoining pressure are present, shrinkage is explained by changes in surface energy. It is well known that a liquid drop is under hydrostatic pressure by virtue of its surface tension (surface free energy). This pressure is described by Eq. (18.4) because the drop is bounded by a meniscus. A solid particle is likewise subjected to a mean presure:

$$P_{sfe} = \frac{2\gamma S}{3} \tag{18.5}$$

where γ is the surface energy in J/m^2 and S the specific surface area of the solid in m^2/g. Since S is a large quantity for C–S–H, P_{sfe} can be large and causes compression in the solid. Adsorption of water on the surface decreases the value γ, and hence of P_{sfe}, by an amount depending on the relative humidity. The change is greatest during adsorption of the first monolayer (up to 20% RH) and is negligible after the second layer is complete above 50% RH. The increase in solid volume with decrease in γ has been observed for other materials and was first determined in experiments on powdered coal conducted by Bangham. It can be described by the Bangham equation:

$$\frac{\Delta \ell}{\ell} = k\Delta\gamma \tag{18.6}$$

where $\Delta\gamma$ is the change in surface free energy during adsorption and k is a constant of proportionality. $\Delta\gamma$ is dependent on ln RH, but in the range 20 to 40 % RH, the change is approximately linear. Above 40% RH, changes in γ can be neglected.

Irreversible Shrinkage

It can be seen from Figure 18.8 that a considerable part of the observed length change on first drying is irreversible, although the mechanisms described above would predict reversible behavior. Figure 18.10 shows that irreversibility occurs only on first drying and that deformations on subsequent wetting and drying cycles are essentially reversible. It has been demonstrated that increases in drying shrinkage that attend the use of some admixtures are due to increases in the irreversible component on first drying and not the reversible portion. Further, Figure 18.12

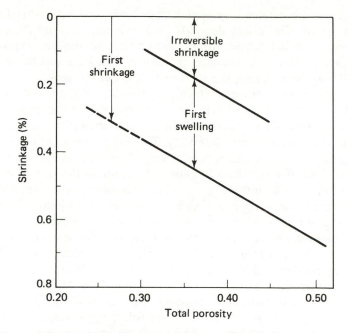

Figure 18.12 Effect of porosity on irreversible shrinkage of cement paste. (From R. A. Helmuth and D. H. Turk, *Journal of the Portland Cement Association, R & D Labs,* Vol. 9, 1967, pp. 8–21.)

shows that total shrinkage on first drying increases as the porosity of the paste increases. If swelling on rewetting is measured, it is found that only the irreversible component is affected by porosity. Since the w/c ratio and degree of hydration control porosity, these parameters also affect irreversible shrinkage. These facts can be explained in part by changes in the distribution of pores within the paste. On first drying the permeability of the paste increases nearly twofold, and this has been taken as evidence that a continuous network of capillaries is formed. The development of capillary stress is then much reduced. Stresses are greatest in the smallest capillaries, and it has been shown that admixtures which increase drying shrinkage have a greater fraction of pores in the diameter range 3 to 10 nm. On drying, the volume of pores in this size range is reduced.

The temperature of curing also influences the drying shrinkage of cement pastes. If pastes are exposed to elevated temperatures during moist curing, irreversible shrinkage is reduced but reversible shrinkage is unaffected. The decrease in shrinkage depends on the maximum temperature to which the paste is exposed; at 65°C (150°F) irreversible shrinkage can be reduced by about two-thirds and total shrinkage by about one-third. The effect depends on the length of time the paste is maintained at higher temperatures. However, the time of exposure at high temperatures that is required to reduce shrinkage can be relatively short and may be less than the total specified curing time.

Irreversibility is strongly affected by the drying history of the paste. If the paste is dried quite slowly, by being conditioned at progressively lower relative humidities, total shrinkage is less than if the paste is dried directly to the lowest relative humidity. This is an important observation, since drying of concrete can be a relatively slow process in thick sections where diffusion of water from the concrete controls the rate of moisture loss. The time at which the paste is held at a low relative humidity also has a strong influence on irreversible shrinkage (see Figure 18.13). The slow, continued increase in total shrinkage that has been observed after all of the water has effectively been removed from the specimen is almost all due to an increase in irreversible shrinkage (see Figure 18.14).

We do not know the exact causes of irreversible changes in cement paste. However, it is a widely held view that C–S–H, as it is formed during cement hydration, is an unstable material and that irreversibility can be the result of one or more of the following changes: changes in pore-size distribution, changes in bonding between C–S–H particles, and permanent rearrangements in the packing of C–S–H particles due to changes in the distribution of water within C–S–H. It has been suggested that changes in the *w/c* ratio, or the use of admixtures, can change the arrangement of C–S–H particles and hence the bonding between them,

Figure 18.13 Effect of drying time on irreversible shrinkage of cement paste. (Based on data by R. A. Helmuth and D. H. Turk, *Journal of the Portland Cement Association, R & D Labs,* Vol. 9, 1967, pp. 8–21.)

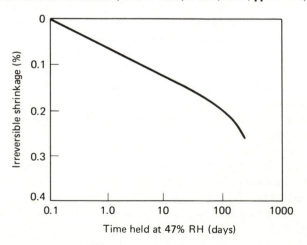

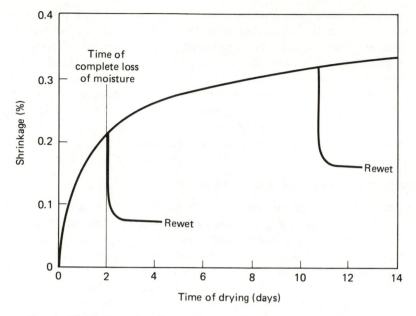

Figure 18.14 Increase of total and irreversible shrinkage of cement paste.

as well as changes in pore size distribution. Exposure to high-temperature curing apparently increases the amount of chemical bonding between C–S–H particles. It is useful to think of an aging process, which is time- and temperature-dependent and which stabilizes the C–S–H so that it is less deformed on loss of moisture.

Feldman and Sereda have postulated that the interlayer regions of C–S–H play an important role in shrinkage, particularly irreversible shrinkage. They consider that loss of water from between the layers of C–S–H has a dominant effect on shrinkage, especially at low relative humidities. This cause is thought to be much greater than the effects of changing surface free energy or disjoining pressure. Regain of interlayer water on rewetting occurs in a different way from removal, and this is responsible for some irreversibility. However, much of the irreversibility occurs by the creation of new interlayer regions between C–S–H particles during first drying and rewetting.

Cement Composition

The composition of the cement can influence the shrinkage of the paste, and hence of the concrete, although the effect is not large. Exact relationships between composition and shrinkage are not well known.

Correlations have been found between shrinkage and C_3A content, from which it might be concluded that the sulfoaluminates contribute to shrinkage. There is also an "optimum gypsum content" for minimum shrinkage for each cement. It has recently been observed that the shrinkage of pastes of pure C_3S and C_2S is less than that of a portland cement paste. Since the bulk of the shrinkage is believed to originate in the C–S–H and its associated porosity, it would seem that in cement pastes the presence of minor components changes the properties of C–S–H. Calcium hydroxide apparently has little effect on shrinkage.

Behavior of Concrete: Effect of Aggregate

The drying shrinkage of concrete will be less than that of pure paste because of the restraining influence of aggregate. With rare exceptions aggregates are dimensionally stable under changing moisture conditions. The amount of restraint provided by aggregate depends on the amount of aggregate in the concrete (Figure 18.15), its stiffness (Figure 18.16), and the maximum size of the coarse aggregate. The stresses at the cement paste–aggregate interface due to drying shrinkage increase as the maximum aggregate size increases. These higher internal stresses will increase the amount of cracking in the interfacial region. Lightweight

Figure 18.15 Influence of aggregate content on shrinkage of concrete.

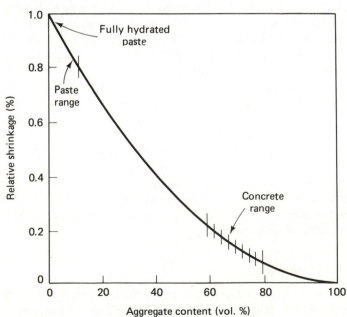

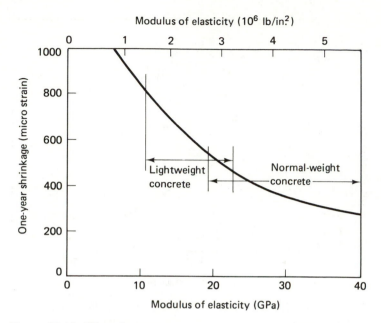

Figure 18.16 Effect of aggregate modulus on concrete shrinkage.

aggregates are generally dimensionally stable, but their low modulus of elasticity means that lightweight concretes can be expected to have higher shrinkages than normal-weight concretes. The combined effects of aggregate content and stiffness are embodied in the empirical equation (18.7):

$$\epsilon_{con} = \epsilon_p (1 - A)^n \qquad (18.7)$$

where ϵ_{con} and ϵ_p are the shrinkage strains of concrete and paste, respectively, A is the aggregate content, and n varies between 1.2 and 1.7.

Specimen geometry. The size and shape of a concrete specimen will determine the rate of moisture loss and hence the rate and magnitude of drying shrinkage. The length of the diffusion path has a strong influence on the rate of moisture loss. In mass concrete it will take approximately 1 month for the outer 75 mm of concrete to reach moisture equilibrium, 1 year for 225 mm, and 10 years for 600 mm (24 in.) to reach equilibrium. Thus, volume-to-surface area ratios will be important, and a T-beam section will dry more rapidly than will a solid beam of equal width and height (Figure 18.17), and exhibit slightly less ultimate shrinkage. The rate of moisture loss depends on total surface area and the average length

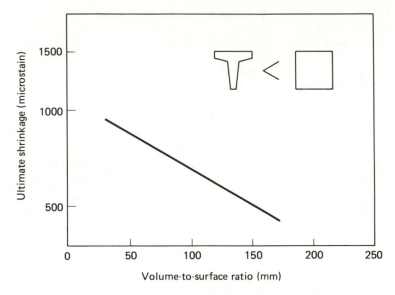

Figure 18.17 Effect of volume-to-surface ratio on the shrinkage of concrete.

of the diffusion path. The smaller the initial cross section of a member, the faster the initial rate of shrinkage, but the lower the magnitude of shrinkage at later times. Similarly, with reduced surface areas, the lower rates of early shrinkage extrapolate to large ultimate shrinkage. Thus, there is an inverse relationship between rate of early shrinkage and ultimate shrinkage. This is opposite to the effect observed for pastes. This apparent contradiction is, in part, due to the increased amount of hydration of the slower drying specimens, but is primarily caused by restrained shrinkage. During drying of a concrete specimen, the situation soon arises where a dry, contracting exterior surrounds a still moist core that has not begun to shrink (see Figure 18.18). Thus, tensile stresses will be set up in the outer part of the prism and microcracking is likely to occur unless the stress is relieved by tensile creep. Either creep or microcracking will act to reduce the observed shrinkage. Furthermore, the moist core will be subject to a compressive stress under which it will creep, and this may also affect the magnitude of later drying shrinkage.

Carbonation Shrinkage

Hardened cement paste will react chemically with carbon dioxide. The amount present in the atmosphere ($\sim$0.04%) is sufficient to cause considerable reaction with cement paste over a long period of time. This

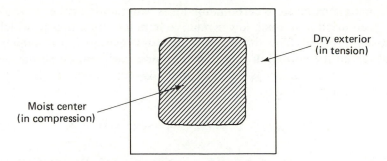

Figure 18.18 Restraint of shrinkage in a partially dried concrete member.

is accompanied by shrinkage, and hence it is called *carbonation shrinkage*. The extent to which cement paste can react with carbon dioxide, and hence undergo carbonation shrinkage, is a function of relative humidity (see Figure 18.19) and is greatest around 50% RH. At high humidities, carbonation is low because the pores are mostly filled with water and CO_2 cannot penetrate the paste very well. At very low humidities an absence of water films is believed to lower the rate of carbonation. Carbonation shrinkage is greatest when carbonation occurs after drying, rather than during drying, except at low humidities.

Figure 18.19 Shrinkage of mortar bars during carbonation. (From G. J. Verbeck, in *Papers on Cement and Concrete*, ASTM STP 205, 1958, pp. 17–36. Reprinted, with permission, from the American Society for Testing and Materials, 1916, Race Street, Philadelphia, PA 19103. Copyright.)

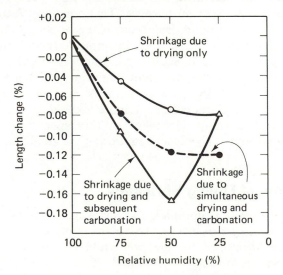

Concrete exposed to carbonation loses water and behaves as though it has been dried to a much lower relative humidity than that to which it is actually exposed. The shrinkage–water loss relationship is similar to that observed for normal drying. Furthermore, carbonation shrinkage is wholly irreversible and it is believed that CO_2 reacts with C–S–H with concomitant loss of water:

$$C-S-H + \overline{C} \rightarrow C-S-H + C\overline{C} + H \qquad (18.8)$$

The C/S ratio decreases during this reaction. Carbonation of C–S–H is known to change the bonding characteristics of the material, which could account for the irreversible nature of the accompanying shrinkage. Thus, carbonation can be viewed as promoting changes in C–S–H that normally only occur at much lower relative humidities. Calcium hydroxide will form calcium carbonate by reacting with atmospheric CO_2.

Carbonation shrinkage can be important from a practical point of view. Advantage can be taken of its irreversible nature for precast concrete. For example, by exposing concrete block to CO_2-rich air (to hasten the carbonation process), the block can be made much more dimensionally stable to subsequent wetting and drying. On the other hand, carbonation can be detrimental to cast-in-place concrete. Since it is much less porous than block, carbonation can occur only near the outside, precisely where the maximum rate of drying is also occurring. Thus, carbonation shrinkage is likely to be maximized and when added to drying shrinkage may cause severe shrinkage cracking. Concrete may be damaged by carbonation in the winter months when it is cured in enclosed areas heated with oil burners. This condition leads to excess plastic shrinkage due to rapid drying out if the relative humidity is not maintained, but higher levels of CO_2 in improperly vented enclosures may also cause carbonation. Such a condition can lead to crazing or dusting of the surface. In the long term, the slightly higher levels of CO_2 that can exist in enclosed parking garages may eventually lead to carbonation damage. Carbonation is more extensive the greater the surface area, and in small specimens or thin specimens carbonation shrinkage may become a significant fraction of the total shrinkage.

Prediction of Drying Shrinkage

If some allowance is not made for shrinkage of concrete in design, unacceptable cracking may occur. There are many empirical equations that have been developed for the prediction of shrinkage. The most common approaches are those of ACI and CEB.

ACI Method

ACI Committee 209 recommends a set of empirical equations that allow shrinkage to be estimated as a function of time of drying and relative humidity [Eqs. (18.9) and (18.10)]. The shrinkage, $(\epsilon_{sh})_t$, at any time t after age 7 days for moist cured concrete is given by

$$(\epsilon_{sh})_t = \frac{t}{35 + t}(\epsilon_{sh})_{ult} \qquad (18.9)$$

where $(\epsilon_{sh})_{ult}$ is the value of ultimate shrinkage for drying at 40% RH. The effect of relative humidity is given by a correction factor (C.F.):

above 80% RH $\qquad$ (C.F.)$_H$ = 3.00 − 0.03H $\qquad$ (18.10a)

below 80% RH $\qquad$ (C.F.)$_H$ = 1.40 − 0.01H $\qquad$ (18.10b)

where H is the relative humidity expressed as a percentage. Note that it is necessary to assume a value of $(\epsilon_{sh})_{ult}$. Values for concretes will fall mostly in the range 415 to 1070 × 10^{-6}, and ACI suggests a figure of 730 × 10^{-6} be used where no data are available, but recommends that specific values be used wherever possible. This recommendation is obviously sensible since shrinkage depends on the w/c ratio, degree of hydration, presence of admixtures, and cement content. There is also the problem of determining the magnitude of $(\epsilon_{sh})_{ult}$, since it must be extrapolated from short-term test data. Further, ASTM C157, used to determine drying shrinkage, is an arbitrary test using one specimen size (25 × 25 × 285 mm or 1 × 1 × 11¼ in.) under one set of drying conditions, 23°C, 50% RH. We have seen that specimen size and the drying regime affect the ultimate shrinkage. Nevertheless, some correction for shrinkage, even if not accurate, is better than no correction at all. The ACI equations also do not take into account the irreversible shrinkage on first drying, but if this irreversible shrinkage can be safely accommodated, the strains that occur on moisture cycling will be greatly reduced.

CEB –FIP Method

In Europe the shrinkage that develops in an interval of time $(t - t_o)$ is estimated by

$$\epsilon_s(t, t_o) = \epsilon_{so}[\beta_s(t) - \beta_s(t_o)] \qquad (18.11)$$

where $\epsilon_s(t, t_o)$ is the shrinkage strain at time t after drying commences at time t_o; ϵ_{so} is the basic shrinkage coefficient, which depends on relative humidity and specimen dimensions; and β_s is a function corresponding to the change of shrinkage with time, which also depends on specimen dimensions. Rather than using a direct measure of time, corrections are made for temperature changes during curing as follows:

$$t = \frac{1}{30} \Sigma \left[(T(t_m) + 10) \, \Delta t_m \right] \qquad (18.12)$$

where Δt_m is the number of days when the mean daily temperature has assumed a value of $T°C$. Thus, Equation (18.12) is a measure of maturity. ϵ_{so} and β_s are determined using standard tables or graphs.

18.3 CREEP OF CONCRETE

Although creep is observed for all materials, creep of fired ceramics and metals is negligible at room temperature. The fundamental origins for creep of concrete must be quite different, since significant volume changes occur at ambient temperatures and the presence of moisture in the material plays a major role. This discussion will concentrate on the compressive creep strains that occur under axial loading. Flexural, tensile, and torsional creep will be considered more briefly.

Relationship between Creep and Shrinkage

It is commonly stated that creep and shrinkage are interrelated phenomena because there are a number of similarities between the two. The strain–time curves are very similar, experimental parameters affect creep in much the same way as shrinkage, the magnitudes of the strains are the same, and they include a considerable amount of irreversibility. Like shrinkage, creep is a paste property and the aggregate in concrete serves to act as a restraint. However, we do not really understand the processes of creep, and there is mounting evidence that creep and shrinkage occur by quite different mechanisms. The origins of creep are believed to reside in the response of C–S–H to stress. Since very few creep data are available on pure pastes, it is necessary to rely on data from concretes to assess the influence of experimental variables. Table 18.3 lists the various experimental variables that can be expected to affect creep, which are the same parameters that influence drying shrinkage.

Definition of Terms

A typical creep curve is given in Figure 18.20. When a specimen is unloaded, the instantaneous recovery is approximately the same as the instantaneous strain on first application of the load, but creep recovery, although it occurs more rapidly than creep, is by no means complete: a considerable portion of the total creep is irreversible. Under typical

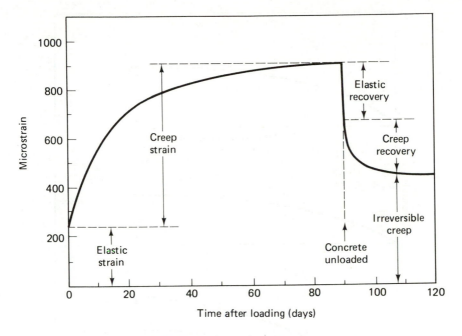

Figure 18.20 Typical creep curve for plain concrete.

service conditions concrete is most likely to be drying while under load, and it has been found that under such conditions creep deformations are greater than if the concrete is dried prior to loading. Terminology has been developed to take this fact into account and is shown diagrammatically in Figure 18.21. If the free shrinkage (ϵ_{sh}) (determined while the specimen is unloaded, but subjected to the same drying conditions) and basic creep (ϵ_{bc}) (determined while specimen is loaded but not drying) are added together, their sum is less than the total strain (ϵ_{tot}) determined from simultaneous loading and drying. The excess deformation is called *drying creep* (ϵ_{dc}). Total creep strain (ϵ_{cr}) is the sum of ϵ_{bc} and ϵ_{dc}. It is common practice, however, to ignore this distinction, and creep is simply considered as the deformation under load in excess of free shrinkage.

Factors Influencing Paste Behavior

Applied Stress

It is generally assumed that creep strains are linearly related to the applied stress, up to a stress of about 50% of the ultimate strength of the concrete. However, recent work has shown that experimental data are more accurately represented by a nonlinear relationship:

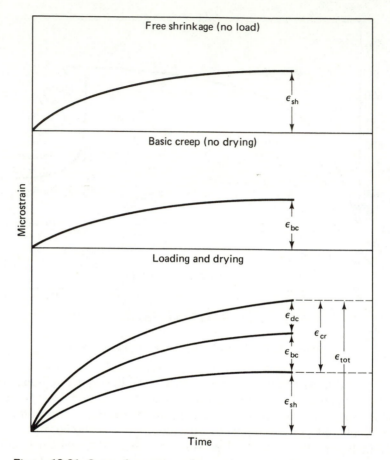

Figure 18.21 Creep of concrete under simultaneous loading and drying.

$$\epsilon_{cr} = C \cdot \sinh\left(\frac{V\sigma}{RT}\right) \tag{18.13}$$

which represents a thermally activated process. (C is a constant, V the activation volume, R the gas constant, T the absolute temperature, and σ the applied stress.) When this function is represented graphically (Figure 18.22), it is found that the creep–stress relationship is approximately linear in the stress range generally used. Thus, from a practical point of view, a linear relationship is justified and the concept of *specific creep* can be used:

$$\text{specific creep } (\phi) = \frac{\epsilon_{cr}}{\sigma} \tag{18.14}$$

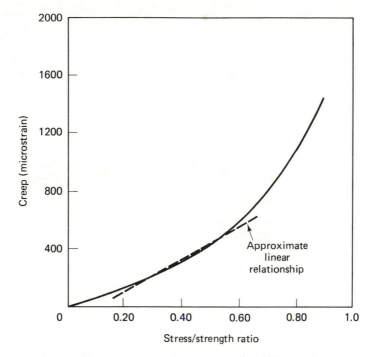

Figure 18.22 Creep –stress relationship for concrete.

Specific creep is useful for comparing creep of different concrete specimens loaded at different stress levels; a typical value is 150×10^{-6}/MPa
$(10^{-6}$/lb/in.$^2)$.

w/c Ratio

A study of the literature provides conflicting data concerning the effect
of *w/c* ratio on creep. This illustrates nicely the difficulties of interpreting
creep data, since it is not possible to change one parameter independently
of others. A change in the *w/c* ratio of concrete means a change in
cement content and concrete strength. If allowances are made for these
factors, it is found that specific creep increases with increasing *w/c* ratio
(Figure 18.23). Thus, it might be anticipated that specific creep of
concrete would be a function of its compressive strength, since that
property is most strongly influenced by *w/c* ratio. An inverse relationship
between the ultimate specific creep and the compressive strength of the
concrete does indeed exist.

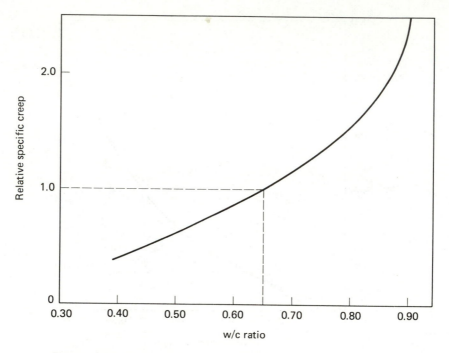

Figure 18.23 Effect of *w/c* ratio on creep of concrete. (Adapted from A. M. Neville, *Creep of Concrete: Plain, Reinforced and Prestressed*, North-Holland Publishing Company, Amsterdam, 1970.)

Curing Conditions

The time of moist curing of the concrete at loading affects the magnitude of creep (Figure 18.24). This is to be expected since the degree of hydration is lower at shorter curing times and the porosity of the paste is higher. However, the age effect continues even in more mature concretes, when porosity and strength do not change markedly with time (Figure 18.24). Thus, there seems to be an aging effect that may be caused by changes in bonding within C–S–H, which increases its resistance to stress. Some investigators have postulated that a "maturing creep" component occurs when the concrete is loaded while it is still hydrating rapidly. An increase in the temperature of curing reduces both basic and drying creep. The amount of reduction depends on the temperature and its duration, but even quite short periods at elevated temperatures may cause significant reductions in creep. The observed reductions occur primarily in the irreversible components of creep and are similar in magnitude to those observed for drying shrinkage under the same conditions. These observations can be rationalized by assuming that higher temperatures increase the aging process of C–S–H.

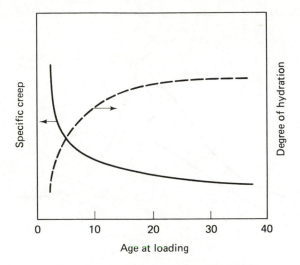

Figure 18.24 Effect of age on concrete creep.

Temperature

If concrete is maintained at elevated temperatures while under load, the amount of creep is increased over that of concrete held at room temperature (Figure 18.25). Creep increases approximately linearly with temperature up to 80°C (175°F) when it is about three times that at

Figure 18.25 Effect of temperature and temperature change on creep.

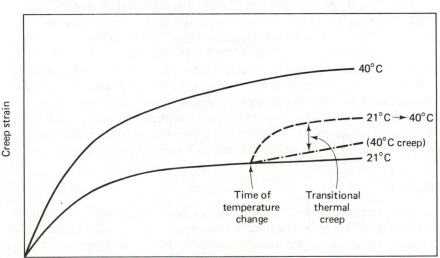

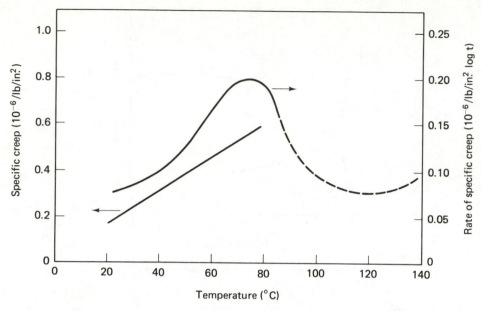

Figure 18.26 Effect of temperature on the rate of creep. (Based on data from H. Geymayer, in *Concrete for Nuclear Reactors* Vol. 1, SP-34, American Concrete Institute, Detroit, Mich., 1972, pp. 565–589.)

ambient temperature. However, above 80°C there is some uncertainty as to the exact relationship: some investigators report a continued linear variation, but there are some reports of a maximum at 70 to 80°C (155 to 175°F). The increased creep is a result of an increase in the rate of creep (Figure 18.26), and most investigators report a maximum in the creep rate somewhere between 50 and 90°C (120 and 195°F). The reason for these conflicting data and the existence of a rate maximum is not clear. It has been suggested that it may be a result of different preloading treatments (time and temperature), and if aging processes are indeed important this would be expected. If temperature increases occur while concrete is under load, an additional creep strain component has been observed that has been called *transitional thermal creep* (Figure 18.25).

Moisture

Moisture appears to be a necessary condition for creep, although the exact role of water in the creep processes is still under dispute. By preconditioning at a lower relative humidity before applying an external

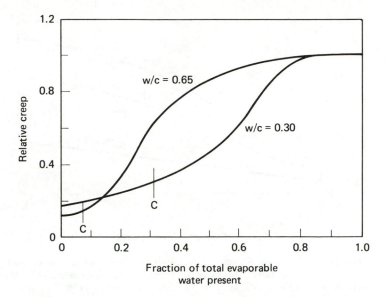

Figure 18.27 Creep as a function of nonevaporable water content. (*C* represents the approximate point at which capillary water is removed.) (Adapted from W. Ruetz, in *The Structure of Concrete*, ed. A. E. Brooks and K. Newman, Cement and Concrete Association, London, 1968, pp. 146–153.)

load, the amount of creep is reduced. Creep is thus a function of the evaporable water in the concrete (i.e., the water lost on D-drying) and falls to zero when no evaporable water is present. The greatest decrease in evaporable water, and hence in creep, occurs on drying to 40% RH while water is being lost from capillary pores. Thus, the creep vs. water content relationship depends on the *w/c* ratio (Figure 18.27).

The amount of creep that occurs when concrete is dried as well as loaded depends on relative humidity and time, as can be seen in Figure 18.28. Drying creep may be considerably more than the basic creep strain. Creep occurs during moisture change irrespective of the direction of the change. Thus, dry concrete will start to creep if it is resaturated while under load, but an appreciable rate of creep occurs only during exposure above 40% RH. Like drying creep, "wetting creep" depends on the RH change and is greatest when concrete is completely resaturated. It is believed that the disjoining pressure that causes C–S–H particles to become separated by a water film at high relative humidities (and contributes to shrinking and swelling) plays an important role in creep.

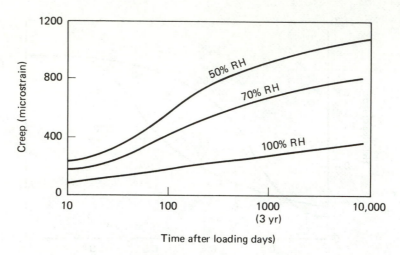

Figure 18.28 Creep of concrete during drying (RH change commenced on application of load). (From G. E. Troxell, J. M. Raphael, and R. E. Davis, *Proceedings, ASTM,* Vol. 58, 1958, pp. 1101–1120. Reprinted, with permission, from the American Society for Testing and Materials, 1916 Race Street, Philadelphia, PA 19103. Copyright.)

Cement Composition

There appears to be some relationship between the composition of a portland cement and creep, similar to that observed with shrinkage. Increasing the C_3A content or decreasing the effective C_3S content seems to increase creep. Type I cement concrete is reported to creep more than Type III, and a low-heat concrete containing blast-furnace slag or pozzolan may have even higher creep. However, exact relationships are not known, as creep tests with different cements are not always comparable. Creep of pastes of the pure silicates C_3S and C_2S is less than for Type I cement pastes, and thus the presence of minor components in C–S–H must have important implications for creep. There is also an "optimum gypsum content" at which creep is minimized for each cement; this tends to be higher than the optimum gypsum content at which drying shrinkage is a minimum.

Admixtures

It is generally stated that admixtures that increase drying shrinkage also increase creep. Calcium chloride is a common example, but water-reducing and set-retarding admixtures have also been implicated, particularly lignosulfonate-based formulations. However, it has been pointed out that changes in mix design that often accompany the use of water reducers (lower w/c ratio or lower cement content) can be used to offset possible increases in creep. Where creep is of concern, admixtures proposed for use should be tested to evaluate their influence on creep.

Creep Recovery

Only a relatively small proportion of the total creep strain is reversible when concrete is unloaded. The recoverable creep generally is in the range 10 to 20% of total creep after loading for 200 days. Thus, irreversible creep dominates creep behavior, and it is of importance in predicting the behavior of concrete under variable stress.

Creep recovery is strongly affected by time, temperature, and relative humidity. The proportion of irreversible creep increases with time under load. Experimental data suggest that after about the first 30 days under load, additional creep strain is largely irreversible (see Figure 18.29). It also appears that the decrease in total creep that occurs when concrete is aged before loading is a result of a decrease in irreversible creep, creep recovery being little affected. Similarly, the decrease in creep that is observed when the paste is exposed to a higher temperature is the result of a decrease in irreversible creep only. Thus, it would seem that irreversible creep can be considered as being the result of aging processes under stress; if the C–S–H has been "aged" by prior treatment, further aging is reduced. The additional drying creep that occurs

Figure 18.29 Influence of time on creep and creep recovery.

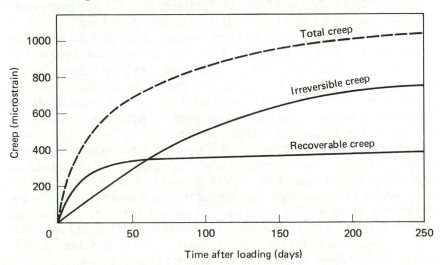

when concrete is dried while under load is largely irreversible. Similarly, transitional thermal creep that occurs when the temperature is changed while the concrete is under load is also irreversible. Transitional thermal creep may be the thermal analog of drying creep; and these changes may accelerate the aging process. Increasing the applied stress also increases irreversible creep but does not change creep recovery. Therefore, to a good approximation it can be said that the external factors that affect creep affect primarily the irreversible component. However, creep recovery may depend on mix parameters such as *w/c* ratio and aggregate content.

Creep under Different States of Stress

Creep in tension is of considerable interest in estimating cracking potential due to stresses imposed by moisture or thermal changes. Reduction of tensile stresses by tensile creep can minimize cracking in water-retaining structures and thin-shell roofs, where impermeability is important. However, it is difficult to measure accurately because of the low tensile strength of concrete, so that it is not easy to draw quantitative comparisons between creep in compression and in tension. It appears, however, that the initial rate of creep is higher in tension and that tensile creep is therefore greater for relatively short durations of load, although at longer times the reverse may hold. Again it can be said qualitatively that creep in torsion is affected by experimental variables in a way similar to creep under compressive stresses. Flexural creep is complicated by the fact that part of concrete is in compression and part in tension. Creep in the tensile and compressive fibers is not necessarily the same, and can be affected to different extents by drying.

Creep also occurs under dynamic loading, although it is difficult to separate out those time-dependent strains due to creep and those due to progressive microcracking under the changing stress conditions (which leads to fatigue failure). It is generally found that dynamic creep is greater than static creep compared under the same maximum stress. Creep strains appear to depend on the range of stress, the frequency of loading, and the duration of dynamic loading.

Under a uniaxial compressive stress some lateral creep also occurs. Poisson's ratio for creep is approximately the same as for instantaneous loading, but it may appear to be less if corrections are not made for simultaneous shrinkage. Under multiaxial loading it is of interest to know whether the creep strains from each stress act independently of one another; that is, can creep strains be superimposed? It appears that under multiaxial loading, the creep Poisson's ratio is less than it is in uniaxial compression. Axial creep of concrete confined by lateral stress is less than for unconfined concrete after allowance is made for Poisson creep. A linear relationship is found between volumetric creep and the mean normal stress [$\frac{1}{3}(\sigma_1 + \sigma_2 + \sigma_3)$]. Even under hydrostatic compression there is considerable creep.

Mechanisms of Creep

Numerous theories have been advanced to explain creep of cement paste and hence of concrete. It is not possible to review all of the approaches, and the discussion will be limited to three different approaches which have attracted considerable interest in recent years and which can be considered to be developments of ideas and concepts presented by earlier investigators. It should be noted that there are elements of similarity among the three approaches, and perhaps the different approaches treat the same phenomena in different ways and with different emphases.

Thermally Activated Creep

This approach takes as its basic premise the assumption that the time-dependent strains are the result of thermally activated processes that can be described by rate process theory. This approach is exactly analogous to that used for other materials, and the differences between cement paste and metals, for example, lie in the different types of processes that can operate. Creep strains will originate through deformation of a microvolume of paste, designated a "creep center." The creep center will undergo deformation to a lower energy configuration under the influence of energy added to the system by external sources. This deformation can only occur by going through an energy barrier (Figure 18.30) in the form of an intermediate, high-energy state. The ability for a creep center to cross the barrier depends on the height of the energy barrier and the inputs of energy from external sources: temperature, stress (strain energy), changes in moisture content, and changes in temperature. There is a time dependence which reflects the reduction in the number of centers as the activation barrier is overcome. Thus, creep (or rate of creep) can be expressed by the function of time, temperature, applied stress, and changes in moisture content or temperature while under load.

This view of microscopic deformation automatically indicates irreversible annihilation of creep centers. The question is, thus: How can a reversible process occur? On the removal of stress, the material can behave as if a negative stress is applied. The new stress pattern will activate a new set of creep centers, causing deformations in the opposing

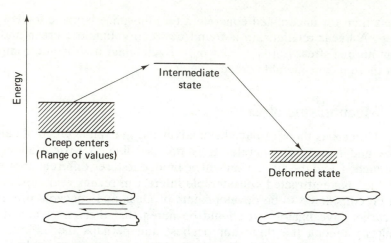

Figure 18.30 Schematic representation of temperature-dependent deformation of creep centers in cement paste.

direction, but the total deformations will be smaller, since they are accompanied by a relaxation of the stress distribution to zero. Hence, the observed creep recovery is less than the original creep.

Another question is: What is the nature of the creep center, and what process is operating? The most prevalent view involves a shear strain (slip) between adjacent particles of C–S–H. The ease and extent of slip depends on the forces of attraction between the particles. If the particles are chemically bonded, no slip can occur, but if only van der Waals' interactions are operating, slip is theoretically possible. It appears that measurable slip occurs only when a sufficient thickness of water exists between the particles. The water can reduce the van der Waals forces sufficiently to allow slippage more readily; it can be thought of as an analogy to lubrication.

Distribution of Adsorbed Water

The thermal activation approach views water as a necessary, but not sufficient, condition for creep. An opposing view explains creep in terms of the diffusion of adsorbed water under stress. These ideas were first developed qualitatively by Powers in the 1960s (the so-called "seepage theory") and quantitative thermodynamic treatment has been derived by Bazant in the 1970s.

The basic ideas are shown schematically in Figure 18.31. Water films exist on the surfaces of C–S–H particles due to adsorption. The

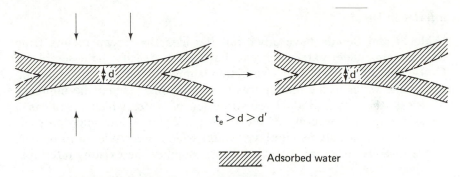

$$t_e > d > d'$$

///// Adsorbed water

Figure 18.31 Schematic representation of creep due to changes in disjoining pressures.

thickness of these films depends on the relative humidity with which the system is in equilibrium. In a saturated paste (100% RH) the equilibrium thickness (t_e) is about five water molecules thick (about 1.3 nm). If two adjacent C–S–H particles are closer than $2t_e$, the equilibrium films cannot be attained without forcing the particles apart. If the particles are fixed, a disjoining pressure is developed. This phenomenon has been termed *hindered adsorption* and can occur only in the very fine pores where capillary effects do not take place (i.e., the micropores). The equilibrium state of water in the micropores is thus determined by a combination of stress and thickness (interactions with the solid surface).

When an applied external stress is applied, the stress exerted on the water in the micropores is increased. To maintain equilibrium the thickness of the adsorbed layer must be decreased to compensate for the effective increase in disjoining pressure that has been created. Thus, water diffuses from the micropores to the larger capillary pores where no stress exists. The amount of water actually involved in diffusion is considered to be a very small part of the total water in the cement paste, so that creep occurs even in saturated specimens. Water movement will occur also when the external RH is lowered, since this change reduces t_e. Water will therefore diffuse from the micropores to maintain equilibrium, thereby reducing the disjoining pressure and causing a bulk deformation (see the discussion in Section 18.2). Thus, this can be considered to be a unified creep and shrinkage theory. Drying creep is explained by the fact that the reduction of t_e generally occurs much more rapidly than does creep. There is thus a temporary diminution of the stressed volume of water which raises the effective stress under which diffusion occurs.

Interlayer Theory

Feldman and Sereda have advanced the idea that creep results from deformations under external stress that tend to densify the assembly of C–S–H particles. The basic assumption (similar to the adsorption model) is that the water in the micropores is load-bearing. Under the influence of stress, there is a gradual redistribution of water which results in a densification and ordering of the C–S–H. Water in the micropores is believed to exist only as interlayer water which is considered to be part of the C–S–H structure. External stress modifies the existing interlayer spacings.

Other Processes

Various other processes have been proposed to explain the irreversibility of creep strains. It has been suggested, for example, that diffusion of *solid material* as well as water can occur. Increases in interlayer regions can be viewed as a change in the binding energy between adjacent C–S–H particles. Other types of bonding, such as formation of new silicate bonds, would have a similar effect and would prevent slip between adjacent particles. Increases in silicate bonding nicely explains the aging effects of C–S–H. At high stresses internal microcracking would lead to additional irreversible strain. Crack growth has been cited as an important mechanism in tensile creep.

Creep in Concrete

Influence of Aggregate

The foregoing discussion has not been concerned specifically with the role of aggregate in creep, except in passing. The role of aggregates in creep is similar to that in shrinkage: they act as a restraint to reduce the potential deformations of the paste. Thus, the aggregate content (Figure 18.32) and modulus of elasticity (Figure 18.33) are the most important parameters affecting creep of concrete. Aggregate size, grading, and surface texture have little influence. There is an empirical relationship similar to that found for drying shrinkage [Eq. (18.7)].

Specimen Geometry

When drying occurs while under load, factors such as specimen size and shape become important. Thus, the volume-to-surface ratio and specimen thickness affect total creep in much the same way as drying shrinkage is affected (Figure 18.34).

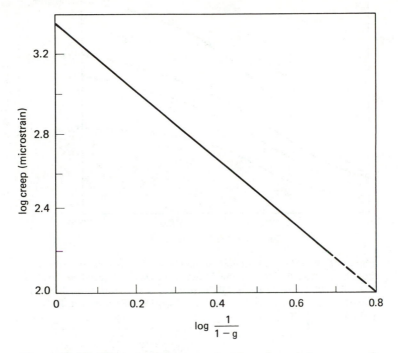

Figure 18.32 Effect of aggregate content by volume (*g*) on creep of concrete. (From A. M. Neville, *Magazine of Concrete Research,* Vol. 16, No. 46, 1964, pp. 21–30. Reproduced with permission of Cement and Concrete Association.)

Prediction of Creep

A large number of empirical equations have been developed for the prediction of creep in concrete. Under various conditions of loading and drying, these equations may range from very simple equations to quite complex relationships that require a computer for their analysis. There is considerable disagreement concerning the different mathematical approaches to creep prediction because of differences in the choice of function used to represent different aspects of creep. Two approaches to creep prediction have been developed by ACI and by CEB. These will be briefly described. Both methods use simple equations amenable to hand calculation.

Two serious deficiencies of both methods are the failure to correct for the effects of temperature (either before or during loading) and the inability to predict creep recovery. Both these effects could be readily included in either approach. Neither method is likely to prove adequate when concrete is subject to complex sequences of load, relative humidity,

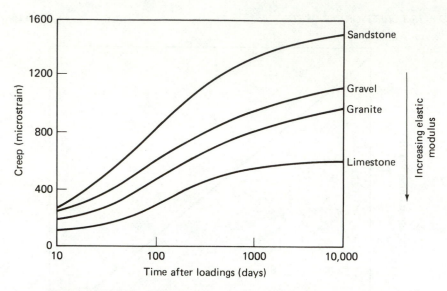

Figure 18.33 Effect of aggregate modulus on creep of concrete. (Adapted from G. E. Troxell, J. M. Raphael, and R. E. Davis, *Proceedings, ASTM,* Vol. 58, 1958, pp. 1101–1120. Reprinted, with permission, from the American Society for Testing and Materials, 1916 Race Street, Philadelphia, PA 19103. Copyright.)

Figure 18.34 Effect of volume-to-surface ratio on creep. (Adapted from T. C. Hansen and A. H. Mattock, *Journal of the American Concrete Institute,* Vol. 63, No. 2, 1966, pp. 267–290.)

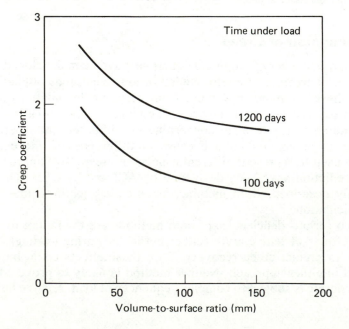

and temperature, particularly under conditions not commonly encountered. For example, both temperature cycling and load cycling will increase the magnitude of the observed creep.

ACI Method

The basic creep equation is of the form

$$C_t = \frac{t^{0.6}}{B + t^{0.6}} \cdot C_{\text{ult}} \tag{18.15}$$

where C_t is the creep coefficient at time t, C_{ult} the ultimate creep coefficient, and B is a constant (it is equal to 10 if the concrete is more than 7 days old before loading). The creep coefficient is the creep strain (ϵ_{cr}) divided by the instantaneous strain (ϵ_e):

$$C_t = \frac{\epsilon_{cr}}{\epsilon_e} \tag{18.16}$$

Equation (18.15) is very similar to the corresponding expression for drying shrinkage [Eq. (18.9)]. Again, the problem lies in selecting a suitable value for C_{ult}. When concrete is drying under load at 40% RH, C_{ult} will be in the range 1.30 to 4.15, and ACI recommends an average value of 2.35 if actual experimental data are not available for an estimation of C_{ult}. Correction factors can be used to adjust C_{ult} for different conditions: relative humidity [Eq. (18.17)] and age at loading [Eq. (18.18)]:

$$(\text{C.F.})_H = 1.27 - 0.0067H, H \geq 40\% \tag{18.17}$$

where H is the relative humidity as a percentage, and

$$(\text{C.F.})_{t_L} = 1.25t_L^{-0.118} \tag{18.18}$$

where t_L is the age of the concrete at time of loading in days.

CEB –FIP Method

The basic equation used for prediction of creep is

$$\epsilon_c(t, t_o) = \psi(t, t_o)\frac{\sigma_o}{E_{c28}} \tag{18.19}$$

where $\epsilon_c(t, t_o)$ is the creep strain at any time t [corrected according to Eq. (18.12)] after applying a constant stress σ_o at time t_o, E_{c28} the longitudinal modulus of elasticity at 28 days, and $\psi(t, t_o)$ the creep coefficient. The coefficient is determined from Eq. (18.20):

$$\psi(t, t_o) = \beta_o(t_o) + \psi_d\,\beta_d(t - t_o) + \psi_f[\beta_f(t) - \beta_f(t_o)] \qquad (18.20)$$

$\beta_o(t_o)$ represents the irreversible part of the deformation, which develops in the first few days after the load has been imposed. $\psi_d\beta_d(t - t_o)$ represents the recoverable part of the delayed elasticity. ψ_d is assumed to be constant and can be taken as 0.4. $\psi_f[\beta_f(t) - \beta_f(t_o)]$ represents the irreversible delayed deformation (flow), which is strongly influenced by the value of t_o, specimen dimensions, and relative dimensions. The quantities β_o, β_d, β_f, and ψ_f are determined from standard graphs or tables.

Loss of Prestress

Creep has been considered in this chapter as a problem of time-dependent strain, but it can also be manifested as a stress relaxation. Creep (and shrinkage) will thus reduce compressive stress in concrete induced by prestressing and enhance the tendency for cracking. In fact, creep of concrete frustrated early attempts at prestressing. It is thus essential to estimate the magnitude of loss of prestress due to creep and shrinkage to ensure an adequate residual prestress. It is beyond the scope of this book to consider the effects of creep and shrinkage on loss of prestress. Suffice it to say that pretensioning and post-tensioning are quite different problems. The time of pretensioning will be critical. Tension is applied while the concrete is quite young, and creep and shrinkage losses become a large part of total prestress losses. On the other hand, post-tensioning of precast elements generally involves quite mature concrete, which has had an opportunity to dry. In such cases prestress losses due to creep and shrinkage are quite small.

Bibliography

Elastic Deformations

HANSEN, T. C., "Influence of Aggregates and Voids on Modulus of Elasticity of Concrete, Cement Mortar, and Cement Paste," *Journal of the American Concrete Institute,* Vol. 62, No. 2, pp. 193–216 (1965).

RAO, C. V. S. KAMESWARA, R. N. SWAMY, AND P. S. MANGAT, "Mechanical Behavior of Concrete as a Composite Material," *Materials and Structures (Paris),* Vol. 7, No. 40, pp. 256–271 (1974).

Mechanisms of Shrinkage and Creep

BAZANT, Z. P., "Thermodynamic Theory of Deformations of Concrete with Explanation of Drying Creep," in SP-27: *Designing for Effects of Creep, Shrinkage, Temperature in Concrete Structures*, pp. 411–420. American Concrete Institute, Detroit, Mich. 1971.

FELDMAN, R. F., "Mechanism of Creep of Hydrated Portland Cement Paste," *Cement and Concrete Research*, Vol. 2, No. 5, pp. 521–540 (1972).

FELDMAN, R. F., AND P. J. SEREDA, "A New Model for Hydrated Portland Cement and Its Practical Implications," *Engineering Journal* (Canada), Vol. 53, No. 8/9, pp. 53–59, 1970.

GAMBLE, B. R., AND J. M. ILLSTON, "Rate of Deformation of Cement Paste and Concrete during Regimes of Variable Stress, Moisture-Content and Temperature," in *Hydraulic Cement Pastes: Their Structure and Properties*. Cement and Concrete Association, Slough, U.K., 1976.

POWERS, T. C., "Mechanism of Shrinkage and Reversible Creep of Concrete," in *The Structure of Concrete*, eds. A. E. Brooks and H. Newman. Cement and Concrete Association, Slough, U. K., 1968.

WITTMANN, F. H., "On the Action of Capillary Pressure on Fresh Concrete," *Cement and Concrete Research*, Vol. 6, No. 1, pp. 49–56 (1976).

WITTMANN, F. H., "The Structure of Hardened Cement Paste—A Basis for Better Understanding of the Materials Properties," in *Hydraulic Cement Pastes: Their Structure and Properties*. Cement and Concrete Association, Slough, U.K., 1976.

General References on Creep and Shrinkage

ACI COMMITTEE 209, "Prediction of Creep, Shrinkage and Temperature Effects in Concrete Structures," in SP-27: *Designing for Effects of Creep, Shrinkage, Temperature in Concrete Structures*, pp. 51–93. American Concrete Institute, Detroit, Mich. 1971.

GEYMAYER, H. G., "Effect of Temperature on Creep: A Literature Review," in *Concrete for Nuclear Reactors*, SP-34, Vol. 1, pp. 547–564. American Concrete Institute, Detroit, Mich., 1972.

HELMUTH, R. A., AND G. VERBECK, "Structures and Physical Properties of Cement Pastes," *Proceedings, Fifth International Symposium on the Chemistry of Cement, Tokyo, 1968*, Vol. 3, pp. 1–32. Cement Association of Japan, Tokyo, 1969.

International Recommendations for the Design of Concrete Structures: Principles and Recommendations. Cement and Concrete Association, Slough, U.K., 1970.

NEVILLE, A. M., *Creep of Concrete: Plain, Reinforced and Prestressed*. North-Holland Publishing Company, Amsterdam, 1970.

ZIA, P., H. K. PRESTON, N. L. SCOTT, AND E. B. WORKMAN, "Estimating Prestress Losses," *Concrete International: Design and Construction,* Vol. 1, No. 6, pp. 32–38 (1979).

Problems

18.1 Given that the compressive strength of a concrete is 6000 lb/in.^{2}, compute its modulus of elasticity if the unit weight is 150 lb/ft^3.

18.2 If a concrete has a compressive strength of 30 MPa and a density of 1600 kg/m^3, calculate the modulus of elasticity according to (a) the ACI code; (b) the CEB recommendations, and (c) the British Code of Practice.

18.3 Estimate the rate of evaporation of water from fresh concrete at 85°F that is exposed to an ambient temperature of 90°F, a relative humidity of 60%, and a wind velocity of 15 mi/h. Will the concrete need protection to avoid plastic shrinkage?

18.4 What are the underlying causes of drying shrinkage?

18.5 Suggest a strategy for reducing the drying shrinkage of precast concrete elements when in service. Could this approach also be used for cast-in-place concrete?

18.6 What are the effects of carbonation shrinkage, and when may it occur?

18.7 Estimate the drying shrinkage of a concrete beam after 3 months of drying at an average relative humidity of 75%.

18.8 What are the approximations involved in predicting shrinkage?

18.9 Define drying creep, specific creep, and basic creep.

18.10 How do curing conditions affect creep?

18.11 What material parameters influence creep?

18.12 Estimate the creep coefficient of a concrete column after (a) 30, (b) 60, (c) 90, and (d) 180 days, for which C_{ult} has been determined to be 3.05. The concrete is loaded after 7 days' curing and is exposed to 80% RH while under load.

19

other properties of concrete

19.1 THERMAL PROPERTIES

The properties of concrete, like those of other materials, are affected by changes of temperature, as discussed in the appropriate chapters. This section is concerned with the thermal properties of concrete: thermal expansion, thermal conductivity, and specific heat, as well as a general discussion of the response of concrete when exposed to extreme temperatures. These are of special concern in the design of structures where the concrete may be (1) exposed to thermal stresses or sustained high temperatures, (2) used for thermal insulation purposes, (3) used for fire protection, or (4) used for cryogenic applications. These categories cover a wide variety of practical applications. The thermal properties are more complex than they are for most materials, because not only is concrete a composite material whose components have different thermal properties (see Table 19.1), but its properties depend on moisture content and porosity.

Thermal Expansion

Thermal expansion is an important factor in all types of structures where differential heating may occur, either from environmental effects, such as the solar heating of pavements and bridge decks, or from service conditions, as in nuclear-reactor pressure vessels or furnace installations. Failure to allow for thermal expansion, or for thermal stresses resulting from differential expansion, will cause failure. The differential expansion

Table 19.1

Thermal Properties of Concrete Constituents

	Thermal Conductivity W/m·K (Btu/ft·h·°F)		Specific Heat, J/kg·°C (Btu/lb·°F)		Coefficient of Linear Expansion, 10^{-6}/°C (10^{-6}/°F)	
Aggregate						
Granite	3.1	(1.8)	800	(0.19)	7–9	(4–5)
Basalt	1.4	(0.8)	840	(0.20)	6–8	(3.3–4.4)
Limestone	3.1	(1.8)	—		6	(3.3)
Dolomite	3.6	(2.1)	—		7–10	(4–5.5)
Sandstone	3.9	(2.3)	—		11–12	(6.1–6.7)
Quartzite	4.3	(2.5)	—		11–13	(6.1–7.2)
Marble	2.7	(1.6)	—		4–7	(2.2–4)
Cement paste[a]						
$w/c = 0.4$	1.3	(0.75)	—		18–20	(10–11)
$w/c = 0.5$	1.2	(0.7)	—		18–20	(10–11)
$w/c = 0.6$	1.0	(0.6)	1600	(0.38)	18–20	(10–11)
Concrete	1.5–3.5	(0.9–2.0)	840 –1170 (0.2 –0.28)		7.4–1	(4.1–7.3)
Water	0.5	(0.3)	4200	(1.0)	—	—
Air	0.03	(0.02)	1050	(0.25)	—	—
Steel	120	(70)	460	(0.11)	11–12	(6.1–6.7)

[a]Saturated condition.

that can occur between cement paste and aggregate will give rise to high internal stresses, which may be critical in cases where there are large temperature changes.

Cement Paste

The coefficient of thermal expansion, in the range −10 to 100°C (14 to 212°F), is not a unique value but depends on the moisture content of the paste (Figure 19.1), the w/c ratio, and the age of the paste. There is an unusual moisture dependency, in which the coefficient increases considerably at intermediate relative humidities. This has been explained by considering that internal rearrangement of water takes place between capillary pores and gel pores without a change in the total water content of the paste. This can be called *hygrothermal expansion,* and it is dependent on w/c ratio and age, because these factors determine the porosity characteristics of the paste. The following relationship holds:

$$\alpha_{\text{actual}} = \alpha_{\text{true}} + \alpha_{\text{hygro}} \tag{19.1}$$

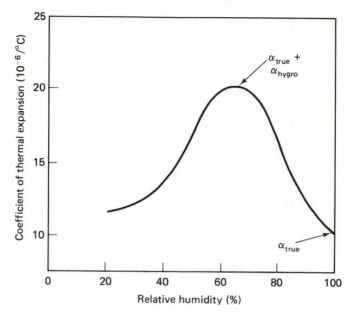

Figure 19.1 Variation of the coefficient of thermal expansion with moisture content of cement paste (equilibrated prior to measurement). (Adapted from N. G. Zoldners, in *Temperature and Concrete*, SP-25, American Concrete Institute, Detroit, Mich., 1971, pp. 1–31.)

where α_{actual} is the measured coefficient of thermal expansion, α_{true} is that measured in the absence of hygrothermal change, depending (as with other materials) on kinetic molecular movement, and α_{hygro} is the coefficient of hygrothermal expansion. It is believed that α_{true} is independent of paste properties, averaging about $10 \times 10^{-6}/°C$ ($5.5 \times 10^{-6}/°F$), and can be measured in a saturated paste.

If a moist paste is heated, loss of moisture should be accompanied by shrinkage. The amount of shrinkage that can occur will depend on the duration of heating, the permeability of the paste, and the thickness of the specimen. Below 100°C, shrinkage due to loss of moisture will be compensated for by an increase in the thermal coefficient at lower water contents, and a net expansion may occur (Figure 19.2), but if heating is not too slow, a net contraction will be observed at higher temperatures until all moisture has been lost from the paste. Much greater shrinkage is observed for a given amount of moisture loss above 100°C than for the same loss below 100°C, because structural breakdown of the hydration products is occurring.

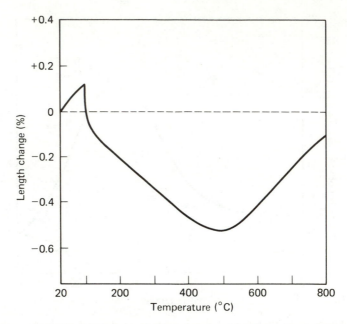

Figure 19.2 Length change of cement pastes heated to elevated temperatures. (Adapted from N. G. Zoldners, in *Temperature and Concrete*, SP-25, American Concrete Institute, Detroit, Mich., 1971, pp. 1–31.)

Aggregates

It can be seen from Table 19.1 that the coefficient of thermal expansion for common rocks varies with the mineralogical composition. Within each type a range of values occurs which depends on chemical composition, structure, and porosity; rocks are themselves composite materials. An average value for most rocks is in the vicinity of 6 to 8 $\times$ 10^{-6}/°C, which is less than α_{true} for cement paste and considerably less than α_{actual} (Figure 19.1). Quartz has the highest coefficient of thermal expansion (12 $\times$ 10^{-6}/°C) of any common mineral, and the coefficients of various rocks are related to their quartz content. Rocks with a high quartz content (quartzites, sandstones) have coefficients similar to quartz; those containing no quartz (limestones, marbles) have coefficients around 5 $\times$ 10^{-6}/°C.

The coefficient of thermal expansion of rocks also depends on moisture content, but to a much lesser degree than cement pastes. Air-dry rocks may expand about 10% more than water-saturated rocks. Temperature has a much greater effect on the thermal coefficient (Table 19.2), which increases markedly, but nonlinearly, with temperature. In

Table 19.2

Effect of Temperature on the Average Coefficient of Thermal Expansion of Some Aggregates ($10^{-6}/°C$)[a]

Temperature Range (°C)	Rock Type			
	Sandstone	Limestone	Granite	Anorthosite
20–100	10	3	4	4
100–300	15	9	14	9
300–500	22	17	26	10
500–700	25	33	48	13

[a]$°C \times \,^9/_5 + 32 = °F$; $10^{-6}/°C \times \,^5/_9 = 10^{-6}/°F$.

contrast, the thermal coefficient of cement paste does not change appreciably. However, at high temperatures other changes may also take place. Rocks may lose integrity due to chemical decomposition (limestones, basalts), or phase changes (quartzite). Anorthositic rocks show the best thermal stability, being relatively stable up to 1000°C (1830°F).

Concrete

The coefficient of thermal expansion will be a variable quantity depending on the mix design and the type of aggregate used. Since aggregates make up the bulk of concrete, their properties will largely determine the concrete properties. In a mortar the coefficient of thermal expansion will be a linear function of the volume of sand and cement (Figure 19.3a). The coefficient of thermal expansion of concrete can be estimated from the volumes of mortar and coarse aggregate (Figure 19.3b). Because cement paste has a high thermal expansion, the coefficient will also depend on the cement content, although the variation over the normal range of cement contents may not be as great as changing the type of aggregate.

Thermal Conductivity

Thermal conductivity can be defined as the ratio of heat flux to temperature gradient. The coefficient of thermal conductivity, k, represents the uniform flow of heat through a unit thickness of material between two faces of unit area that are subjected to a unit temperature difference. It is an important property in the design of concrete installations exposed to heat since it determines the rate of penetration of heat into the concrete and hence the magnitude of temperature gradients and

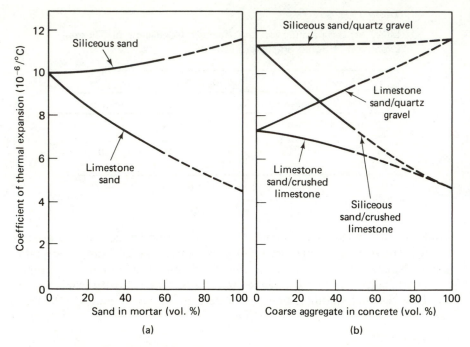

Figure 19.3 Effect of aggregate content on the thermal expansion of (a) mortar; (b) concrete. (Dotted lines are an extrapolation of experimental data.) (Adapted from N. G. Zoldners, in *Temperature and Concrete*, SP-25, American Concrete Institute, Detroit, Mich., 1971, pp. 1–31.)

thermal stresses. The thermal conductivity of concrete will be determined by the values for cement paste and aggregate and their relative proportions in the mix. The thermal conductivities of paste and aggregate are strongly influenced by porosity and moisture contents, and so too is that of concrete. This is not surprising, since the medium contained in the pore system also contributes to heat flow. Reference to Table 19.1 shows the thermal conductivity of water and air to be markedly different and in both cases substantially lower than the value for a solid material.

Neat cement pastes have thermal conductivities in the range 1.0 to 1.5 W/m·K, the value depending on the *w/c* ratio (Table 19.1). The degree of saturation has a greater influence than the total porosity; however, in strongly dried pastes, where all pore water has been removed, the value of the coefficient drops to about half that of the saturated paste. The thermal conductivity of most rocks is approximately 3.0 W/m·K (Table 19.1). Notable exceptions are quartz and dolomite. The moisture content may also have a significant effect on the thermal conductivity of rocks.

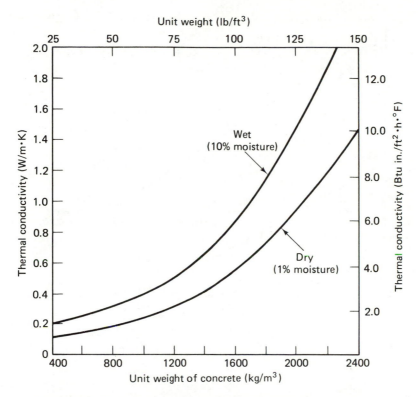

Figure 19.4 Thermal conductivity as a function of density and moisture content. (Based on data from A. Short and W. Kinniburgh, *Lightweight Concrete*, 3rd ed., Applied Science Publishers, Ltd., London, 1978.)

Thermal conductivity is related to the density of concrete (Figure 19.4); lightweight concretes that have high porosities have very low thermal conductivities, because of the large volume of air voids. This accounts for their good thermal insulating properties; it is immaterial whether the porosity is contained within the aggregate (lightweight aggregates) or within the paste (foamed or aerated concretes). Since the thermal conductivity of both paste and aggregate depends on the degree of saturation, the same is true for concrete (Figure 19.4). The actual effect that a change of moisture content has on thermal conductivity depends on the initial degree of saturation of the concrete. The thermal conductivities of paste, aggregate, and concrete are independent of temperature within the normal climatic range. Above 100°C, however, thermal conductivity decreases linearly with temperature (Figure 19.5). This has been attributed to changes in moisture content on drying.

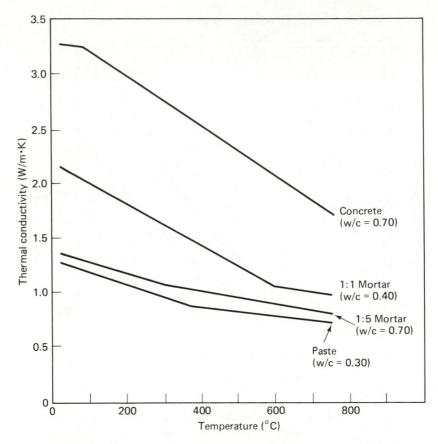

Figure 19.5 Variation of thermal conductivity with temperature. (Based on data from T. Harada et al., in *Concrete for Nuclear Reactors*, Vol. I, SP-34, American Concrete Institute, Detroit, Mich., 1972, pp. 377–406.)

Specific Heat

Specific heat, or heat capacity, is little affected by the type of aggregate, since the specific heats of rocks do not change much with mineralogical type. The specific heat of cement paste, however, is strongly dependent on porosity (*w/c* ratio), water content, and temperature (Table 19.3). The specific heat of concrete also depends on these factors. The common range of values for concrete is 800 to 1200 J/kg·°C.

Thermal Diffusivity

The thermal diffusivity measures the rate at which temperature changes take place in the concrete and is a function of both thermal conductivity and specific heat:

Table 19.3

Specific Heats of Pastes, Concretes and Mortars

Material	a/c Ratio	w/c Ratio	Temperature (°C)	Specific Heat J/kg·°C (Btu/lb·°F)
Neat paste	—	0.25	21	1140 (0.27)
	—		65	1680 (0.40)
	—	0.60	21	1600 (0.38)
	—		65	2460 (0.58)
Mortar	1:1	—	21	1720 (0.41)
	1:2	—	21	1180 (0.28)
	1:6	—	21	1100 (0.26)
Concrete	—	—	—	800–1200 (0.20–0.28)

$$D = \frac{k}{c\rho} \tag{19.2}$$

where D is the diffusivity constant, k the thermal conductivity, c the specific heat, and ρ the density. Diffusivity is often measured experimentally for the determination of thermal conductivity. Obviously, factors that affect conductivity and specific heat will also affect diffusivity. Typical values for concrete range from 0.002 to 0.007 m²/h (0.02 to 0.08 ft²/h).

Exposure to High Temperatures

From the foregoing discussions it can be concluded that within the normal environmental temperature range, the thermal properties of a concrete can be considered to be constant, *provided that there is no change in moisture content.* However, at elevated temperatures these properties change because of changes in the moisture content of the concrete components and because of progressive deterioration of the paste and in some cases of the aggregate. These processes depend on the conditions of exposure: the rate of temperature rise, the maximum temperature, and the time at elevated temperatures. The response of concrete will also depend on its initial properties and those of its constituents. Therefore, prediction of concrete behavior at elevated temperatures is a difficult problem.

Strength

Unless large temperature differentials are allowed to develop (as in rapid heating), the compressive strength of concrete at elevated temperatures is usually maintained up to about 300°C (570°F), but above this temper-

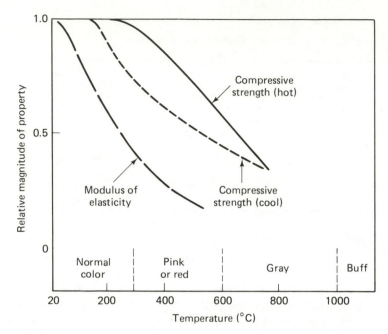

Figure 19.6 Effect of heating on strength and modulus of concrete.

ature significant decreases can be anticipated (Figure 19.6). The magnitude of the decreases depends on the nature of the aggregate and the initial moisture content of the specimen. The changes in both strength and modulus have been attributed to a combination of decomposition of the hydrated pastes, deterioration of the aggregates, and thermal incompatibilities between paste and aggregate leading to stress concentrations and microcracking. The effect on flexural strengths is more marked, and this would be anticipated since flexural strength is more sensitive to the internal microcracking that would be expected to occur. When concretes are cooled back to room temperature before testing, the strength is less than that found if the concrete is tested while hot. This may be due in part to the imposition of additional thermal stresses and also to the rehydration of those hydration products which have been partially dehydrated, thereby causing *in situ* expansions.

If concretes are heated in a sealed condition to prevent loss of moisture, or when loss of moisture is slow, the cement paste is effectively subjected to high-temperature autoclaving conditions. Over short periods of time, additional reaction of the cement may result in a slight increase in strength, but eventually considerable loss of strength is observed due to the formation of crystalline calcium silicate hydrates, which is accompanied by an increase in porosity (see Chapter 11).

Deformations

The modulus of elasticity also decreases with temperature (Figure 19.6), and the change is more marked than in the case of compressive strength. The decrease is greater than would be expected from a change in bonding energies, and it is believed that internal microcracking at the paste–aggregate interface contributes to the change in modulus.

As mentioned earlier, drying shrinkage occurs when concrete is heated to elevated temperatures due to additional loss of moisture from the paste (Figure 19.2). The drying shrinkage at 100°C is four to five times higher than that at 21°C under comparable conditions. Shrinkage continues to increase at higher temperatures (Figure 19.7) as structural decomposition of the hydration products continues, and much of this is irreversible. The rate of shrinkage depends on the rate of moisture loss from the concrete and thus depends on such factors as water content, w/c ratio, aggregate content, specimen geometry, and drying conditions at the surface.

Creep will also increase with increasing temperature. Between 50 and 140°C (120 and 285°F) there are conflicting data concerning the magnitude of specific creep and the rate of creep (see Chapter 18). At higher temperatures the rate of creep can be expected to increase, but the amount of creep should depend on the extent of moisture loss from the concrete.

Figure 19.7 Effect of temperature on magnitude and rate of shrinkage. (From N. G. Zoldners, in *Temperature and Concrete*, SP-25, American Concrete Institute, Detroit, Mich., 1971, pp. 1–31.)

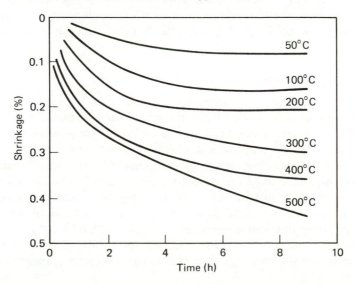

Fire Resistance

Compared to structural steel, concrete has excellent fire-resistant properties and is often used to protect steel from the effects of fire. Concrete has a lower thermal conductivity and a higher specific heat than metals; indeed, its properties are typical of ceramic materials. However, as we have seen, concrete is damaged by exposure to high temperatures and will suffer loss of strength, cracking, and spalling. The advantages of concrete in exposure to fire are:

1. A low rate of penetration.

2. Retention of strength if exposure is not too long.

3. Absence of toxic fumes.

The low rate of heat penetration is a consequence of the low thermal conductivity; hence, lightweight concrete has better fire resistance than ordinary concrete, but other factors can also slow down the rise of temperature in concrete. Water has a high specific heat and considerable amounts of heat are consumed in the evaporation of water, so that a moist concrete may heat more slowly, although concomitant shrinkage may lead to surface spalling and cracking. Under certain circumstances, aggregates may improve fire resistance. For example, dolomitic aggregates lose carbon dioxode when heated to about 600°C; this process consumes heat and also forms a layer of porous calcined material which can insulate the interior of thick cross sections of concrete.

Concretes made with limestone or siliceous aggregates show a color change on heating (Figure 19.6). This color change persists on cooling and so can be a useful guide to the extent of fire damage. Concrete that has passed beyond the pink stage is likely to be severely damaged.

Refractory Concretes

By a suitable choice of materials, concretes can be made to perform satisfactorily at high service temperatures in furnace installations and as a substitute for refractory brick where awkward shapes are required or monolithic construction is desirable. Both cements and aggregates should be chosen for their refractory properties (Table 19.4). Calcium aluminate cements should be used for refractory concretes. High-alumina cement is a rather impure mixture of calcium aluminate containing much iron and silica as well as CA. A much purer calcium aluminate cement containing only calcium aluminates—CA, C_3A_5, and $C_{12}A_7$—should be used for very high temperatures. Aggregates should also be chosen for their refractory properties, using increasingly refractory materials as the

Table 19.4

Types of Refractory Concrete

Type	Service Temperature Limit (°C)[a]	Cement	Aggregate
Structural	300	Portland	Common rocks
Heat resisting	1000	High alumina	Fine-grained basic igneous rocks (basalt, dolerite); heat-treated porous aggregates (bricks, scoria, expanded shale)
Ordinary refractory	1350	High alumina	Firebrick
Super-duty refractory	1450	Calcium aluminate	Firebrick
	1350–1600	High alumina	Refractory (magnesite, silicon carbide, bauxite, fired clay)
	—	Calcium aluminate	Refractory

[a]°C × $9/5$ + 32 = °F.

service temperature rises. Reinforcement should be omitted or used sparingly in the form of mesh rather than bars, since increased elastic deformations and possibilities for corrosion can cause problems at high temperatures. Refractory concretes have a high compressive strength when cast due to the hydraulic bond. On heating, this bond breaks down in the range 800 to 1000°C (1470 to 1830°F), and the strength may fall to about 20% of the initial value. On further heating, the strength rises again, due to the formation of a ceramic bond.

Methods of Test

Many different tests are available for the determination of thermal properties of materials. Two tests (ASTM C177 and C135) are available for the measurement of thermal conductivity of insulating concretes. The U.S. Corps of Engineers has standard tests for specific heat and thermal diffusivity in its *Handbook for Cement and Concrete*. ASTM E119 is the only established test for fire resistance for concrete, although other special tests may well be used for particular applications. ASTM E119 requires a fixed time–temperature curve (Figure 19.8) to be developed by the furnace used to heat the sample. The test can be applied to various specimens representing structural or nonstructural elements of a building. The success of a test can depend on all or any of the following

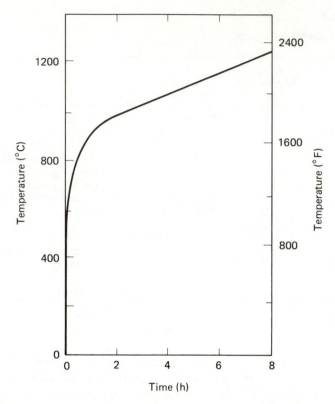

Figure 19.8 Standard time –temperature curve specified by ASTM E119. (Reprinted, with permission, from the American Society for Testing and Materials, 1916 Race Street, Philadelphia, PA 19103. Copyright.)

criteria: ability to sustain the applied load, prevention of the passage of flame or hot gases through the sample, or limitation of a specified peak temperature on the face not exposed to heat. Fire resistance is expressed in terms of the time during which the specimen meets the criteria under the conditions of test. Fire-resistance testing is a specialized undertaking and only two or three laboratories in North America are equipped for the job. Special tests may be used depending on the particular requirements.

Cryogenic Applications

Concrete is also being used at very low temperatures, such as in tanks for the storage of liquefied propane gas. If the concrete is cooled slowly and not exposed to thermal cycling, it will perform satisfactorily. The freezing of capillary moisture will allow concrete that has been

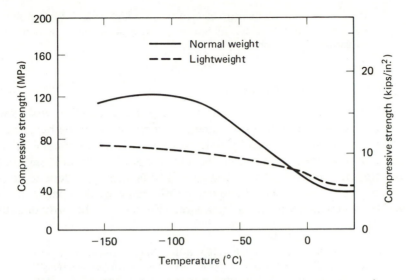

Figure 19.9 Effect of cryogenic temperatures on concrete strength. (Adapted from H. Woods, *Durability of Concrete Construction,* Monograph No. 4, American Concrete Institute, Detroit, Mich., 1968.)

previously well cured to attain much higher strengths (two to three times higher) than the same concrete tested at room temperature (Figure 19.9). Dry concrete will be only slightly stronger when frozen. The modulus of elasticity is also increased at lower temperatures, while the coefficient of thermal expansion is decreased somewhat. Concrete will undergo very large and irreversible losses in strength after only one or two cycles of freezing to very low temperatures (below about −70°C) and then thawing at ordinary room temperature. However, there is some evidence that if the concrete is dried before the temperature cycling begins, losses in strength may be relatively small.

19.2 RESISTANCE TO WEAR

In certain applications, severe wearing of concrete surfaces may lead to service problems. Three distinct types of wear have been distinguished:

1. *Abrasion.* Wearing by repeated rubbing or frictional processes (attrition). This term is used in connection with traffic wear on pavements and industrial floors.

2. *Erosion.* Wearing by the abrasive action of fluids and suspended solids. Erosion is a special case of abrasion and occurs in water-supply installations: canals, conduits, pipes, and spillways.

3. *Cavitation.* Impact damage caused when a high-velocity liquid flow is disturbed. It will occur at spillways and sluiceways in dams and irrigation installations.

Abrasion Resistance

Cement paste generally does not have good abrasion resistance, and the performance of concrete depends a great deal on the hardness of the aggregates used. Especially hard aggregates such as emery or iron shot have been used in abrasion-resistant toppings for industrial floors. High-strength concretes with low *w/c* ratios are less dependent on aggregate type, and the use of a low *w/c* ratio can provide a dense, strong concrete which is resistant to abrasion (Figure 19.10), provided that

Figure 19.10 Influence of *w/c* ratio on the abrasion resistance of concrete. (From F. L. Smith, in *Papers on Cement and Concrete,* ASTM STP 205, 1958, pp. 91 – 105. Reprinted, with permission, from the American Society for Testing and Materials, 1916 Race Street, Philadelphia, PA 19103. Copyright.)

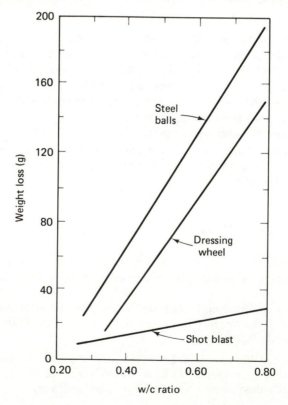

correct finishing techniques are used. To provide a dense uniform finish, the surface should be steel-trowelled; more than one pass over the surface can be used. Special chemicals can be used to provide additional surface hardening. Such treatments involve impregnating the surface layers with a liquid, which subsequently reacts to deposit solid material in the pores of the paste. Many formulations are based on sodium silicate, which reacts with calcium hydroxide to form C–S–H, or on soluble fluorosilicates, which form insoluble calcium fluorosilicate.

It is difficult to assess the abrasion resistance of concrete, since the damaging action depends on the exact cause of wear. Wear on pavements and floors includes scraping damage and impact loading as well as frictional attrition. Several tests have been prepared using different kinds of wear, but no one test is satisfactory for evaluating all conditions of abrasion. Indeed, the most difficult part of an abrasion test may be interpreting the results. ASTM C799 gives three optional methods for testing the abrasion resistance of horizontal concrete surfaces. The *steel-ball abrasion test* applies a load to a rotating head containing steel balls, while the abraded material is washed away by water to expose fresh surfaces. The *dressing wheel test* applies a load through rotating steel dressing wheels. The *revolving disk test* uses revolving flat steel disks in conjunction with a silicon carbide abrasive grit. In each of the tests the depth of wear is measured after a set time of test; weight loss is also a common measure of wear. These tests are meant to simulate wear from wheeled and heavy foot traffic, and apply various amounts of frictional abrasion, together with impact and scuffing such as might occur in practice. The *sandblast test* (ASTM C418) is probably a better measure of resistance of concrete to erosion, although it can be used as an abrasion test also. This test tends to abrade the less-resistant components of the concrete more severely. In addition, there are tests for the abrasion resistance of aggregates, such as the *Los Angeles Machine test,* which was discussed in Chapter 6.

Special tests for aggregates may be required for particular applications; for example, skid resistance of pavements is an important consideration. The gradual polishing of aggregates will eventually cause a smooth skid-prone surface to develop. Hard, friable aggregates, which fracture rather than polish, may be the most desirable to maintain a skid-resistant surface. Aggregates which are made up of minerals that have different rates of polishing will also give a more skid-resistant surface. There are various ASTM tests designed to measure the skid resistance of pavement surfaces, including concrete. ASTM E274 and ASTM E303 are designed for field testing. ASTM E274 uses a specified, full-scale automotive tire, while ASTM E303 uses a special rubber slider in contact with the surface. ASTM E303 can be adapted for laboratory testing or ASTM E510 can be used for this purpose. Both ASTM E303 and ASTM

E510, as well as ASTM E451, can be used to study the polishing characteristics of the pavement.

Erosion Resistance

The erosion of concrete will depend very much on the quantity and properties of suspended solids. Erosion by waters free of sediment is not likely to be great unless chemical attack can occur. Again, resistance to erosion is primarily determined by the compressive strength (or *w/c* ratio) of the concrete. The use of large aggregate particles seems to be advantageous. As in the case of abrasion, it is only the concrete at the surface that is subjected to erosion. Thus, particular attention should be paid to ensuring that a high-quality surface is attained during placement. The sandblast test (see above) can be used to measure resistance to erosion.

Cavitation Damage

Damage by cavitation is a far more serious problem, and even good-quality concrete may suffer severe attack. Cavitation occurs when a high-velocity flow of water (or any other fluid) suffers an abrupt change in direction or velocity. This causes a zone of low pressure to occur at the surface immediately downstream from the change, and this drop in pressure may be sufficient to allow pockets (or cavities) of vapor to form. These pockets of vapor collapse catastrophically on leaving the low-pressure zone, and their collapse causes a localized high-energy impact on the concrete surface. The impact from the implosion of vapor cavities may generate localized pressures as high as 700 MPa (1,000 lb/in.²), which is sufficient to affect even the strongest metals. Clearly, concrete can be severely affected, and perhaps the best defense against cavitation is to try to eliminate its possibility in design. The elimination of irregularities in flow or the reduction of flows to less than 12 m/s (40 ft/s) will do much to help. Deliberate air entrainment in water is said to cushion the effect of cavitation and hence reduce damage.

Cavitation cannot always be avoided, however, and the use of very good materials should be considered in minimizing damage. Experience has shown that the use of a high-strength, low-*w/c*-ratio concrete with a small aggregate size (not greater than 20 mm) and a good paste–aggregate bond will give the best resistance. The use of polymer-impregnated concrete (for high strength and good bond) or fiber-reinforced concrete (for good impact resistance) has been advocated for protection against cavitation damage. These newer materials (see Chapter 22) are now being tested in the field.

19.3 OTHER PROPERTIES

Radiation Shielding

Concrete has useful properties as a radiation-containment material, its most visible application at the present time being in prestressed concrete pressure vessels for nuclear reactors. Normal concrete can be used for shielding but may require excessive thicknesses. The choice of suitable materials will improve the attenuation of radiation and reduce the required thicknesses. The principal types of radiation that need to be considered in the design of radiation shielding are neutrons and gamma rays.

Gamma radiation is highly penetrating, electromagnetic radiation with wavelengths much shorter than X-rays. Gamma rays are attenuated primarily through elastic collision with electrons (the Compton scattering effect). High-density materials are good attenuators, and the use of special high-density aggregates is desirable. Attenuation of fast neutrons also requires a high density, while the presence of hydrogen atoms will cause moderate and slow neutrons to be absorbed. For normal-density concrete, a hydrogen content of 0.45% by weight is required, which is reached when the concrete contains about 4% by weight of water. Fully dried concrete will not contain sufficient water, so that an aggregate containing some hydrogen may also be desirable (Table 19.5). When hydrogen absorbs slow neutrons, high-energy gamma-rays are produced, which must also be attenuated. Boron is a more effective absorber of neutrons and produces gamma rays with a much lower energy which are more readily shielded. Thus, boron-containing aggregates are useful for neutron shielding. The addition of water-soluble boron compounds is not advised, however, since they may retard the hydration of cement. Siliceous aggregates have a high proportion of oxygen, which also helps to moderate neutrons. Aggregates for radiation shielding are covered by ASTM C637 and ASTM C638, and in BS 4619. Heavy aggregates are more expensive and require special placing and mix design considerations. They should be used only when space limits the thickness of the concrete. Heavy aggregates tend to be harsh, with a tendency to segregate. Therefore, more fine aggregate is needed than usual; the use of fine sands and high cement contents will give workable, cohesive concretes. (See Chapter 21 for mix design of heavyweight concrete.)

Radiation Damage

It appears that concrete properties are unaffected by exposure to gamma rays or neutrons at levels generally encountered. This statement is equivocal because it is difficult to separate the effects of radiation from

Table 19.5

Aggregates Recommended for Radiation Shielding

Aggregate Type	Specific Gravity	Shielding Capability
Natural		
Bauxite	~2.0	Fast neutrons (H)[a]
Serpentine	~2.5	Fast neutrons (H)
Goethite	~3.5	Fast neutrons (H)
Limonite	~3.5	Fast neutrons (H)
Borocalcite	~2.5	Neutrons (B)[b]
Colemanite	~2.5	Neutrons (B)
Barite	~4.2	Gamma rays
Magnetite	~4.5	Gamma rays
Illmenite	~4.5	Gamma rays
Hematite	~4.5	Gamma rays
Synthetic		
Heavy slags	~5.0	Gamma rays
Ferrophosphorus	~6.0	Gamma rays
Ferrosilicon	~6.7	Gamma rays
Steel punching or shot	~7.5	Gamma rays
Ferroboron	~5.0	Neutrons (B)
Boron carbide	~2.5	Neutrons (B)
Boron frit	~2.5	Neutrons (B)

[a](H), moderation by hydrogen.

[b](B), moderation by boron.

the effects of temperature. The energy of absorbed radiation is converted to heat and results in quite large temperature rises in concrete—100°C (180°F) or more. The temperature of concrete in reactor shields may attain several hundred degrees unless steps are taken to lower temperatures by removal of heat. The safe operating temperature depends on many factors: the thermal stability of aggregates, the presence of temperature cycling (which is more damaging than a steady temperature), the size of temperature gradients, and the minimum shielding requirements.

Acoustic Properties

Two acoustic properties of building materials are of interest: sound absorption and sound transmission. Sound absorption is primarily concerned with controlling noise levels within a room. The *sound absorption coefficient* depends strongly on porosity; thus, lightweight concrete gives better sound absorption than does normal-weight concrete. Yet texture is also important and a concrete made with porous, lightweight aggregate which has an irregular, interconnected porosity absorbs sound better than a foamed concrete of the same porosity which has discrete air bubbles.

The main factor influencing *sound transmission* is the density per unit area of wall; the type of material used is less critical. Low-density concretes are both good transmitters and absorbers of sound; thus, if both high absorption and low transmission are required simultaneously, some solution must be found, or a compromise must be made. The presence of cavities increases transmission loss, so the use of cavity walls or the use of sandwich panels with a dense interior may provide an acceptable solution. The use of gypsum board and furring strips to provide a cavity adjacent to a concrete wall will reduce sound transmission.

Sound transmission depends on the frequency of sound. Measurements of sound transmission loss are made according to ASTM E90 for a series of frequencies. From these measurements a single-figure rating, the sound transmission class, is determined according to ASTM E413. Where thin partitions are used, the bending stiffness of the wall may be important. Under the right conditions the air-wave frequency imposed on the wall may equal its structural resonance frequency, and this condition will enhance sound transmission.

Electrical Properties

Electrical Conductivity

Dry concrete is a good insulator with a resistivity of about 10^{13} ohm·m. However, moist concrete has a much lower resistivity, about 10^6 ohm·m, which is in the range for semiconductors. The increase in conductivity is due to the presence of water in the capillary pores, which contains dissolved salts and so acts as an electrolyte. Therefore, the resistivity of concrete will be controlled by the concentration of the electrolytic solution. Factors that increase the concentration are the alkali content of the cement (alkali salts predominate in pore solutions), the presence of electrolytic admixtures such as calcium chloride, and the amount of capillary water, which is controlled by the degree of hydration and the *w/c* ratio. These effects are shown in Table 19.6.

The discussion above applies to alternating current. Direct current could have a polarizing effect, but at 50 Hz no difference is observed between the resistance of concrete to ac or dc. Concrete offers a high resistance to the passage of current to or from reinforcing steel, believed to be due to polarizing effects.

Table 19.6

Effect of Alkali Content on Resistivity of Concrete[a]

Na₂O Content	w/c Ratio	Resistivity (at 1000 Hz, 4 V) (ohm·m)		
		7 Days	28 Days	90 Days
0.2%	0.4	103,000	117,000	157,000
	0.5	79,000	88,000	109,000
	0.6	53,000	70,000	76,000
1.0%	0.4	123,000	136,000	166,000
	0.5	82,000	95,000	120,000
	0.6	72,000	73,000	79,000

[a]Adapted from G. E. Monfore, *Journal of the Portland Cement Association, R & D Laboratories,* Vol. 10, No. 2, pp. 35–48 (1968).

Other Properties

The capacitance of concrete decreases with age and increasing frequency. The dielectric strength of concrete is apparently independent of moisture content, but for a composite material such as concrete, it should depend on the microstructure of the material. The dielectric properties of cement pastes have been studied in the radiofrequency range, and microwave absorption has been used to study the behavior of water in cement paste.

Bibliography

Behavior of Concrete Under Temperature Extremes, SP-39, American Concrete Institute, Detroit, Mich. 1973.

Concrete for Nuclear Reactors, SP-34, Vol. 1. American Concrete Institute, Detroit, Mich., 1972.

HILSDORF, H. K., J. KROPP, AND H. J. KOCH, "The Effects of Nuclear Radiation on the Mechanical Properties of Concrete," pp. 223–251 in SP-55, *Douglas McHenry International Symposium on Concrete and Concrete Structures,* American Concrete Institute, Detroit, Mich., 1978.

LITVAN, A., AND H. W. BELLISTON, "Sound Transmission Loss through Concrete and Concrete Masonary Walls," *Journal of the American Concrete Institute,* Vol. 75, No. 12, pp. 641–646 (1978).

NEVILLE, A. M., *Hardened Concrete: Physical and Mechanical Aspects,* Monograph 6. American Concrete Institute, Detroit, Mich., 1971.

Refractory Concrete, SP-57. American Concrete Institute, Detroit, Mich., 1978.

ROSTÁSY, F. S., U. SCHNEIDER, AND G. WIEDEMANN, "Behavior of Mortar and Concrete at Extremely Low Temperatures," *Cement and Concrete Research,* Vol. 9, No. 3, pp. 365–76 (1979).

Significance of Tests and Properties of Concrete and Concrete-Making Materials, ASTM STP 169B. American Society for Testing and Materials, Philadelphia, Pa., 1978.

Temperature and Concrete, SP-25. American Concrete Institute, Detroit, Mich., 1977.

TURNER, F. H., *Concrete and Cryogenics,* Cement and Concrete Association, Wexham Springs, Great Britain (1979).

WHITING, D., A. LITVIN, AND S. E. GOODWIN, "Specific Heat of Selected Concretes," *Journal of the American Concrete Institute,* Vol. 75, No. 7, pp. 299–305 (1978).

WOODS, H., *Durability of Concrete Construction,* Monograph No. 4. American Concrete Institute, Detroit, Mich., 1968.

YAMANE, S., H. KASAMI, AND T. OKUNO, "Properties of Concrete at Very Low Temperatures," pp. 207–221 in SP-55, *Douglas McHenry International Symposium on Concrete and Concrete Structures,* American Concrete Institute, Detroit, Mich., 1978.

Problems

19.1 What is the significance of the fact that concrete aggregates and cement paste have different coefficients of thermal expansion?

19.2 Why are lightweight concretes good thermal insulators?

19.3 How would you determine whether a structure has been structurally damaged after exposure to fire?

19.4 Why is concrete more fire-resistant than structural steel?

19.5 What factors affect the wear resistance of concrete?

20

durability

Concrete is inherently a durable material. If properly designed for the environment to which it will be exposed, and if carefully produced with good quality control, concrete is capable of maintenance-free performance for decades without the need for protective coatings, except in highly corrosive environments. However, concrete is potentially vulnerable to attack in a variety of different exposures (Table 20.1) unless certain precautions are taken. Deterioration of concrete can be caused by the adverse performance of any one of the three major components: aggregate, paste, or reinforcement, and can be due to either chemical or physical causes. Although one particular environmental factor may initiate distress, other factors may contribute and aggravate the situation.

A major difficulty in studying durability is predicting concrete behavior several decades in the future on the basis of short-term tests. Much of our understanding of durability has come through a direct study of actual field problems. A major challenge facing the modern concrete technologist is the more accurate prediction of concrete durability under a variety of service conditions.

20.1 PERMEABILITY OF CONCRETE

The single parameter that has the largest influence on durability is the w/c ratio. As the w/c ratio is decreased, the porosity of the paste is decreased and the concrete becomes more impermeable (Figure 20.1). Clearly, the permeability of concrete plays an important role in durability because it controls the rate of entry of moisture that may contain

Table 20.1

Durability of Concrete

Chemical attack
   Leaching and efflorescence (P)[a]
   Sulfate attack (P)
   Alkali–aggregate reaction (A) (Chapter 6)
   Acids and alkalis (P)
   Corrosion of metals (R)
Physical attack
   Freezing and thawing (P, A)
   Wetting and drying (P) (Chapter 18)
   Temperature changes (P, A) (Chapter 19)
   Wear and abrasion (P, A) (Chapters 6, 19)

[a]Letter(s) in parentheses indicates the concrete
component most affected, in order of importance: A, aggregate; P, paste; R, reinforcement.

aggressive chemicals and the movement of water during heating or
freezing. The w/c ratio has a dual role to play in concrete durability,
since a lower w/c ratio also increases the strength of concrete and hence
improves its resistance to cracking from the internal stresses that may
be generated by adverse reactions.

Figure 20.1 Influence of w/c ratio on the permeability of: (a) cement
paste [from T. C. Powers, L. E. Copeland, J. C. Hayes, and H. M. Mann,
Journal of the American Concrete Institute, Vol. 51, No. 3, 1954, pp.
285–298]; (b) concrete [adapted from *Concrete Manual*, 8th ed., U.S.
Bureau of Reclamation, Denver, Colo., 1975].

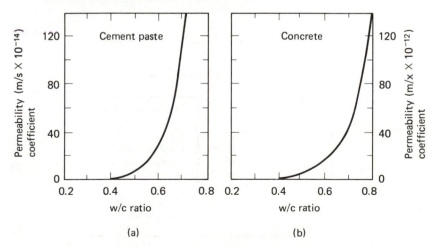

The permeability of concrete can be measured by determining the rate of flow of moisture through a concrete slab. Since the porosity of concrete resides in the paste, the permeability of concrete should be controlled by the paste, but it is modified by internal cracking at the cement–aggregate interface. It should be noted that the flow of water through concrete is of interest in construction aside from considerations of durability. Impermeable concrete is required for water-retaining structures and construction below grade. The term *watertightness,* which is commonly used by the construction industry, is synonymous with impermeability. The flow of water through cement paste obeys *D'Arcy's law* for flow through a porous medium:

$$\nu = K_p \frac{h}{x} \tag{20.1}$$

where ν is the rate of flow of water, h the head of water (hydraulic pressure), x the thickness of the specimen, and K_p the permeability coefficient. K_p is not a constant for cement paste, being dependent on the w/c ratio and the age of the paste, as can be seen in Table 20.2. This is because K_p is strongly dependent on the capillary porosity of the paste, which is controlled by both w/c ratio and degree of hydration. There is justification for this point of view, since it can be seen from Table 20.2 and Figure 20.1 that K_p varies over several orders of magnitude.

In a mature paste the permeability coefficient is very small, even though the total porosity is high and is of the same order as that observed

Table 20.2

Effect of Age of Cement Paste on Its Permeability Coefficient ($w/c = 0.51$)

Age (days)	K_p (m/s)	
Fresh paste	10^{-5}	Independent of w/c
1	10^{-8}	
3	10^{-9}	
4	10^{-10}	Capillary pores
7	10^{-11}	interconnected
14	10^{-12}	
28	10^{-13}	
100	10^{-16}	Capillary pores
240 (maximum hydration)	10^{-18}	discontinuous

Table 20.3

Curing Time Required to Produce
a Discontinuous System of Capillaries (Assuming Continuous Moist
Curing)

w/c Ratio	Curing Time (days)
0.40	3
0.45	7
0.50	28
0.60	180 (6 months)
0.70	365 (1 year)
> 0.70	Not possible

for low-porosity rocks. Thus, it can be concluded that water does not easily move through the very small gel pores and that permeability is controlled by an interconnecting network of capillary pores. As hydration proceeds, the capillary network becomes increasingly tortuous as interconnected pores are blocked by the formation of C–S–H. This is accompanied by a continuous decrease in K_p, and the time at which complete discontinuity of the capillary pores occurs is a function of the w/c ratio (Table 20.3). In concretes with a w/c ratio greater than 0.70, complete discontinuity of capillary pores can never be achieved, even with continuous moist curing, and concretes will have relatively high permeabilities. Even after the capillary pores have been completely isolated by regions of C–S–H and its attendant gel porosity, K_p continues to decrease by several more orders of magnitude. This is due not only to an increase in thickness of C–S–H between capillaries but also to the fact that calcium hydroxide is continuing to grow within the residual capillary pores, thus forming impermeable regions. A limiting value of K_p should occur when all capillary porosity has been eliminated, and this is believed to be less than 10^{-22} m/s.

The foregoing discussion refers to well-compacted pastes continuously moist-cured. If pastes are allowed to dry and then are rewetted, the permeability coefficient is higher. This may be due to changes in pore-size distributions that occur on shrinkage and which allow capillary pores to become partially interconnected again. The effect is even more marked in concrete, since cracking at the paste–aggregate interface will create further opportunities for water flow. Even in saturated concretes, permeability will be increased by imperfect consolidation or excessive segregation of materials, which can create bleeding channels within the paste. Typical values of the permeability of mass concrete used in some dams range from 8 to 35 $\times$ 10^{-12} m/s (2.6 to 11.5 $\times$ 10^{-11} ft/s). The U.S. Bureau of Reclamation has adopted $K_p = 15 \times 10^{-12}$ m/s (4.8 $\times$ 10^{-11} ft/s) as a maximum permeability for some of its work.

20.2 CHEMICAL ATTACK

Leaching and Efflorescence

Efflorescence occurs quite frequently on the surface of concrete when water can percolate through the material, either continuously or intermittently, or when an exposed face is alternately wetted and dried. Efflorescence consists of deposited salts that are leached out of the concrete and are crystallized on subsequent evaporation of the water or interaction with carbon dioxide in the atmosphere. Typical salts are sulfates and carbonates of sodium, potassium, or calcium, the major constituent being calcium carbonate.

Efflorescence, in itself, is an aesthetic rather than a durability problem, but it does indicate that substantial leaching is occurring within the concrete. Extensive leaching causes an increase in porosity, thereby lowering the strength of the concrete and increasing its vulnerability to aggressive chemicals. Calcium hydroxide is the hydration product that is most readily leached from concrete; C–S–H is essentially insoluble and is only decomposed under exposure to severe leaching conditions for long periods of time. Thus, pastes that have a high content of calcium hydroxide are likely to be more prone to leaching and efflorescence and to have a greater potential for deterioration in unfavorable conditions. The rate of leaching is dependent on the amount of dissolved salts contained in the percolating water. Soft waters, such as rainwater, are the most aggressive, while hard waters containing large amounts of calcium ions are less dangerous. The temperature of the water is also a consideration, since calcium hydroxide is more soluble in cold water than it is in warm water. Leaching is, of course, most prevalent when the water can seep through the concrete, particularly under pressure. Concrete is not significantly leached by water flowing over its surface unless accompanied by physical abrasion from suspended solid matter.

Sulfate Attack

Perhaps the most widespread and common form of chemical attack is the action of sulfates on concrete. Sulfates are often present in groundwaters, particularly when high proportions of clay are present in the soil, and seawater has sulfates as a major constituent. Groundwaters

Figure 20.2 Concrete in a sewage disposal plant damaged by sulfate attack. (Photograph courtesy of Bryant Mather.)

may have local concentrations of sulfates in the vicinity of industrial wastes such as mine tailings, slag heaps, and rubble fills. Sulfates present in rainwater from air pollution, or produced by biological growths, may cause slow deterioration even in concrete above ground (Figure 20.2).

Mechanisms of Sulfate Attack

Sulfate attack is actually a rather complex process which may involve a number of secondary processes. However, laboratory and field experiences have definitely established a correlation between the C_3A content of a portland cement and its susceptibility to sulfate attack. As discussed in Chapter 4, the major cause is *sulfoaluminate corrosion,* in which ettringite is formed from monosulfoaluminate:

$$C_4 A \overline{S} H_{12} + 2C \overline{S} H_2 + 16H \rightarrow C_6 A \overline{S}_3 H_{32} \qquad (4.5)$$

This is accompanied by a very large increase in solid volume (Table 20.4), which causes a volume expansion within the paste and which generates accompanying internal stresses and ultimately leads to cracking.

Table 20.4

Processes Involved in Sulfate Attack

Process	Equation	Controlling Variable	Molar Volume Expansion (cm³)
Basic sulfate attack:			
1. Penetration of sulfate ions		K_p and K_d	—
2. Gypsum corrosion	20.2	Concentration of $\overline{S}$, removal of OH^-	41
3. Sulfoaluminate corrosion	4.5	Concentration of $C_4A\overline{S}H_{12}$	254
Additional magnesium attack:			
1. Magnesium corrosion	20.3 20.4	Concentration of $M\overline{S}$	>200 >100
2. Magnesium–gypsum	20.5	Concentration of $\overline{S}$	65.5

Sulfoaluminate attack is initiated by an initial reaction between sulfate ions and calcium hydroxide:

$$CH + SO_4{}^{2-}(aq) \rightleftharpoons C\overline{S}H_2 + 2OH^-(aq) \qquad (20.2)$$

This reaction can be described as *gypsum corrosion,* since it is also an expansive reaction, but gypsum corrosion only contributes directly to concrete deterioration at concentrations of sulfate ions above 1000 ppm and remains of secondary importance until much higher levels are reached (in excess of 4000 ppm). At low levels of sulfate concentration, the importance of Eq. (20.2) is that it encourages the penetration of sulfate ions into the concrete and concentrates them in a form in which they can react directly with monosulfoaluminate.

Sulfate attack can therefore be considered as a sequence of three processes, as outlined schematically in Table 20.4:

1. The first process is the diffusion of sulfate ions into the pores of the concrete, which is controlled by the permeability coefficient (K_p) and the diffusion coefficient of the sulfate ions (K_d).

2. In its initial stages, gypsum corrosion may actually be beneficial, since gypsum is more soluble than calcium hydroxide and the dissolution–crystallization reaction will allow gypsum first to crystallize without expansion. The extent of gypsum formation will depend on whether the hydroxide ions are removed, either by water moving through the concrete (in which case gypsum will also be leached out) or by precipitating as another insoluble compound.

3. As sulfoaluminate corrosion causes internal cracking, the effective K_p of the concrete will be increased, thereby accelerating further sulfate attack.

Effect of Different Sulfates

With relatively small concentrations of sulfate ions, the nature of the accompanying cations does not affect the process of sulfate attack. With increasing concentrations of sulfate ions, this is no longer true, although in all cases the rate of sulfate attack has a less than linear relationship with sulfate concentration. At intermediate concentrations (1000 to 2000 ppm) calcium sulfate may appear more aggressive because of a faster rate of gypsum corrosion, but its effect is limited by the relatively low solubility of gypsum. At the increasingly higher sulfate concentrations that can be achieved with more soluble sulfates, gypsum corrosion has an increasing contribution. Magnesium sulfate can be even more aggressive because of the possibility of additional corrosive reactions due to the presence of magnesium ions, which can decompose both C–S–H and the calcium sulfoaluminates:

$$C_3S_2H_3 + 3M\overline{S}(aq) \rightarrow 3C\overline{S}H_2 + 3MH + 2SH_x \qquad (20.3)$$

$$C_4A\overline{S}H_{12} + 3M\overline{S}(aq) \rightarrow 4C\overline{S}H_2 + 3MH + AH_3 \qquad (20.4)$$

The silica gel formed, as in Eq. (20.3), may react slowly with MH to form a crystalline magnesium silicate which has no cementing properties. At very high concentrations of $MgSO_4$, sulfoaluminate corrosion is totally replaced by magnesium corrosion.

Equations (20.3) and (20.4) proceed because of the insolubility of magnesium hydroxide. Precipitation of MH also increases the rate of gypsum corrosion:

$$CH + M\overline{S}(aq) \rightarrow C\overline{S}H_2 + MH \qquad (20.5)$$

The increase in solid volume is greater in Eq. (20.5) than it is in Eq. (20.2), but the formation of MH is such that it often forms in the existing pores without causing disruptive expansions. When this happens, it tends to seal the concrete and hinder further penetration of sulfates.

Action of Seawater

The protective nature of magnesium hydroxide is one reason why seawater is less corrosive than might be expected from a consideration of typical salt concentrations (Table 5.2). Another reason is that gypsum and ettringite are more soluble in solutions containing the chloride ion, another major constituent of seawater, and this reduces deleterious expansions. Thus, concretes remain intact, although strength may decrease slowly, due to leaching. Deterioration by seawater is greatest in the intertidal zone, where additional destructive mechanisms can occur. Frequent wetting and drying will aggravate the effect of sulfate attack, while the crystallization of sea salts in the concrete on evaporation may also contribute to expansive forces. Further damage may be caused by frost action and is aggravated by wave impact, and abrasion by floating debris. Furthermore, the chloride ion will cause severe corrosion of the reinforcing steel if this is not controlled by the use of low-permeability concrete.

Control of Sulfate Attack

ACI has no standard defining critical sulfate levels, but those specified by the Canadian Standards Association (Table 9.7) are also widely used in the United States. Under the provisions of Table 9.7, seawater would be classed "severe," but when concrete is totally immersed the protection given by the formation of magnesium hydroxide allows it to be classed as "mild." If the sulfate content is less than 150 ppm in groundwater or 0.10% in soils, no special precautions are taken. The w/c ratio is an important factor in controlling sulfate attack because it determines the permeability of the concrete, and Table 9.7 gives allowable limits of w/c ratio for different sulfate levels, as well as recommendations on cement type.

The use of admixtures such as pozzolans and blast-furnace slag can improve sulfate resistance. Type IP or IS cements can be used instead of a Type II cement (see Chapters 3 and 7). Use of Type VP or VS cements gives superior protection under very severe sulfate conditions where even Type V may not give satisfactory protection. Under such extreme conditions, supersulfated slag cements, if available, are an alternative (see Chapter 3). Since ettringite is the stable sulfate phase in the mature paste, the reaction described by Eq. (4.5) cannot occur.

Another way of reducing sulfate attack that can be used for precast concrete products is high-pressure steam curing. Curing at elevated temperatures below 100°C may not improve sulfate resistance, and may even decrease it. Sulfate resistance improves with an increase in curing temperature above 100°C (Table 20.5). Steam curing, with the use of a silica addition, removes calcium hydroxide from the hydrated pastes, and calcium sulfoaluminate hydrates are no longer stable phases. The alumina is incorporated into C–S–H, and may form some C_3AH_6, which is much more resistant to attack by sulfate ions. More stable crystalline calcium silicate hydrates are formed at higher temperatures.

Table 20.5

Sulfate Resistance of Autoclaved Mortars (Time to Reach Given Expansion)[a]

Solution	Linear Expansion (%)	Temperature of Curing (24 h)				
		$21°C^b$	$100°C$	$125°C$	$150°C$	$175°C$
2.1% NA_2SO_4	0.02	3 days	10 days	1 yr	1 yr	> 1 yr
Sat'd $CaSO_4 \cdot 2H_2O$	0.02	7 days	18 days	1 yr	1 yr	> 1 yr
1.8% $MgSO_4$	0.20	6 days	32 days	75 days	130 days	> 1 yr

[a]From F. M. Lea. *The Chemistry of Cement and Concrete*, 3rd ed., Chemical Publishing Co., New York, 1971.
[b]$°C \times {}^9/_5 + 32 = °F$.

Although calcium hydroxide does play an important role in sulfate attack, the susceptibility of a cement to sulfate attack depends on how much ettringite can potentially form. This is usually related to the amount of monosulfoaluminate in the hydrated paste, which relates back to the C_3A content. Expansive cements are generally susceptible to sulfate attack because the ettringite that causes early expansion eventually converts to monosulfoaluminate during later hydration. However, if ettringite can be maintained as the stable sulfoaluminate hydrate, as can occur in a Type K cement, which is made by blending $C_4A_3\overline{S}$ with a Type V cement, sulfate resistance will be good. The form of alumina that occurs in some pozzolans and slags can also react with sulfates to form ettringite, and this should be kept in mind when contemplating the use of such admixtures.

Tests for Sulfate Attack

At present, no standard ASTM test exists for assessing the sulfate resistance of a concrete, and predictions of performance cannot be made with certainty. However, intelligent use of Table 9.7, with due allowance for special circumstances, such as percolation of groundwater through concrete, acid sulfate waters, and so on, should ensure adequate resistance in most cases. There is still no satisfactory, rapid test for sulfate attack, and many different approaches have been used. ASTM C452, which is proposed for research purposes only (Chapter 3), uses an abnormally high gypsum content in the cement to test for sulfate resistance. It is a test for the cement only, not for concrete. This procedure has several drawbacks, the major objections being that the test is too slow for use as a routine specification test, that the behavior of concrete does not necessarily parallel the behavior of mortar bars, and that realistic exposure conditions are not reproduced. Furthermore, tests that are used with portland cements do not always adequately assess the behavior of cements blended with pozzolans or slags.

Attack by Acids and Bases

Hydrated cement paste is an alkaline material and therefore specific attack by alkaline materials will not normally be encountered. High concentrations of alkaline materials that may come in contact with concrete in industrial processes cause deterioration by processes other than direct chemical reaction with hydroxide ions. The situation is entirely different for acidic solutions, which will readily attack basic materials such as concrete. Generally, naturally occurring acidic ground-waters are not common, being confined to marshy or peaty regions, where extensive decomposition of organic matter occurs. Acidic waters may also occur in, or adjacent to, landfilled areas and in places where mining operations and stockpiling of mine tailings have occurred. Highly acidic conditions may exist in agricultural and industrial wastes, partic-ularly from the food- and animal-processing industries.

The hydrogen ion will accelerate the leaching of calcium hydroxide:

$$Ca(OH)_2 + 2H^+ \rightarrow Ca^{2+} + 2H_2O \tag{20.6}$$

and, if highly concentrated, C–S–H may also be attacked, forming silica gel:

$$3CaO \cdot 2SiO_2 \cdot 3H_2O + 6H^+ \xrightarrow{H_2O} 3Ca^{2+} + 2(SiO_2 \cdot nH_2O) + 6H_2O \tag{20.7}$$

The nature of the anion that accompanies the hydrogen ion may further aggravate the situation. Both sulfuric acid and carbonic acid are common constituents of acid groundwaters. The sulfate ion will obviously partic-ipate in sulfate attack, and thus sulfuric acid is particularly corrosive. Carbonic acid can also be very corrosive because of the formation of soluble calcium bicarbonate:

$$Ca(OH)_2 + H_2CO_3 \rightarrow Ca(HCO_3)_2 + H_2O \tag{20.8}$$

Any acid that can form soluble calcium salts in a like manner will be particularly aggressive, while if the acid forms an insoluble calcium salt, its precipitation during acid attack can protect the concrete against further deterioration. Some examples are given in Table 20.6.

Sugar, although not an acid, is another substance that dissolves more than just calcium hydroxide; both C–S–H and the calcium alumi-nate hydrates will be slowly attacked. For this reason sugar in solution

Table 20.6

Acid Attack of Concrete

Acid	Formula	Likely Occurrence
Aggressive Acids that Form Soluble Calcium Salts		
Hydrochloric acid	HCl	Chemical industry
Nitric acid	HNO_3	Fertilizer manufacture
Acetic acid	CH_3CO_2H	Fermentation processes
Formic acid	$H \cdot CO_2H$	Food processing and dyeing
Lactic acid	$C_2H_4(OH) \cdot CO_2H$	Dairy industry
Tannic acid	$C_{76}H_{52}O_{46}$	Tanning industry, peat waters
Acids that Form Insoluble Salts		
Phosphoric acid	H_3PO_4	Fertilizer manufacture
Tartaric acid	$[CH(OH) \cdot CO_2H]_2$	Winemaking

is very aggressive to concrete, and should not be allowed to come in direct contact for more than brief periods.

It should be noted that corrosive chemicals can only attack concrete when water is present. Thus, concrete can be used to store dry chemicals.

Crystallization of Salts

We have been concerned hitherto with chemical attack of concrete, but salts can also cause damage to concrete through the development of crystal growth pressures that arise through physical causes. Corrosion of this type occurs when concrete is placed in contact with water containing considerable quantities of dissolved solids. These salts will permeate into the concrete and will crystallize in the pores as the salts are concentrated by evaporation. Repeated or continuing evaporation will cause salt deposits to build up to the point where they will cause cracking. This problem is observed in regions between fluctuating water levels, for example in the intertidal region in marine exposures, but it can also occur when concrete is in contact with groundwaters rich in salts. In such cases, capillary effects can cause the salt-laden water to rise several hundreds of millimeters above the water level. If the water table is close to the ground surface, the capillary rise zone may extend above ground, where evaporation will take place most readily. Thus, damage generally occurs just above ground level. Damage of this type is found to be severe in certain areas of the Middle East, where high salt levels and high rates of evaporation make the problem acute. Capillary rise will also aggravate crystallization problems that occur with fluctuating water levels above ground. This problem can be corrected by sealing the concrete, either to prevent ingress of moisture or subsequent evaporation. A barrier can be built into the structure to prevent capillary effects from occurring. Alternatively, concrete below ground may be surrounded by an impervious clay fill to keep salts from coming in contact with the concrete.

Corrosion of Sewer Pipes

Concrete sewer pipes may be exposed to severe corrosion conditions. Not only may the pipes be corroded by substances in the groundwater (sulfates or acids), but they are also more prone to corrosion from the sewage itself. Furthermore, the thinness of the concrete pipe wall can make the effect of corrosion critical in a much shorter time. If an unbalanced water pressure exists, flow of water through the pipe wall may cause problems of leaching or salt deposition if wall thicknesses are less than 300 to 400 mm (12 to 16 in.). The *w/c* ratio of the concrete will obviously be an important parameter.

Domestic sewage or animal manures are usually harmless to concrete unless the reinforcing steel is inadequately protected. Acids or organic chemicals in industrial wastes can cause problems, as discussed above. However, the greatest problem in sewer pipes is corrosion caused by bacteria. The metabolic activity of many bacteria forms acids that may attack concrete, but the most harmful result of bacterial action in sewers is the formation of hydrogen sulfide in the sewage from the reduction of sulfate compounds by anaerobic bacteria. The hydrogen sulfide gas dissolves in water films in the upper part of the pipe, where there will be sufficient air to allow oxidation back to sulfate compounds by aerobic bacteria. In this way, high concentrations of sulfuric acid can be formed locally which can corrode the upper part of the pipe very rapidly.

Protective treatment of sewer pipe surfaces will prolong the life of installations. Coatings of bitumens, tars, and resins have been used. Chemical treatments of the concrete surfaces are another possibility, and several are available which, when exposed to the concrete, precipitate insoluble salts to fill the pores or provide a resistant surface coating. Examples are water glass (sodium silicate), insoluble soaps, fluoride salts, and iron compounds. The surfaces can also be treated with carbon dioxide gas or silicon tetrafluoride vapor (the Okrat process). These gases form highly insoluble $CaCO_3$ or CaF_2, respectively, which seal the surfaces.

Corrosion of Metals

Concrete reinforcement is invariably made from mild steel which is susceptible to corrosion under the right conditions. The formation of rust is an expansive reaction that will lead to cracking and spalling of the concrete above the rusting steel. Corrosion of other metals may also be of importance and will be discussed briefly.

Mechanism of Corrosion

Rusting is an electrochemical process that requires a flow of electrical current for the chemical corrosion reactions to proceed. For electrochemical (galvanic) corrosion to occur, two dissimilar metals must come in electrical contact in the presence of oxygen and moisture (Figure 20.3a). Oxidation–reduction processes can then take place at the surfaces of the two metals, and under normal conditions iron becomes the *anode* and oxidation of iron takes place:

$$\text{anode reaction:} \quad Fe - 2e \rightarrow Fe^{2+} \qquad (20.9a)$$

At the other metal surface, the cathode, reduction of oxygen occurs:

$$\text{cathode reaction:} \quad 2H_2O + O_2 + 4e \rightarrow 4OH^- \qquad (20.9b)$$

Figure 20.3 Corrosion of iron: (a) General conditions for electrochemical corrosion; (b) rusting of an isolated steel bar.

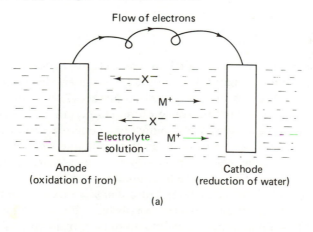

(a)

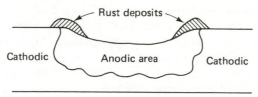

(b)

Oxidation and reduction can occur only if electron transfer can take place, so that the net redox reaction [Eq. (20.10a)] can proceed with the precipitation of ferrous hydroxide:

$$2Fe + 2H_2O + O_2 \rightarrow 2Fe(OH)_2 \tag{20.10a}$$

The spontaneous oxidation of ferrous oxide to hydrated ferric oxide (rust) occurs rapidly:

$$2Fe(OH)_2 \xrightarrow{O_2,\ H_2O} 2Fe(OH)_3 \rightarrow Fe_2O_3 \cdot nH_2O \tag{20.10b}$$

A separate cathodic metal is not required for corrosion of steel to occur, since an isolated bar will rust spontaneously. This is because different areas of the bar may develop "active sites" with different *electrochemical potentials* (i.e., different tendencies for oxidation), and thus set up anode–cathode pairs *(galvanic couples)*, as illustrated in Figure 20.3b. Corrosion occurs in localized anodic areas. The development of anodic and cathodic areas can be caused by a variety of conditions, such as different impurity levels in the iron, different amounts of residual strain, or different concentrations of oxygen or electrolyte in contact with the metal.

Corrosion in Concrete

For corrosion of steel embedded in concrete to occur, the following conditions must all be met: (1) the provision of an anode–cathode couple with at least part of the steel acting as an anode, (2) the maintenance of an electrical circuit, (3) the presence of moisture, and (4) the presence of oxygen. In a good-quality, well-compacted concrete, reinforcing steel should not be liable to corrosion, even though active sites resulting from inhomogeneities in the steel, or nonuniform concentrations, are likely to be prevalent. This is because the high-alkaline conditions present within concrete (a pH of about 12 to 12.5) cause a passive oxide film to form on the surface of the iron and prevent corrosion. This is analogous to the passive layer of alumina that prevents aluminum from corroding rapidly under normal conditions of use. The passive iron oxide layer is destroyed when the pH is reduced to about 11.0 or below, causing the normal, porous oxide layer (rust) to form during corrosion. This critical reduction of pH occurs when calcium hydroxide, which maintains the high pH in cement paste, is converted to calcium carbonate (calcite) by atmospheric carbonation. In a well-cured concrete with a low *w/c* ratio, the depth of the carbonated zone is unlikely to exceed 25 mm, and therefore a concrete cover of 25 to 40 mm over reinforcing bars should provide adequate protection from corrosion in most instances. Where more severe conditions of exposure are encountered or concrete with a fairly high permeability is used, the cover should be increased to at least 50 mm. Penetration of CO_2 can be accelerated by the presence of micro-cracks in the concrete due to the effects of external or internal stress. Destruction of the passive film at the base of the crack initiates corrosion, which may occur well beyond the actual area of exposed steel. The rate of penetration of CO_2 depends on the width of the crack. If the crack can be kept below 0.025 mm (0.001 in.), by suitable spacing of reinforcement, then CO_2 diffusion will be quite slow.

Even in fully carbonated concrete, where the reserve of alkalinity has been lost, corrosion of the reinforcing steel may not be a problem. The same applies to concretes made with slag cements, calcium aluminate cements, and portland–pozzolan cements, where CH is absent or much reduced in quantity. Under such conditions the rate of corrosion depends on the availability of oxygen and water at the surface of the steel. Concretes with a low permeability will limit the rate of corrosion by limiting the rate of diffusion of oxygen. If the concrete pores are filled with water, diffusion of oxygen will be further reduced. Thus, very low or very high moisture contents will prove beneficial to corrosion, while intermediate moisture contents may be dangerous.

Effect of Chloride Ions

Chloride ions have the special ability to destroy the passive oxide film of steel even at high alkalinities. The amount of chloride required to initiate corrosion depends on the pH of the solution in contact with the steel (Figure 20.4), and comparatively small quantities are needed to offset the basicity of portland cement. Chlorides may enter into concrete from three major sources: from $CaCl_2$ added as an accelerating admixture; from de-icing salts used on pavements and bridge decks; and from seawater or salt spray. Prestressed elements should not be exposed to chloride without adequate protection, since when steel is under tensile stress, corrosion may proceed more readily, owing to rupture of the passive film, and *stress corrosion* may occur. Thus, the chloride ion may accelerate corrosion at lower concentrations than usual (see Chapter 7). The corrosion of reinforcing steel in bridge decks that are regularly treated with de-icing salts is one of the most acute durability problems in the United States. De-icing salts are generally mixtures of NaCl and $CaCl_2$, and much of the salt will penetrate into the pores of the concrete and slowly diffuse down to the reinforcement. Little chloride is lost once it enters the concrete, so that there is a steady buildup of chloride ion

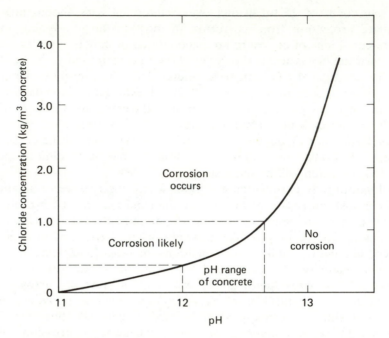

Figure 20.4 Effect of chloride on the corrosion of steel reinforcement.

until critical concentrations (0.6 to 1.2 kg Cl^-/m^3 or 1.0 to 2.0 lb Cl^-/yd^3 of concrete) are reached at the rebar level. Corrosion of reinforcing steel in structures exposed to seawater or sea spray is another severe problem. It has been shown that the high concentration of chloride ion in seawater is such that it will penetrate even into high-quality concrete beyond the depth of cover normally specified. Whether severe corrosion will occur depends on other factors, such as the availability of oxygen at the corrosion site and the extent of other forms of deterioration that can take place in seawater.

Protection against Corrosion

There are several strategies that have been developed to combat corrosion of reinforcing steel, based largely on work that has been done in recent years on the bridge-deck problem. The possible strategies can be divided into four categories: (1) reduction of permeability of the concrete, (2) protective coatings on the concrete, (3) protective coatings on the steel, and (4) suppression of the electrochemical process. The use of high-quality, impermeable concrete with a low *w/c* ratio and adequate concrete cover can do much to alleviate the problem. Although chloride ions will eventually fully penetrate concrete regularly exposed to de-icing salts or seawater, the rate of penetration can be considerably reduced. The rate of diffusion of oxygen will also be reduced and its availability may limit the rate of corrosion even when the critical chloride ion concentration has been reached. Furthermore, concrete with a low *w/c* ratio has better resistance to cracking and spalling because of its higher strength. Nevertheless, even greater reduction of permeability could be advantageous on bridge decks, and the use of partial or full polymer impregnation or integral sealing with wax beads incorporated in the concrete has been suggested.

Protective coatings are also commonly used on bridge-deck surfaces. Treatments with water-repellant materials, such as linseed oil in kerosene and asphalt emulsions, have traditionally been used, but these are eventually destroyed by traffic wear and weathering. Overlays such as asphalt, polymer concretes, or polymer latex-modified concrete, can be very effective. Generally, such overlays are placed over deteriorated concrete and their performance is often limited by the state of the existing concrete and the efficiency of the repair operation, but there is no reason why overlays could not be applied to new bridge decks. Overlays of this type offer low permeability, good crack resistance, and good bonding properties. A waterproof membrane between the normal concrete and the overlay is needed to serve as an additional barrier to penetration of chloride salts.

Protective coatings can be applied to the reinforcing steel. Metal coatings, such as zinc galvanizing, are often used to protect steel from corrosion, and galvanized reinforcing bars have been used successfully in concrete. Other metals, such as cadmium, nickel, and copper, have also been tested. Zinc and cadium provide *sacrificial anodic protection* because they are more prone to corrosion and act as anodes, making the iron cathodic. Although such coatings provide protection, they may be insufficient in the presence of chloride ions, which accelerate the corrosion of the coatings. The corrosion products may themselves cause distress in the concrete, and protection ceases when all the metal is used up. Nickel and copper act as inert coatings at high pH, although they, too, can be susceptible to corrosion in the presence of the chloride ion. Thus, they act more like other inert coatings that have been used, such as epoxy resins, asphalt, and synthetic rubbers. Coatings can provide good protection if they provide a continuous film without holes and weak spots; a break in a film may cause severe localized corrosion at that point. The effectiveness of coatings depends on how well they retain their integrity during fabrication (including bonding to the steel), transportation, handling during erection of the structure, and under the service environment.

Suppression of the electrochemical corrosion of iron is the basis of *cathodic protection,* which is used to protect steel structures and appears to be another promising method of protecting reinforced concrete structures. A current is applied to the rebars (which must be connected electrically) in the opposite direction to the current flow in spontaneous corrosion, using an inert anode. If the current is sufficiently large, the iron is made cathodic and corrosion is prevented. A sacrificial anode can be used instead of galvanizing. The rebars are connected electrically to zinc or magnesium bars buried close to the structure. The iron is now cathodic with respect to these metals, which corrode without damaging the structure and can be replaced as necessary.

Various chemicals are known to inhibit corrosion of steel. Those that have been used in concrete are chromates, nitrites, benzoates, phosphates, stannous salts, and ferrous salts. It appears that such chemicals may act either to suppress the anodic reaction by stabilizing the passive oxide film or forming a new insoluble coating (chromate, phosphates), or to suppress the cathodic reaction by scavenging oxygen (nitrites, benzoates, stannous and ferrous salts).

Corrosion of Other Embedded Metals

Aluminum can be severely corroded by continued exposure to moist concrete, since its passive film is destroyed at high alkalinities. Corrosion is aggravated in the presence of chloride ions (as is the case for steel) or if the aluminum is in contact with reinforcing steel (since it then acts as a sacrificial anode). Spalling of concrete and destruction of aluminum conduit have been reported in some instances. Lead and zinc behave similarly, although the latter is more resistant to corrosion. Copper and copper alloys, such as brass and bronze, have good resistance to corrosion unless chlorides are present.

20.3 PHYSICAL ATTACK

Freezing and Thawing

Porous materials containing moisture are susceptible to damage under repeated cycles of freezing and thawing (frost attack). Hardened cement paste, which has a high porosity, is particularly susceptible to such conditions, and concrete may be destroyed in a single winter in northern climates. Fortunately, as discussed in Chapter 7, air entrainment has proved to be an effective and reliable means of protecting concrete from frost attack. It is of interest to examine the mechanism by which damage occurs on repeated freezing and thawing and thereby determine the critical factors affecting frost resistance and the reasons why air entrainment is effective.

Freezing of Cement Paste

When a saturated concrete is cooled to below 0°C, immediate freezing of most of the water in the cement paste does not occur. It must be remembered that paste contains a wide spectrum of pore sizes, and it can be shown thermodynamically that water in capillary pores will not freeze until the temperature is lowered below 0°C by an amount that depends on the diameter of the pore. For example, water in pores of 10-nm diameter will not freeze until −5°C (23°F), and in pores of 3.5-nm diameter water will not freeze until −20°C (−4°F). Furthermore, water adsorbed on the surfaces of C–S–H, which forms the surfaces of capillary pores and also creates micropores within the paste, will never freeze, although it may migrate to the capillary pores, where it can freeze. It has been shown that in an unprotected paste severe dilation accompanies freezing (Figure 20.5), which leads to internal tensile stresses and cracking. In an air-entrained paste very little dilation occurs and considerable shrinkage during freezing is observed at low temperatures.

Mechanism of frost attack. Several different processes can contribute to paste behavior during freezing. These are generation of hydraulic pressure by ice formation, desorption of water from C–S–H, and segregation of ice. A 9% volume increase occurs as water turns to ice, but this change is insufficient to account for all of the dilation observed in cement paste and in concrete. Some of the dilation that occurs on first freezing is probably due to direct expansion of ice in microcracks, and this will increase if progressive microcracking occurs during continued freezing and thawing.

However, the major dilation that occurs was attributed by Powers to the generation of *hydraulic pressure*. As ice forms in a capillary, the accompanying volume increase causes the residual water to be compressed. This pressure can be relieved if the water can escape from the capillary to a free space by diffusing through unfrozen pores, but if the water has too far to move to an escape boundary, the capillary will tend to dilate and the surrounding material will come under stress (see Figure 20.6). The superposition of pressure from adjacent capillaries will eventually cause the tensile strength of the paste to be exceeded and rupture will occur. As the temperature is progressively lowered, more capillary water is involved in freezing, increasing the hydraulic pressures and

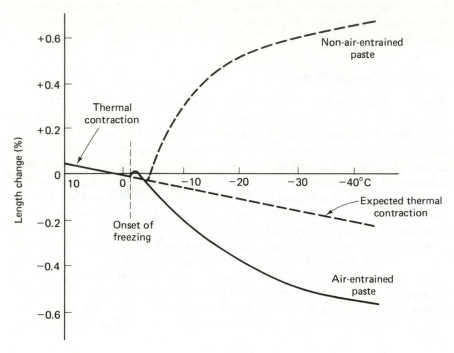

Figure 20.5 Volume changes occurring in cement pastes as the temperature is lowered. (Adapted from T. C. Powers and R. A. Helmuth, *Proceedings of the Highway Research Board,* Vol. 32, 1953, pp. 285–297.)

Figure 20.6 Creation of hydraulic pressure in frozen cement paste: (a) non-air-entrained paste; (b) air-entrained paste.

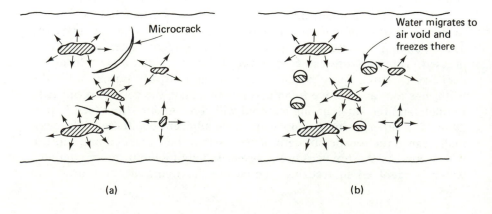

(a)   (b)

thereby increasing microcracking and dilation. In a saturated non-air-entrained paste, the only free space is the exterior of the specimen, and the diffusion of water to the outside is too slow to relieve the hydraulic pressure. Thus, the inclusion of entrained air provides empty space within the paste to which the excess water can move and freeze without damage. The bubbles act as "safety valves" and the spacing factor (see Chapter 7) determines the average distance the water must travel to reach the free space. This distance must not be too great if hydraulic pressure is to be relieved; hence the requirement of a critical spacing factor.

When properly air-entrained, concrete does not dilate on freezing; rather, a contraction is observed. This contraction is greater than predicted by thermal change alone and is believed to be due to the fact that less water is held by the gel at temperatures below freezing. Powers considered that this was a result of water in gel pores diffusing to large capillaries or voids to form microscopic ice lenses, since the thermodynamic basis for this phenomenon is the fact that the vapor pressure in equilibrium with ice is lower than it is over supercooled water at the same temperature. Thus, the effective relative humidity is lowered, and part of the water adsorbed on the gel surfaces is removed to achieve equilibrium again. The removal of water from the gel further increases the hydraulic pressure caused by freezing and explains why dilation can continue when all freezable water has frozen. Litvan, however, suggests that this desorption of water is solely responsible for creating hydraulic pressure within the porous body, that ice formation is not necessary, and that ice will not form in the small capillaries found in the cement paste.

Using the mechanisms described above, it can be shown that the resistance to freezing and thawing depends on the permeability, the degree of saturation of the paste, the amount of freezable water, the rate of freezing, and the average maximum distance from any point in the paste to a free surface where ice can form safely.

Partially dry concrete will not suffer freeze–thaw damage, because the larger capillaries are now empty and provide the necessary free space throughout the paste. The critical degree of saturation has been found to be close to 0.80 of the total freezable water and about 0.85 if expressed as a fraction of the total evaporable water (which includes the gel pore water, which does not freeze). Air entrainment is thus required for protection only for concrete that is frozen in near-saturated conditions. The entrained air bubbles do not readily fill with water except when the concrete is immersed for long periods of time. It can be assumed that the surfactants that cause air entrainment in the first place are concentrated at the surfaces of the air bubbles and make it difficult for water to enter. Once the air bubbles are filled with water, however, the concrete has a greater amount of freezable water and will become highly susceptible to frost damage.

The buildup of ice in particular parts of the structure may aggravate frost damage in concrete, although it is not the basic cause. The formation of *ice lenses,* as this phenomenon is called, occurs in soils and is responsible for frost heave. The development of ice lenses in concrete will require the presence of large voids, which could be microcracks, or voids occurring under reinforcement or large aggregate particles due to segregation or poor compaction. An external supply of water that can readily migrate to these voids is also needed.

It can be shown that concrete should develop a certain strength in order to withstand frost damage (i.e., fresh concrete will be damaged by freezing even when air entrained). Furthermore, theory predicts that there will be a *w/c* ratio (about 0.36) below which air entrainment is not needed if the paste is completely hydrated, since there is little freezable water in the concrete. However, it is generally advisable to air-entrain all concretes, even those made with low *w/c* ratios.

Freezing of Aggregates

Certain rocks are also susceptible to damage on freezing and thawing, and if present in aggregate may contribute to concrete damage. The same concept of development of hydraulic pressure can be applied to rocks (as to all porous materials). There is thus a critical size of aggregate above which the aggregate is liable to fracture. This critical size is a measure of the maximum distance water must flow to reach the outside surface in order to relieve hydraulic pressure. This size depends on freezing rate, degree of saturation, permeability, and tensile strength (i.e., the same factors as for cement paste). For most rocks the critical size is greater than the maximum size used for coarse aggregates. Most rocks (limestones and granites, for example) have very little porosity, and since their permeability may be low, they do not saturate readily in concrete. Therefore, they have internal voids available to relieve hydraulic pressures. Very porous rocks, such as sandstones or synthetic lightweight aggregates, have a high permeability, so that water can readily escape during freezing and a high degree of saturation is not critical. The most critical situation is rocks containing fine pores but a relatively high absorption (total porosity) combined with a low permeability. Once saturated, the critical size of the rocks can become less than the size present in the aggregate and fracture of the aggregate can occur. This can occur in some fine-grained rocks, such as cherts and shales, and causes pop-outs at the surface of concrete. Rapid damage is most

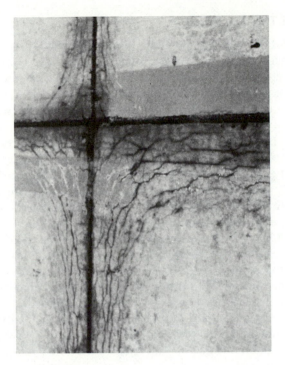

Figure 20.7 D-cracking in a concrete pavement. (Photograph courtesy of Bryant Mather.)

prevalent at the surface because the aggregate is more prone to saturation at this point and freezing and thawing occurs more frequently. However, D-cracking (see Table 20.9 and Figure 20.7) is a more serious consequence of freezing and thawing damage due to aggregates in midwestern states. Deterioration usually starts in the lower parts of concrete slabs adjacent to joints, where moisture accumulates, and eventually progresses through the entire slab. When D-cracking is observed at the surface, deterioration is well advanced in the lower part of the slab, and this generally occurs within 10 to 15 years.

Even though aggregates with high absorptions may not be damaged by frost action, the water that is forced from the aggregate must either be accommodated in the entrained air bubbles of the surrounding paste or be expelled at the surface of the concrete. If too much water is expelled, the paste may not be able to accommodate it all and thus suffers damage. Water expelled from aggregates could also contribute to the formation of ice lenses under special circumstances.

Freezing and Thawing Tests

A number of different tests have been developed to assess the long-term frost resistance of concrete. These involve subjecting concrete to different freeze–thaw cycles and measuring the progressive internal damage by monitoring weight loss, length change (dilation), decrease in strength, or dynamic modulus of elasticity. Tests differ in the nature of the freeze–thaw cycle used and the condition of the specimen during the test, but all tests are arbitrary, since they involve only one set of conditions. As we have seen, the potential for frost damage depends on rate of freezing, w/c ratio, time of moist curing, and degree of saturation, and these may vary widely in actual service conditions. Thus, predictions from freeze–thaw tests are not always a good indication of actual service performance, and results should be treated with caution.

The standard method of test described in ASTM C666 prescribes a cooling and heating cycle between 4.4 and $-17.8°C$ (40 to 0°F) at a rate of cooling in the range of 12.2 to 14.4°C/h (22 to 26°F/h). Freezing can be done while the specimen is in air (procedure A) or immersed in water (procedure B), but it is thawed in water in both cases. The test is continued (either continuously or intermittently) until 300 cycles or until the dynamic modulus has reached 60% of its initial value, whichever occurs first. A durability factor (D.F.) can then be calculated:

$$D.F. = \frac{P \times N}{300} \qquad (20.11)$$

where P is the percentage of the initial dynamic modulus after N cycles. There are no definite values of the durability factor for acceptance or rejection of the concrete subjected to this test. A value of less than 40 suggests that the concrete may be unsatisfactory, whereas above 60 it is likely to perform well, but there is no guarantee that this prediction will hold true.

A major objection to ASTM C666 is the unrealistically high rate of freezing, since in the field rates exceeding 3°C/h (5°F/h) are rarely encountered. ASTM C671 overcomes this objection by specifying one freeze–thaw cycle every 2 weeks, employing this rate of freezing, with a rapid thawing regime. Damage is monitored by measuring linear expansion during the tests, which is continued until critical dilation (i.e., until dilation begins to increase sharply) or the desired number of cycles is completed. This test allows the user to specify the curing history of the specimen and the exposure conditions to most nearly match the expected service conditions. ASTM C682 for freezing and thawing of aggregates was mentioned in Chapter 6. This test involves casting 3×6 in. (75×150 mm) concrete cylinders, made with the aggregate in question, and having an air content of 6%. They are then tested and the results analyzed in accordance with ASTM C671.

Salt Scaling

Concrete that is adequately air-entrained for frost resistance may nevertheless be damaged by repeated application of de-icing salts. We have already discussed damage that can arise from accelerated corrosion of reinforcing steel caused by the chloride salts commonly used, but even properly air-entrained high-quality concrete may be damaged directly by de-icing chemicals that contain no chloride. Concrete that has suffered salt scaling becomes roughened and pitted due to the spalling of small pieces of mortar.

Mechanism of Scaling

The exact causes underlying salt scaling are not known, but they probably involve more than one process. It has been suggested that the consumption of heat required to melt ice when de-icer is applied causes a rapid drop in the temperature of the concrete just below the surface, which may cause damage either from the effects of rapid freezing or stress caused by differential thermal strains. The additional free moisture now present at the surface of the concrete may encourage the growth of microscopic or macroscopic ice lenses near the surface, where ice formation can still occur. Osmosis has also been suggested as a mechanism of salt scaling. De-icing chemicals can accumulate in the concrete just below the surface of the concrete to form relatively concentrated solutions. When rainwater accumulates on the surface, the phenomenon of osmosis occurs, whereby water flows to equalize concentration differences. Considerable osmotic (hydraulic) pressures can be created by this effect, causing rupture of pastes. The most damage usually occurs in the late winter at places where rainwater can accumulate and remain for some time.

Prevention of Scaling

Scaling is most likely to occur on surfaces that have been overvibrated, trowelled too early and too long, subjected to plastic shrinkage, or where excessive bleeding occurred. Such surfaces tend to have a weak layer of paste or mortar either at the surface or just below, and may have

Table 20.7

Minimum Moist-curing Times to Develop Salt Scaling Resistance[a]

Cement Type	Acceleration (% $CaCl_2$)	Minimum Curing Period (days)		
		23°C (73°F)	4°C (39°F)	−4°C (25°F)
I	0	7	15	>60
	2	7	7	30
II	0	7	12	35
	2	7	7	28
III	0	7	7	24

[a]From H. Woods, *Durability of Concrete Construction,* Monograph No. 4, American Concrete Institute, Detroit, Mich., 1968.

microcracks or bleeding channels that can transport surface solutions to lower levels. Careful attention to mix design, placing, and finishing should eliminate many potential problems. If adequate moist curing is followed by a period of drying before de-icing chemicals are applied, scaling should not be a problem. Table 20.7 indicates recommended curing times for air-entrained pavements. ASTM C672 is used for assessing scaling resistance of concrete surfaces. A solution of $CaCl_2$ is ponded on the surface of the specimens, which are then subjected to freezing–thawing cycles. (The freeze–thaw cycles are not the same as in the freeze–thaw durability tests.) Visual examination of the surface is made every five cycles.

20.4 CRACKING IN CONCRETE

Throughout this book we have continually referred to potential cracking of concrete. Cracking may be caused by many different situations and may range from very small internal microcracks that occur on the application of modest amounts of stress, through to quite large cracks caused by undesirable interactions with the environment. In extreme cases the structural integrity of the concrete may be seriously affected. In many other instances, however, cracks do not affect the ability of the concrete to carry load, but may affect the durability of the concrete by providing easy points of access to the body of the concrete for aggressive agents that might otherwise not seriously affect the material. In this section we briefly review the causes of cracking and discuss methods of control.

Causes of Cracking

Table 20.8 summarizes the kinds of cracking that can occur by interactions involving the materials of concrete and its surroundings. In most instances cracking originates internally, forming a network of microcracks throughout the concrete. Internal damage may be considerable before cracks are visible at the exterior surfaces. In other cases, such as humidity and temperature changes, localized large cracks may occur in the structure. Cracking may be used to help predict the cause of deterioration of concrete, since in many cases characteristic cracking patterns are produced (see Table 20.9). Concrete that will resist cracking under normal environmental conditions cannot be assured of remaining intact under catastrophic conditions, such as fire.

Control of Cracking

In most instances cracking can be avoided by proper specification of materials provided that the potential problem has been anticipated through a careful assessment of the expected environment. For example, unsoundness of cement should never be encountered when ASTM C150 is adhered to, and proper testing of groundwaters should enable severe sulfate attack to be avoided through choice of an appropriate cement and w/c ratio.

Chemical attack of concrete involves ingress of moisture either as a carrier for aggressive agents or as a participant in destructive reactions. Thus, precautions in mix design and construction practices that prevent the entry of water into, and passage through, concrete should improve durability. Concrete of low permeability can be assured by the use of sufficient cement contents and low w/c ratios, proper placement, consolidation and finishing, and adequate moist curing. Provision of proper drainage and the use of watertight construction joints can be very beneficial in some situations.

The use of surface sealing compounds can be of value in severe environments if they are properly applied. These can be of two types:

1. Materials that are applied to the surface to provide a waterproof, impervious coating.

2. Materials that penetrate the surface of the concrete and block the capillary pores.

The first type is exemplified by paints, but is more likely to be thicker, more robust coatings or overlays based on asphalt, epoxy resin,

Table 20.8

Causes of Cracking in Concrete

Component	Type	Cause of Distress	Environmental Factor(s)	Variables to Control
Cement	Unsoundness	Volume expansion	Moisture	Free lime and magnesia
	Temperature cracking	Thermal stress	Temperature	Heat of hydration, rate of cooling
Aggregate	Alkali–silica reaction	Volume expansion	Supply of moisture	Alkali in cement, composition of aggregate
	Frost attack	Hydraulic pressure	Freezing and thawing	Absorption of aggregate, air content of concrete, maximum size of aggregate
Cement paste	Plastic shrinkage	Moisture loss	Wind and temperature	Temperature of concrete, protection of surfaces
	Drying shrinkage	Moisture loss	Relative humidity	Mix design, rate of drying
	Sulfate attack	Volume expansion	Sulfate ions	Mix design, cement type, admixtures
	Thermal expansion	Volume expansion	Temperature change	Temperature rise, rate of change
Reinforce-ment	Electro-chemical corrosion	Volume expansion	Oxygen, moisture	Adequate concrete cover

synthetic rubber, or other polymeric materials. Such coatings may also be used to keep moisture in and reduce drying shrinkage. Coatings may themselves contain cement and aggregate, examples being latex-modified mortars, polymer concretes, and asphaltic concrete. Materials that penetrate and block the capillary pores can be used to seal the surface of

Table 20.9

Types of Cracking in Concrete Structures

Nature of Crack	Cause of Cracking	Remarks
Large, irregular	Inadequate support, overloading	Slabs on ground, structural concrete
Large, regularly spaced	Shrinkage cracking, thermal cracking	
Coarse, irregular "map cracking"	Alkali–silica reaction	Extrusion of gel
Fine, irregular "map cracking" (crazing)	Excessive bleeding, plastic shrinkage	Finishing too early, excessive trowelling
Fine cracks roughly parallel to each other on surface of slab	Plastic shrinkage	Perpendicular to direction of wind
Cracks parallel to sides of slabs adjacent to control joints (D-cracking)	Excessive moisture contents, porous aggregates	Deterioration of concrete at base of slab due to destruction of aggregates by frost
Cracking along rebar placements	Rebar corrosion	Aggravated by the presence of chlorides

good-quality concrete, but they are not effective with poor-quality concretes.

Cracking due to drying shrinkage and thermal expansion is caused by tensile stresses which are created by differential strains that occur under nonuniform drying, temperature rise, or uneven restraint. Thus, shrinkage and thermal cracking resemble flexural cracking and can be controlled by suitable location of reinforcement, which will reduce the amount of cracking and will cause several fine cracks rather than a single large crack. The finer the crack, the less likely it is that it will contribute to durability problems. Recommended crack widths are given in Table 20.10, but crack widths of less than 0.10 mm (0.004 in.) would be desirable in cases where severe exposure is anticipated.

The use of properly located construction joints will allow free expansion and contraction. Rigid joints should be adequately designed to accommodate additional stress that may be caused by moisture or temperature changes. Shrinkage cracking in walls and slabs usually cannot be completely avoided, and shrinkage-control joints are required. These are grooves cast into the concrete, or sawn after hardening, which provide planes of weakness that will crack preferentially. In this way

Table 20.10

Recommended Crack Widths in Reinforced Concrete*a*

Exposure Condition	Maximum Crack Width mm (in.)
Dry air, Protective coating	0.40 (0.016)
High humidity and moisture	0.30 (0.012)
De-icing chemicals	0.175 (0.007)
Seawater and spray, Wetting and drying	0.15 (0.006)
Water-retaining structures	0.10 (0.004)

*a*Data from Report of ACI Committee 224, "Control
of Cracking in Concrete Structures," *Journal of the
American Concrete Institute,* Vol. 69, No. 12, pp.
717–753 (1972).

random cracking is avoided and the cracks are located in a manner that
allows them to be sealed against moisture. More shrinkage control joints
must be cut than are actually needed because the amount of restraint is
not known. Cracked joints must also be regularly maintained. For this
reason strategies to eliminate shrinkage-control joints in pavements have
been sought; the use of expansive cements, continuously reinforced or
prestressed pavements, or fiber-reinforced concrete are possible solutions.

Cracking may also be controlled if the magnitude or rate of
environmental changes can be reduced. In this way tensile stresses will
be lower and may be further reduced by tensile creep and, in the early
life of a structure, by an increase in tensile strength (see Figure 20.8).
Changes can be made by temporary protection of concrete or the use of
reflective or impermeable coatings.

20.5 REPAIR AND MAINTENANCE OF CONCRETE

Once cracking or other manifestations of deterioration are visible, the
concrete is more susceptible to further damage, which may eventually
render it unsuitable for further use. Although it is obvious that economic

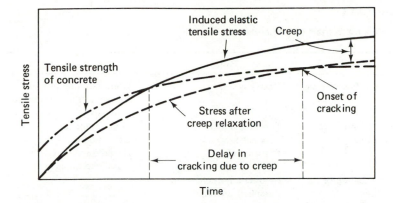

Figure 20.8 Relief of tensile stresses in concrete by creep. (Adapted from A. M. Neville, *Properties of Concrete,* 2nd ed., John Wiley & Sons, Inc., New York, 1973.)

and political considerations will be important in decisions on maintenance and repair, from a technical point of view early maintenance is desirable for maintaining the integrity of the concrete. Preventive maintenance (regular inspection and restoration of sealed joints, drainage systems, etc.) will play an important role in the durability of concrete structures. Early attention to sealing of cracks and restoration of waterproof joints may eliminate the need for more costly repairs later. In cases where more extensive deterioration has occurred or earlier measures have not controlled a potential problem, an investigation should be made to determine the cause and steps taken to counteract the situation during the repair operations.

Materials for Repair

Modern technology has made available many kinds of materials for repair and maintenance of concrete. These range from low-viscosity polymers for the sealing of very fine cracks, very rapid setting cements for repairs in the presence of flowing or seeping water, special concretes for overlays, to portland cement mortar and concrete itself. A selection of materials is given in Table 20.11, but this can be used only as a guide, since the diversity of materials is so great. The engineer will be faced with an array of potential materials to choose from, requiring a specialized knowledge for proper evaluation. A final selection will depend on many factors, such as properties during repair, mechanical response, long-term durability, cost, and prior field experience.

Table 20.11

Materials for Repair of Concrete

Repair Operation	Material	Comments
Sealing of fine cracks	Epoxy resins	Good bonding properties even in the presence of moisture
Sealing of large cracks and joints	Portland cement mortar	Well-compacted
	Polymer mortar	Good bonding properties
	Putties and caulks	Based on synthetic polymers and tars
General sealing of surfaces	Synthetic polymers and asphalt coatings	
Localized patching of surfaces	Concretes or mortars using portland cement	
	Rapid-setting cements	Calcium aluminate and regulated-set cements
	Polymer resins	Epoxies, polyesters; good bonding
Overlays and shotcreting	Portland cement concrete	Quick-setting admixtures
	Fiber-reinforced concretes	Resistance to cracking
	Latex-modified concrete	
	Polymer concretes	Good bonding
	Asphaltic concrete	

Techniques of Repair

It is not possible in this book to provide more than a general discussion of the varied techniques that can be used in the repair and restoration of concrete. Nevertheless, there are two considerations that should be kept in mind. These are that efficient, durable repairs depend on (1) proper selection of materials, and (2) proper preparation of the damaged concrete prior to restoration. We have already discussed the former and the latter will be discussed in the ensuing sections. However, the following general points should be remembered:

1. All damaged materials should be removed until a sound surface is reached.

2. Where possible, the cavity should be prepared to ensure good bonding between the concrete and the repair material, and to ensure proper consolidation.

3. Measures should be taken to remove aggressive materials or to prevent their reentry.

Repair of Cracks

Fine cracks, as small as 0.05 mm (0.002 in.), have been successfully sealed by the injection of low-viscosity polymeric grouts. Such materials should be capable of forming a solid polymer *in situ* after injection.

Epoxies are a popular choice, and many proprietary formulations are commercially available. The epoxy is injected under pressure in order to penetrate the very fine and tortuous crack pattern that may exist. The success of pressure grouting depends on proper application to ensure that all cracks are sealed. Epoxy grouting can restore structural integrity as well as seal cracks against seepage. It has been used underwater or in the presence of seeping moisture with good success, but the choice of suitable materials is important.

Larger cracks and joints may be sealed using cementitious mortars and caulks or putties. For a durable and successful seal, the crack should be cleaned out and cut back to form a V-shaped groove into which the sealant can be well compacted. A good-quality portland cement mortar is satisfactory (and cheap) for larger cracks. In the presence of moisture, quick-setting admixtures should be used or the portland cement replaced by a quick-setting proprietary cement (inorganic or organic). For finer openings under dry conditions, caulks and putties based on organic polymers can be used. Lead wool is often used in connection with a conventional sealant. Sealants are of many types and properties; the selection should depend on considerations of anticipated service conditions, such as applied loads, conditions of exposure, and the like. Once cracks and joints are repaired, a general protective coating is both beneficial and aesthetic. In cases where little seepage is likely to occur, a general protective coating may be sufficient.

Localized patching. This may involve filling of tie holes, bolt holes, prestressing ducts, and so on. The simplest approach is to use dry-pack mortar for shallow holes and conventional "replacement mortar" for deeper cavities or for filling around rebars. Pressure grouting may be required for deep and narrow cavities, such as prestressing ducts. Mortar should be as dry as possible consistent with good compaction and pumpability. The use of admixtures to improve flow characteristics and to avoid shrinkage on subsequent drying may be advisable where the creation and maintenance of a good bond is important. For large cavities, replacement concrete may be used for economy; in difficult situations the use of prepacked aggregate may be advisable, with the mortar grouted in subsequently.

The use of replacement mortar and concrete, or equivalent materials, such as asphalt or polymer concrete, is common practice in localized repair of pavements and floors. All too often repairs are not durable because insufficient attention is paid to initial preparation. All unsound, unbonded concrete should be removed, since a good bond is required between the old concrete and patching material. If reinforcement is corroded, the surface rust should be removed and where possible it is advisable to completely expose the outer layer of reinforcement, to provide additional "interlocking." A mechanical bond will be improved by roughening the surface and shaping the hole to provide a mechanical key. Priming with cement mortar or a polymer bonding agent will help develop additional chemical bond between the old and the new concrete. Alternatively, the use of materials such as polymer concrete or latex-modified concrete will in themselves give a good bond.

Overlays. The foregoing considerations will apply equally well to general overlays of surfaces where extensive deterioration does not warrant localized patching. Resurfacing of pavements and bridge decks is a common application. If overlays are to be bonded directly to the underlying, sound concrete, a strong bond should be developed and the overlay should match the underlying concrete in thermal properties or have good crack-resistant properties. If this is not the case, differential movements will cause cracking in the overlay, which will destroy its integrity and protection. In such circumstances a nonbonded overlay should be used which can move independently of the base concrete.

On vertical surfaces pneumatic application of concrete is often used. This is particularly important where considerable structural damage has occurred and additional reinforcement is required or where a reasonably thick layer of new concrete is desirable. The application of thinner coatings (or stuccos) can be done by spraying or hand application. Conventional portland cement concrete using special quick-setting admixtures is most commonly used, but the use of modified concretes, such as fiber-reinforced, regulated-set cement or latex additions, has been advocated.

Bibliography

General

ACI COMMITTEE 201, "Guide to Durable Concrete," *Journal of the American Concrete Institute,* Vol. 74, No. 12, pp. 573–609 (1977).

BICZOK, I., *Concrete Corrosion and Concrete Protection.* Chemical Publishing Co., New York, 1967.

Durability of Concrete, SP-47. American Concrete Institute, Detroit, Mich., 1975.

LEA, F. M., *The Chemistry of Cement and Concrete.* 3rd ed. Chemical Publishing Co., New York, 1971.

Significance of Tests and Properties of Concrete and Concrete-making Materials, ASTM STP 169B. American Society for Testing and Materials, Philadelphia, Pa., 1978.

WOODS, H., *Durability of Concrete Construction,* Monograph No. 4. American Concrete Institute, Detroit, Mich., 1968.

Permeability

VERBECK, G. J., AND R. A. HELMUTH, "Structures and Physical Properties of Cement Paste," *Proceedings, Fifth International Symposium on the Chemistry of Cement, Tokyo, 1968,* Vol. 3, pp. 1–32. Cement Association, Japan, Tokyo, 1969.

Chemical Attack

SWENSON, E., ed., *Performance of Concrete.* University of Toronto Press, Toronto, Canada, 1968.

Physical Attack

GORDON, W. A., *Freezing and Thawing of Concrete—Mechanisms and Control,* Monograph 3. American Concrete Institute, Detroit, Mich., 1966.

LITVAN, G. G., "Mechanism of Frost Action in Hardened Cement Paste," *Journal of the American Ceramic Society,* Vol. 55, No. 1, pp. 38–42 (1972).

POWERS, T. C., "Freezing Effects in Concrete," in *Durability of Concrete,* SP-47, pp. 1–12. American Concrete Institute, Detroit, Mich., 1975.

Corrosion of Reinforcement

Chloride Corrosion of Steel in Concrete, ASTM STP 627. American Society for Testing and Materials, Philadelphia, Pa., 1977.

Corrosion of Metals in Concrete, SP-49. American Concrete Institute, Detroit, Mich., 1975.

"Halting Deck Deterioration on Existing Bridges," *Civil Engineering,* Vol. 43, No. 4, pp. 80–86 (1973).

Cracking and Repair of Concrete

ACI COMMITTEE 224, *Causes, Mechanism and Control of Cracking in Concrete,* Bibliography No. 9. American Concrete Institute, Detroit, Mich., 1971.

ACI COMMITTEE 224, "Control of Cracking in Concrete Structures," *Journal of the American Concrete Institute,* Vol. 69, No. 12, pp. 717–753 (1972).

Causes, Mechanism, and Control of Cracking in Concrete, SP-20. American Concrete Institute, Detroit, Mich., 1968.

Concrete Manual, 8th ed. U.S. Bureau of Reclamation, Denver, Colo., 1975.

Epoxies with Concrete, SP-21. American Concrete Institute, Detroit, Mich., 1968.

Problems

20.1 What is the effect of permeability on durability?

20.2 What does the appearance of efflorescence on the surface of a structure imply?

20.3 Why is concrete highly porous but relatively impermeable?

20.4 What is the most effective way of preventing sulfate attack?

20.5 Describe the mechanism of sulfate attack of concrete.

20.6 Why is seawater not as corrosive as groundwater with the same concentration of sulfate ions?

20.7 Why is concrete cured by high-pressure steam less susceptible to sulfate attack?

20.8 Why do acids attack concrete?

20.9 What conditions must be present for the corrosion of steel to occur?

20.10 What strategies could you adopt to minimize the corrosion of reinforcing steel in concrete bridge decks?

20.11 Why do de-icing salts cause severe deterioration of concrete bridge decks?

20.12 Why is concrete damaged by repeated freezing and thawing cycles?

20.13 How does air entrainment minimize the effects of freezing and thawing?

20.14 Why are some aggregates susceptible to damage from repeated freezing and thawing?

20.15 What problems are caused by the cracking of concrete?

20.16 How might cracks be repaired?

21

special concretes

21.1 LIGHTWEIGHT CONCRETES

As discussed in Chapter 6, lightweight concretes can be divided into structural lightweight concretes and ultra-lightweight concretes used for nonstructural purposes (see Table 6.12). ACI Committee 213 makes three divisions (Figure 21.1) on the basis of strength and unit weight: low-density, low-strength concrete used for insulation; moderate-strength lightweight concrete used for concrete block and other applications where some useful strength is desirable; and structural lightweight concrete.

The various types of lightweight aggregates were discussed in Chapter 6 and are summarized in Figure 21.1. However, lightweight concretes can also be obtained by other means. Low densities are achieved by a high porosity within the concrete, and this porosity need not be confined to the aggregate. Aerated (foamed or cellular) concrete has a uniform distribution of air voids throughout the paste or mortar, while "no-fines" concrete or lightly compacted concretes also contain large, irregular voids. Aerated concrete is reviewed in this section; no-fines concrete is discussed later in the chapter as a special kind of gap-graded concrete. Table 21.1 summarizes the various kinds of lightweight concretes.

The use of lightweight concretes in the United States dates back to the early 1900s. Expanded clays and shales were developed commercially by S. H. Hayde (the Haydite process) and were used for shipbuilding during World War I. The Park Plaza Hotel in St. Louis was an early example of lightweight concrete construction in the 1920s. Clinker

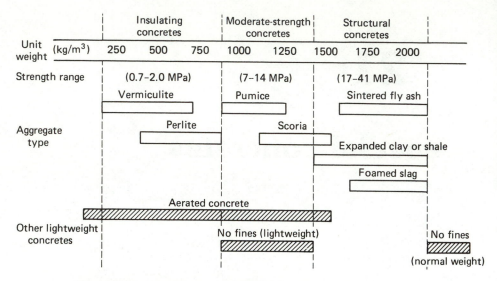

Figure 21.1 Classification of lightweight concretes (kg/m³ × 0.062 = lb/ft³; MPa × 145 = lb/in.²).

aggregate was also developed prior to World War I and used mostly in lightweight block, and foamed slag was produced commercially in the late 1920s. Since the 1950s, lightweight concrete has been used on many multistory buildings and other large structures around the country. Some of the more notable examples are the Busch Memorial Stadium, St. Louis; the Watergate Apartments, Washington, D.C.; the Lake Point Towers, Chicago (Figure 21.2a); and the Assembly Hall, University of Illinois at Urbana–Champaign (Figure 21.2b).

Structural Lightweight Concrete

Structural lightweight concretes are made with both coarse and fine lightweight aggregates, but it is common with the higher-strength concretes to replace all or part of the fine fraction with normal-weight sand. Such a replacement will increase the unit weight of the concrete by as much as 320 kg/m³ (20 lb/ft³). Although synthetic lightweight aggregates are generally more expensive than normal aggregates, the increased strength-to-weight ratio offers sufficient overall saving in materials, through the reduction of dead-weight loadings to more than offset the higher aggregate cost per cubic meter of concrete. Lower total loads mean reduced supporting sections and foundations, and less reinforcement.

Table 21.1

Properties of Different Types of Lightweight Concrete

Type of Lightweight Concrete	Type of Aggregate	Density of Aggregate $(kg/m^3)^a$	Cube Crushing Strength at 28 days $(MPa)^a$	Density of Concrete $(kg/m^3)^a$
Aerated concrete	—	—	1.4–4.8	400–600
Partially compacted	Expanded vermiculite and perlite	64–240	0.5–3.4	400–1120
	Pumice	480–880	1.4–3.8	720–1120
	Foamed slag	480–960	1.4–5.5	960–1520
	Sintered pulverized-fuel ash	640–960	2.8–6.9	1120–1280
	Expanded clay or shale	560–1040	5.5–8.3	960–1200
	Clinker	720–1040	2.1–6.9	720–1520
No-fines concrete	Natural aggregate	1360–1600	4.1–13.8	1600–1920
	Lightweight aggregate	480–1040	2.8–6.9	880–1200
Structural lightweight aggregate concrete	Foamed slag	480–960	10.3–41.4	1680–2080^b
	Sintered pulverized-fuel ash	680–960	13.8–41.4	1360–1760^b
	Expanded clay or shale	560–1040	13.8–41.4	1360–1840^b

^{a}kg/m³ × 0.062 = lb/ft³; MPa × 145 = lb/in.²

^{b}These heavier concretes are obtained by replacing some of the lightweight fines by a natural sand.

Engineering Properties

The engineering properties of lightweight concretes depend to a large extent on the materials used in mix design. As seen in Figure 21.1, with some lightweight aggregates there is no difficulty in obtaining concrete strengths up to 41 MPa (6000 lb/in.²) in spite of the high porosity and inherent weakness of the aggregate. A relationship between strength and density exists for lightweight concrete, but depends on the particular aggregate used (see Figure 21.3) and the amount of normal-weight sand. Low w/c ratios are required to achieve the higher strengths, but the high absorption of most lightweight aggregates makes it difficult to calculate the w/c ratio of the paste exactly. The need for a lower w/c ratio in the paste to obtain high strengths means that generally higher cement contents are needed for structural lightweight concretes compared to normal-weight concretes of the same strength (see Table 21.2). In

(a)

(b)

Figure 21.2 (a) Lake Point Towers, Chicago. (Photograph courtesy of Portland Cement Association.); (b) Assembly Hall, University of Illinois at Urbana-Champaign. (Photograph courtesy of University of Illinois.)

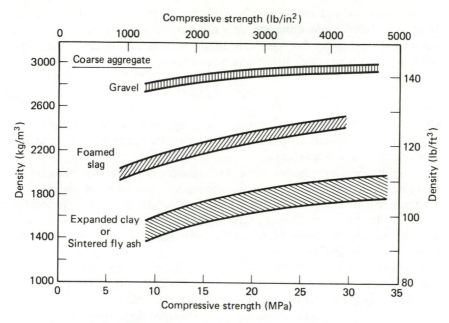

Figure 21.3 Relationship between density and compressive strength (6-in. test cubes). (Adapted from A. Short and W. Kinniburgh, *Lightweight Concrete*, 3rd ed., Applied Science Publishers, Ltd., London, 1978.)

addition, the physical characteristics of lightweight aggregates are such that more paste is often required to provide good workability.

The fracture of lightweight concrete is somewhat different from that of ordinary concrete. Failure commonly occurs through, rather than around, the aggregate. The strengths of the aggregate and the cement paste are more nearly equal, and in the weaker concretes, the paste

Table 21.2

Cement Contents of Lightweight and Normal-Weight Concretes

Compressive Strength MPa (lb/in.²)	*Cement Content, kg/m³ (lb/yd³)*	
	Lightweight	*Normal-Weight*
17 (2500)	255–420 (425–700)	210–330 (350–550)
21 (3000)	285–450 (475–750)	210–360 (350–600)
28 (4000)	330–510 (550–850)	240–420 (400–700)
35 (5000)	390–570 (650–950)	300–450 (500–750)

strength may exceed the aggregate strength. It has been found that the strength of lightweight concrete depends on the volume fraction of the lightweight aggregate. For concrete made with normal-weight sand, the following empirical relationship has been found:

$$\sigma_c = \sigma_a{}^n \cdot \sigma_m{}^{(1-n)} \tag{21.1}$$

where σ_c, σ_a, and σ_m are the compressive strengths of the concrete, aggregate, and mortar, respectively; and n is the volume fraction of the lightweight coarse aggregate.

Lightweight aggregates have low moduli of elasticity because of their high porosity. Consequently, the elastic modulus of lightweight concrete will be lower than that of normal-weight concrete. Generally, values are in the range 10 to 17 GPa (1.5 to 2.5 × 10⁶ lb/in.²), about one-third to two-thirds those of normal-weight concrete. The exact value depends on the nature of the aggregates used, and thus the range of variation is somewhat greater for a given ultimate compressive strength. The equation used in the ACI Building Code to estimate the elastic modulus (Equation 18.1a), may not approach the actual value closer than ±20%. The lower elastic modulus of lightweight aggregates would also offer less restraint to time-dependent deformations such as drying shrinkage and creep. On the average, creep or shrinkage strains of lightweight concrete tend to be greater than for normal-weight concrete (Figure 21.4). It must be remembered that there is considerable variation in creep and shrinkage among concretes of a given density, the magnitude depending on the cement content, w/c ratio of the paste, modulus of elasticity of the aggregate, and the rate of moisture loss.

Physical Properties and Durability

The coefficient of thermal expansion of lightweight concrete is much the same as that for normal-weight concrete, but its thermal conductivity is considerably lower because of the large amount of void space. The thermal conductivity depends on unit weight (Figure 19.4). The lower thermal conductivity means that lightweight concretes are generally more fire-resistant than are normal-weight concretes.

Lightweight aggregates are more friable than most rocks, so that lightweight concretes are generally not suitable for heavy wear. However, many synthetic aggregates have quite hard surfaces and lightweight concretes can perform just as well as normal-weight concretes under less rigorous conditions of wear. The use of natural sand and the development of high compressive strengths improve the abrasion resistance.

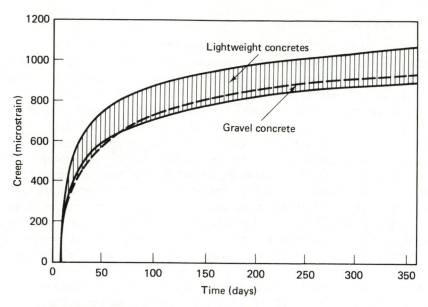

Figure 21.4 Comparison of creep of lightweight and normal-weight concrete. (From A. Short and W. Kinniburgh, *Lightweight Concrete*, 3rd ed., Applied Science Publishers, Ltd., London, 1978.)

The freeze–thaw resistance of lightweight concrete is similar to that of ordinary concrete. Air entrainment should be used whenever the concrete will be exposed to freezing and thawing. The moisture content of the aggregates can be critical, because when aggregates are close to saturation, freezing of water in the aggregate pores will force water out of the aggregate particle into the surrounding paste. The resulting hydraulic pressure may cause tensile failure if sufficient entrained air is not present to accommodate the excess water. To avoid this situation, the aggregates should have as low a moisture content as is practical during mixing, or the concrete should have ample time to dry out before being exposed to freezing temperatures. The resistance to de-icer salt scaling is also similar to normal-weight concrete. Resistance is improved by air entrainment, low *w/c* ratios, adequate curing, and a period of drying before service. Otherwise, there are no special durability problems associated with lightweight concretes. The tendency for such concretes to have a greater absorption of water bears no relation to their durability. The higher absorption is a reflection of the porous aggregates and does not indicate a higher concrete permeability, since the latter is controlled by the porosity of the cement paste.

Fresh Concrete Properties

Lightweight concrete has essentially the same properties in the plastic state as does normal-weight concrete. Mixes tend to be somewhat harsher than ordinary concrete mixes because of the nature of the synthetic aggregates. Thus, air entrainment may be desirable solely to obtain improvements in workability. The addition of sand may be needed to improve finishability if the lightweight aggregate is deficient in fines passing the No. 30 (600-μm) sieve. Slump loss can be a severe problem when the aggregates continue to absorb large quantities of water after mixing. This condition can generally be avoided by batching the aggregate in a damp condition or by mixing the aggregate with about two-thirds of the mix water before adding the cement and the balance of the water.

Slump should be limited to a maximum of 100 mm, because higher slumps tend to cause segregation of the lightest coarse aggregate particles rather than of the paste. It must be remembered that lightweight concretes will tend to have slightly lower slumps for a given workability because of their lower density. When placing and finishing the concrete, care should be taken to avoid segregation by using practices recommended for normal-weight concrete.

Mix Design of Structural Lightweight Concrete

The particular properties of lightweight aggregates pose special problems in calculating proportions for lightweight concrete. The absolute volume method, which is the basis of the ACI method of proportioning normal-weight concrete, cannot be used with confidence for lightweight concrete. This is due to two major factors: (1) variations in bulk specific gravity (BSG), and (2) changes in moisture content.

The BSG of normal-weight aggregates is essentially independent of particle size. Thus, when the grading curve is determined on a weight basis (by sieve analysis), it can be converted directly to a volume basis, since it is really volume relationships that are important in proportioning. However, the constancy of the BSG does not hold with lightweight aggregates because the amount of porosity varies with particle size. The variation is particularly marked in the fine-aggregate fraction (see Table 21.3) and means that the grading curve on a volume basis is different from that on a weight basis. In the example in Table 21.3, the fineness modulus determined by weight is 3.03, whereas by volume it is 3.23. Therefore, the lightweight sand is really coarser than the value 3.03 indicates, and a greater weight of material should be retained on the finer sieve sizes to provide the same volumetric grading as a normal-weight sand.

Table 21.3

Typical Sieve Analysis of Lightweight Fine Aggregate

Sieve Size	BS G[a]	Percent Retained		Cumulative Percent Retained	
		By Wt.	*By Vol.*	*By Wt.*	*By Vol.*
No. 4	1.40	0	0	0	0
No. 8	1.55	22	26	22	26
No. 16	1.78	24	25	46	51
No. 30	1.90	19	19	65	70
No. 50	2.01	14	13	79	83
No. 100	2.16	12	10	91	93
Passing No. 100	2.40	10	7	100	100
Fineness modulus	—	—	—	3.03	3.23

[a]SSD basis.

Because of the excessive absorption of many lightweight aggregates (absorption capacity 10 to 20%), the BSG will also vary markedly with changing moisture content. Furthermore, the amount of absorption that occurs during mixing is not easily determined accurately, and thus the calculation of the net w/c ratio cannot be used as a basis for proportioning. The amount of water absorbed by the aggregate depends both on its initial moisture content and the time of exposure to water in the fresh paste. It is thus necessary to adopt a trial-and-error procedure for mix design. ACI Recommended Practice 211.2 provides a general guide for estimating the proportions for the first mix and procedures for adjusting quantities for the subsequent mixes. The recommendations must of necessity be very general because of the wide range of properties among lightweight aggregates. The supplier of the aggregate can generally suggest more precise values for a particular aggregate.

First estimation of quantities. The cement content is estimated on the basis of the desired strength requirement, as given in Table 21.4. It can be seen that these estimates have very wide limits and if more precise information is not available, a series of trial mixes will need to be made to determine the optimum cement content. The strength is not estimated from the w/c ratio because that quantity is not determined. Sufficient water is added to provide the workability for proper placing, consolidating, and finishing without segregation and consistent with strength requirements. Slump requirements for different placements are similar to those for normal-weight concretes. No requirements are laid down for durability, but air entrainment should be used for freeze–thaw protection.

Table 21.4

First Estimation of Material Quantities for Lightweight Concrete

Material	*General Requirements*
Cement	250–420 kg/m³ for 17 MPa (420–700 lb/yd³ for 2500 lb/in.²)
	280–450 kg/m³ for 21 MPa (470–750 lb/yd³ for 3000 lb/in.²)
	330–500 kg/m³ for 28 MPa (550–850 lb/yd³ for 4000 lb/in.²)
	390–560 kg/m³ for 35 MPa (650–950 lb/yd³ for 5000 lb/in.²)
Water	Sufficient to obtain the desired slump: 180–300 kg/m³
	(300–500 lb/yd³)
Entrained air	4–8% with 19 mm (¾ in.) max. size
	5–9% with 9.5 mm (⅜ in.) max. size
Aggregates	1.04–1.19 m³/m³ (28–32 ft³/yd³) concrete (dry-loose basis)
	40–60% fine aggregate

Air contents should be slightly higher than for normal-weight concrete because cement contents are generally higher and maximum aggregate sizes are usually lower. Air contents in the range 5 to 9% are usual. The total volume of aggregates required is in the range 1.04 to 1.19 m³/m³ measured as the sum of the uncombined volumes of coarse and fine aggregates on a dry-loose basis. (This figure is greater than 1 m³ because it includes the space that will be occupied by the paste, as well as some additional void space because the aggregates are not compacted.) The exact volume depends on grading, shape, texture, and surface porosity and will be least (i.e., lowest paste requirements) with well-distributed size gradation, well-rounded shape, and low surface porosity, but it also depends on the required cement, air, and water contents. The percentage of fine aggregates is generally somewhat higher than in normal-weight concrete and also depends on grading, shape, and porosity. The use of normal-weight sand allows the proportion of fine aggregate to be decreased.

Adjustment of mixes. Once suitable proportions, consistent with required strength, air content, and slump, have been developed through trial-and-error procedures, it may still be necessary to make adjustments in the field or laboratory to compensate for variations in materials (e.g., moisture content of the aggregate) or to make desired changes in the concrete properties. If these changes are small, the rules of thumb given in Table 21.5 show the approximate amount of change required without affecting other properties of the concrete. These values are approximations for guidance only. When such changes are made, the fine-aggregate content should be adjusted to ensure that yield is maintained.

Table 21.5

Guidelines for Adjustment of Mixes of Lightweight Concrete

Change Required	Amount of Change	Changes in Other Quantities per m³ (yd³) of Concrete
Proportion of fine aggregate	+1%	+2 kg (4.4 lb) water
		+1% cement
Air content	+1%	−3 kg (6.6 lb) water
Slump	+25 mm (1 in.)	+6 kg (13.2 lb) water
		+3% cement

To calculate these adjustments precisely, a procedure based on the method of absolute volumes is recommended. This involves calculating the effective volume that the lightweight aggregates displace in the concrete, which requires knowledge of the total moisture content of the aggregate and the *specific gravity factor* (SGF). SGF is defined as

$$\text{SGF} = \frac{\text{weight of aggregates added}}{\text{effective volume of aggregates}} \tag{21.2}$$

and is determined by the volume displacement of water in a pycnometer. This determination allows water to be absorbed by the aggregate before the displacement volume is measured. Thus, SGF is not a true specific gravity, since its value incorporates compensation for absorption of free water by the aggregates, but it is used in exactly the same way to calculate volume relationships. SGF depends on the initial moisture content of the aggregate (Figure 21.5), and this relationship must be determined for the aggregates being used.

An example for calculating a change in cement factor of a concrete mix is given in Table 21.6. In this example the strength of the trial mix was too low, and therefore the cement was raised from 415 kg/m³ to 445 kg/m³. It can be seen in these calculations that the BSG of cement is used to calculate its effective volume in the normal way and that the SGF of the fine aggregate at the appropriate moisture content is used for the same purpose. The fine-aggregate content is changed to keep the yield constant on the dry-mix basis. When adjusting for stockpile moisture contents, the effective volumes of both the coarse and fine aggregates change because the SGF changes. The water content is then adjusted to keep the yield constant.

Table 21.6

Adjustment for Change in Cement Factor (Metric Units)[c]

	Original Mix Dry Basis[a] $M_c = 0\%$ $S_{co} = 1.34$	$M_f = 0\%$ $S_{fo} = 1.99$
	(1) Weight[b] (kg)	(2) Effective Displaced Volume[b] (m³)
Cement	415	$\dfrac{415}{1000 \times 3.15} = 0.132$
Air (5.5%)	—	0.055
Coarse aggregate	351	$\dfrac{351}{1000 \times 1.34} = 0.262$
Fine aggregate	549	$\dfrac{549}{1000 \times 1.49} = 0.276$
Added water	275	0.275
Total	1590	1.000

	Adjusted Mix Dry Basis $M_c = 0\%$ $S_{co} = 1.34$	$M_f = 0\%$ $S_{fo} = 1.99$	Adjusted Mix Damp Basis $M_c = 1.5\%$ $S_{c1\frac{1}{2}} = 1.35$	$M_f = 4\%$ $S_{f4} = 1.97$
	(4) Weight (kg)	(3) Effective Displaced Volume (m³)	(5) Weight (kg)	(6) Effective Displaced Volume (m³)
Cement	445	$\dfrac{445}{1000 \times 3.15} = 0.141$	445	0.141
Air (5.5%)	—	0.055	—	0.055
Coarse aggregate	351	0.262	$351 \times 1.015 = 356$	$\dfrac{356}{1000 \times 1.35} = 0.264$
Fine aggregate	0.267×1000 $\times 1.99 = 531$	$1.000 - 0.733 = 0.267$	$531 \times 1.04 = 552$	$\dfrac{552}{1000 \times 1.97} = 0.280$
Added water	275	0.275	$1000 \times 0.260 = 260$	$1.000 - 0.740 = 0.260$
Total	1602	1.000	1613	1.000

[a]M_c, M_f = moisture contents of fine and coarse aggregate; S_{co}, $S_{c1\frac{1}{2}}$ = specific gravity factor for coarse aggregate, dry, or with a moisture content of 1.5%, respectively; S_{fo}, S_{f4} = specific gravity factor for fine aggregate, dry, or with a moisture content of 4%, respectively.

[b]kg × 2.2 = lb; m³ × 35.3 = ft³.

[c]The sequence of calculations is determined by observing the following sequence. In the original mix, the effective volumes (column 2) are calculated from batch weights (column 1), and then the mix is adjusted by volume (column 3) to correct the cement content. Batch weights on a dry basis (column 4) are calculated from these volumes and then adjusted for changes in moisture content (column 5). The effective volume (column 6) is calculated to determine the water content. (Adapted from ACI 211.2; reproduced by permission.)

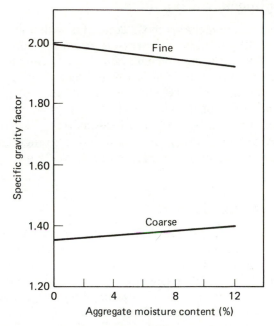

Figure 21.5 Variation of specific gravity factor with moisture content. (From ACI 211.2-69 (Revised 1977). *Recommended Practice for Selecting Proportions for Structural Lightweight Concrete,* American Concrete Institute, Detroit, Mich., 1977.)

Moderate-Strength Lightweight Concretes

This classification is intermediate between structural lightweight concrete and insulating concrete. Such concrete combines additional insulating capacity with some useful strength. Lightweight concrete block is a common application for this class of concrete. Typical aggregates are pumice, scoria, or expanded clay or shale. The aggregates are selected or manufactured with lower unit weights than are needed for structural lightweight concretes, to provide better thermal insulation. "No-fines" concrete, using either normal-weight or lightweight aggregates, also fits into this category, but is discussed under gap-graded concrete.

Ultra-lightweight Concretes

This classification includes those concretes with unit weights of less than 1100 kg/m³ (70 lb/ft³) and compressive strengths below about 7 MPa (1000 lb/in.²). With ultra-lightweight concretes, strength requirements are minimal and the concrete is used for its other properties, principally thermal insulation and light weight; such concretes are used primarily for insulating fills on roof decks, metal decking or floors, underground conduit linings, fire walls, and non-load-bearing fill over structural concrete. Thermal conductivity and compressive strength are dependent on concrete density, as shown in Figures 19.4 and 21.6.

Vermiculite and Perlite Concretes

In the United States, expanded (exfoliated) vermiculite and expanded perlite (Chapter 6) are most commonly used as aggregates in insulating concrete. The expanded materials are very durable and do not interact chemically with the cement paste, but the concretes have high water requirements because of the high absorption of these aggregates. Strength is dependent on the cement content and the use of normal-weight sand; rich mixes have strengths around 7 MPa (1000 lb/in.²) but a greater density (Figure 21.6) and higher thermal conductivity. The concretes are made with a high slump and a high air content (25 to 30%) for easy handling and finishing. The high air contents prevent segregation, and the high slump means that generally only screeding and darbying are required to give a satisfactorily smooth surface. Since the thermal conductivity is strongly dependent on moisture content, insulating concretes should be allowed to dry after curing before being sealed. Insulating materials based on vermiculite and perlite are covered by ASTM C196 and ASTM C610, respectively.

Aerated Concretes

The lightest kind of aggregate possible is, of course, air, and this is the basis of aerated or cellular concrete. The "aggregate" is a uniform cellular structure of air voids distributed throughout a matrix of cement paste or mortar, so that the term "concrete" is not strictly correct. Aerated concretes are more popular in England and Europe than in the United States, where vermiculite and perlite concretes are more extensively used. Aerated concrete can be either formed *in situ* or precast; in the latter case high-pressure steam curing can be used, which gives a stronger product with better dimensional stability.

The nature of the void structure is similar to that of entrained air (i.e., discrete, nearly spherical bubbles), except that the voids are one or two orders of magnitude larger in diameter (0.1–1 mm), large enough to be seen with the naked eye. Thus, the concrete will have good frost resistance, although it is of low strength, and also will not show rapid absorption of moisture, since the bubbles are not interconnected. The void system is formed either by generating a foam within the concrete during mixing, or by blending a preformed foam with the paste or mortar. Preformed foams are generally made from liquid concentrates, based on hydrolyzed protein or synthetic detergents, which are agitated with compressed air. A foam generator supplies foam at a standard rate, and this is blended with a slurry of cement and water to provide the finished product. It is also possible to add the concentrate to the cement slurry in a high-speed mixer which provides a high shearing action that will cause foaming. In either case the foam must be sufficiently stable to remain intact during handling and until the concrete has hardened. The other method of generating a foam during mixing is to form a gas within the concrete by a chemical reaction (hence the term "gas-concrete," which is sometimes used). Although other chemicals have been proposed, the material generally used is a finely divided aluminum powder (sometimes powdered zinc is used). The aluminum reacts with the soluble alkalis in the cement slurry to generate small bubbles of hydrogen.

$$Al + 2OH^- + 2H_2O \rightarrow Al(OH)_4^- + H_2 \uparrow \tag{21.3}$$

The rate and amount of hydrogen generation determines the amount and character of the air-void system.

Typical properties of insulating cellular concretes produced commercially are: density, 300 to 1100 kg/m^3 (20 to 80 lb/ft^3); compressive strength, 0.3 to 7.0 MPa (50 to 1000 lb/in.^{2}); thermal conductivity, 0.1 to 0.3 W/m·K. In most cases these are low-density, low-strength materials suitable for nonstructural insulating concretes, but it is possible to produce moderate-strength lightweight concretes from aerated concrete. The higher-strength materials have a lower void content and use mortar as the matrix. Density and thermal conductivity are a function of unit weight (Figures 21.6 and 19.4) in the same range as perlite or vermiculite concretes. Since the "aggregate" is air, the concretes have a very low modulus of elasticity and high drying shrinkage.

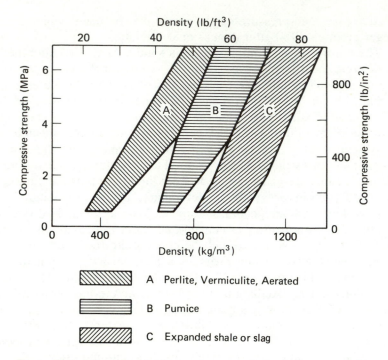

Figure 21.6 **Influence of density on the strength of insulating concretes.** (From *Engineering Bulletin*, 11th ed., Portland Cement Association, Skokie, Ill., 1968.)

21.2 CONCRETES FOR RADIATION SHIELDING

The desirable properties of heavyweight aggregates and the factors affecting attenuation of radiation by concrete were discussed in Chapters 6 and 19, respectively. In this section we discuss the proportioning and properties of heavyweight concretes in general, and the particular requirements of concrete for nuclear reactors.

Heavyweight Concrete

Most heavyweight concrete produced today is used as a shielding material to protect workers from the damaging effects of gamma rays and neutrons produced by radiation sources. Where space is not a limitation, concrete of normal density can be used most economically, but concrete of high density, >3200 kg/m³ (>200 lb/ft³), has better attenuating properties and can be used in thinner sections for the same amount of shielding.

Heavyweight concrete can be proportioned using the ACI method of absolute volumes developed for normal-weight concrete. Thus, it is necessary to determine the bulk specific gravities of both the fine and coarse aggregate, the fineness modulus of the fine aggregate, and the unit weight and maximum size of the coarse aggregate. Selection of quantities of cement, water, and coarse aggregate are the same as for normal-weight concrete. However, heavyweight aggregates tend to give a harsh mix because of their shape and texture, so that it is desirable to use fine aggregates with a lower fineness modulus or a higher ratio of fine to coarse aggregate. Cement contents are therefore generally quite high, greater than 350 kg/m³ (600 lb/yd³). This helps to improve the shielding characteristics of the concrete because of the high bound water content of the paste.

Heavyweight concrete can be pumped, but over much shorter distances, because of its high density. Also, pressures on formwork are much greater and formwork must be constructed more sturdily. Because of the high density, segregation can be more pronounced and it is recommended that both fine and coarse aggregates be high-density materials to minimize undue segregation. The preplaced aggregate method is often used to avoid problems of segregation and to produce concrete of uniform density. Preplacing the aggregate may also assist difficult placements where reinforcement is rather congested or where there are many items to be embedded in the concrete.

The properties of heavyweight concrete are similar to those of normal-weight concrete. Strengths can be estimated from the same *w/c* ratio relationship, and the physical properties will depend to a large extent on the properties of the aggregate. Thermal conductivity is about the same as for normal-weight concrete when magnetite is used, but is up to 50% lower when barite is the aggregate; it is higher if steel aggregate is used.

Concrete for Nuclear Reactors

The use of prestressed concrete reactor pressure vessels (PCRV) for nuclear power generation places much greater demands on heavyweight concrete than other forms of radiation shielding. Pressure vessels operate at higher temperatures and stress levels than conventional structures, and the concrete may be under appreciable thermal and moisture gradients. The concrete must maintain its integrity in both the short and long terms under these more severe operating conditions, so that the thermal properties of the concrete are an important consideration in design.

PCRVs can be expected to operate at bulk vessel temperatures of about 70°C (160°F) under relatively high multiaxial stresses, and the vessel will be subjected to some thermal cycling during its lifetime. Inelastic deformations, such as creep and shrinkage, should be minimized since they can cause microcracking of the concrete and loss of prestress. Similarly, thermal incompatibility between aggregate and cement paste can also cause microcracking within the concrete, particularly if temperature cycling occurs. Microcracking not only reduces the predictability of vessel response to prestress and pressure forces, but also reduces concrete strength and increases permeability.

In many cases knowledge of the appropriate concrete properties under rather complex conditions is insufficient for proper analysis, and much research in the past 20 years has been directed toward learning more about the behavior of concrete under multiaxial stresses at elevated temperatures and conditions of temperature cycling. We do, however, know enough about the properties of concrete materials to make informed choices and to recognize the compromises that have to be made. These considerations are summarized in Table 21.7.

Concrete strengths should be higher than for conventional concrete to ensure adequate long-term sustained strength. For example, it has been suggested that a compressive cylinder strength of 45 MPa (6500 lb/

Table 21.7

Material Properties for Concrete for Nuclear Reactors

Material		*Concrete Properties*
Cement	High content	Good workability
	Low content	Low creep and shrinkage, low heat of hydration
Water	Low content	High strength, low creep and shrinkage, low permeability
	High content	Good workability, good shielding
Aggregate	High density	Good shielding
	High modulus	Low creep and shrinkage
	High thermal expansion	Low internal stresses
	Low thermal expansion	Low thermal expansion

in.^{2}) would correspond to a long-term structural strength of 21 MPa (3000 lb/in.^{2}). A high strength requirement means that the use of low w/c ratios is desirable, but high workabilities are required in order to place and consolidate around high concentrations of reinforcement. Therefore, a high cement content is needed, and this can possibly result in thermal cracking if steps are not taken to control the heat of hydration. High cement contents can also lead to larger creep and shrinkage strains than would be desirable. These can be minimized by the use of aggregates with a high modulus of elasticity to restrain paste deformations.

A low-thermal-expansion concrete is desired to minimize thermal stresses. On the other hand, stress losses that can occur on heating to typical working temperatures are believed to be due primarily to thermal incompatibility between the coarse aggregate and the cement paste, leading to bond failure. Using an aggregate with a high siliceous content increases the thermal coefficient of the aggregate so that it more nearly matches that of the paste, but this increases the thermal expansion of the concrete.

Moisture loss in service is also of considerable importance in reactor design, since it can lead to differential shrinkage strains and the generation of internal stresses. Loss of moisture also decreases the thermal conductivity of the concrete and thus may aggravate thermal stresses during temperature cycling. Furthermore, the water content of the concrete affects its shielding capabilities, since water is a good attenuator of slow neutrons. In the cases where hydrous aggregate has been used to improve neutron-shielding properties, the maximum service temperature must be limited to avoid dehydration of the aggregate.

The use of low w/c ratios to attain high strength is also beneficial in producing concrete of low permeability (assuming good compaction). This is of advantage in minimizing leakage in the event of failure of the impermeable liner that is used in pressure vessels. Low permeability also reduces the rate of moisture loss, thereby reducing shrinkage gradients.

The early use of concrete in prestressed nuclear reactor pressure vessels drew on the limits of our knowledge about concrete. Subsequent research and field evaluation of operating vessels has greatly expanded our knowledge, but considerably more understanding of the material is still required. The application is a very interesting materials problem requiring a considerable range of service requirements. Further improvement could well be achieved by drawing on the most recent technology (superplasticizing admixtures, fiber reinforcement, polymer composites, etc.) and by considering the advantages of composite construction. This would require a more careful analysis of the exact requirements for different parts of the structure and appreciation of the role fundamental material properties will play in achieving these requirements.

21.3 HIGH-STRENGTH CONCRETE

There is a trend toward the use of higher-strength concrete in conventional structures, with 28-day compressive strengths in excess of 55 MPa (8000 lb/in.²). Concrete in the range 55 to 70 MPa (8000 to 10,000 lb/in.²) can be supplied by a capable ready-mixed producer and strengths up to 100 MPa (15,000 lb/in.²) have been obtained in the laboratory. The use of high-strength concrete has advantages in the precast and prestressed concrete industries, where it can result in a more rapid output of components and less product loss during handling. In high-rise construction, advantage can be taken of reduced dead loads, which allows thinner concrete sections and longer beam spans. A disadvantage of high-strength concrete is that it behaves in a more brittle fashion. It must be remembered that we are concerned with concretes that attain high strengths at 28 days or later, and that the task of obtaining high early strengths (1 to 7 days) is quite a different problem. It is true that high-strength concrete will show somewhat higher strengths even as early as 1 day, but even greater 1-day strengths can be obtained without large increases in the 28-day strength.

The biggest single factor affecting strength is the porosity of the concrete, which is controlled primarily by the *w/c* ratio of the paste,

Table 21.8

Production of High-Strength Concretes

Type	w/c Ratio	28-Day Strength MPa (lb/in.²)	Remarks
Normal consistency	0.35–0.40	35–80 (5000–12,000)	50–100 mm (2–4 in.) slump; high cement contents
No-slump	0.30–0.45	35–50 (5000–7000)	Less than 25 mm (1 in.) slump; normal cement contents
Very low *w/c* ratios	0.20–0.35	100–170 (15,000–25,000)	Use of admixtures
Compacted	0.05–0.30	70–240 (10,000–35,000)	Compaction pressure 70 MPa (10,000 lb/in.²) and greater

although consolidation methods also play a part. High-strength concretes fail in a more brittle fashion because the paste–aggregate bond is also strengthened. Much less progressive microcracking occurs on loading and failure passes through both aggregate and paste. In this discussion we consider the role of these factors in the production of high-strength concrete of normal workabilities, concretes of very low workabilities, specially compacted concretes, and concretes with very low *w/c* ratios. These classifications are summarized in Table 21.8.

Design of High-Strength Concrete

The production of high-strength concretes requires the supplier to optimize the three aspects of concrete that affect strength: (1) cement paste, (2) aggregates, and (3) cement–aggregate bond. To do this it is necessary to pay careful attention to all aspects of concrete production (i.e., selection of materials, mix design, handling, and placing). It cannot be emphasized too strongly that quality control is an essential part of the production of high-strength concrete and requires full cooperation among the materials or ready-mixed supplier, the engineer, and the contractor.

Cement Paste

The major factor affecting the strength of the paste component is the *w/c* ratio, and this should be as low as possible, consistent with adequate workability. There is thus a tendency to produce very stiff concretes ("no-slump" concretes), because this lowers the water requirements. The special considerations of no-slump concrete are discussed in the next section. However, high-strength concretes with *w/c* ratios in the range 0.32 to 0.40 can be made with slumps of 50 to 100 mm using high cement contents and water-reducing admixtures. In this regard the newer high-range water-reducing admixtures (superplasticizers) have considerable potential for reducing *w/c* ratios to less than 0.35 while maintaining adequate workability. This strategy will be discussed later.

There is no need to use a rapid-hardening cement (e.g., Type III), since high early strength is not the objective. Indeed, a Type I cement is generally preferable, since slightly higher ultimate strengths are obtained. Pozzolans are often used to improve long-term strength. The choice of cement can affect the strength of the concrete (see Figure 21.7). A high C_3S content and a moderate fineness (350 to 400 m^2/kg Blaine) seem to be desirable characteristics, but are probably not the sole determining factors. A suitable cement should be chosen through testing and in

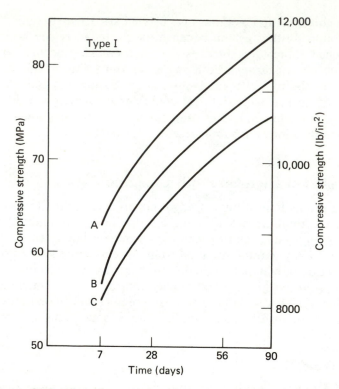

Figure 21.7 Effect of cement brand on concrete strength. (From *High Strength Concrete,* National Crushed Stone Association, Washington, D.C., 1975.)

consultation with the supplier. Uniformity of cement performance is as important as the potential strength gain, if not more so.

Aggregates

Aggregates should consist of strong, dense material with an absence of porous and weak impurities. Since aggregate particles are broken during fracture, the strength of the aggregate is very important. In many areas, the quality of the aggregate may be the principal limitation to achieving high strength. Particle shape should be compact and angular and the aggregate should have a grading within the limits prescribed by ASTM C33. The optimum grading for minimum cement content should be established and this grading maintained uniformly during the job. The fine aggregate can generally contain less material passing the No. 50 and No. 100 sieves because of the higher cement content. Recommended limits are 5 to 20% passing the No. 50 (300-μm) sieve and 0 to 5% passing the No. 100 (150-μm) sieve. The proportion of fine aggregate should also be somewhat less than that used for normal-strength concrete.

Paste –Aggregate Bond

Since the bond between paste and aggregate is a weak point in concrete (Chapter 13), attention should be given to improving its contribution to overall concrete strength. Crushed stone has a rougher surface texture than gravel and gives a better bond and, therefore, better strength (Figure 21.8). Also, crushed stone has a greater surface-to-volume ratio than does rounded gravel. Special care should be taken to ensure that aggregate surfaces are clean. To increase the total surface area still more (and thereby improve the total bond contribution), the maximum aggregate size is generally held below 19 mm. This means that the paste content has to be increased to provide sufficient workability. The combination of low w/c ratio and small maximum aggregate size means that cement contents will be quite high, generally in the range 400 to 600 kg/m³ (650 to 1000 lb/yd³). Too high a cement content can adversely

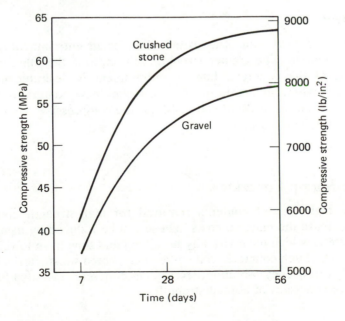

Figure 21.8 Influence of aggregate type on concrete strength (same grading). (From *High Strength Concrete*, National Crushed Stone Association, Washington, D.C., 1975.)

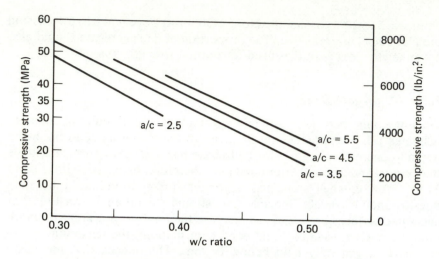

Figure 21.9 Effect of cement content on concrete strength (*a/c* = aggregate/cement ratio). (Acknowledgement is made to McGraw-Hill Book Company, Australia Ppy. Limited for the use of Fig. 2.9 from W. H. Taylor, *Concrete Technology and Practice,* 3rd ed., 1969.)

affect strength, however (Figure 21.9), and the cement content should be the minimum consistent with workability and *w/c* ratio requirements.

Air Content

There is some disagreement as to the need for air entrainment of high-strength concrete. The greater strength will improve the inherent frost resistance of the concrete, but air entrainment is desirable in most applications where moderate or severe freeze–thaw conditions will be encountered. For each 1% of entrained air, the compressive strength will be reduced by about 5%.

No-Slump Concrete

The high cement contents required for high-strength concretes greatly increase the material costs. These can be reduced by using very stiff concretes which have virtually no slump and thus have lower paste requirements. Such concretes can still be compacted satisfactorily by the use of high frequency vibration. (No-slump concretes can also be used to produce concretes of normal strength.)

Table 21.9

Workability of No-Slump Concretes[a]

Description	Slump mm (in.)	Vebe (s)	Compacting Factor	Drop Table (rev)
No slump				
Extremely dry	—	32–18	—	112–56
Very stiff	—	18–10	0.70	56–28
Stiff	0–25 (0–1)	10–5	0.75	28–14
Plastic				
Stiff plastic	25–50 (1–2)	5–3	0.85	14–7
Plastic	75–100 (3–4)	3–0	0.90	7–0
Flowing (very wet)	150–175 (6–7)	—	0.95	—

[a]Adapted from ACI 211.3; reproduced by permission.

Since the slump test cannot differentiate between different no-slump concretes, other workability tests must be used. The ACI Recommended Practice for Selecting Proportions for No-Slump Concrete (ACI 211.3) suggests the use of the Vebe apparatus, the compacting factor apparatus, or the Thaulow drop table (Table 21.9). These tests are described in Chapter 8. The Vebe and drop-table tests more closely measure the behavior of no-slump concrete under vibration, which is an appropriate feature since heavy vibration is needed for good compaction. No-slump concretes are frequently used in modern precasting operations, where heavy vibration can be more easily provided, but can also be used for cast-in-place concrete. They are particularly suited for slip-forming operations, such as slip-formed pavements, where the concrete must maintain its shape unassisted before setting occurs. The proportioning of no-slump concretes described below is designed for cast-in-place concrete.

Proportioning

The ACI Recommended Practice for Proportioning No-Slump Concrete is basically the method of absolute volumes used for concretes of normal consistencies modified to produce the stiff concretes. The reduction in workability is obtained by lowering the amount of water used and increasing the proportion of coarse aggregate (Table 21.10). The *w/c* ratio is chosen on the basis of strength or durability requirements. The criteria for durability and strength are the same as for normal concrete (see Tables 9.3 and 9.4). The method of calculation of batch weights is otherwise identical to normal proportioning.

Table 21.10

Approximate Proportions of Water and Coarse Aggregate for No-Slump Concrete[a]

Consistency	*Relative Water Content (%)*[b]	*Relative Coarse Aggregate Content (%)*[b]				
		9.5 mm (⅜ in.)	*12.5 mm (½ in.)*	*19 mm (¾ in.)*	*25 mm (1 in.)*	*39 mm (1½ in.)*
Extremely dry	78	190	170	145	140	130
Very stiff	83	160	145	130	125	125
Stiff	88	135	130	115	115	120
Stiff plastic	100	108	106	104	106	109

[a]Adapted from ACI 211.3; reproduced by permission.
[b]Relative to the amount of material required per m³ (yd³) to attain a slump of 75–100 mm (3–4 in.) according to recommendations in ACI 211.

Concretes with Very Low *w/c* Ratios

The approximate practical limits for *w/c* ratios of concretes consistent with adequate workabilities are about 0.35 for normal consistencies and about 0.30 for no-slump concretes. These low values can only be achieved by the use of high cement contents, carefully graded aggregates, and the use of water-reducing agents. However, it is possible to reduce the *w/c* ratios still further, to values well below 0.30. Such low *w/c* ratios could lead to potentially much higher strengths than have hitherto been achieved. The attainment of very low *w/c* ratios can be achieved by two distinct approaches:

1. The casting of fluid concretes using special admixtures.

2. The mechanical compaction of very dry, unworkable mixes.

When very low *w/c* pastes are used, the gel/space ratio relationship must be modified. The intrinsic strength at "zero porosity" estimated by Powers (see Chapter 15) was obtained by extrapolation of data in well-hydrated pastes using *w/c* ratios in the normal range 0.40 to 0.70. In that case the intrinsic strength reflected the strength of "cement gel" (C–S–H and CH) with a certain minimum intrinsic porosity. In fluid pastes at very low *w/c* ratios, complete hydration cannot occur, and it is thought that the intrinsic strength may be higher due to a contribution from the remnants of unhydrated cement present even in a mature paste. In compacted pastes, this contribution can be still greater, and it is likely that the intrinsic porosity of the "cement gel" may also be reduced, increasing the potential strength still further. Thus, concretes or mortars with very low *w/c* ratios have the potential for attaining strengths greater

Table 21.11

Use of Superplasticizing Admixtures to Obtain Low-w/c-Ratio Concretes[a]

Cement Content, kg/m³ (lb/yd³)	w/c Ratio	Compressive Strength, MPa (lb/in.²)	
		7 Days	28 Days
500 (850)	0.30	72 (10,400)	84 (12,200)
500 (850)	0.27	73 (10,600)	88 (12,800)
600 (1000)	0.27	77 (11,200)	87 (12,600)
600 (1000)	0.25	79 (11,500)	92 (13,300)

[a]Slumps in the range 70 to 95 mm (2¾ to 3¾ in.). Based on data from "Superplasticizing Admixtures in Concrete," Cement and Concrete Association, 1976.

than those predicted by Powers, and this has been borne out in laboratory tests.

Superplasticizing admixtures have the potential for reducing the *w/c* ratio without the problems of excessive retardation. After 28 days, compressive strengths greater than 85 MPa (13,000 lb/in.²) have been obtained (Table 21.11), and subsequent long-term strength gains are similar to those of normal concretes. These low-*w/c*-ratio concretes have very high early strengths (Table 21.12) even under ambient curing conditions, and when steam curing is used further gains can be made.

Table 21.12

Early Strength Development of Superplasticized Concrete[a]

w/c Ratio	Conditions	Compressive Strength MPa (psi)		
		1 Day	7 Days	28 Days
0.32	Ambient	49 (7,100)	76 (11,000)	—
0.33	LP-steam	42 (6,100)	—	62 (9,000)
0.33	Autoclaved	62 (4,000)	—	64 (9,300)
0.30	LP-steam	55 (8,000)	—	70 (10,200)
0.30	Autoclaved	89 (12,900)	—	92 (13,300)
0.27	LP-steam	60 (8,700)	—	78 (11,300)
0.27	Autoclaved	104 (15,100)	—	107 (15,500)

[a]Slumps in the range 50 to 60 mm (2 to 2½ in.) Based on data from "Superplasticizing Admixtures in Concrete," Cement and Concrete Association, 1976.

The data given in Tables 21.11 and 21.12 refer to concretes with normal consistencies, generally using very high cement contents. There is still potential for additional strength increases, or for using lower cement contents by reducing workabilities to the no-slump range. It is known that some superplasticizers give unusual workabilities with enhanced thixotropic properties, but this changed behavior could most probably be used to advantage in no-slump concretes. Thus, it is likely that superplasticized concretes with compressive strengths exceeding 140 MPa (20,000 lb/in.²) will be produced in the future.

Compacted Concrete

Consolidation by compaction is already used in industry. Concrete masonry blocks are compacted by both pressure and vibration. In this way very dry mixes can be used and reasonably high strengths can be obtained in conjunction with a high void content, which reduces the unit weight. High-pressure compaction is used to produce precast products such as curbstones; excess water is squeezed out of the concrete so that the net w/c ratio is less than 0.30. Spinning of concrete pipe is, in essence, compacting under centrifugal forces, which also results in low w/c ratio concretes.

High-pressure compaction is limited by the amount of water contained in the concrete or mortar. If very low water contents are used (w/c ratio < 0.20), higher compaction pressures can be used before water will start to be squeezed from the paste. The compressive strength of compacted pastes or mortars cured under normal conditions is primarily a function of the initial compaction pressure (Figure 21.10). Strength gain is most rapid in the first 24 h of moist curing, but strength continues to increase with age.

21.4 OTHER TYPES OF CONCRETES

Gap-graded Concrete

In Chapter 6, during the discussion of aggregate, mention was made of gap grading, in which certain aggregate sizes are omitted. Gap-graded concrete is not inherently better than concrete made with continuously graded aggregate. Gap grading of coarse aggregate gives better exposed aggregate finishes. It will improve the workability of undersanded mixes, but is still susceptible to segregation. In fact, segregation during handling and consolidation is a major drawback to gap-graded concrete, and special care must be taken in mix design and in handling techniques.

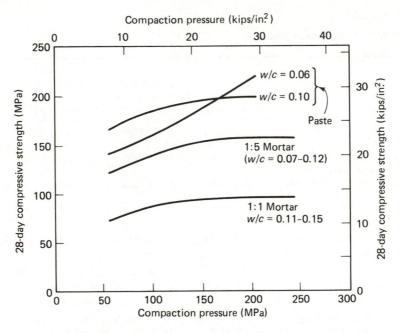

Figure 21.10 Effect of compaction pressure on the compressive strength of compacts. (Based on data by J. P. Skalny and A. Bajza, *Journal of the American Concrete Institute,* Vol. 67, No. 3, 1970, pp. 221–227.)

"No-Fines" Concrete

A special type of gap-graded concrete is "no-fines" concrete, where the fine aggregate is omitted entirely and a uniform size of coarse aggregate is used. This type of concrete can be viewed as an assembly of coarse aggregate particles cemented together by a layer of hardened cement paste at their points of contact. The material has an open structure with a high void content and a fairly low cement content. Thus, no-fines concrete has a low unit weight, low strength, low shrinkage, and a low thermal conductivity (Table 21.13). It is suited for applications where strength requirements are not great but lightweight and insulating properties are of particular importance. When the aggregate is itself a lightweight material, it has excellent insulating properties and is an ultra-lightweight material.

Since no-fines concrete is very porous, it is highly permeable to air and water and thus should not be used in foundations unless an effective moisture barrier is provided. However, there is little tendency for water to be drawn through in the absence of a pressure head (e.g., in an

Table 21.13

Properties of No-Fines Concrete

a/c Ratio	w/c Ratio	Cement Content kg/m³ (lb/yd³)	Unit Weight kg/m³ (lb/ft³)	Compressive Strength, 28 days MPa (lb/in.²)	Thermal Conductivity (W/m·K) (Btu/ft·h·°F)	Shrinkage (%)
6:1	0.38	259 (432)	2000 (125)	14.3 (2070)	—	—
8:1	0.41	193 (322)	1915 (120)	9.4 (1400)	—	0.018
10:1	0.45	155 (258)	1860 (116)	7.0 (1000)	0.74 (0.45)	0.019
Conventional concrete						
6:1	0.40	250 (417)	2550 (159)	35 (5000)	1–4 (0.6–2.4)	0.035

exterior wall exposed to rain), because there is very little capillary action. Water can only pass through by vapor transmission and moisture penetration is only two to three times the diameter of the aggregate particles under conditions of high humidity and no air movement.

When designing no-fines concrete the paste content should be just sufficient to thoroughly coat each aggregate particle. The *w/c* ratio should not be too high or the paste will separate from the aggregate. When properly designed, the mix will not easily segregate during handling. It should be placed with light rodding; vibration or ramming should not be used. In spite of the high porosity of no-fines concrete, it must be air-entrained for frost resistance.

Concretes for Mass Structures

We have from time to time discussed the special requirements of mass concrete. It is appropriate in this section to summarize the problems associated with the placing of mass concrete and the changes in material properties that can be tolerated. Since the majority of mass concrete is placed in dams, it is convenient to discuss the topic with this application in mind.

Strength requirements for the structure as a whole are not high and this can be used to advantage, because it is generally possible to make use of a local aggregate source even when it is not of high quality. Aggregate durability problems will not be a major concern in the core of the dam, with the exception of the alkali–aggregate reaction. Large aggregates, up to 150 mm in diameter, are used by the U.S. Bureau of Reclamation. The use of large aggregates requires the use of higher-capacity equipment than is normally employed in placing concrete. In some cases very large stones or "plums" (up to 0.03 m³ or 1 ft³) have been used, although the savings in material costs may be offset by extra labor costs. There should be no more than 20 to 30% of the total volume as "plums."

Large aggregate sizes allow quite low cement contents (less than 150 kg/m³, 250 lb/yd³) to be used in the interior of a dam. This helps to reduce the heat of hydration, which can be further reduced by the use of a Type IV cement or a pozzolan as a replacement for some of the cement. Internal heating of the structure is also reduced if the concrete temperature is kept below 16°C (60°F) at the time of placing. Even so, it is usually necessary to cool the concrete during curing by pumping water through a system of pipes embedded in the concrete at the top of each lift. Thermal cracking occurs not on heating but on subsequent cooling, when thermal stresses become greatest. When heat generation is controlled, the proper provision of contraction joints should prevent random cracking. The joints may be filled with a waterproof joint sealant or grouted when the structure has cooled.

More care should be taken in the quality of concrete used for exposed faces. Concretes should have higher cement contents, lower w/c ratios, and good-quality aggregates to ensure adequate durability. Air entrainment will be needed for frost resistance, and on spillway sections special measures for improved abrasion and cavitation resistance are required. Special care should be taken to prevent excessive moisture loss or temperature changes at the surfaces after the concrete has been newly placed, to avoid the possibility of surface cracking.

Grouts

Grouts can be defined as cementitious slurries that are injected into cracks, ducting, and other voids and fissures in concrete, or adjacent to concrete structures, to provide an impermeable barrier. The cementitious material used in a grout may be based on an organic or inorganic cement, but generally a portland cement is used when appreciable strength is also required. Grouts usually contain other materials besides cement and water. Sand may be used when a considerable volume of void space with a relatively open structure is to be filled. Mineral admixtures, such as fly ash and bentonite, are often used to help provide good fluid properties, without segregation occurring during injection. Fly ash provides additional cementitious action; bentonite is inert but results in high permeability because it absorbs large amounts of water with a concomitant large increase in volume. Chemical admixtures are used to reduce the water contents of a fluid slurry, to provide better cohesiveness, or to control setting times. Admixtures may also be added to counteract possible subsequent shrinkage of the grout. Aluminum powder is commonly used because the hydrogen generated by the alkaline corrosion causes expansion during setting. A sulfoaluminate-based expansive component may also be used, which causes expansion during the first few days of curing. These admixtures ensure that the grout fully fills the available space and counteracts subsequent shrinkage cracking. The important parameter of a grout is not initial or final set, but pumping time, which is a measure of the length of time the grout remains fluid enough to be properly injected. This is usually less than the time of initial set. Impermeability (or watertightness) are more important considerations than strength. The grout's major purpose is to prevent entry of moisture or to prevent corrosion of steel, so that injection procedures should be such as to ensure proper filling of the total void space.

One application where grout plays an important structural role is in the placing of concrete by the prepacked aggregate method. The void space in the aggregate bed is filled with a grouted mortar, which should be made with a *w/c* ratio low enough to satisfy strength and durability requirements and should be air-entrained if exposure conditions warrant it. Water-reducing admixtures and pozzolans help to attain the required flowability at the proper *w/c* ratio; pozzolans also help to minimize bleeding and at the same time improve later strength and permeability. A large amount of grouting is done by drilling companies for oil and gas wells. Grouts are pumped down to form an impermeable barrier around well casings. Fortunately, the strength of such grouts need not be very high, but there are other quite critical requirements that have to be met. In deep wells, placement of the grout takes a considerable length of time and the grout must remain pumpable. Since the temperature of the wells can be quite high [generally over 200°C (390°F) and in the deepest wells over 400°C or 750°F], special cements have been developed that have long setting times (Chapter 3). Long pumping times increase the chance of segregation during injection, so that special attention must be paid to the cohesiveness of the grout. Since the grout displaces the drilling mud used to cool the rig during the sinking of the well, mixing of grout and mud should be kept to a minimum. Generally, the sand used in the grout is finer than that used to make conventional concrete, with a fineness modulus in the range 1.3 to 2.3. Cement/sand ratios are usually in the range 1:1 to 1:2. The grout is mixed in special high-speed mixers, the sand being added to the cement–water slurry.

Bibliography

Lightweight Concrete

ACI COMMITTEE 213, "Guide to Structural Lightweight Aggregate Concrete," *Journal of the American Concrete Institute,* Vol. 64, No. 8, pp. 433–469 (1967).

LEWIS, D. W., "Lightweight Concrete and Aggregates," in *Significance of Tests and Properties of Concrete and Concrete-making Materials,* ASTM STP 169B, pp. 503–524. American Society for Testing and Materials, Philadelphia, Pa., 1978.

Lightweight Aggregate Concrete, CEB/FIP Manual of Design and Technology. The Construction Press, Lancaster, U.K., 1977.

Lightweight Concrete, SP-29. American Concrete Institute, Detroit, Mich., 1971.

Recommended Practice for Selecting Proportions for Structural Lightweight Concrete, ACI 211.2 (rev. 1977). American Concrete Institute, Detroit, Mich., 1977.

SHORT, A., AND W. KINNIBURGH, *Lightweight Concrete,* 3rd ed. Applied Science Publishers Ltd., London, 1978.

Heavyweight Concrete

Concrete for Nuclear Reactors, SP-34, 3 vols. American Concrete Institute, Detroit, Mich., 1973.

POLIVKA, M., AND M. S. DAVIS, "Radiation Effects and Shielding," in *Significance of Tests and Properties of Concrete and Concrete-making Materials,* ASTM STP 169B, pp. 420–434. American Society for Testing and Materials, Philadelphia, Pa. 1978.

Recommended Practice for Selecting Proportions for Normal and Heavyweight Concrete, ACI 211.1. American Concrete Institute, Detroit, Mich., 1977.

High-Strength Concrete

High Strength Concrete, Special Note 3p. National Crushed Stone Association, Washington, D.C., 1975.

PERENCHIO, W., *High Strength Concrete,* Bulletin RD014. Portland Cement Association, Skokie, Ill., 1973.

PERENCHIO, W., AND P. KLIEGER, *Some Physical Properties of High Strength Concrete,* Bulletin RD056.01T. Portland Cement Association, Skokie, Ill., 1978.

Recommended Practice for Selecting Proportions for No-Slump Concrete, ACI 211.3. American Concrete Institute, Detroit, Mich., 1974.

SCHMIDT, W., AND E. S. HOFFMAN, "9000 psi Concrete—Why? Why Not?" *Civil Engineering,* pp. 52–55 (May 1975).

Special Concretes

ACI COMMITTEE 207, "Mass Concrete for Dams and Other Massive Structures," *Journal of the American Concrete Institute,* Vol. 67, No. 4, pp. 273–309 (1970).

MALHOTRA, V. M., "No-Fines Concrete—Its Properties and Applications," *Journal of the American Concrete Institute,* Vol. 73, No. 11, pp. 628–643 (1976).

RAPHAEL, J. M., "The Nature of Mass Concrete in Dams," pp. 133–159 in SP-55, *Douglas McHenry International Symposium on Concrete and Concrete Structures,* American Concrete Institute, Detroit, Mich. 1978.

Symposium on Mass Concrete, SP-6. American Concrete Institute, Detroit, Mich., 1963.

Problems

21.1   What are the problems underlying the mix design of lightweight aggregate concretes?

21.2   Discuss the freeze–thaw resistance of concretes made with lightweight aggregates.

21.3   Why do lightweight concretes tend to show greater deformations in service than do normal-weight concretes?

21.4   Compare the properties of normal-weight and heavyweight concretes.

21.5   What special problems arise in the placement of heavyweight concretes?

21.6   Discuss the special material requirements needed to produce concrete with a compressive strength in excess of 60 MPa.

21.7   Why does high-strength concrete behave in a more brittle manner than does concrete of normal strength?

21.8   What are the advantages and disadvantages of no-slump concrete?

21.9   What are the implications of making concrete with a w/c less than 0.3?

21.10 Discuss the properties of (a) no-fines, and (b) gap-graded concrete.

21.11 What are the special requirements of concretes used in massive structures?

21.12 Describe the production and use of grouts.

22

modern developments

The brittle nature of concrete is an inherent property of the material and one that is overcome by the use of reinforcing materials. The very high porosity of concrete is also a disadvantage, severely limiting the strength of the material and limiting its durability under severe service conditions. Several approaches have been taken to improving concrete properties, resulting in quite different materials. Three different kinds of materials are discussed in this chapter.

1. *Polymer-impregnated concrete (PIC)* involves filling the capillary pores of hardened concrete with a polymer.

2. *Latex-modified concrete (LMC)* is made by incorporating a polymer latex with fresh concrete, which improves the tensile properties of concrete.

3. *Fiber-reinforced concrete (FRC)* is made by adding fibrous materials (usually steel or glass) to the fresh concrete, which improves the crack-resisting properties.

22.1 POLYMER-IMPREGNATED CONCRETE

When concrete is prepared and cured in the normal way, considerable capillary porosity is present in the cement paste, even in well-cured concretes made with quite low *w/c* ratios. It must be remembered that some of this porosity (below 10-nm diameter) is considered to be a part

Table 22.1

Typical Properties of Plain Concrete and PIC

Properties	*Plain*	*PIC (Methyl methacrylate)*	*PIC (Styrene)*
Compressive strength (28 days), MPa (lb/in.²)	37 (5300)	140 (20,300)	70 (10,000)
Tensile strength (28 days), MPa (lb/in.²)	2.8 (400)	11 (1600)	5.8 (840)
Flexural strength (28 days), MPa (lb/in.²)	5.2 (750)	18 (2600)	—
Modulus of elasticity, GPa (lb/in.²)	24 (3.5×10^6)	44 (6.3×10^6)	44 (6.3×10^6)
Permeability of water, m/s (ft/yr)	5.3×10^{-4} (5.3×10^4)	1.4×10^{-8} (1.4×10^4)	1.5×10^{-8} (1.5×10^4)
Water absorption (%)	6.4	0.3	0.7
Thermal coefficient of expansion, $10^{-6}/°C$ ($10^{-6}/°F$)	10.0 (5.5)	9.5 (5.3)	9.0 (5.0)
Bond to steel, MPa (lb/in.²)	1.7 (250)	3.8 (550)	4.1 (600)

of the C–S–H. If water present in the capillaries could be removed and replaced by some solid material, this would greatly improve the strength and durability of the concrete, by eliminating much of this void space. This is the basis of PIC. The problem of incorporating a solid into the pore system is solved by using a liquid *monomer* (the molecule that is the repeating unit of the polymer) to impregnate the concrete, and subsequently polymerizing the monomer to form the solid polymer *in situ* within the pores. Typical improvements in concrete properties are given in Table 22.1.

Materials

Concrete Materials

Any type of concrete can be successfully impregnated, regardless of the type of cement, admixtures, or aggregates used. Variations in mix design and materials will alter the amount of polymer needed for impregnation and the ease with which impregnation can occur. The amount of polymer needed for complete impregnation depends on the porosity of the concrete, which is determined by the *w/c* ratio, the amount of curing, the porosity of the aggregate, and so on. A more porous concrete (e.g., a concrete with a high *w/c* ratio) will require more polymer, but full impregnation will generally be achieved more easily and rapidly. The strength of the resulting PIC will be largely independent of the strength of the initial concrete, provided that full impregnation is achieved. Thus, the choice of initial materials can be determined largely by economic considerations.

This is not the case when partial impregnation is used, since the quality of the remaining unimpregnated concrete is obviously now of importance. The strength of PIC is still dependent on the final porosity of the material. Thus, if the pores of lightweight aggregate cannot be impregnated, lightweight PIC will not be very strong. Obviously, complete impregnation of porous, lightweight aggregate will require high polymer loadings.

Monomers and Polymers

Many different monomers have been successfully used to produce PIC. The desirable properties of a monomer include low viscosity, relatively high boiling point, low toxicity, ease of polymerization, low cost, and availability. Methyl methacrylate (MMA)—the monomer of plexiglass—and styrene are the most suitable of the common monomers. Their very low viscosities make them ideal liquids for penetrating the tortuous pore system of hardened concrete. Polyesters are too viscous to be successfully used alone, but when blended with styrene their viscosities can be reduced sufficiently to be used for partial impregnation of concrete. The high volatility and toxicity of another common monomer, vinyl chloride, effectively rules out its use in PIC.

Polystyrene and polymethyl methacrylate (PMMA) do not maintain their mechanical properties at elevated temperatures. If PIC is to be used in environments above ambient temperatures, monomers that provide polymers with improved performance would be required. Special cross-linking agents can be used to give polystyrene and PMMA better properties at higher working temperatures.

Process Technology

Production of PIC requires the following sequence of operations once the concrete has been cast and cured (Figure 22.1): (1) dry the concrete, (2) impregnate the concrete with monomer, and (3) polymerize the monomer.

Drying requirements. For full impregnation of precast specimens, the removal of as much evaporable water as possible (2 to 4% by weight of concrete) is desirable if the optimum performance is to be achieved. A temperature of 150°C (300°F) is recommended for typical structural concrete. The time of drying depends on the drying temperature, the

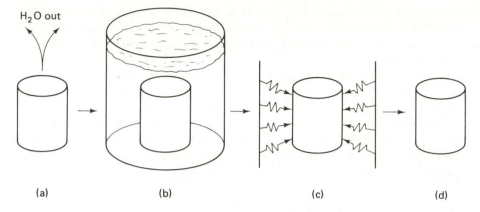

Figure 22.1 Process technology for polymer-impregnated concrete: (a) heat and dry concrete; (b) impregnate concrete; (c) polymerize concrete; (d) PIC.

w/c ratio of the concrete, and the thickness of the specimen. At 150°C a 100-mm-thick concrete pipe will take about 24 h to dry. A rapid rate of drying may induce microcracking, but this is of little consequence since the cracks will also be impregnated with polymer. However, when only partial impregnation is desired, such as for the upper surace of a slab, cracking could be serious and careful control of heating and cooling rates and the maximum drying temperature will be needed. Lower drying temperatures can be tolerated in many applications of partial impregnation.

Impregnation. Saturation of the concrete pore system depends on the viscosity of the monomer, the porosity and pore-size distribution of the concrete, the hydrostatic pressure, and time. Complete impregnation of good-quality dense concrete can be achieved after about 1 h of immersion if the dried concrete is first evacuated, the monomer then introduced into the impregnation chamber under vacuum, and the monomer pressurized to about 70 kPa (10 lb/in.²). This procedure can obviously only be used in a precasting plant.

Partial impregnation can be achieved satisfactorily by soaking at atmospheric pressure without prior evacuation. For example, 70 to 80% of total impregnation with MMA can be expected by soaking unevacuated specimens overnight. The depth of penetration will depend on soak times and the viscosity of the monomer. Depth of penetration is not linear with time and has been variously described as proportional to the logarithm of time or square root of time, thereby indicating a diffusion process. Low-viscosity monomers may penetrate the concrete faster, but tend to give less uniform impregnation. Bridge decks have been impregnated to a depth of 35 to 50 mm after soaking 8 to 12 h with MMA. The monomer is ponded on the surface and retained in a layer of sand to minimize losses from evaporation. Various methods of pressure impregnation for *in situ* impregnation are being developed.

Polymerization. The ultimate success of the whole operation depends on proper polymerization. Monomers used in PIC can be polymerized in two ways: (1) by exposing them to gamma radiation (e.g., from a cobalt-60 source), or (2) by the use of a catalyst and heat (thermal-catalytic polymerization). Radiation-induced polymerization can take place at room temperature and occurs uniformly through relatively thick concrete sections, but the health hazards associated with the use of gamma radiation effectively rule out its use in the field. Alternatively, a catalyst may be used to initiate polymerization. The catalyst–monomer mixture needs to be heated to 80 to 100°C (195 to 212°F) for polymerization to proceed. Heating breaks down the catalyst into free radicals which are needed for polymerization. There are also chemicals called promotors which break down the catalyst at ambient temperatures. But thermal–catalytic methods are often preferred over the use of a promotor–catalytic system because polymerization will proceed faster at elevated temperatures and is more predictable. Steam, hot water, or infrared heating can readily be used as a heat source. When thermal–catalytic polymerization is used, care must be taken to minimize loss of monomer by evaporation. Wrapping the specimen in a plastic sheet or aluminum foil is convenient for smaller specimens; for large specimens, polymerization under hot water is most practical. The water provides the heat for curing and also prevents evaporation.

Properties of PIC

Mechanical Properties

Table 22.1 shows an approximate fourfold increase in compressive, flexural, and tensile strength obtained from a fully impregnated PIC. It should be noted, however, that the relationships between the different strengths remain unchanged and PIC behaves as a more brittle material. The stress–strain curve in compression remains linear up to about 75% of ultimate load (Figure 22.2), and the departure from linearity is never very great. Consequently, failure under compression is sudden and the specimen shatters. Abrupt failure can be prevented by adding a monomer that gives a tough, flexible copolymer. Fracture mechanics studies indicate that fracture energies and the fracture toughness are higher for

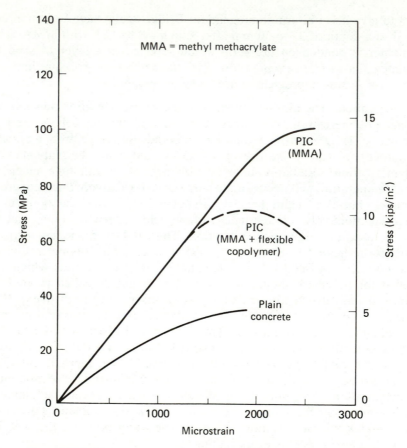

Figure 22.2 Stress–strain curves for PIC compared to plain concrete.

PICs than they are for plain concrete. Polymer impregnation strengthens the cement paste and improves the cement–aggregate bond. Fracture is therefore likely to occur through the aggregate, which may determine the strength limit of PIC. Weak, porous aggregates will also be strengthened by impregnation, although they will still retain weight advantages over normal aggregates to some extent.

    PIC still retains some porosity due to incomplete impregnation and/or incomplete polymerization and to the fact that some shrinkage will occur during polymerization. Both the strength and the elastic modulus of PIC are dependent on its porosity. Radiation-cured PIC may show higher strengths than thermal–catalytically polymerized PIC (see Figure 22.3), due to more complete polymerization, particularly with monomers

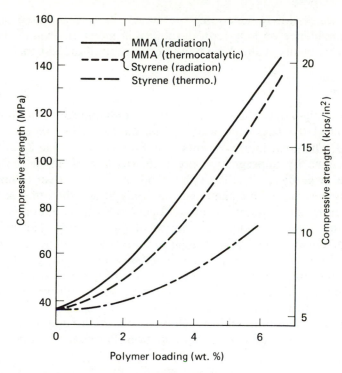

Figure 22.3 Effect of polymer loading and polymerization on strength of PIC. (From "Concrete-Polymer Materials, First Topical Report," Brookhaven National Laboratories, Upton, N.Y., 1968.)

that polymerize less readily (e.g., styrene). The strength of PIC will fall off quite markedly with temperature, owing to the softening of the polymer. A cross-linking agent will improve the performance in this regard; PICs containing cross-linking polystyrene can be safely used up to 180°C (355°F).

Creep and Shrinkage

Creep of PIC is reported to be very small, perhaps one-tenth of the creep of plain concrete. This is not surprising, because the concrete was strongly dried before impregnation; thus, no free water remains to cause creep of the cement paste. At higher temperatures, some increase in creep is observed as the polymer begins to soften. Similarly, no drying shrinkage would be anticipated, since the concrete has already been dried. An important question is whether deleterious expansions will

occur if water slowly penetrates the matrix, since this would obviously
have an influence on long-term durability in moist environments. However, no problems of this kind have been reported in field trials.

Durability

Many potential applications of PIC require only improved durability and
do not need the large increases in mechanical properties that can be
achieved. The great improvements in durability (see Table 22.2) that are
also obtained by impregnation are primarily the result of the marked
decrease in permeability. Since the capillary pore system is now filled
with polymer, aggressive chemicals can only attack the outer surface of
the concrete and cannot penetrate deeply. Further, there is often a
"skin" of polymer sealing the exterior of the concrete. Partial impregnation of concrete will give similar improvements in durability, since
partial impregnation will effectively seal the surface layer of the concrete,
even though improvements in mechanical properties are lower. When
partial impregnation is used, care must be taken to ensure the development of a uniform impregnated layer.

Improvement in durability will also be partially due to the fact that
the concrete was dried before impregnation. For example, the improved

Table 22.2

Durability of PIC

	Plain	PIC (MMA)	PIC (Styrene)
Freeze–thaw			
Number of cycles	740	2650	5440
Weight loss (%)	25	2	21
Sulfate attack			
Number of days	48	720	630
Expansion (%)	0.488	0.006	0.003
Acid resistance (15% HCl)			
Number of days	105	805	805
Weight loss (%)	27	9	12
Abrasion resistance[a]			
Abrasion depth, mm (in.) per 1000g of shot	1.25 (0.050)	0.38 (0.015)	0.93 (0.037)
Total weight loss (g) of specimens	14	4	9

[a]Shot blast test, using a Ruemelin abrasion machine. (Based on data from "Polymers
in Concrete: First Topical Report," Dept. BNL S0134, Brookhaven National Laboratory, 1968.)

freeze–thaw resistance would be expected, because there is no longer freezable water in the concrete. Since the entrained air voids will be mostly impregnated, air-entrained concrete need not be used. However, if partial impregnation is planned, air entrainment will be needed to protect the unimpregnated section. Although the durability measurements given in Table 22.2 look impressive, there are still no adequate long-term data available concerning exposure over several years. Field and laboratory tests will eventually provide this information, which is necessary for a proper evaluation of PIC.

Fire Resistance

Although polymers are flammable materials, they are not able to support combustion when distributed in PIC. Nevertheless, they can char and produce noxious fumes; fire retardants can be used to reduce these problems. More serious is the potential decrease in mechanical properties that can occur when PIC is exposed to higher temperatures. This will be a severe limitation on the use of PIC in high-strength applications.

Applications

Most applications that have been suggested for PIC take advantage of the tremendous improvements in durability. Since the production of PIC is expensive, because of the high cost of the monomer compounds and the impregnation process, the high cost of PIC must be justified in terms of a long maintenance-free service life or for applications where replacement would be difficult and costly.

Full impregnation can only be used on precast products. Applications that have been the subject of testing programs include concrete sewer pipe, precast segments for tunnel linings, underground support systems for mining operations, railroad ties, and precast pilings for wharves and jetties. Partial impregnation has advantages in its greater potential for field use. There has been a great deal of interest in the use of polymer impregnation to solve the bridge-deck problem (Chapter 20), since surface impregnation can provide an effective barrier against de-icing salts. Techniques for field impregnation are being explored, and this field promises to be the focus of considerable inventiveness and ingenuity. A precast PIC bridge-deck system has also been proposed. Impregnation of floor slabs used in corrosive industrial environments is another possibility.

Although the use of PIC is still relatively limited, there is no doubt that its use will grow. The high cost of production will undoubtedly limit its use to a variety of specialty applications where the initial investment can be justified in terms of improved performance. PIC will not become a general-purpose construction material.

22.2 LATEX-MODIFIED CONCRETE

The second method of combining concrete and polymer is to add a polymer latex at the mixer. A latex is a colloidal suspension of polymer in water. Polymer latexes are the vehicles of water-based latex paints, but can be specially formulated for use in concrete. The combination is called latex-modified concrete (LMC) or *polymer portland cement concrete*. The interaction between polymer and concrete in LMC is quite different from PIC. Whereas in PIC the polymer phase fills the capillary porosity, in LMC it is believed to form a continuous polymer film within the paste. This film can effectively coat the walls of the capillary pores. Also, the high tensile strength of the polymer film can offset the brittle nature of the paste and inhibit crack propagation under stress.

Materials

The materials used in LMC are the same as those used in normal concretes, apart from the polymer latex. Most commercial latexes (Table 22.3) are based on thermoplastic (vinyl-type polymers) or elastomeric (rubberlike) polymers which form coherent films readily when the latex is dried. They are generally copolymer systems which incorporate more

Table 22.3

Some Commercial Polymer Latexes Developed for Use as Admixtures

Chemical Name	Type	Use	Wet Strength
Polyvinyl acetate	Thermoplastic	Bonding aid	Low
Polyvinylidene chloride– polyvinyl chloride copolymer (Saran)	Thermoplastic	Overlays and patching	Good
Styrene–butadiene copolymer	Elastomer	Overlays and patching	Moderate
Poly acrylate copolymer	Thermoplastic	Patching	Moderate
Epoxy	Thermosetting	Overlays and patching	Good

than one type of polymer in the formulation, to optimize film formation and flexibility. Thermosetting resins, such as epoxies, polyesters, and polyurethanes, have also been added to concrete. Water-soluble or dispersible epoxy resins probably act similarly to other latexes. Materials containing cement and polyester or cement and polyurethane appear to behave more like resins filled with hydrated cement.

Effects on Properties

Fresh Concrete

The addition of polymer latex generally improves the properties of fresh concrete. The very small ($\sim$0.01 to 1 μm in diameter) spherical polymer particles that make up the latex act much as entrained air bubbles to improve the workability and decrease the bleeding of the paste. Usually the latex also entrains considerable amounts of air, due to the action of the emulsifying surface-active agent that stabilizes the latex. Indeed, it may be necessary to suppress the air entrainment by the addition of an antifoaming agent to keep the entrained air within reasonable limits. Thus, the addition of latex allows a reduction in the w/c ratio of the concrete, and a good rule of thumb is to replace some of the mix water with an equal quantity of latex. It must be remembered that a latex generally contains only about 50% by weight of solids, and typically 10 to 25% polymer by weight of cement is added to the concrete. At higher polymer loadings there may be some tendency for the latex to bleed and form a skin on the surface.

Hardened Concrete

Optimum properties of LMC are developed only when the dispersed latex is converted to a continuous polymer film within the concrete. The removal of water in the paste by hydration of the cement will help film formation to occur, but this will not happen if continuous moist curing is employed. Many suppliers recommend only 1 to 2 days of moist curing followed by drying at 50% RH to encourage film formation, although this does not allow the full strength of the paste to develop. Furthermore, with some polymers (e.g., polyvinyl acetates and some polyacrylates), some hydrolysis of the polymer can occur on prolonged exposure to the moist, alkaline environment of saturated concrete, leading to a loss of strength. For stable polymers, exposure to a relative humidity in the range 80 to 95% is said to give optimum results, allowing both slow, continued hydration of the cement, and coalescence of the latex to a continuous film.

Mechanical properties. Improvements in compressive, tensile, and flexural strengths occur (see Table 22.4) but are proportionally greater for tension and flexure. It is thought that the polymer film inhibits the propagation of microcracks under tensile stress because the high flexibility of the polymer will relieve the stresses at the crack tip. Thus, LMC will have a high strain at failure (Figure 22.4), as is reflected in a lower modulus of elasticity. Increases in strength are proportionally lower when the moist-curing period is relatively brief. If a latex is unstable in the

Table 22.4

Mechanical Properties of Latex-modified Mortars[a,e]

	Plain Concrete Control		Styrene–Butadiene	Saran[b]	Acrylic	PVAc	Epoxy
Compressive strength							
MPa	31	41[c]	33	61	32	26	52
lb/in.²	4500	5800 [c]	4800	8400	4700	3700	7500
Tensile strength							
MPa	2.2	3.7[c]	4.3	6.3	5.8	4.8	5.0
lb/in.²	310	535[c]	620	910	835	700	730
Flexural strength							
MPA	4.2	7.4[c]	9.9	12.9	12.7	12.7	11.3
lb/in.²	610	1070[c]	1430	1820	1835	1840	1640
Modulus of elasticity							
GPa	23	—	10.8	15.5	—	—	18
10⁶ lb/in.²	3.40	—	1.56	2.25	—	—	2.7
Shear bond strength							
MPa	0.35–1.4	—	>4.5[d]	>4.5[d]	>4.5[d]	>4.5[d]	—
lb/in.²	50–200	—	>650[d]	>650[d]	>650[d]	>650[d]	—
Impact strength							
m·kg	0.07	0.08[c]	0.22	—	0.25	0.18	—
in. lb	6	7[c]	19	—	22	16	—
Abrasion resistance (% wear)	24	5[c]	2.5	—	1.7	5	—

[a]Sand/cement = 3; polymer/cement = 0.20; dry-cured 28 days at 50% RH.

[b]Vinyl chloride–vinylidene chloride copolymer.

[c]Moist-cured 28 days.

[d]Failure occurred through mortar rather than at the interface.

[e]Adapted from "Polymers in Concrete," ACI Committee 548 Report, American Concrete Institute, Detroit, Mich., 1977.

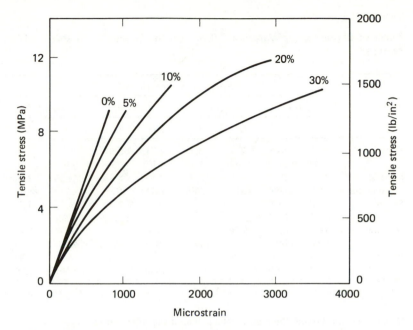

Figure 22.4 Stress–strain curves for latex modified concretes containing polyvinyl acetate (polymer loadings in percent by weight of cement). (From V. R. Riley and I. Razl, *Composites*, Vol. 5, No. 1, 1974, p. 28.)

presence of alkali, considerable loss of strength can be anticipated if the LMC is in contact with water for prolonged periods (Table 22.5). LMCs also have greatly improved bond strengths both to old concrete and to reinforcing steel. Invariably, failure will occur through the old concrete. Reduction in the w/c ratio will lead to decreased drying shrinkage and creep. The presence of the polymer phase may tend to increase creep, but this is only likely to be significant at high temperatures. In most applications, creep is of little consequence and the high flexibility of LMC prevents shrinkage cracking.

Durability. LMC also shows considerable improvement in durability over plain concrete. This is due in part to the improved resistance to tensile cracking. When cracks do form, they are kept very small and thus do not become a point of weakness for further environmental attack. It is thought that polymer films bridge the cracks and hold them closed. Also, it is believed that a polymer film lining the pores will tend to reduce the permeability of the concrete and prevent the entry of aggressive agents. The air-entraining properties of the latex will naturally provide frost resistance.

Table 22.5

Effect of Immersion in Water for 7 Days on the Strength of Latex-modified Mortars[a,e]

	Strength in MPa (lb/in.²)							
	Compressive		Tensile		Flexural		Shear Bond	
	Dry[c]	Wet[c]	Dry	Wet	Dry	Wet	Dry	Wet
Control	16.5	30.5	2.0	2.1	4.21	5.07	0.3	1.0
	(2390)	(4420)	(300)	(310)	(610)	(735)	(40)	(140)
Styrene–	34.2	28.3	4.1	2.4	9.83	6.38	>4.5[b]	2.4
butadiene	(4950)	(4100)	(600)	(350)	(1425)	(925)	(>650)[b]	(350)
Saran[d]	58.2	49.3			9.11	7.59	>4.5[b]	4.5
	(8430)	(7150)	—	—	(1320)	(1100)	(>650)[b]	(650)
Acrylic	39.3	27.7	5.76	3.38	12.6	7.25	>4.5[b]	2.4
	(5690)	(5460)	(835)	(490)	(1835)	(1050)	(>650)[b]	(340)
PVAc	25.9	9.0	4.8	0.35	12.7	2.2	>4.5[b]	0.9
	(3750)	(1300)	(700)	(50)	(1840)	(320)	(>650)[b]	(130)

[a]Sand/cement = 3; polymer/cement = 0.2; cured 28 days at 50% RH prior to immersion.

[b]Failure occurs through the mortar rather than at the interface.

[c]Dry, no immersion after curing; wet, immersed 7 days after curing.

[d]Vinyl chloride—vinylidene chloride copolymer.

[e]Adapted from "Polymers in Concrete," ACI Committee 548 Report, American Concrete Institute, Detroit, Mich., 1977.

Applications

The high cost of the polymer latexes must be offset by improved performance; thus applications take advantage of improved durability, flexibility, and bonding abilities. LMCs are ideal for repair work, because of their excellent bonding ability and good durability. The best LMCs are ideal for overlays for pavements and bridge decks and for stuccos on walls, since they can be applied in thin layers, have good flexural strength, resist drying shrinkage cracking, have good abrasion resistance and frost resistance, and show excellent resistance to de-icing salts. Latex additions improve the workability and bonding ability of masonry mortars.

22.3 FIBER-REINFORCED CONCRETE

As we have already seen (Chapter 15), the tensile strength of concrete is very low compared to the compressive strength. Because of the highly statistical nature of the strength of brittle materials, even this low strength cannot be relied upon, and therefore concrete structures are designed to minimize tensile stresses. However, such stresses invariably do occur in flexure, diagonal tension, or due to differential strains. Where they do occur, tensile stresses are expected to be carried entirely by steel reinforcing bars. Since one of the consequences of tensile stresses in concrete is cracking, which can lead to a decrease in durability, any technique that can provide concrete with greater tensile strength and "ductility" would be very valuable. One relatively new development in this direction has been the development of fiber-reinforced concrete (FRC).

Types of Fibers

Fiber-reinforced concrete may be defined as concrete made from portland cement (with or without aggregates of various sizes) and incorporating discrete fibers. A number of different types of fibers have been found suitable for use in concrete: steel, glass, organic polymers, ceramics, and asbestos are the most common. These fibers vary considerably in both cost and effectiveness. Some typical fiber properties are shown in Table 22.6, and a brief description of some of these fiber types follows.

Steel fibers may be produced either by cutting wire, shearing sheets, or from a hot-melt extract; they may be smooth, or deformed in a variety of ways to improve the bond. They will rust at the surface of the concrete, but appear to be very durable within the concrete mass. *Glass fibers* are generally available as "chopped strand," where each strand may consist of 100 to 400 separate filaments. Ordinary glass is not suitable for use in concrete as fibers, since attack by the highly alkaline

Table 22.6

Typical Properties of Fibers and Cement Matrix

Fiber	Diameter (μm)	Specific Gravity	Modulus of Elasticity (GPa)[a]	Tensile Strength (GPa)[a]	Elongation at Break (%)
Asbestos	0.02–20	2.55	165	3–4.5	2–3
Glass	9–15	2.60	70–80	2–4	2–3.5
Graphite	8–9	1.90	240–415	1.5–2.6	0.5–1.0
Steel	5–500	7.84	200	0.5–2.0	0.5–3.5
Polypropylene	20–200	0.91	5–77	0.5–0.75	20
Kevlar	10	1.45	65–133	3.6	2.1–4.0
Sisal	10–50	1.50	—	0.8	3.0
Cement matrix	—	2.50	10–45	$3–7 \times 10^{-3}$	0.02

[a]GPa $\times$ 0.145 = 10^6 lb/in.².

environment will rapidly reduce the strength of the fibers. Glass fibers have been produced (containing significant amounts of ZrO_2) which are highly alkali-resistant. The long-term durability of these materials, however, is still being investigated. Naturally occurring *asbestos fibers* have long been used with cement and water in the manufacture of pipe and other building components because of their high corrosion and abrasion resistance. However, there are significant health hazards associated with the production and handling of asbestos fiber. Most *polymeric fibers,* such as nylon and propylene, have lower elastic moduli than concrete. They therefore cannot increase the strength of the composite material and may indeed reduce it. However, they are effective in increasing the impact and shatter resistance of the concrete. A new fiber, *Kevlar,* which is an aromatic polyamide, has both a high tensile strength and a high modulus of elasticity, and shows considerable promise as reinforcement, but it is currently very expensive. *Carbon fibers* also have a very high elastic modulus, tensile strength, and cost. Like organic fibers, they are not attacked chemically by cement and are most effective when used in continuous lengths. *Natural organic fibers* such as sisal, coir, and jute are based on cellulose. They may not be very suitable for use, as they have low tensile strengths and elastic moduli and tend to deteriorate in damp or alkaline environments.

Mechanics of Fiber Reinforcement

Definitions

First, several parameters that will be used in the following discussion are defined:

  1. *Aspect ratio* = (fiber length/equivalent fiber diameter), where the equivalent diameter is the diameter of a circle having the same cross-sectional area as the fiber.

  2. *Minimum effective length,* l_m = minimum length at which the fibers have any effect on the *first-crack strength* of the concrete matrix.

  3. *Critical length,* l_c = length above which the fibers will fracture rather than pull out when the crack intersects the fiber at its midpoint. It has been shown that the critical length is approximated by

$$l_c = \frac{d}{2\tau} \sigma_f \qquad (22.1)$$

where d is the fiber diameter, τ the interfacial bond stress, and σ_f the fiber strength.

4. *Orientation factor,* or *fiber efficiency factor* = efficiency with which randomly oriented fibers can carry a tensile force in any one direction. Assuming perfect randomness, this can be shown to be $0.41l$, where l is the fiber length, but with different assumptions (e.g., orientation effects near the surface), values from about $0.33l$ to $0.65l$ have been obtained.

5. *Spacing factor:* if the fibers are close enough together, the first cracking strength is higher than that of the matrix alone because the fibers effectively reduce the stress intensity factor, which controls fracture. A typical expression for the spacing factor, s, is

$$s = 13.8d \frac{\sqrt{l}}{p} \tag{22.2}$$

where d is the fiber diameter and p the percent fiber (by volume). Other, similar expressions, have also been developed.

A typical load-deflection curve for fiber-reinforced concrete in flexure is shown in Figure 22.5. Point A represents the load at which the matrix begins to crack, referred to as the "first-crack strength." Usually, this is at about the same stress at which cracking occurs in nonreinforced concrete, and thus the segment OA is about the same for both plain and fiber-reinforced concrete. Once the matrix has cracked, all of the load

Figure 22.5 Typical load−deflection curve for fiber-reinforced concrete in flexure.

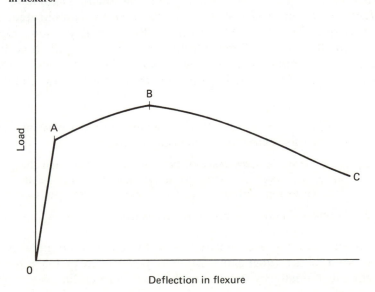

must be carried by the fibers bridging the crack. The segment AB represents the region where there is continued cracking of the matrix and some debonding and pulling out of the fibers. The maximum load (point B) depends on the fiber content and geometry. It should be noted that during this part of the debonding and pulling-out process, the fiber stress is generally substantially less than the yield stress of the fibers, so yielding of the fibers does not occur. In the declining portion of the curve, BC, matrix cracking and fiber pull-out continue; if the fibers are long enough to maintain their bond, they may eventually fail by yielding or fracture in this region of the curve. A reasonably good empirical equation has been developed by ACI Committee 544, Fiber Reinforced Concrete, for the ultimate strength S_c:

$$S_c = AS_m(1 - V_f) + BV_f \, l/d \qquad (22.3)$$

where S_m is the ultimate stress of the matrix, l/d the aspect ratio, V_f the volume fraction of fibers adjusted for the effect of randomness, and A and B are constants which can be obtained by a plot of $V_f \, (l/d)$ vs. composite strength.

Fiber –Matrix Bond

For a composite system such as fiber-reinforced concrete, the mechanical behavior depends not only on the properties of the fiber and the concrete, but also on the bonding between them. The nature of the interface in cement-based systems is particularly complicated, since there may be a chemical reaction between the cement and some types of fiber. Also, the nature of the interface may change with time as the cement matures or undergoes time-dependent volume changes. However, the general form of the bond is fairly well known for the different classes of fibers:

1. *Steel:* a combination of adhesion, friction, and mechanical interlocking, although some chemical reactions may also occur.

2. *Glass:* there is some reaction between the cement and the glass; in particular, alkali attack tends to weaken the fiber reinforcement, although to a much lesser extent with the alkali-resistant glasses.

3. *Organics:* the bond is primarily mechanical interlocking.

As mentioned above, most fiber-reinforced concrete failures occur due to bond failure (fiber pull-out). It is possible to increase the bond strength substantially by deforming the fibers in various ways so as to

Table 22.7

Typical Fiber-Matrix Pull-out Strengths

Matrix	Fiber	Pull-out Strength, MPa (lb/in.²)
Cement paste	Asbestos	0.8–3.2 (115–460)
	Glass	6.4–10.0 (930–1450)
	Polycrystalline alumina	5.6–13.6 (810–1970)
	Steel	6.8–8.3 (990–1200)
Mortar	Steel	5.4 (780)
Concrete	Steel	3.6 (520) (first crack)
		4.2 (610) (failure)
	Nylon	0.14 (20)
	Polypropylene	1.0 (150)

increase the end anchorage. Large changes in the bond strength are not reflected by similar changes in the concrete strength, but will improve the post-cracking behavior. A very good bond may increase the tensile strength, while a poor bond may increase the energy absorption. Table 22.7 shows the pull-out strengths for a number of different fibers in various matrices.

Fabrication of Fiber-reinforced Concrete

There are a number of ways of introducing fibers into a concrete mix. For short fibers (normally steel or glass) which are supposed to achieve a random orientation, standard concrete mixers may be used. The problem is to avoid "balling up" of the fibers, particularly steel. This can most easily be overcome by mixing all the other ingredients and then adding the fibers gradually; steel fibers may have to be vibrated through a coarse screen or otherwise separated as well. Also, workability aids may have to be added to the mix. With this method, only about 3% by volume of fibers can be added.

For glass fibers, a "spray-up" method is often used to produce thin sheets. Using a special pump and spray gun, the fibers are chopped and combined with a cement slurry and then sprayed onto a mold. This method can also be used with carbon and other organic fibers. About 10% by volume of fibers can be introduced by this method. For obtaining a more efficient fiber orientation, the "winding" process may be used with continuous fibers or filaments. The fibers are passed through a cement slurry, then wound on a frame; additional slurry and chopped

Table 22.8

Typical Proportions for Normal-Weight Fiber-reinforced Concrete

Cement	325–560 kg/m³ (550–950 lb/yd³)[a]
w/c Ratio	0.4–0.6
Fine aggregate/total aggregate	0.5–1.0
Maximum aggregate size	10 mm (⅜ in.)
Air content	6–9%
Fiber content	0.5–2.5% by volume[b]

[a]Pozzolan is often used to replace some of the cement when high cement contents are needed.

[b]Steel fiber: 1% = 78 kg/m³ (132 lb/yd³); glass fiber: 1% = 25 kg/m³ (42 lb/yd³).

fiber may then be sprayed on to achieve the desired thickness. With this method, up to 15% by volume of fibers can be achieved.

It should be noted that the mix proportions for fiber-reinforced concrete (FRC) are not the same as those for ordinary concrete. Higher cement contents are generally used for FRC to provide enough paste to coat the fibers. Pozzolans are a useful way of increasing the paste content without increasing the cement. Higher air contents are needed because of the high paste content, but the percentage of air in the paste need not be higher. The maximum aggregate size is generally 10 mm, and a fairly high ratio of fine to coarse aggregate is used. Typical mix proportions are given in Table 22.8. The aspect ratio of the fibers is generally in the range 50 to 150.

Rheology of Fiber-reinforced Concrete

The rheological properties of FRC depend on the size and type of fiber and on the method of production. Since fibers tend to have relatively large surface areas, they have a large water requirement, as well as exhibiting a tendency to interlock or "ball." Glass fibers may have a particularly high water requirement, because water is absorbed between the filaments making up each individual fiber. In addition, the *w/c* ratio and ratio of fine to coarse aggregate must be considered, as with conventional concrete. As a general rule, the workability is decreased as the fiber content increases, as the aspect ratio of the fibers increases, or as the coarse aggregate content increases. It is, however, difficult to define a satisfactory method of testing the workability; at present, the Vebe test is considered to be most suitable. Some results indicating the effect of aspect ratio and type of matrix on the workability of steel fiber concrete are shown in Figures 22.6 and 22.7.

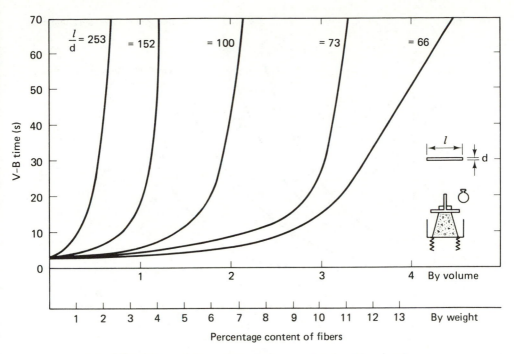

Figure 22.6 Effect of fiber aspect ratio on V-B time of fiber-reinforced mortar. (From J. E. Edgington, D. J. Hannant, and R. I. T. Williams, BRE Current Paper CP 69/74, Building Research Establishment, Garston, Watford, U.K., 1974. British Crown Copyright, HMSO.)

Figure 22.7 Workability against fiber content for matrices with different maximum aggregate size. (From J. E. Edgington, D. J. Hannant, and R. I. T. Williams, BRE Current Paper CP 69/74, Building Research Establishment, Garston, Watford, U.K., 1974. British Crown Copyright, HMSO.)

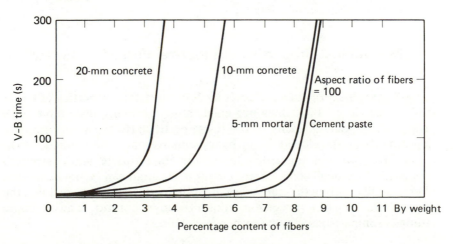

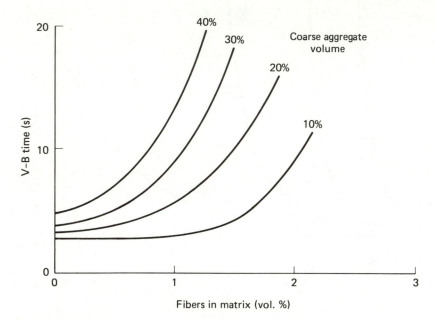

Figure 22.8 Aggregate –fiber interaction on the compactibility of steel fiber concrete. (From R. N. Swamy, *Materials and Structures (Paris)*, Vol. 8, No. 45, 1975, pp. 235 –254.)

Apart from difficulties with workability, it is also harder to compact FRC, although again no very satisfactory test is available. However, it is known that increases in the coarse aggregate content can greatly decrease the compactibility. This is shown in Figure 22.8. For most fiber mixes, external vibration is preferred; however, it may not be practical in the field. In general, fibers tend also to reduce the bleeding and improve the cohesion of a mix.

Mechanical Properties of Fiber-reinforced Concrete

As yet, a set of standard tests to determine the properties of either fresh or hardened FRC does not exist, although many tests have been proposed. In general, many of the tests, particularly those for strength, that have been developed for ordinary concrete may be applied. However, FRC has not been developed for its ultimate static strength properties. Although some improvement in strength can be obtained with fibers, similar strengths can often be obtained simply by making the appropriate changes in cement content and *w/c* ratio. Thus, simple strength comparisons may be rather misleading.

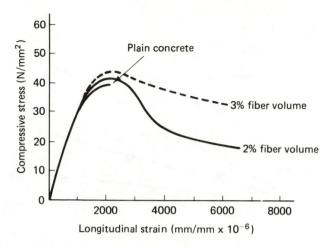

Figure 22.9 Stress–strain deformation in compression of steel-fiber concrete. (From report by RILEM Technical Committee 19-FRL, *Materials and Structures (Paris)*, Vol. 10, No. 56, 1977, pp. 103–120.)

For adequately compacted specimens, the addition of fibers has very little effect on the compressive strength of FRC. Typical compressive stress–strain curves for steel fiber concrete are shown in Figure 22.9. Moreover, it would appear that fiber reinforcing has little effect on the elastic modulus of FRC. The direct tensile strength of FRC can be increased considerably by the addition of high modulus fibers. The increase is, however, dependent on the aspect ratio of the fibers, as shown in Figure 22.10. It appears that the tensile strength can be adequately predicted by the composite materials "law of mixtures" [see Eq. (13.1)]:

$$\sigma_t = \sigma_m(1 - V_f) + 2\tau(l/d)V_f \qquad (22.4)$$

where σ_t and σ_m are tensile strengths of the composite and the matrix, respectively; V_f the percent of fibers by volume; l/d the aspect ratio; and τ the average interfacial bond strength. This equation is very similar to Eq. 22.3. Tensile strengthening thus occurs at all fiber contents as long as $2\tau(l/d) > \sigma_m$. However, it should be noted that some investigations have shown very little increase in direct tensile strength due to fiber additions. Torsional strength of FRC is also very little affected by fiber additions.

The effects of fiber additions on the flexural strength of FRC are not clear. Some investigators have found an increase in both the first-crack strength and in the ultimate strength, the latter increase being up to 3 times the strength of plain concrete, as shown in Figure 22.11.

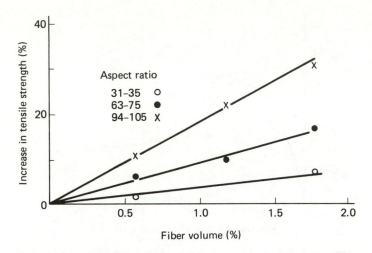

Figure 22.10 Relation between fiber volume and tensile strength. (Adapted from C. D. Johnston and R. A. Coleman, in *Fiber Reinforced Concrete,* SP-44, American Concrete Institute, 1974, pp. 177–193.)

Figure 22.11 Load-deflection curves of steel fiber composites (mortar matrix) in flexure. (From R. N. Swamy, *Materials and Structures (Paris),* Vol. 8, No. 45, 1975, pp. 235–254.)

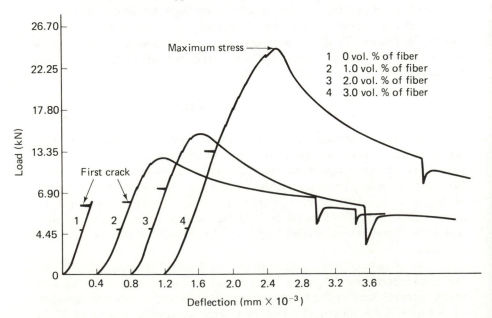

However, other investigations have shown little or no improvement in flexural strength; much seems to depend upon the details of the tests carried out, both the coarse aggregate volume and methods of fabrication being important. The real advantage of FRC would appear to be that a certain amount of flexural strength can be relied upon, even after some cracking of the matrix occurs.

Although few data are available, it appears that steel fiber reinforcement can improve the shear strength of concrete. Some research has been carried out in replacing conventional shear reinforcement in reinforced concrete beams with steel fibers. Although the fibers were found to be less effective than conventional shear reinforcement, the increase in shear, moment, and energy capacity of beams with fiber reinforcement suggests that they may have structural applications, particularly for earthquake-resistant structures. Much more work in this area is required.

The greatest advantage of using FRC is that fiber additions improve the toughness (the total energy absorbed in breaking a specimen). That is, fiber additions give concrete a considerable amount of apparent ductility. However, there are no standard tests for ordinary concrete to measure the toughness, and to date no satisfactory test has been developed for FRC, although several have been suggested. It seems likely that some measure of the area under the stress–strain curve is necessary to characterize the effectiveness of the fibers. If we define toughness as the area under the stress–strain (or load-deflection) curves, it may be seen from Figure 22.12 that increasing the fiber content has little effect on ultimate strength but vastly increases the toughness. Basically, toughness refers to the ease with which cracks can propagate within a material and is a particularly significant property of brittle materials such as concrete. Since plain concrete is limited by its brittle behavior, anything that can be done to improve its toughness will be useful.

However, it is important to emphasize the point that the toughness referred to above must not be confused with the fracture toughness, K_{IC}, discussed in Chapter 14. There is now considerable evidence that linear elastic fracture mechanics cannot be applied to fiber-reinforced concrete. Thus, tests which show that K_{IC} does not change much with fiber additions are not very meaningful.

Related to the concept of toughness is the impact resistance of FRC. Many studies have shown that the impact resistance of concrete can increase dramatically (by more than an order of magnitude) with the addition of fibers. Low-modulus fibers like nylon and polypropylene seem to be particularly effective in this regard. The fiber effectiveness for impact resistance is also related to the bond characteristics; fibers with different shapes can give quite different results, as shown in Figure

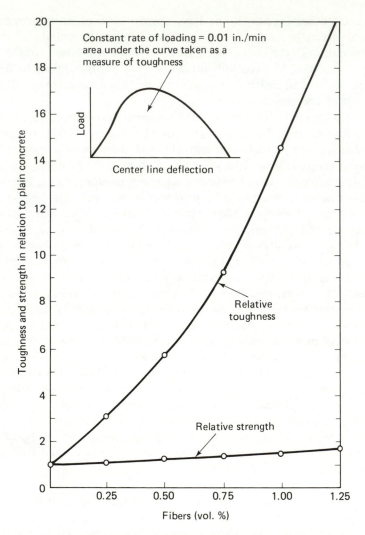

Figure 22.12 Effect of volume of fibers in flexure. (From S. P. Shah and B. V. Rangan, *Journal of the American Concrete Institute,* Vol. 68, No. 2, 1971, pp. 126–135.)

22.13. Unfortunately, no satisfactory test for the impact strength of FRC has yet been developed which makes it impossible to compare the results obtained in different investigations. Finally, fibers will not only improve the impact strength of FRC but also help prevent the shattering of the mass into fragments under shock loading. In addition, there are indications that fiber reinforcement improves the ability of concrete to withstand abrasion and cavitation damage.

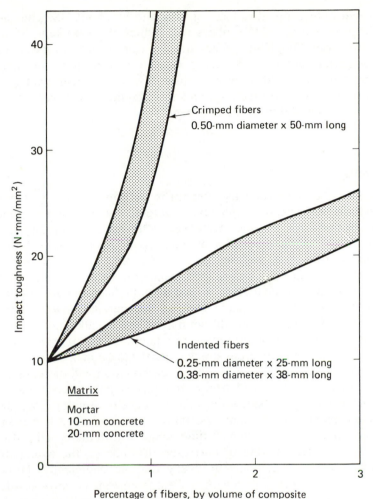

Figure 22.13 Impact toughness of fiber-reinforced mortar and concrete. (From J. Edgington, D. J. Hannant, and R. I. T. Williams, BRE Current Paper 69/74, Building Research Establishment, Garston, Watford, U.K., 1974. British Crown Copyright, HMSO.)

Fatigue resistance is also related to the ability of cracks to propagate. There is some evidence to suggest that the fatigue strength of concrete in flexure increases with increasing fiber content; however, fibers have little effect on fatigue in compression.

Fibers appear to have no effect on compressive creep. However, they can reduce tensile creep at least to some degree. Very stiff carbon fibers have been shown to reduce flexural creep very considerably. Steel fibers will reduce the shrinkage of concrete by about 10%, but not the shrinkage of mortar. Glass fibers will reduce the shrinkage of cement paste by about 20%. Thus, fibers simply seem to act as rigid inclusions in the matrix, without producing much effect because of their low volume. However, and more important, fibers have been very effective in reducing shrinkage cracking, and this should have a very beneficial effect on the durability of the concrete.

Durability

Durability is as important as strength in determining the suitability of concrete for any specific application. Durable concrete should generally be dense and impermeable. However, although the porosity of FRC appears to be higher than that of plain concrete, probably due to the difficulties in fully compacting the mix, the effects of fibers on permeability have not yet been determined. Since the study of FRC is a fairly recent development, many long-term durability data are not yet available.

If the steel FRC is made with an appropriate paste content and w/c ratio for the exposure condition in question, the alkaline environment provides adequate protection for the fibers *in the uncracked state*. However, once the concrete cracks, particularly in a marine or acid environment, the rate of corrosion and carbonation will increase considerably. Surface rusting will occur, but this is only a cosmetic effect. Ordinary glass cannot be used in FRC because of attack by the alkaline portland cement. Special alkali-resistant glasses are better, but even they may show a loss of strength with time, coupled with increased brittleness of the composite; the rate of corrosion depends on the availability of moisture. Asbestos cements are very durable chemically. However, because of the short fiber length, asbestos cement is fairly brittle and subject to damage in handling. Carbon and polymeric fibers can also be expected to be durable, but natural organic fiber may suffer from alkaline, bacterial, or fungus degradation.

Thermal Properties

The thermal conductivity of FRC increases with the amount of steel fiber. For glass fibers, the thermal conductivity is controlled simply by the density of the concrete. The thermal expansion of steel FRC is about 10.4 to 11.1 $\times$ 10^{-6}/°C; (5.8 to 6.2 $\times$ 10^{-6}/°F); with glass FRC, it is in the range 7 to 12 $\times$ 10^{-6}/°C (3.9 to 6.7 $\times$ 10^{-6}/°F).

Applications of Fiber-Reinforced Concrete

The controlling factor in the use of FRC is not its material properties but the cost; fibers are still expensive. However, where the additional material cost can be justified, FRC can be used successfully in many applications. Steel FRC has been used for pavements, and highway and runway overlays, to reduce cracking and thickness. It has been used in the spillways of large dams to reduce cavitation damage. It has also been used to reduce construction damage, for nuclear reactor shielding, in refractory concrete, and in shotcrete applications for tunnel linings. Asbestos cement has long been used for pipes, fire-resistant boards, and flat sheets. Glass FRC has been used as spray-on cladding on structures, to provide both architectural and structural properties. It has also been used to make thin, precast panels. In summary, the use of FRC is still in its infancy. New applications are being found wherever the energy-absorbing characteristics of the material are important. It seems clear that these materials will be much more widely used in the future, particularly if the economics of the material becomes more favorable.

Bibliography

Polymers

General

ACI Committee 548, *Polymers in Concrete*. American Concrete Institute, Detroit, Mich., 1977.

Polymers in Concrete (International Symposium), SP-58. American Concrete Institute, Detroit, Mich., 1978.

Polymers in Concrete, Proceedings, 2nd International Congress on Polymers in Concrete, 1978. The University of Texas at Austin, 1978.

Proceedings, First International Congress on Polymers in Concrete, 1975. The Construction Press, Lancaster, U.K., 1976.

Polymer-Impregnated Concrete

CLIFTON, J., AND G. FROHNSDORFF, "Polymer-impregnated Concretes," in *Cements Research Progress, 1975*, pp. 173–196. American Ceramic Society, Columbus, Ohio, 1976.

Introduction to Concrete-Polymer Materials, Rept. No. FHWA-RD-75-507. Federal Highway Administration, U.S. Department of Transportation, Washington, D.C., 1975. 152 pp.

Latex-Modified Concrete

POPOVICS, S., *Proceedings ASCE, Journal of the Construction Division,* Vol. 100, No. CO3, pp. 469–487 (1974).

RILEY, V. R. AND I. RAZL, "Polymer Latex Modified Mortars—A Review," *Composites,* Vol. 5, No. 1, pp. 27–33 (1974).

Fiber-Reinforced Concrete

EDGINGTON, J., D. J. HANNANT, AND R. I. T. WILLIAMS, *Steel Fibre Reinforced Concrete,* BRE Current Paper CP 69/74. Building Research Establishment, Garston, U.K., 1974.

Fiber Reinforced Concrete, SP-44. American Concrete Institute, Detroit, Mich., 1974.

"Fibre Concrete Materials," report prepared by RILEM Technical Committee 19-FRC, *Materials and Structures (Paris),* Vol. 10, No. 56, pp. 103–120 (1977).

HANNANT, D. J., *Fibre Cements and Fibre Concretes.* Wiley-Interscience, Chichester, U.K., 1978.

HOFF, G. C., C. M. FONTENOT, AND J. G. TOM, *Selected Bibliography on Fiber-Reinforced Cement and Concrete,* Miscellaneous Paper C-76-6. U.S. Army Corps of Engineers Waterways Experiment Station, Vicksburg, Miss., 1976 (and supplement No. 1, 1977, supplement No. 2, 1979).

NEVILLE, A., ed., *Rilem Symposium 1975: Fibre Reinforced Cement and Concrete,* 2 vols. The Construction Press Ltd., Lancaster, U.K., 1976.

SWAMY, R. N., "Fiber Reinforcement of Cement and Concrete," *Materials and Structures (Paris),* Vol. 8, No. 45, pp. 235–254 (1975).

Testing and Test Methods of Fibre Cement Composites, R. N. Swamy, ed. The Construction Press, Lancaster, U.K., 1978. (RILEM Symposium 1978.)

Problems

22.1 Describe the different polymer concretes that can be made.

22.2 What effect do different polymers have on the performance of polymer-impregnated concrete?

22.3 What effect does the porosity of the concrete have on the amount of polymer used in impregnation?

22.4 What are the different ways in which monomers can be polymerized?

22.5 What advantages would polymer-impregnated concrete have in bridge construction?

22.6 How do polymers affect the properties of concrete (a) in polymer-impregnated concrete, (b) in latex-modified concrete?

22.7 What difficulties may be associated with field use of polymer latexes?

22.8 What is the effect of moist curing, or subsequent exposure to moisture, on the performance of latex-modified concrete?

22.9 How do fibers affect the ultimate strength of concrete in (a) tension; (b) compression; (c) flexure; and (d) impact loading?

22.10 What is the significance of the aspect ratio of fibers?

22.11 What are the difficulties associated with the production of fiber-reinforced concrete?

22.12 What problems are associated with the use of glass fibers in concrete?

22.13 How would fibers be expected to change the cracking behavior of concrete?

APPENDIX

TESTS AND SPECIFICATIONS FOR CONCRETE AND ITS CONSTITUENTS

ASTM stands for American Society for Testing and Materials.

CSA stands for Canadian Standards Association.

BSI stands for British Standards Institution.

This is not an exhaustive list. All specifications refer to the latest editions.

Designation	ASTM	CSA	BSI
Specifications			
Cements and Mortars			
Portland cement	C150	CAN3-A5	BS 12
			BS 1370
			BS 4027
Natural cement	C10		
Blended cements	C595	A362-M1977	BS 146
			BS 4246
Cementitious slag		A363-M1977	
Expansive cement	C845–76T		
High-alumina cement			BS 915, Part 2
Tests			
Optimum SO_3 in portland cement	C563		
Sampling cement	C183	CAN3-A5	BS 4550, Part 1
Chemical analysis of cement	C114	CAN3-A5	BS 4550, Part 2
Density of cement by air permeability	C188		BS 4550, Part 3, §3-2
Fineness of cement (air permeability) (Blaine)	C204		BS 4550, Part 3, §3-3
Fineness of cement (Wagner turbidimeter)	C115	CAN3-A5	
Fineness of cement—minimum passing 75-μm sieve			
Mechanical mixing of cement pastes and mortars	C305	CAN3-A5	
Flow table	C230	CAN3-A5	
Time of setting of cement by Vicat needle	C191	CAN3-A5	
Time of setting of cement by Gillmore needles	C266		
Time of setting of mortar by Vicat needle	C807		BS 4550, Part 3, §3-6
Air content of mortar	C185		
Normal consistency of cement	C187	CAN3-A5	BS 4550, Part 3, §3-5
Tensile strength of mortars	C190		
Flexural strength of mortars	C348		

Designation	ASTM	CSA	BSI
Cements and Mortars			
Compressive strength of mortars (cube test)	C109	CAN3-A5	BS 4550, Part 3, §3-4
Compressive strength of mortars (prism test)	C349		
Bleeding of cement pastes and mortars	C243		
Heat of hydration of cement	C186		BS 4550, Part 3, § 3-8
Autoclave expansion of cement (soundness test)	C151	CAN3-A5	BS 4550, Part 3, §3-7
Length change of hardened mortar and concrete	C157		
Potential sulfate expansion of mortars	C452	CAN3-A5	
Restrained expansion of expansive cement mortar	C806		
Drying shrinkage of mortar	C596		
Aggregates			
Specifications			
Concrete aggregate	C33	CAN 3-A23.1	BS 882 and 1201
Aggregates for radiation-shielding concrete	C637		BS 4619
Lightweight aggregates for structural concrete	C330		BS 3797, BS 877, BS 1165
Tests (sampling aggregates)	D75	A23-2-1A	
Petrographic examination	C295	A23-2-15A CAN3-A23-2 (Appendix B)	BS 812, Part 1, BS 3681
Sieve analysis	C136	A23-2-2A	BS 812, Part 1
Total moisture content	C566		
Surface moisture in fine aggregate	C70	A23-2-11A	BS 812, Part 2
Unit weight	C29	A23-2-10A	BS 812, Part 2
Specific gravity and absorption of coarse aggregate	C127	A23-2-12A	BS 812, Part 2

	ASTM	CSA	BS
Specific gravity and absorption of fine aggregate	C128	A23-2-6A	BS 812, Part 2
Soundness of aggregates	C88	A23-2-9A	
Evaluation of frost resistance of coarse aggregates in air-entrained concrete	C682		
Impact and crushing	C131		BS 812, Part 3
Resistance to abrasion (Los Angeles machine)	C535		BS 812, Part 3
Organic impurities in sands	C40	A23-2-8A	
Effect of organic impurities on strength	C87	A23-2-8A	
Materials finer than No. 200 (75-μm) sieve	C117	A23-2-5A	BS 812, Part 1
Lightweight pieces	C123	A23-2-4A	
Flat and elongated particles		A23-2-13A	BS 812, Part 1
Clay lumps and friable particles	C142	A23-2-3A	
Potential alkali reactivity of cement–aggregate combinations (mortar bar method)	C227		
Alkali–aggregate reaction (concrete prisms)		A23-2-14A	
Potential reactivity of aggregate (chemical method)	C289		
Potential alkali reactivity of carbonate rocks	C586		
Potential volume change of cement–aggregate combinations	C342		
Control of alkali–aggregate reaction using mineral admixtures	C441		
Sulfate ion content in groundwater	D516	A23-2-2B	
Total or water soluble sulfate ion content of soil		A23-2-3B	
Chloride content of aggregate			BS 812, Part 4

Admixtures

Specifications			
Chemical admixtures	C494	CAN3-A226-2	BS 5075, Part 1
Air-entraining admixtures	C260	CAN3-A266-1	

	Designation	ASTM	CSA	BSI
Admixtures				
	Fly ash and raw or calcined natural pozzolans	C618	CAN3-A266-3	
Tests				
	Air-entraining admixtures	C233	CAN3-A266-1	
	Sampling and testing fly ash or natural pozzolans	C311	CAN3-A266-3	
Water				
Tests for water				BS 3148
Fresh Concrete				
Specifications				
	Ready-mixed concrete	C94	CAN3-A23-1	BS 1926
Tests				
	Sampling	C172	A23-2-1C	BS 1881, Part 1
	Making and curing concrete test specimens in the laboratory	C192	A23-2-2C	BS 1881, Part 3
	Making and curing concrete test specimens in the field	C31	A23-2-2C	BS 1881, Part 3
	Slump test	C143	A23-2-5C	BS 1881, Part 2
	V-B consistometer test			BS 1881, Part 2
	Ball penetration in fresh concrete (Kelly ball)	C360		
	Compacting factor test			BS 1881, Part 2
	Unit weight, yield, and air content (gravimetric)	C136	A23-2-6C	BS 1881, Part 2
	Air content of fresh concrete (volumetric method)	C173	A23-2-7C	
	Air content of fresh concrete (pressure method)	C231	A23-2-4C	BS 1881, Part 2
	Analysis of fresh concrete			BS 1881, Part 2
	Bleeding	C232		
	Time of setting by penetration resistance	C403		
	Early volume changes	C827		
	Water retention by concrete curing materials	C156		

Tests

Test			
Capping cylinders	C617		BS 1881, Part 3
Compressive strength	C39	A23-2-9C	BS 1881, Part 4
Compressive strength of concrete (prisms)	C116		BS 1881, Part 4
Compression tests of no-slump concrete		A23-2-12C	
Accelerated curing and testing of concrete	C684	A23-2-10C	
Splitting tensile strength	C496		BS 1881, Part 4
Flexural strength	C78	A23-2-8C	BS 1881, Part 4
Static modulus of elasticity and Poisson's ratio	C469		BS 1881, Part 5
Fundamental, transverse, longitudinal, and torsional frequencies	C215		BS 1881, Part 5
Mechanical properties under triaxial loads	C801		
Creep of concrete in compression	C512		
Length change of drilled or sawed specimens	C341		BS 1881, Part 5
Specific gravity, absorption, and voids	C642	A23-2-11C	BS 1881, Part 5
Microscopical determination of the air-void system	C457		
Cement content of hardened concrete	C85		BS 1881, Part 6
Resistance to rapid freezing and thawing	C666		
Critical dilation of concrete subjected to freezing	C671		
Abrasion resistance by sandblasting	C418		
Abrasion resistance of horizontal surfaces	C779		
Scaling resistance of concrete exposed to de-icing chemicals	C672		
Bond developed with reinforcing steel	C234		
Examining and sampling of concrete in constructions	C823		
Testing drilled cores and sawed beams	C42	A23-2-14C	BS 1881, Part 4
Rebound number	C805-75T		BS 4408, Part 4
Penetration resistance	C803-75T		
Pulse velocity	C597		BS 4409, Part 5
Pull-out strength	C800-78T		
Analysis of hardened concrete			BS 1881, Part 6
Electromagnetic cover measuring devices			BS 4408, Part 1
Gamma radiography			BS 4408, Part 3

solutions to numerical problems

Chapter 3

3.1 (a) $C_3S = 46.6$, $C_2S = 27.5$, $C_3A = 8.0$, $C_4AF = 11.0$
(b) $C_3S = 50.4$, $C_2S = 23.2$, $C_3A = 8.0$, $C_4AF = 11.0$
(c) $C_3S = 42.6$, $C_2S = 30.6$, $C_3A = 11.5$, $C_4AF = 8.0$
(d) $C_3S = 42.3$, $C_2S = 34.3$, $C_3A = 0$, $C_4AF = 18.4$ (case B)

3.2 (a) 258.1 J/g; (b) 189.2 J/g

3.3 (a) Type III; (b) Type I; (c) Type V; (d) Type IV

Chapter 4

4.9 (a) $V_g = 0.54$ cm³/g; (b) $P_c = 0.16$ cm³/g; (c) $X = 0.77$

4.10 $\alpha = 0.58$; (a) $V_g = 0.39$ cm³/g; (b) $P_c = 0.21$ cm³/g; (c) $X = 0.64$

4.11 Complete hydration is not possible, $\alpha \not> 0.75$
(a) $V_g = 0.51$ cm³/g; (b) $P_c = 0$; (c) $X = 1.0$

4.12 (a) $V_g = 0.68$ cm³/g in each case; (b) $P_c = 0, 0.04, 0.14, 0.24$ cm³/g;
(c) $X = 1.0$ ($\alpha = 0.75$), 0.94, 0.83, 0.74

Chapter 6

6.3 (a) 1.96; (b) 13.52

6.4 (a) 2.91; (b) 2.73

6.5 2.72 g/cm³

6.6 2.59 g/cm³

6.7 $W_{SSD} = 1007$ g, E.A. = 0.7%

6.8 $W_{SSD} = 479$ g, S.M. = 4.38%

6.9 1.52%

6.10 T.M. = 0.81%, E.A. = 0.71%

6.11 1.01%

6.12 T.M. = 5.37%, S.M. = 4.36%

6.13 14 lb less water

6.16 27.8%

Chapter 7

7.6 (a) 0.54; (b) 0.49; (c) 0.45

Chapter 9

9.2 $w/c = 0.41$, $\sigma = 5970$ lb/in.²; 0.48, 5160 lb/in.²; 0.57, 4780 lb/in.²; 0.68, 3400 lb/in.²; 0.82, 2540 lb/in.²

9.3 $g = 0.33$, 25 mm = 100%, 20 mm = 93%, 17.5 mm = 80%, 10 mm = 74%, 5 mm = 59%, 2.5 mm = 47%, 1.25 mm = 37%, 0.6 mm = 30%, 0.3 mm = 24%, 0.15 mm = 18%
$g = 0.5$, 25 mm = 100%, 20 mm = 89%, 12.5 mm = 71%, 10 mm = 63%, 5 mm = 45%, 2.5 mm = 32%, 1.25 mm = 22%, 0.6 mm = 16%, 0.3 mm = 11%, 0.15 mm = 8%
$g = 0.67$, 25 mm = 100%, 20 mm = 86%, 12.5 mm = 63%, 10 mm = 54%, 5 mm = 34%, 2.5 mm = 21%, 1.25 mm = 13%, 0.6 mm = 8%, 0.3 mm = 5%, 0.15 mm = 3%

9.4 2-3 in. slump, air content = 5.0%, $w/c = 0.44$, cement type II, or V (ACI permits Type I). Water 293 lb/yd³, cement 666 lb/yd³, coarse aggregate (SSD) 1813 lb/yd³, fine aggregate (SSD) 1133 lb/yd³. Corrected for moisture contents of aggregates: water 241 lb/yd³, cement 666 lb/yd³, coarse aggregate 1831 lb/yd³, fine aggregate 1167 lb/yd³.

9.5 Water = 348 lb/yd³, cement = 666 lb/yd³, coarse aggregate (OD) = 1786 lb/yd³, fine aggregate (OD) = 1105 lb/yd³. Corrected for moisture contents: water = 297 lb/yd³, cement = 666 lb/yd³, coarse aggregate = 1804 lb/yd³, fine aggregate = 1138 lb/yd³.

9.6 50-75 mm slump, air = 1.0%, $w/c = 0.49$. Cement Type I. Water = 168 kg/m³, cement = 343 kg/m³, coarse aggregate (SSD) = 1230 kg/m³, cement = 343 kg/m³, coarse aggregate = 123.5 kg/m³, fine aggregate = 778 kg/m³.

9.7 Water $= 183$ kg/m^3, cement $= 343$ kg/m^3, coarse aggregate (OD) $= 1224$ kg/m^3, fine aggregate (OD) $= 762$ kg/m^3.

9.8 Assumed crushed aggregate, 35 MPa $= f_m$, slump 30–60 mm, and $w/c = 0.60$. Water $= 190$ kg/m^3, cement $= 317$ kg/m^3, coarse aggregate $= 1378$ kg/m^3, fine aggregate $= 590$ kg/m^3.

Chapter 13

13.1 Eq. (13.3) $E_c =$ (a) 6.5×10^6 lb/in.^{2}; (b) 2×10^6 lb/in.^{2}; (c) 8.8×10^5 lb/in.^{2}
Eq. (13.4) $E_c =$ (a) 4.6×10^6 lb/in.^{2}; (b) 2×10^6 lb/in.^{2}; (c) 6.2×10^5 lb/in.^{2}
Eq. (13.5) $E_c =$ (a) 5.4×10^6 lb/in.^{2}; (b) 2×10^6 lb/in.^{2}; (c) 7.2×10^5 lb/in.^{2}
Eq. (13.6) $E_c =$ (a) 5.3×10^6 lb/in.^{2}; (b) 2×10^6 lb/in.^{2}; (c) 7.7×10^5 lb/in.^{2}

13.2 Eq. (13.3) $E_c =$ (a) 84 GPa; (b) 20 GPa; (c) 7.2 GPa
Eq. (13.4) $E_c =$ (a) 56 GPa; (b) 20 GPa; (c) 4.8 GPa
Eq. (13.5) $E_c =$ (a) 67 GPa; (b) 20 GPa; (c) 5.7 GPa
Eq. (13.6) $E_c =$ (a) 66 GPa; (b) 20 GPa; (c) 6.2 GPa

13.3 (a) $E_c =$ (i) 5.3×10^6 lb/in.^{2}; (ii) 2×10^6 lb/in.^{2}; (iii) 7.6×10^5 lb/in.^{2}
(b) $E_c =$ (i) 66.0 GPa; (ii) 20 GPa; (iii) 6.1 GPa

Chapter 14

14.11 Approximately 10 MPa

Chapter 15

15.2 $P = 0.25$, $S = 20,600$ lb/in.^{2}; $P = 0.5$, $S = 12,500$ lb/in.^{2}
$P = 0.75$, $S = 7600$ lb/in.^{2}; $P = 1.0$, $S = 4600$ lb/in.^{2}

15.3 $n = 2.6$ (a) $w/c = 0.4$, $\sigma = 18,840$ lb/in.^{2}; $w/c = 0.5$, $\sigma = 12,920$ lb/in.^{2};
$w/c = 0.6$, $\sigma = 9290$ lb/in.^{2}; $w/c = 0.7$, $\sigma = 6930$ lb/in.^{2}
(b) $\alpha = 0.2$, $\sigma = 2090$ lb/in.^{2}; $\alpha = 0.5$, $\sigma = 6060$ lb/in.^{2}; $\alpha = 0.7$,
$\sigma = 11,430$ lb/in.^{2}; $\alpha = 0.7$, $\sigma = 17,620$ lb/in.^{2}
$n = 3.0$ (a) $w/c = 0.4$, $\sigma = 17,200$ lb/in.^{2}; $w/c = 0.5$, $\sigma = 11,130$ lb/in.^{2};
$w/c = 0.6$, $\sigma = 7610$ lb/in.^{2}; $w/c = 0.7$, $\sigma = 5430$ lb/in.^{2}
(b) $\alpha = 0.3$, $\sigma = 1363$ lb/in.^{2}; $\alpha = 0.5$, $\sigma = 4650$ lb/in.^{2}; $\alpha = 0.7$,
$\sigma = 9660$ lb/in.^{2}; $\alpha = 0.9$, $\sigma = 15,930$ lb/in.^{2}

15.6 5 days

15.8 (a) $\sim$7 MPa; (b) $\sim$7 MPa; (c) $\sim$4 MPa

Chapter 16

16.6 (a) (1) $>$; (2) $>$; (3) no effect
(b) (3) $>$; (2) $>$; (1) no effect

Chapter 17

17.3 Eq. (17.9a) (a) 4810 lb/in.^{2}; (b) 5160 lb/in.^{2}; (c) 6050 lb/in.^{2}

Chapter 18

18.1 4.7×10^6 lb/in.^{2}

18.2   (a) 15 GPa; (b) 19 GPa; (c) 12 GPa

18.3   0.26 lb/ft^2/h, yes

18.7 295×10^{-6}

18.12 (a) 0.97; (b) 1.20; (c) 1.33; (d) 1.54

index

Z